Electrical Machines

Electrical Machines

Imtiaz Ashraf
Electrical Engineering Department
Z.H. College of Technology
A.M.U. (Aligarh)

M.A. Mallick
Electrical and Electronics Engineering Department
Integral University
Lucknow

TECHSAR™
NEW DELHI

Published and Printed by

Techsar Pvt. Ltd.
4435-36/7, Ansari Road, Daryaganj,
New Delhi - 110 002 (India)
E-mail: tech_info@techsarworld.com
Website: www.techsarworld.com

ISBN 978-93-80026-09-1

Reprint 2025

Dedicated to
Our Parents

Preface

The contribution of electrical machines is enormous in the present technological world. The application ranges from small watch and toys to the most recent concept of electric ship propulsion, electric and hybrid vehicles and 'more electric' aircraft. Almost seventy per cent of the total energy generation in the world is consumed by rotating electrical machines. Since the inception of electrical machine almost one and half century ago, a continuous effort is being put worldwide to enhance the performance of machine by developing new designs and control algorithms. A number of new variants of basic machines have been developed in the past years and new topologies have emerged such as permanent magnet machine, reluctance machine, brushless d.c. machines and linear machines. Apart from the design and basic structure of machines, their control algorithm is another aspect where effort is being made worldwide. Nevertheless the basic underlying principle of operation remains more or less same for all types of machines. It is thus fundamental concept where emphasis is being put in the present textbook.

The book contains altogether twelve chapters. The fundamental concept of rotating machine is illustrated in the first two chapters which helps in better understanding of the subsequent chapters. Chapter 3, elaborates the properties of various magnetic materials and their behaviour with current carrying coils. Chapter 4 discusses the principle of electromechanical energy conversion and it further elaborates the principle of operation of generators and motors. Chapter 5 deals with transformers, their types and the underlying operating principle is covered in it. Chapter 6 and 7 deal with the d.c. machines and their types. The operating principle in the motoring and generating mode is discussed at length with the necessary mathematical analysis. The operation of machines under no load and full load condition is dealt with in the chapters with the relevant mathematical expression. In chapters 8 to 10 the a.c. machines and their types are discussed with the necessary characteristic. The constructional differences and the starting behaviour of both types of machines are discussed at length with the operating characteristics. In the eleventh chapter, the single phase a.c. machines are discussed. The different types of single phase machines are shown pictorially with their starting characteristics. The last chapter is very important from the point of view of industrial applications. Different types of a.c. and d.c. drives are discussed with relevant figures showing the output at different stages. It is regretted, however that despite careful scrutiny of the proofs, printing mistakes may have nevertheless remained in the book. If such mistakes are pointed out by the readers, it will be gratefully entertained. The authors sincerely hope that readers of the text will be benefited from this book.

In the last, the authors feel highly indebted to all those who had helped in some way or the other in bringing this book.

Imtiaz Ashraf
M. A. Mallick

Acknowledgements

It would be impossible to complete this book without the strength which is given by the Almighty Creator of Universe, the ultimate source of power, who is controlling the lives of all and had given the man with immense potential which could be used in the positive or negative direction. We would like to express our gratitude to the teachers at AMU, Aligarh Dr. Atif Iqbal (TPO), ZHCOET, AMU, Aligarh. We special thanks are due to the V.C., Prof. S.W. Akhtar. Integral University, Lucknow, PVC, Prof. S.M. Iqbal, I.U., Lucknow and Dean (Engg.), Dr. M.I. Khan, I.U, Lucknow, for their healthy suggestions and support and we would also like to express my thanks to Mr. Arshad Ali, Senior Instructor, EED, I.U., Lucknow for his constant boosting me during me work.

We would like to express our deep hearted gratitude to our parents for their right teaching and moral support. It would be not possible for me to complete the work without sacrifice of my wife, Ruksana Mallick and my son Rahmatullah Mallick and my daughter Aafia Nisa, who laid the patience during this work. I would also like to express my gratitude to my sisters Sahana, Rezina, Salina and brother Asif whose motivation cannot be neglected.

At last, we would like to thank all those who had helped me or supported me in any way in our work till the competition of the manuscript of this book. I would also like to thank Mr. Rajat Garg, proprietor, PTC Meerut for providing the real photos of machines, which will be helpful for the students of engineering.

Imtiaz Ashraf
M. A. Mallick

Contents

Preface *vii*
Acknowledgements *ix*

1. Review of Basic Circuits **1**

1.1 Introduction 1
1.2 Direct-Current Circuits 2
1.3 Alternating Current Circuits 15
1.4 Three-Phase Circuits 19
1.5 Three-Phase Connections 22
1.6 Three-Phase Power 28
1.7 Power Measurement in Single Phase and Three Phase A.C. Circuits 29
- *Solved Questions* *33*
- *Tutorial Questions* *41*
- *Review Questions* *44*

2. Review of Basic Laws **49**

2.1 Introduction 49
2.2 Maxwell's Equations 50
2.3 Self and Mutual Inductances 51
2.4 Transformer EMF 51
2.5 Motional EMF 53
2.6 Faraday's Law of Electromagnetic Induction 54
2.7 Lenz's Law 55
2.8 Force on a Current Carrying Conductor in a Magnetic Field 56
2.9 Torque on a Current Carrying Coil in a Magnetic Field 58
2.10 Flemming's Right Hand (R.H) Rule 60
2.11 Flemming's Left Hand (L.H.) Rule 61
2.12 Ampere's Law 62

2.13 Right Hand Rule 63
- *Solved Questions* *64*
- *Tutorial Questions* *67*

3. Magnetic Circuits **70**

3.1 Introduction 70
3.2 Permanent Magnets 71
3.3 Magnetic Materials 72
3.4 Magnetic Circuit Terminology 75
3.5 Analogy of Magnetic Circuit with the Electric Circuit 78
3.6 Series Magnetic Circuit 78
3.7 Parallel Magnetic Circuit 83
3.8 Fringing 85
3.9 Hysteresis 86
- *Solved Questions* *87*
- *Tutorial Questions* *101*

4. Principle of Electromechanical Energy Conversion **105**

4.1 Introduction 105
4.2 Electric Field as a Medium 107
4.3 Magnetic Field as a Medium 110
4.4 Moving Coil in a Uniform Magnetic Field 112
4.5 Stationary Coil in a Time Varying Magnetic Field 114
4.6 Motoring Action 116
4.7 Generating Action 117
4.8 Singly Excited System 118
4.9 Doubly Excited System 120
4.10 Force/Torques in Systems with Permanent Magnets 121
4.11 Dynamical Equations of Electromechanical Systems 122
- *Solved Questions* *123*
- *Review Questions* *125*
- *Unsolved Questions* *125*

5. Transformers **127**

5.1 Introduction 127
5.2 Constructional Features of Transformers 128
5.3 Types of Transformers 130
5.4 Ideal Transformer 131
5.5 Phasor Diagram of Ideal Transformer 133
5.6 Practical Transformer 137
5.7 Phasor Diagram of Real Transformer 137
5.8 Equivalent Circuit of Transformer 141

5.9	Voltage Regulation of Transformer	145
5.10	Efficiency of Transformer	149
5.11	Transformer Testing	151
5.12	Auto Transformers	156
5.13	Three-Phase Transformer	158
5.14	Harmonics	164
5.15	Special Transformer	165
5.16	Per Unit System	167
	• *Solved Questions*	*168*
	• *Unsolved Questions*	*175*
	• *Exam Questions*	*179*

6. D.C. Generator **184**

6.1	Introduction	184
6.2	Constructional Features of D.C. Machines	185
6.3	Types of D.C. Machines	192
6.4	Principle of Operation	199
6.5	Symbolic Representation	200
6.6	Characteristics of D.C. Generators	201
6.7	EMF Equation of D.C. Generator	208
6.8	Commutation	210
6.9	Armature Reaction	214
6.10	Losses in D.C. Machines	218
6.11	Voltage Regulation of D.C. Generator	220
6.12	Efficiency of D.C. Generator	221
6.13	Applications of D.C. Generators	222
	• *Solved Questions*	*223*
	• *Tutorial Questions*	*229*
	• *Review Questions*	*231*

7. D.C. Motors **233**

7.1	Introduction	233
7.2	Operating Principle of Motor	234
7.3	Torque Equation of Motor	235
7.4	Speed Regulation	236
7.5	Types of D.C. Motor	237
7.6	Characteristics of D.C. Motor	237
7.7	Performance Characteristics of a Motor	243
7.8	Losses in Motor	244
7.9	Efficiency of D.C. Motor	245
7.10	Starting of D.C. Motor	246
7.11	Speed Control of D.C. Motor	253

7.12 Braking of D.C. Motor 262
7.13 Permanent Magnet D.C. Motor (PMDC) 264
7.14 Brushless Motor 265
7.15 Universal Motors 266
7.16 Applications of D.C. Motor 266
- *Solved Questions* *267*
- *Unsolved Questions* *273*
- *Review Questions* *275*

8. Three-Phase Induction Motor **277**

8.1 Introduction 277
8.2 Types of Polyphase IM 278
8.3 Construction of Polyphase IM 279
8.4 Principle of Operation 282
8.5 Rotating Magnetic Field 283
8.6 Equivalent Circuit 287
8.7 Power Flow Diagram 290
8.8 Power Expressions 290
8.9 Torque Expressions 292
8.10 Characteristic of Polyphase IM 296
8.11 Starting of 3-Phase IM 298
8.12 Speed Control of 3 ϕ IM 306
8.13 Cogging and Crawling of Induction Motor 313
8.14 Braking of Induction Motor 314
8.15 Induction Generator 315
8.16 Applications 316
- *Solved Questions* *317*
- *Unsolved Questions* *325*
- *Review Questions* *327*

9. Synchronous Generators **331**

9.1 Introduction 331
9.2 Constructional Features 332
9.3 Winding Factor 'K_w' 336
9.4 EMF Equation 342
9.5 Armature Reaction 343
9.6 Phasor Diagram on Load 346
9.7 Characteristics of Alternators 348
9.8 EMF or Synchronous Impedance Method 351
9.9 MMF Method 353
9.10 Potier's Method 354
9.11 Capability Curve 355

9.12	Blondel's Two Reaction Theory	356
9.13	Parallel Operation of Synchronous Generators	361
9.14	Hunting	368
9.15	Transients in Synchronous Machines	369
9.16	Synchronous Machine Dynamics	370
	• *Solved Questions*	*372*
	• *Unsolved Questions*	*376*
	• *Review Questions*	*380*

10. Synchronous Motors **383**

10.1	Introduction	383
10.2	Constructional Features	384
10.3	Principle of Operation	384
10.4	Starting Methods	385
10.5	*V*-Curve	386
10.6	Torque and Power	388
10.7	Hunting of Synchronous Motor	390
10.8	Braking	392
10.9	Applications	393
	• *Solved Questions*	*393*
	• *Unsolved Questions*	*397*
	• *Review Questions*	*398*

11. Single Phase Induction Motor **401**

11.1	Introduction	401
11.2	Principle of Operation	402
11.3	Types of Single Phase IM	403
11.4	Starting Methods	404
11.5	Equivalent Circuit of SPIM	412
11.6	Testing of SPIM	415
11.7	Stepper Motor	418
11.8	Hysteresis Motor	420
11.9	Linear Induction Motor	421
	• *Solved Questions*	*422*
	• *Unsolved Questions*	*424*
	• *Review Questions*	*424*

12. Solid State Motor Control **427**

12.1	Introduction	427
12.2	Solid State Control Devices	428
12.3	Types of Drive	431
12.4	Chopper Control	457

12.5 Cycloconverter Control 461
12.6 Inverter Control 462
- *Solved Questions* *466*
- *Unsolved Questions* *471*
- *Review Questions* *472*

Appendix-A 473

Appendix-B 476

Appendix-C 478

Appendix-D 480

Appendix-E 482

Experiments 487

Bibliography 505

Index *507*

Chapter

1

Review of Basic Circuits

1.1 Introduction
1.2 Direct-Current Circuits
1.3 Alternating Current Circuits
1.4 Three-Phase Circuits
1.5 Three-Phase Connections
1.6 Three-Phase Power
1.7 Power Measurement in Single Phase and Three-Phase A.C. Circuits
- Solved Questions
- Tutorial Questions
- Review Questions

1.1 INTRODUCTION

The electricity is a need of life, which has made the life simple and comfortable. The electrical energy is converted in one form or other so as to enable different important applications. For example, the a.c. voltage is converted into d.c. voltage and is used for charging the batteries.

The electricity we use is obtained via generator which gets its energy in non-electrical form from the turbines. The turbines get the energy from steam or high-speed water falling on the turbine blades. The steam again is obtained via crushed wet coal heated in boiler or via the potential energy of the water kept at a large height in dams. So we can say that the electric energy we are using in any form does not violate the law of conservation of energy. The question comes in mind then where does the energy come from coal. The energy comes from the hydro-carbons by which the coal is made. These hydrocarbons are present due to the formation process of coal in the form of peat by the compression of plants and animals for millions of years. These plants and animals are a rich source of hydrocarbons. Then again the people get confused as to who has kept the organic compounds and elements in it. The answer to it that the creator of animals and plants had made them a rich source of hydrocarbons so as to fulfil the need of the people who are coming later. Thus, we see one system is dependent on the other. As the generator depends on the turbine and the turbine depends on the velocity of steam and the steam on coal and coal on plants and animals.

In the later chapter, the principle of electromechanical energy conversion will be discussed. Basically, the machines are divided into two broad categories, one is motor and the other is generator. It is immaterial whether the machine is a.c. or d.c., the electromechanical energy conversion process is applicable to all the machines.

In the chapter on transformers, the types of various transformers which are used in the power system are discussed. Since the transformers perform a vital role in the power system by transferring electrical energy at low voltage level to very high voltage level thus making it possible to send the electrical energy over thousands of kilometers efficiently and economically.

But before getting complete understanding of the machines one should brush up the basic laws which are applicable in order to visualize the machine both under static condition and under dynamic condition. One cannot get a clear concept of machines operation without having knowledge of the basic laws of electricity.

The next coming chapters are based over the first two chapters.

1.2 DIRECT-CURRENT CIRCUITS

The phenomenon of electricity is based on the concept of electric charge 'q'. This charge is carried by electrons. The movement of electrons leads to current. A closed path through which current can flow is called an electric circuit.

The process of current flow through a substance is called conduction. Substances which conduct current easily are called conductors. Substances which conduct with difficulty are called insulators. The ability to conduct is usually measured in terms of resistance to current flow. Conductors have a very low resistance, insulators have a very high resistance. A circuit component specifically designed to have a known resistance is called a resistor. A circuit constructed only with resistors is called a purely resistive circuit.

When current flows through a resistor then a potential difference is developed across it. Current flows through a resistor if a source of electrical energy is connected to it, thus causing a potential difference to be developed across it. A potential difference is often called a voltage.

The unit of potential difference or voltage is Volt. The unit of resistance is Ohm. The unit of current is the Ampere.

The most fundamental law of electric circuits is the Ohm's Law. Ohm's law states that the voltage developed across the resistor is the product of the current flowing through it and the value of the resistance.

Mathematically, it is given by $\boxed{V = IR}$ (1.1)

It is important to remember these two ideas when describing the currents and voltages in a circuit. Current flows through a resistor. Voltage is developed across a resistor. It is even more important to remember this distinction between current and voltage when trying to measure the two quantities. In order to discuss these concepts more readily it is helpful to be able to refer to a circuit diagram. A real circuit is made up of a number of components or circuit elements connected together by pieces of copper wire or more commonly nowadays by copper tracks on a printed circuit board. In a circuit diagram, different components are represented by different symbols. The interconnecting wire or track is represented by thin lines.

The d.c. circuit often employs the fundamental laws of circuit theory such as Ohm's law, Kirchhoff's voltage law and current law. The electric circuit becomes more complicated as more active and passive devices are added to it. Under such cases, the other techniques are used. The Maxwell loop analysis and nodal analysis employs the Kirchhoff's laws but the circuits are easily solved for determining the electrical parameters. When only the concern is to determine the response through any branch then the Thevenin's theorem is used.

In the electronic circuits, the power which is transferred from the source to load is also needed to be determined. In that case, the maximum power transfer theorem is used. The circuit laws enable us to determine the desired electrical parameters such as voltage, current, etc. through any branch of the circuit or network. In the coming sections, we discuss the basic circuit laws.

Ohm's Law

A resistor is a conductor having specified resistance 'R' when current 'I' passes through the resistance, then a potential difference of 'V' appears across it. "The Ohm's law asserts that current through the device is always directly proportional to the potential difference applied to the device". It must be noted that a conducting device obeys Ohm's law when the resistance of the device is independent of the magnitude and polarity of the applied potential difference.

The electronic devices such as diodes which form the basis for the rectifier and switching circuits does not obey Ohm's law.

It is often misunderstood that $V = IR$ is a statement of Ohm's law that is not so, but it defines the value of resistance 'R' and it applies to all devices, whether it is linear or non-linear. In the case of non-linear devices which do not obey Ohm's law uses this above equation to determine the resistance 'R' as per the relation given by equation (1.2),

$$\boxed{R = \frac{V}{I}} \tag{1.2}$$

The Ohm's law can be realized only if the plot between 'V' and 'I' are linear as shown in Fig. (1.1).

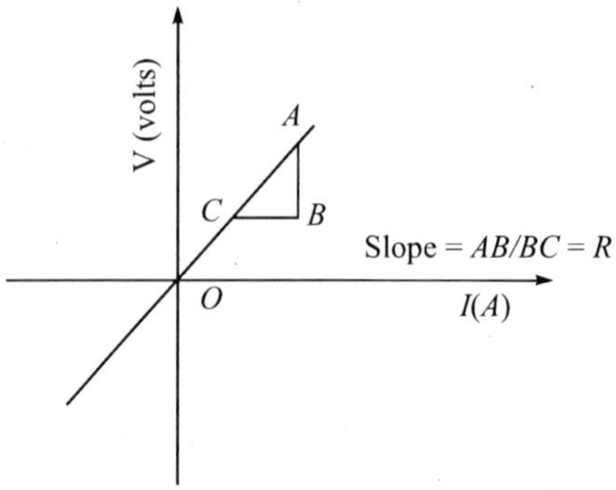

Fig. 1.1 Linear characteristic of resistor.

If the device has a non-linear characteristic as shown in the Fig. (1.2), then it does not obey Ohm's law.

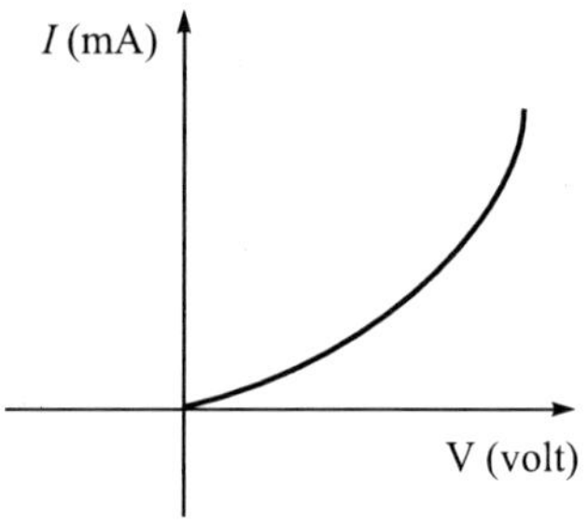

Fig. 1.2 Non-linear characteristic of diode.

The Ohm's law can be expressed in a general way by concentrating on the conducting materials rather than on the basis of conducting devices. The mathematical expression as given by eq. (1.3) gives a general expression of Ohm's law.

$$\boxed{\overline{E} = \rho \overline{J}} \tag{1.3}$$

The equation (2) states that "A conducting material obeys Ohm's law when the resistivity 'ρ' of the material is independent of the magnitude and direction of the applied electric field $\overline{E}$".

Thus, all the homogeneous materials whether they are conductors like copper or semiconductors like pure silicon or impure semiconductors, obeys Ohm's law within some range of values of electric field. If the field $\overline{E}$ is strong enough, then there is departure from Ohm's law in all cases.

Kirchoff's Current Law (KCL)

In the circuits where there are more than two branches carrying currents meeting at a point called junction. Then, "the sum of the currents entering any junction is equal to the sum of currents leaving the junction".

This law is often called Kirchoff's junction rule or Kirchoff's current law (KCL).

It is nothing but the law of conservation of charge for a steady flow of charge, i.e., there is neither increase nor decrease of charge at a junction.

Mathematically, the KCL can be expressed as

$$\boxed{\sum_{i=1}^{n} I_i = 0} \quad (1.4)$$

where, 'n' is the number of branches
'i' is the ith branch
'I_i' is the ith branch current

Illustration: If we consider a circuit having three branches containing active and passive device as shown in the Fig. (1.3).

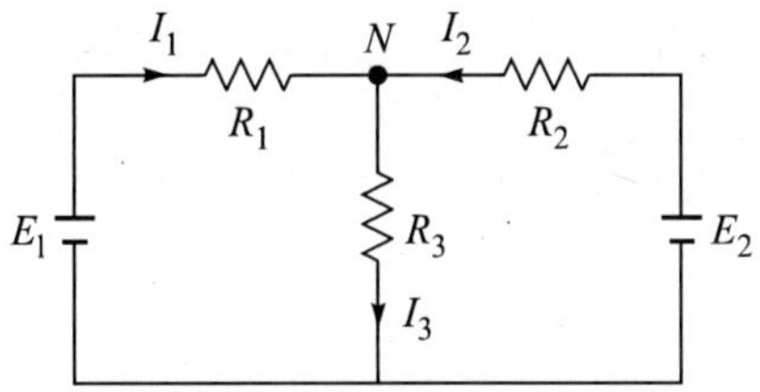

Fig. 1.3 Illustration of KCL.

Applying KCL at junction 'N' of the circuit shown in Fig. (1.3), we have

$$I_1 + I_2 = I_3$$

Sign Convention in 'KCL': The current entering the junction is taken as positive and the current leaving the junction is taken as negative.

EXAMPLE 1: Determine the current 'I' delivered by the battery delivering current of 1 A and 2 A through the branches containing resistances R_1 and R_2 as shown below:

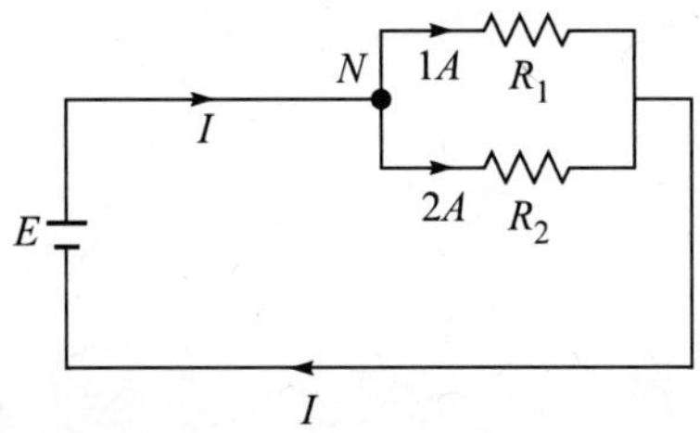

A.1: Applying KCL at junction (N), we get $I = 1 + 2 = 3$ A

Kirchoff's Voltage Law (KVL)

In a mesh of a multiloop circuit, it is seen that the "algebraic sum of the emfs and the potential drops in a mesh is equal to zero". This law is called Kirchoff's voltage law or Kirchoff's loop rule.

Mathematically, KVL can be stated as follows:

$$\boxed{\sum_{i=1}^{n} I_i R_i + \sum_{j=1}^{m} E_j = 0}; \qquad n \neq m \tag{1.5}$$

where, 'I_i' → i^{th} branch current
'R_i' → i^{th} branch resistance
'E_j' → j^{th} branch emf

Sign convention:

(i) The potential drop along the clockwise direction of loop current is taken as positive and vice versa.

(ii) If emfs are present, then in the direction of electronic current from the negative terminal of the battery to the positive terminal of the battery, the sign is taken as negative. That is, the potential rise is taken as positive and vice versa.

Resistance rule: For a move through a resistance in the direction of current; the change in potential is '$-IR$' and in the opposite direction change in potential is '$+IR$' respectively. This rule is illustrated in Fig. (1.4) as shown below.

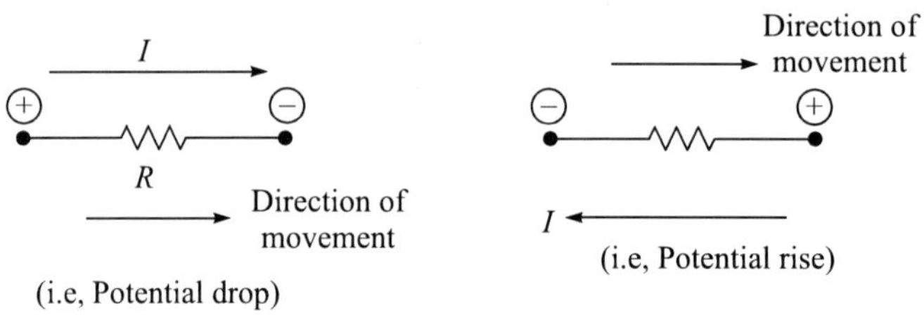

Fig. 1.4 Illustration of resistance rule.

EMF rule: For a move through an ideal emf in the direction of the emf arrow the potential change is '$+E$' and in opposite directions is '$-E$' as shown in Fig. (1.5) below

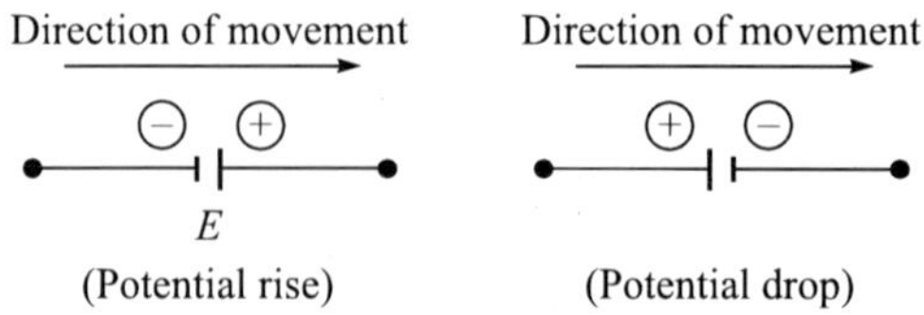

Fig. 1.5 Illustration of EMF rule.

Illustration of KVL: In the Fig. (1.6), which contains a battery having emf 'E' and three resistances R_1, R_2 and R_3 respectively.

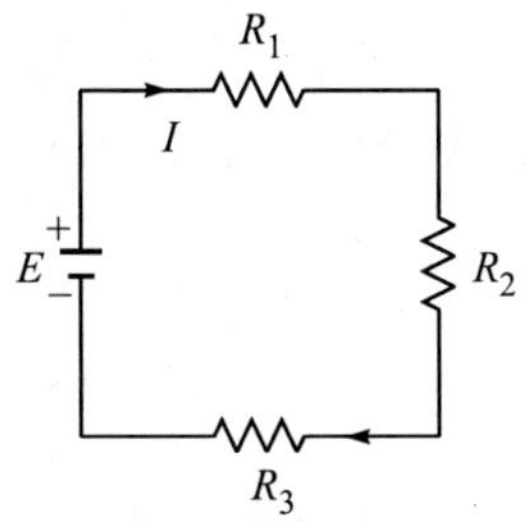

Fig. 1.6 KVL illustration.

The emf 'E' drives a current 'I' through the circuit in the clockwise direction from the positive terminal to the negative terminal of the battery. The direction of the conventional current is always from higher potential to lower potential.

By applying the KVL, we have

$$+ IR_1 + IR_2 + IR_3 - E = 0$$

or,
$$E = IR_1 + IR_2 + IR_3 = I(R_1 + R_2 + R_3)$$

Thus, the emf is equal to the sum of the potential drop which occurs across each resistance. Thus, we see that the response in a given circuit can be determined if the emfs and resistances are known.

EXAMPLE 2: The emfs and resistances in the circuit as shown below have the following values:

$$E_1 = 4 \text{ V}, E_2 = 2 \text{ V}, r_1 = 2\ \Omega, r_2 = 3\ \Omega, R = 5\ \Omega$$

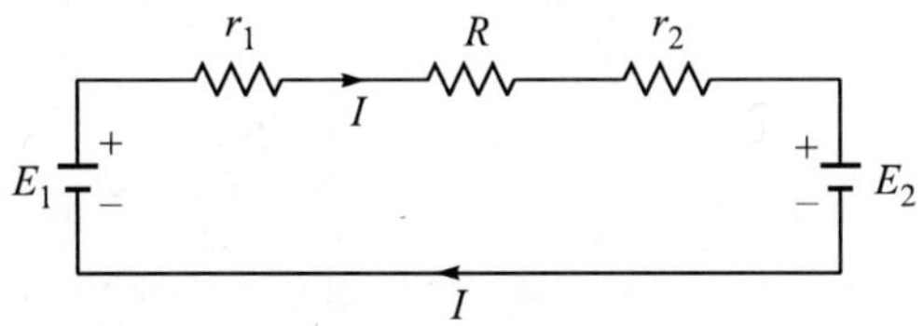

Determine 'I' in the circuit.

A.2: Applying KVL to the circuit, we get

$$r_1 \cdot I + I \cdot R + r_2 \cdot I + E_2 - E_1 = 0$$

or,
$$I(r_1 + r_2 + R) = E_1 - E_2$$

or,
$$I = \frac{E_1 - E_2}{r_1 + r_2 + R}$$

Substituting the values

$$I = \frac{4-2}{2+3+5} = \frac{2}{10} = 0.2 \text{ A}$$

Resistance in Parallel

Consider the circuit shown in Fig. (1.7).

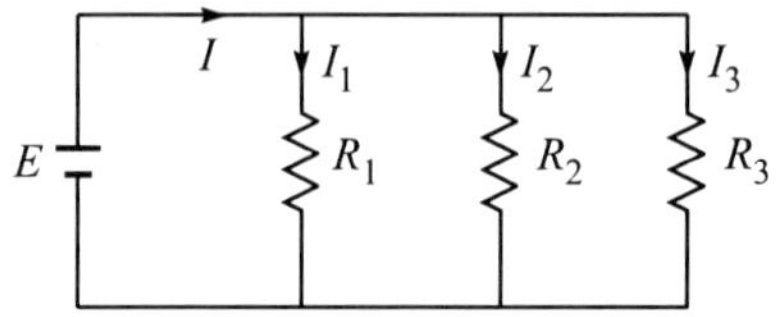

Fig. 1.7 Resistance in parallel.

Parallel resistance means that the potential across each is same and the current through them differs if the values of resistances R_1, R_2 and R_3 are different.

The three resistances as shown in Fig. (1.7) can be replaced by an equivalent resistance 'Req_p' as shown in Fig. (1.8).

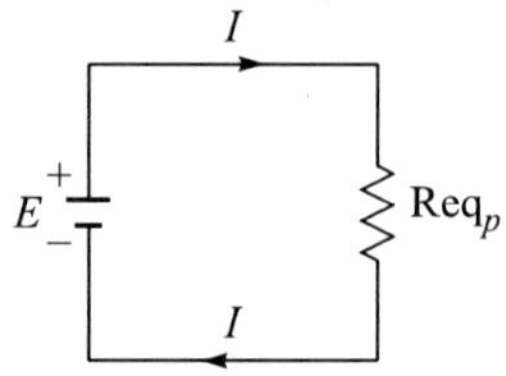

Fig. 1.8 Equivalent circuit with R_1, R_2, R_3, replaced by 'Req_p'.

Applying 'KCL' to the above circuit shown by Fig. (1.7).

$$I = I_1 + I_2 + I_3$$

$$I = \frac{E}{R_1} + \frac{E}{R_2} + \frac{E}{R_3}$$

$$I = E\left[\frac{1}{R_1} + \frac{1}{R_2} + \frac{1}{R_3}\right] \tag{1.6}$$

$\therefore I = E\left(\frac{1}{\text{Req}_p}\right)$ from the equivalent circuit shown in Fig. (1.8).

Thus, by comparing the two above equations, we have

$$\boxed{\frac{1}{\text{Req}_p} = \frac{1}{R_1} + \frac{1}{R_2} + \frac{1}{R_3}} \tag{1.7}$$

For 'n' resistances we have

$$\boxed{\frac{1}{\text{Req}_p} = \sum_{j=1}^{n} \frac{1}{R_j}} \quad \text{('}n\text{' resistances are in parallel)} \tag{1.8}$$

Similarly, for inductors

$$\boxed{\frac{1}{\text{Leq}} = \sum_{j=1}^{n} \frac{1}{L_j}} \qquad \text{('}n\text{' inductances are in parallel)} \tag{1.9}$$

For capacitors

$$\boxed{\text{Ceq} = \sum_{j=1}^{n} C_j} \qquad \text{('}n\text{' capacitances are in parallel)} \tag{1.10}$$

Series Resistance

Consider a circuit containing a battery having emf 'E' and resistances R_1, R_2, R_3 connected in series as shown in Fig. (1.9) below.

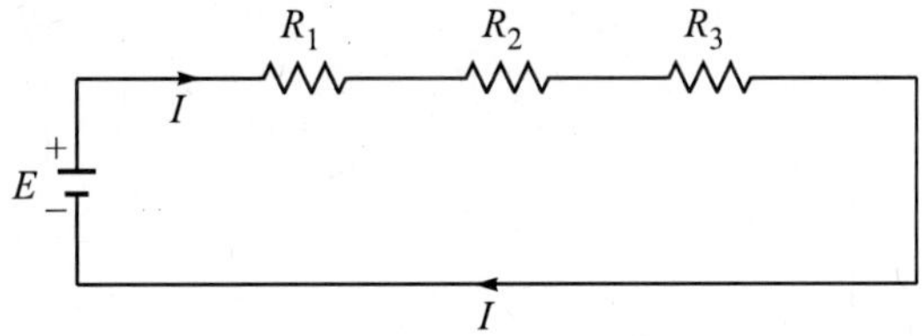

Fig. 1.9 Series resistances.

Consider the current direction as shown, then by applying KVL to the circuit, we have

$$+ I R_1 + I R_2 + I R_3 - E = 0 \tag{1.11}$$

or,
$$E = I (R_1 + R_2 + R_3)$$

or,
$$E = I \, \text{Req}_s \tag{1.12}$$

The resistances can be replaced by a single equivalent series resistance 'Req_s' as shown in the figure below. The value of the equivalent series resistance is given by

$$\boxed{\text{Req}_s = R_1 + R_2 + R_3} \tag{1.13}$$

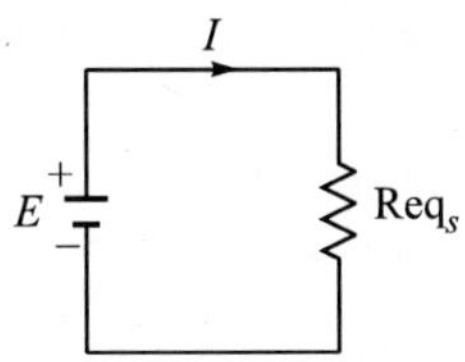

Fig. Equivalent circuit of series circuit containing resistors R_1, R_2, and R_3.

EXAMPLE 3: For the given circuit containing one ideal battery and three resistances with the following values $R_1 = 20\ \Omega$, $R_2 = 20\ \Omega$, $E = 10$ V, $R_3 = 30\ \Omega$

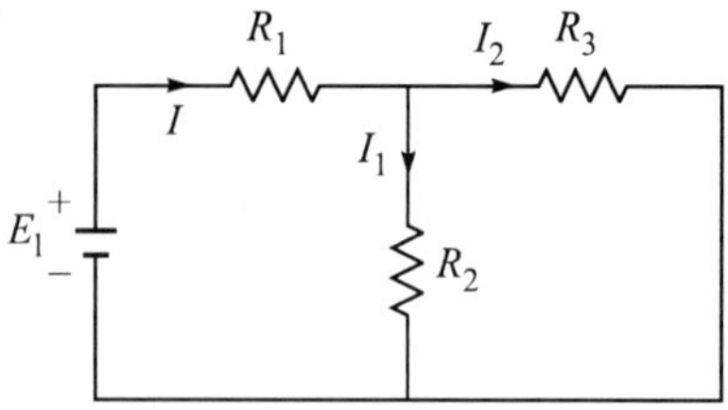

What is the current through the battery?

A.3: The current through the battery is 'I' which is equal to $I_1 + I_2$ as per KCL. The equivalent resistance of the circuit is

$$\text{Req} = R_1 + \frac{R_2 R_3}{R_2 + R_3}$$

$$\text{Req} = 20 + \frac{20 \times 30}{20 + 30} = 20 + \frac{\overset{12}{\cancel{600}}}{\cancel{50}} \; 32\ \Omega$$

So, the circuit reduces to an equivalent circuit given as below

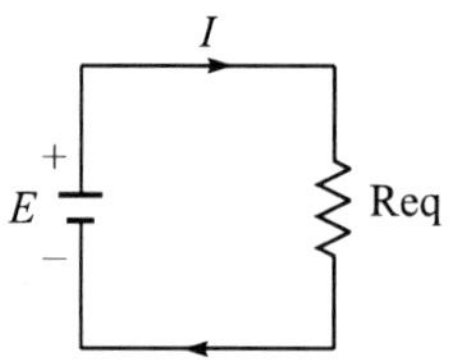

$\therefore$ The current by the battery $I = \dfrac{E}{\text{Req}} = \dfrac{10}{32} = 0.3125$ Amperes.

Thevenin's Theorem

Thevenin's theorem enables to simplify complicated circuits into simple circuit consisting of a single voltage source in series with an equivalent resistance and load resistance.

The Thevenin's theorem states that "a linear bilateral network consisting of active and passive devices can be replaced by a single voltage source called Thevenin's voltage 'V_{th}' and a single equivalent resistance called Thevenin's resistance R_{th}".

The Thevenin's voltage is the open circuit voltage across the terminal about which Thevenin's equivalent circuit is drawn.

The Thevenin's resistance is the equivalent resistance seen across the open circuited terminal with the voltage sources short circuited and their internal resistance considered. If there are current sources then it is open circuited and the internal resistance considered.

Consider the circuit as shown below in Fig. (1.10) Referring to the Fig. (1.11), we see that the circuit given by Fig. (1.10) is simplified.

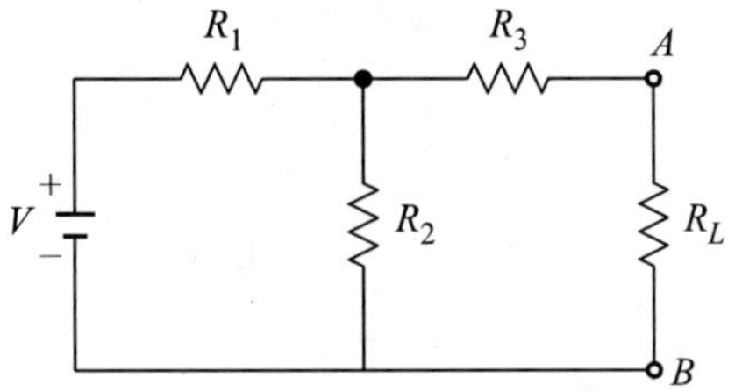

Fig. 1.10 Circuit.

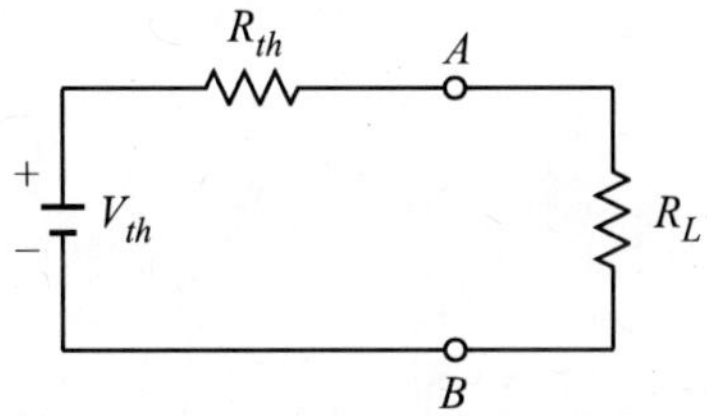

Fig. 1.11 Equivalent Thevenin's circuit.

The Thevenin's voltage 'V_{th}' is determined as follows:

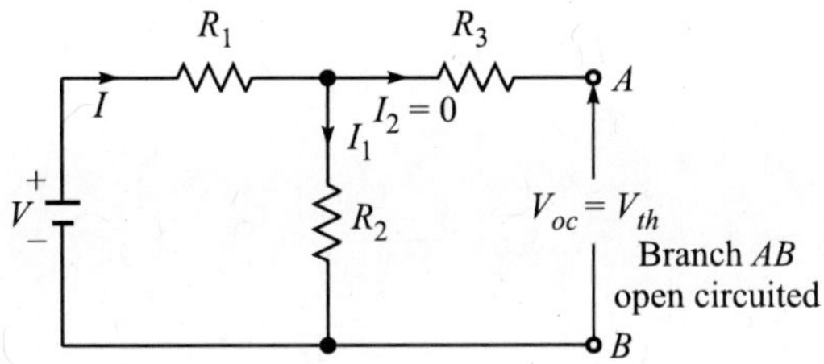

Fig. 1.12 Circuit for determining 'V_{th}".

Then from the fig.

$$V_{th} = V_{oc} = I_1 \cdot R_2, \text{ Volts}$$

∵ 'AB' is open circuited, so, $I_2 = 0$, So $I = I_1 + I_2 = I_1$

Moreover, $$I_1 = I = \frac{V}{R_1 + R_2}, \text{ Amps}$$

Substitute the value of 'I_1' in equation, we get,

$$\boxed{V_{th} = \left(\frac{V}{R_1 + R_2}\right) \cdot R_2} \text{ Volts}$$

Determination of 'R_{th}'

The circuit for obtaining 'R_{th}' is shown in Fig. (1.13).

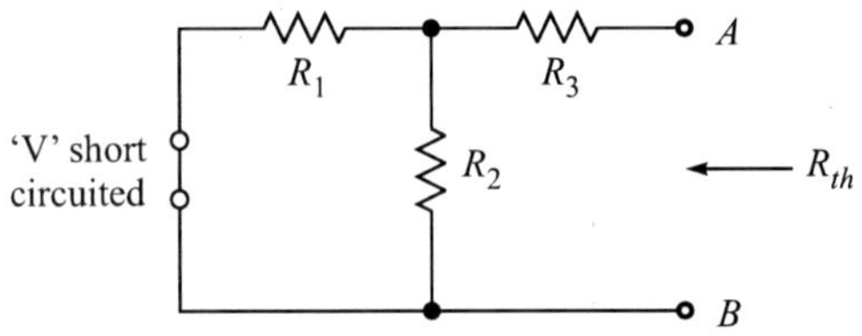

Fig. 1.13 Circuit for calculation of 'R_{th}'.

Thus, from above Fig. (1.12)

$$\boxed{R_{th} = R_3 + \frac{R_1 R_2}{R_1 + R_2}} \text{ Ohms}$$

EXAMPLE 4: Calculate the current through '$R_{5\Omega}$' as shown in the Fig. (1.14) using Thevenin's theorem.

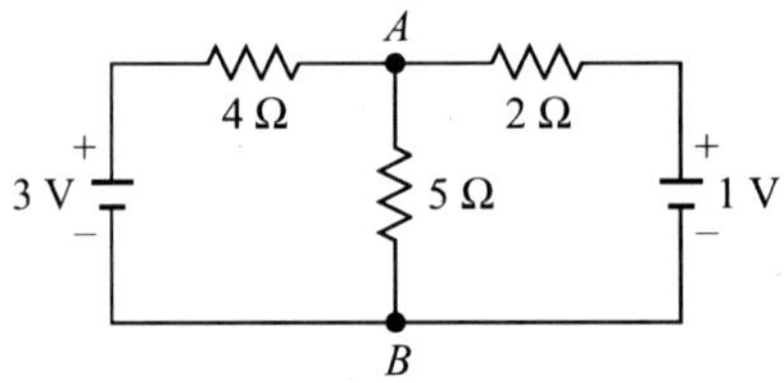

Fig. 1.14

A.4: Applying Thevenin's theorem across '*AB*', we get the Thevenin's equivalent circuit as shown in Fig. (1.15)

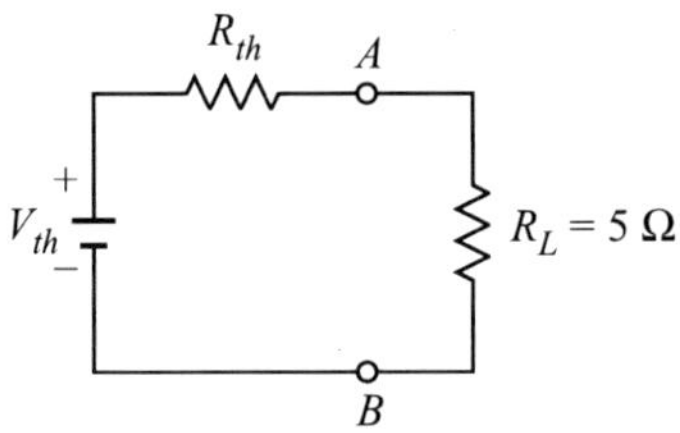

Fig. 1.15 Thevenin's equivalent circuit with load.

V_{th}**:**

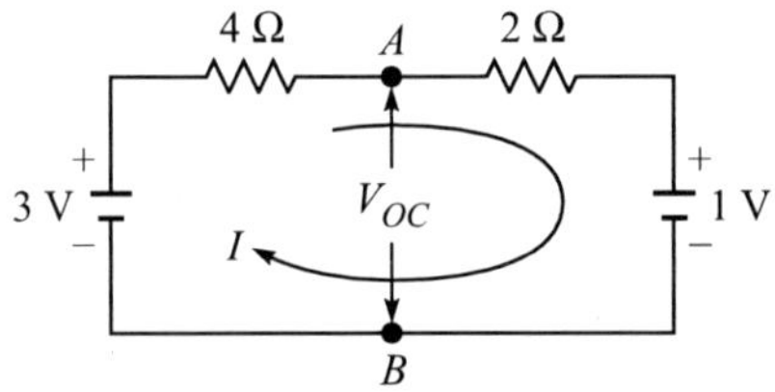

Fig. 1.16 For cal. 'V_{oc}'.

Apply KVL to the above circuit, we get

$$+4I + 2I + 1 - 3 = 0$$

$$6I = 2 \quad \text{or} \quad I = \frac{2}{6} = \frac{1}{3} \text{ Amp} = 0.333 \text{ Amp}$$

Thus, voltage at 'A' is given as $V_A = 3 - I \times 4$

$$V_A = 3 - \frac{1}{3} \times 4 = 3 - \frac{4}{3} = \frac{5}{3} \text{ volts}$$

Since point 'B' is attached to the negative terminal of battery, so $V_B = 0$.

Thus,
$$V_{AB} = V_{oc} = V_A - V_B = \frac{5}{3} - 0 = \frac{5}{3} \text{ volts}$$

So,
$$\boxed{V_{th} = V_{oc} = \frac{5}{3} = 1.67 \text{ volt}}$$

***R_{th}*:**

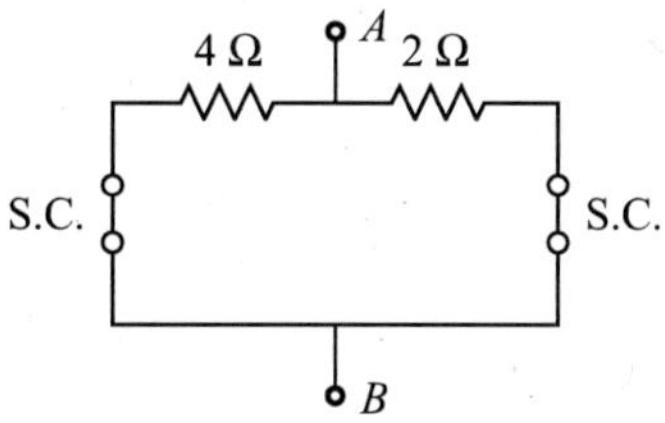

$$R_{th} = R_{AB} = \frac{2 \times 4}{2 + 4} = \frac{8}{6} = \frac{4}{3} = 1.333 \ \Omega$$

$$\boxed{R_{th} = 1.333 \ \Omega}$$

Thus, the load current 'I_L' can be obtained by referring to the Fig. (1.15)

$$I_L = \frac{V_{th}}{R_{th} + R_L} = \frac{1.67}{1.333 + 5} = \frac{1.67}{6.333} = 0.2637$$

$$\boxed{I_L = 0.26 \text{ Amperes}}$$

Maximum Power Transfer Theorem

The maximum power delivered by the source to the load is important in many circuits such as telecommunication circuits and in instrumentation. Sometimes it is applied to the electrical circuits such as equivalent circuit of machines.

The maximum power transfer theorem states that "A circuit containing source having internal resistance 'R_S' and load having resistance 'R_L'. The power delivered from the source to load will be maximum if the source resistance is equal to that of the load resistance".

Alternatively, if the circuit consists of resistances other than source resistance and are connected in series and parallel then the power delivered by the source is maximum if the Thevenin's resistance 'R_{th}' of the circuit is equal to the load resistance 'R_L'.

i.e.,
$$R_L = R_{th} \tag{1.14}$$

EXAMPLE 5: Two batteries of emf 'E' and internal resistance 'r' are connected in parallel across a resistance 'R' as in Fig. (1.16).

(a) For what value of 'R', the power dissipated in 'R' is maximum?
(b) Find the value of maximum power.

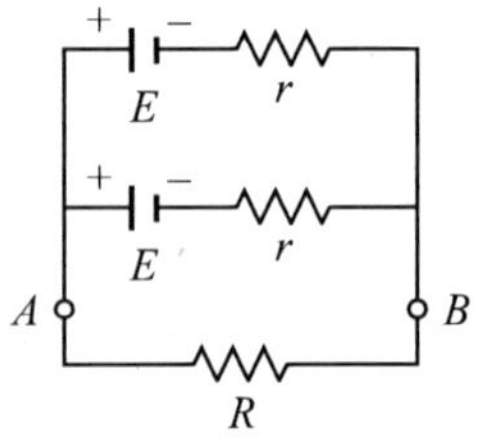

Fig. 1.16

A.5:

(a) Open circuiting the branch containing 'R' and shorting 'E'. The circuit shown by Fig. (1.16) reduces to a simple circuit as shown in Fig. (1.17).

$\boldsymbol{R_{th}}$:
$$R_{th} = \frac{r \times r}{r + r} = \frac{r^2}{2r} = \frac{r}{2}, \text{ ohms}$$

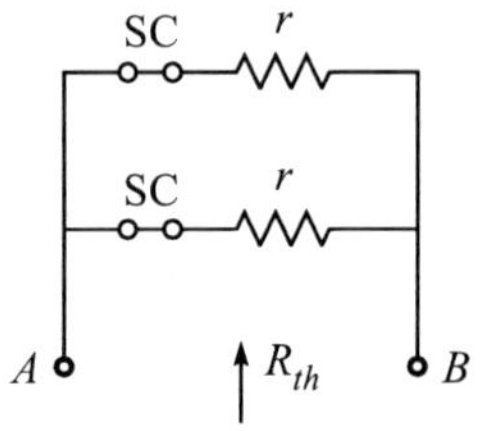

Fig. 1.17 Calculation of 'R_{th}'.

When, $R = R_{th}$ the power delivered to 'R' is maximum.

(b) The maximum power expression is given as

$$P_{R_{\max}} = I_L^2 R = \left(\frac{V_{th}}{R_{th} + R}\right)^2 . R$$

$$= \left(\frac{V_{th}}{2 R_{th}}\right)^2 \times R_{th} = \frac{V_{th}^2}{4 R_{th}^2} \times R_{th}$$

$$P_{R_{\max}} = \frac{V_{th}^2}{4 R_{th}}$$

V_{th} :

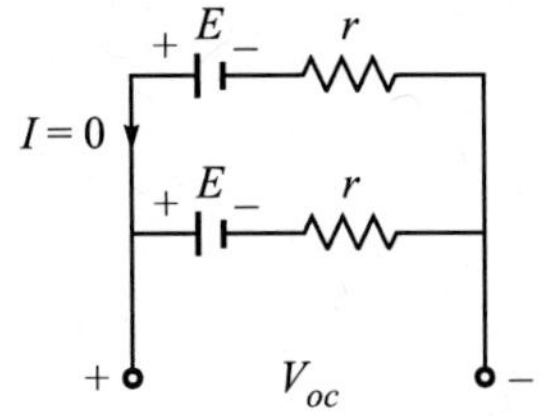

$$V_{oc} = V_{th} = E$$

(Applying parallel throrem)

Fig. For calculation of 'V_{th}'.

$$\therefore \qquad \boxed{P_{R_{max}} = \frac{E^2}{4(r/2)} = \frac{E^2}{2r}} \text{ watt.}$$

1.3 ALTERNATING CURRENT CIRCUITS

The a.c. circuit analysis is important since most of the machines we deal with in the electrical engineering use a.c. power. The popularity of a.c. is due to the easy availability and transmission over large distances from the source centres.

The a.c. excitation or source is sinusoidal in nature. The representation of the sinusoidal voltage in time domain is given by eq. (i)

$$v(t) = V_m \sin \omega t \qquad \text{(i)}$$

where, 'V_m' is the maximum amplitude of the voltage, in Volts.
'ω' is the angular frequency, in radian per sec.
't' is the instantaneous time, in seconds
$v(t)$ is the instantaneous voltage, in Volts.

The representation of the voltage wave is given in Fig. (1.18).

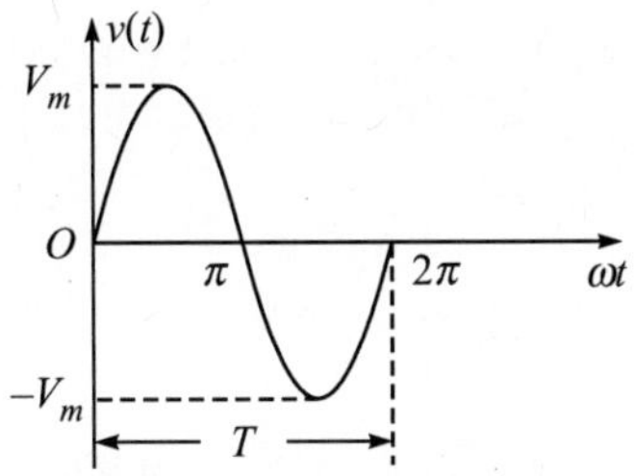

Fig. 1.18 A.C. voltage waveform.

The time period of voltage waveform is given by $T = \frac{1}{f}$, where 'f' is the frequency of voltage wave.

Angular frequency 'ω' = $2\pi f$ rad/s.

so, $$f = \frac{\omega}{2\pi}, \text{ Hertz}$$

Thus, $$T = \frac{2\pi}{\omega}, \text{ second}$$

Average value of the voltage '$v(t)$' is given by

$$V_{av} = \frac{1}{T}\int_0^T v(t) \cdot dt = 0 \tag{1.15}$$

So, the average value of a.c. wave (voltage and current) is zero over the full cycle or one time period.

The r.m.s. value of the voltage is also called the effective value of the a.c. quantity is important from the practical point of view. The rating of the a.c. machines and appliances is given is terms of the root mean square value.

Mathematically, the root mean square value of an a.c. quantity $\{v(t) \text{ or } i(t)\}$ is given as follows:

$$V_{rms} = \sqrt{\frac{1}{T}\int_0^T v^2(t)dt}, \text{ volts} \tag{1.16}$$

and $$I_{rms} = \sqrt{\frac{1}{T}\int_0^T i^2(t)dt}, \text{ Amp} \tag{1.17}$$

By substituting the values of $v(t)$ and $i(t)$ in eq. (1.16) and eq. (1.17), we get

$$V_{rms} = \frac{V_m}{\sqrt{2}} = 0.707\ V_m, \text{ Volts} \tag{1.18}$$

$$I_{rms} = \frac{I_m}{\sqrt{2}} = 0.707\ I_m, \text{ Amperes} \tag{1.19}$$

The r.m.s. value of a.c. quantities are expressed with no subscripts as 'V' and 'I' respectively. The r.m.s value of the voltage and current is 70.7 % of the maximum value.

The r.m.s. value of voltage and current are important as it is marked over different resistive inductive and capacitive loads such as bulbs, inductive, chokes and capacitors used in pumps and fans type loads.

Power: The power rating of the loads is very important from point of view of selection for a particular application. For determining the efficiency, the power required is average or the active power consumed by the load.

The value of average power can be determined from the expression of instantaneous power as given below:

$$p = v(t)\ i(t) \tag{1.20}$$

where, $$v(t) = V_m \sin\omega t$$
$$i(t) = I_m \sin\omega t$$

Substituting the value of $v(t)$ and $i(t)$ in eq (1.20), we get

$$p = V_m I_m \sin^2\omega t$$

$$p = \frac{V_m \cdot I_m}{2}(1 - \cos 2\omega t)$$

$$p = \frac{V_m I_m}{2} - \frac{V_m I_m}{2}\cos 2\omega t \tag{1.21}$$

Taking average of the instantaneous power over one cycle, we get

$$P_{av} = P = \frac{1}{T}\int_0^T p\,dt \tag{1.22}$$

Solving, we get

$$P = \frac{V_m I_m}{2} = \frac{V_m}{\sqrt{2}} \cdot \frac{I_m}{\sqrt{2}} = V I, \text{ watts} \tag{1.23}$$

Thus, we see that the average power or active power or useful power is equal to the product of r.m.s. value of voltage and current in the case when there is no phase difference between the voltage and current as in resistive loads.

For inductive and capacitive loads where there is a phase difference of 'ϕ' between the voltage and current. The expression of the average power is given by

$$P = V I \cos\phi, \text{ watts} \tag{1.24}$$

where, '$\cos\phi$' is the power factor of the load.

Phasor Approach to A.C. Quantities

The phasor approach to a.c. quantities simplifies the analysis, so the concept of phasor should be clear in our mind so as to obtain the clear understanding of the a.c. quantities.

If we consider the sinusoidal voltage and current quantities given by the expression

$$v(t) = V_m \sin\omega t, \text{ volts}$$
$$i(t) = I_m \sin\omega t, \text{ Amperes}$$

Then using Euler's identity,

$$V_m = \sin\omega t = I_m\,[V_m\, e^{j(\omega t+0)}] \tag{1.25}$$

where, $$j = \sqrt{-1}$$
$$V_m = \sqrt{2}\,V, \text{ volts and } V = V_{rms}$$

Since '$\sqrt{2}\,e^{j\omega t}$' is constant for a particular frequency 'ω', so it is not considered in the expression of phasor. The phasor is a vector quantity having magnitude 'V' and direction depending on the phase difference between '$v(t)$' and $i(t)$.

In this case, the phase difference between the voltage and current is equal to zero. So 'ϕ' is equal to zero. The phasor of $v(t)$ represented by an arrow of length 'V' equal to rms value of voltage and direction along 'x' axis. So, mathematically, the phasor '$\overline{V}$' is represented by eq. (1.26),

$$\overline{V} = V\angle 0° \tag{1.26}$$

The phasor of current $i(t)$ is given by (1.27)

$$\bar{I} = I\angle 0° \tag{1.27}$$

where V, I are root mean square value of $v(t)$ and $i(t)$ respectively.

If there is phase difference of 'ϕ' between voltage and current, then the phasors $\overline{V}$ and $\bar{I}$ depends on the type of loads across which the current is flowing and voltage is measured.

For capacitive loads, phasor '$\bar{I}$' leads the phasor $\overline{V}$ by 'ϕ' (say) that is

Let, $$\overline{V} = V\angle 0°\text{(say)} \tag{1.28}$$

then, $$\bar{I} = I\angle +\phi \tag{1.29}$$

Similarly, for inductive loads, the current through the loads lags the voltage across the load by an angle ϕ (say) then,

Let, $$\overline{V} = V\angle 0° \text{ (say)} \tag{1.30}$$

then, $$\bar{I} = I\angle -\phi \tag{1.31}$$

The phasor analysis is possible if the frequency 'ω' of the phasors is not changing. If the frequency 'ω' of two phasors is different, then they cannot be compared.

The phasors addition and subtraction is different from the corresponding scalar addition and subtraction of two a.c. quantities magnitude. Since the phasors are vectors and involve the direction, so, the phasors are represented in the j-plane as given below in Fig. (1.19).

Let $$\overline{V} = V\angle 0°$$

and $$\bar{I} = I\angle -\phi \text{ (say)}$$

then the phasor representation is as follows.

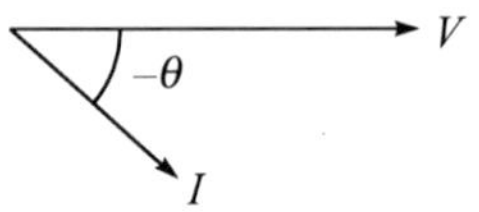

Fig. 1.19 Phasor representation of $\overline{V}$ and $\bar{I}$.

If
$$v_1(t) = V_{m_1} \sin \omega t$$
$$i_1(t) = I_{m_1} \sin 2\omega t$$

then, phasor $\overline{V}_1$ and $\overline{I}_1$ can't be represented in the same phasor diagram as shown in Fig. (1.19) as they differ in frequency.

1.4 THREE-PHASE CIRCUITS

The bulk power is always dealt with three-phase circuits. The power which is generated by an alternator involves three-phase stator windings which is connected such that balanced voltages is obtained across the stator terminals. Three-phase circuits possess certain economic and operating advantages over the single-phase circuits.

The transmission of the three-phase power from the generating end to the utility end is made possible using three-phase transformers whose primary and secondaries are connected in star 'Y' or delta 'Δ'. The knowledge of three-phase circuits is therefore important from the point of view of analysis.

The phasor method enables us to understand the behaviour of the voltages and currents in three-phase circuits. The unbalanced three-phase systems knowledge is necessary because due to improper loading on each phase the system tends to affect the three-phase source. So, if the system gets unbalanced, then it is necessary to know how we can make the system balance.

The power consumed by the load is also to be known. Moreover, the power factor of the load is to be determined accurately so that in case of decrease of the power factor of load, some measure should be taken such as by using capacitor bank to improve the power factor at the receiving end.

Balanced Three-Phase Source

A three-phase source is one which has three equal voltages which are 120° out of phase with one another. Such a source is called balanced three-phase source.

The three voltages are usually generated in the same machine i.e., alternator. The alternator stater phase windings are so designed that across each phase winding we get same voltage magnitude. The individual voltages across each phase winding of an alternator is shown in the Fig. (1.20).

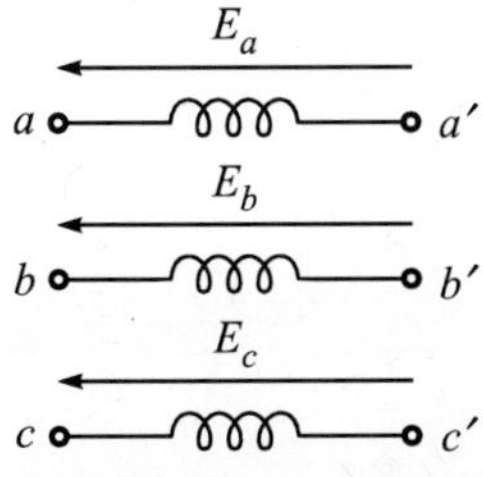

Fig. 1.20 Induced emf in phase windings of an alternator.

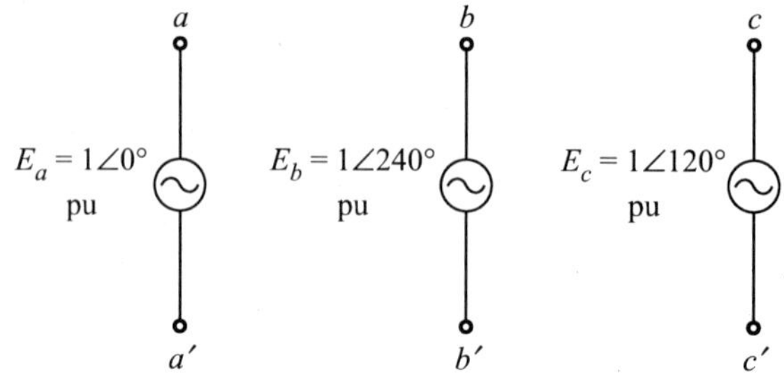

Fig. 1.21 Individual phase voltage in an alternator.

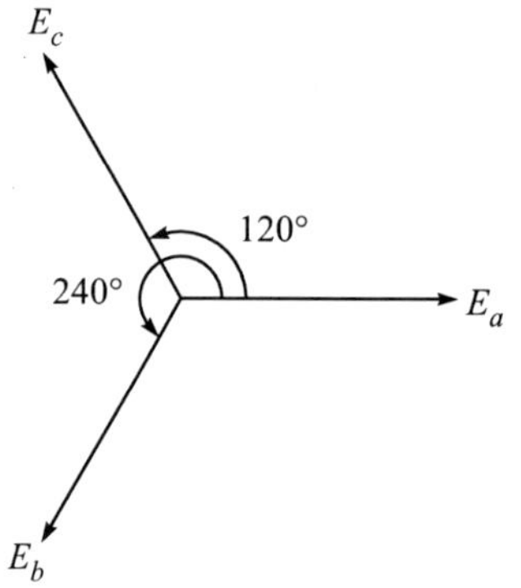

Fig. 1.22 Phasor diagram of three-phase balanced source.

In a three-phase balanced system, the sum of the three voltages is equal to zero. The sum of the phase currents is equal to zero. So the current in a neutral wire is equal to zero i.e., $I_N = 0$. The three-phase balanced source can be obtained by star or delta connections. The star and delta connected three-phase source is shown in the Fig. (1.23).

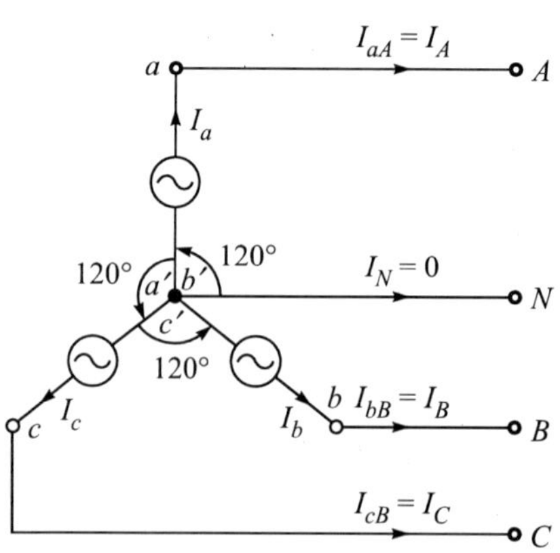

Fig. 1.23 (a) *Y* connected source (3-phase).

There are two manners in which the phase windings of the alternator may be connected either the three phases aa', bb' and cc' are supplying single phase individually by connecting their ends a', b' and c' as shown in Fig. 1.23(a). This type of connection is used in the modern alternators. Thus, the terminal a', b' and c' are interconnected to form the neutral point. To this neutral point, a wire called neutral wire is joined which is connected to earth. This wire carries zero current if the three-phase system is balanced.

Thus, we see that from Fig. (1.23) four wires are coming from the source. So such type of system is called three-phase, four-wire system. Sometimes, the extending of the neutral wire is not necessary. Such system is called three-phase three-wire system.

Another connection of the phase winding is possible by connecting a to b', b to c' and c to a' such that they are connected in Δ fashion. Such a connection is called Δ connection as shown in Fig. 1.23(b).

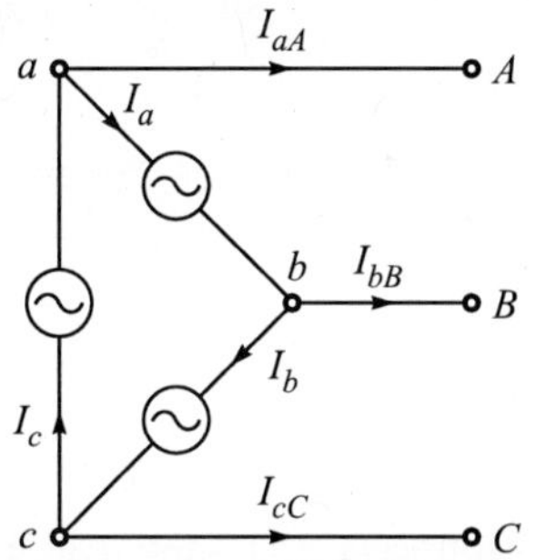

Fig. 1.23 (b) Δ connected source (three-phase).

The basic requirements for three-phase balanced system are as follows:

(i) The three-phase voltages are equal i.e., $E_a = E_b = E_c$.

(ii) The angle between each phase voltages is equal to 120°.

(iii) The impedances of each phase are equal $|\bar{Z}_a| = |\bar{Z}_b| = |\bar{Z}_c|$.

(iv) The phase currents are equal $|\bar{I}_a| = |\bar{I}_b| = |\bar{I}_c|$.

(v) The angle between each phase current is equal to 120°

(vi) The active power flow in each phase is equal

i.e.,
$$P_a = P_b = P_c$$

(vii) The reactive power flow in each phase is equal

i.e.,
$$Q_a = Q_b = Q_c$$

(viii) The current in the neutral wire is equal to zero

i.e.,
$$\bar{I}_N = \bar{I}_A + \bar{I}_B + \bar{I}_C = 0$$

Unbalanced Three-Phase System

The unbalanced system is one in which the phase voltages and currents are not equal in magnitude and direction. The power flow—both active power and reactive power— are also unequal in different phases. The phasor diagram of connected unbalanced three-phase system is shown in the Fig. (1.24).

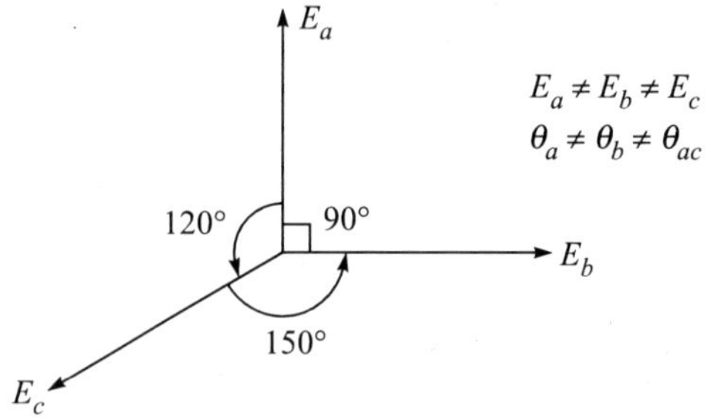

Fig. 1.24 Unbalanced three-phase system.

The unbalanced systems are not dealt with in this book since majority of the practical problems are concerned with balanced systems. We consider that the loads attached to each phase are of equal magnitude.

1.5 THREE-PHASE CONNECTIONS

Star Connected Systems

When the three phases are *Y*-connected as shown in Fig. 1.23(a) the phasor diagram of the phase voltage is shown in Fig. (1.22).

Consider the phase order as '*abc*'. That is, the phase *A* voltage magnitude reaches maximum first then the phase *B* voltage then the phase *C* voltage respectively. This phase order is also called the phase sequence. From Fig. (1.25), we have three-phase voltage E_{oa}, E_{ob} and E_{oc}. The phase voltage is also referred to as line to neutral voltages.

Whereas the voltages E_{ab}, E_{bc} and E_{ca} are called the line voltage or line to line voltages.

Using KVL, the line voltages can be determined as follows.

$$E_{ab} = E_{ao} + E_{ob} = -E_{oa} + E_{ob} = \sqrt{3}\, E_{ob} \angle -30° \tag{1.32}$$

Similarly,

$$E_{bc} = \sqrt{3}\, E_{oc} \angle -30° \tag{1.33}$$

$$E_{ca} = \sqrt{3}\, E_{oa} \angle -30° \tag{1.34}$$

For three-phase balanced system,

$$E_{oa} = E_{ob} = E_{oc} = E_p \text{ (say)} \tag{1.35}$$

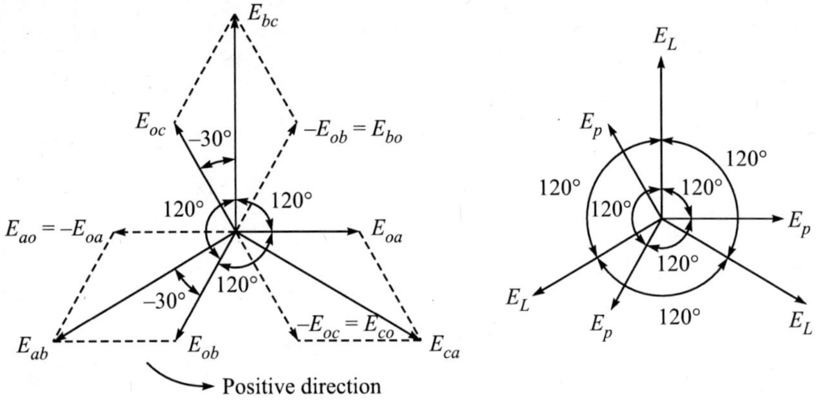

Fig. 1.25 Phasor diagram of *Y*-connection showing line and phase voltages.

and so,
$$E_{ab} = E_{bc} = E_{ca} = \sqrt{3}\, E_p \tag{1.36}$$
$$E_{ab} = E_{bc} = E_{ca} = E_L \text{ (say)} \tag{1.37}$$

Comparing eq. (1.36) and eq. (1.37), we get

$$\boxed{E_L = \sqrt{3}\, E_p} \tag{1.38}$$

From the equation, we see that for a *Y*-connection, the line voltage is $\sqrt{3}$ times the phase voltage or the line to line voltage is $\sqrt{3}$ times that of lines to neutral voltage. The line currents I_{aA}, I_{bB} and I_{cC} are equal and are phase displaced by 120°. This is shown in Fig. (1.26).

So, from the Fig. (1.26), we see that for a *Y*-connection, the line currents and phase currents are equal.

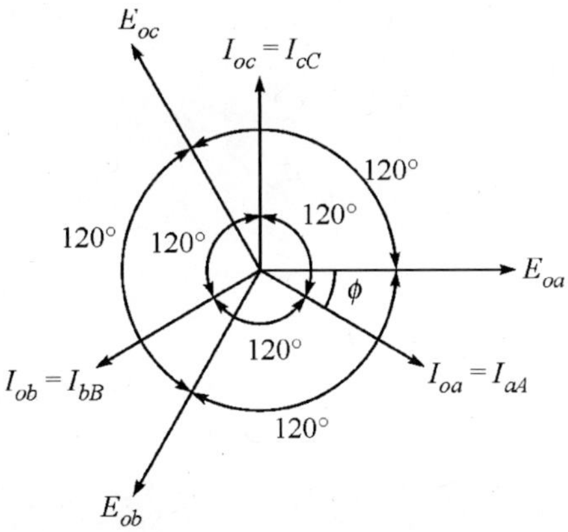

Fig. 1.26 Current phasor diagram (for *Y* connection).

$$I_{oa} = I_{ob} = I_{oc} \tag{1.39}$$

$$I_{aA} = I_{bB} = I_{cC} \tag{1.40}$$

i.e.,

$$\boxed{I_L = I_p}$$

Delta Connected System

The Δ connected system is shown in Fig. 1.23(b).

From the phasor diagram of the voltage as shown in Fig. (1.27).

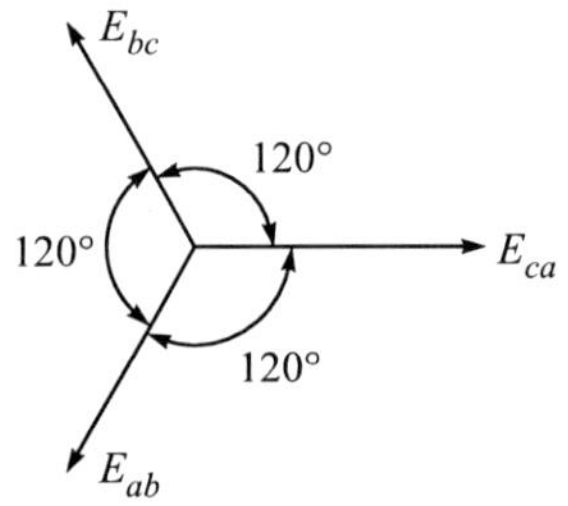

Fig. 1.27 Voltage phasor for Δ system diagram.

From figure

$$E_{ca} = E_{ab} = E_{bc} = E_p \tag{1.41}$$

$$E_{CA} = E_{AB} = E_{BC} = E_L \tag{1.42}$$

Considering negligible drop in the connecting wires. From above eq. (1.41) and eq. (1.42).

$$\boxed{E_L = E_p} \tag{1.43}$$

From the above, we conclude that for 'Δ' connected system, the line voltages and phase voltages are equal. The phasor diagram showing phase currents and line currents are shown in the Fig. (1.28).

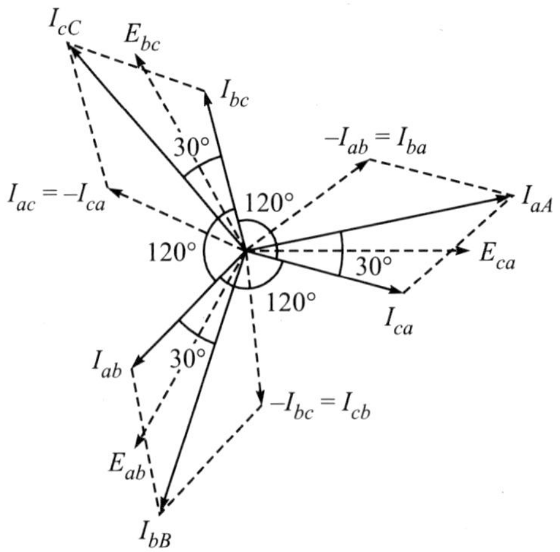

Fig. 1.28 Phasor diagram for 'Δ' connection.

In the Fig. (1.28) I_{ab}, I_{bc}, I_{ca} are the phase currents and the I_{aA}, I_{bB}, I_{cC} are the line currents.

The subscripts indicate the direction of currents.

Using KCL, we can determine the line currents as follows:

$$I_{aA} = I_{ba} + I_{ca} = -I_{ab} + I_{ca} \tag{1.44}$$

$$I_{aA} = \sqrt{3}\, I_{ca} \angle{+30°} \tag{1.45}$$

Similarly,

$$I_{bB} = \sqrt{3}\, I_{ab} \angle{+30°} \tag{1.46}$$

and

$$I_{cC} = \sqrt{3}\, I_{bc} \angle{+30°} \tag{1.47}$$

Thus, we conclude that for a Δ connected system, the line current is $\sqrt{3}$ times the phase current.

Mathematically, we can represent the statement as follows:

$$\boxed{I_L = \sqrt{3}\, I_p} \tag{1.48}$$

where, 'I_L' is the line current, in amperes
'I_p' is the phase current, in amperes.

Three-Phase Load

As we consider the balanced and unbalanced cases when we are discussing the three-phase source, similarly, we have balanced three-phase loads and unbalanced three-phase loads respectively.

The balanced three-phase loads have the following properties as given below.

1. The impedance of each phase of three-phase connected loads is equal in magnitude. Mathematically,

$$|\bar{Z}_{LA}| = |\bar{Z}_{LB}| = |\bar{Z}_{LC}| \tag{1.49}$$

2. The phase angle of the individual impedance of each phase is equal.

$$\theta_A = \theta_B = \theta_C \tag{1.50}$$

where, θ_A, θ_B and θ_C are the phase angles of the impedance $\bar{Z}_{LA}$, $\bar{Z}_{LB}$ and $\bar{Z}_{LC}$ respectively.

3. The current in each phase is equal in magnitude

$$|\bar{I}_A| = |\bar{I}_B| = |\bar{I}_C| \tag{1.51}$$

4. The angle between the voltages and currents is same.
5. The angle between the phase voltages is same.
6. The angle between the phase currents is same.
7. The angle between the line voltages is same.
8. The angle between the phase voltages is same.

The representation of a three-phase balanced load is shown in Fig. (1.29).

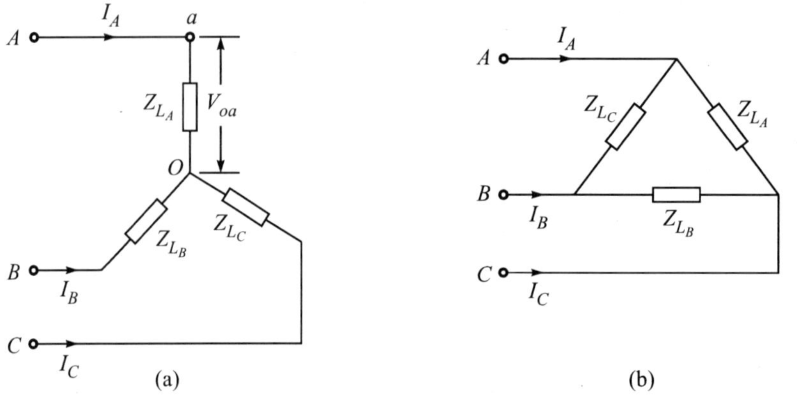

Fig. 1.29 *Y* and Δ connected balanced load.

The three-phase load consumes power and thus current passes through the load impedances.

The phasor of the currents and voltages in the three-phase loads is given in similar manner as that of three-phase source. The phasor representation is given in Fig. (1.30).

V_{oa}, V_{ob}, V_{oc} are the phase voltages.

I_a, I_b, I_c are the phase currents.

θ_a, θ_b, θ_c are the phase angles between V_{oa} and I_a, V_{ob} and I_b V_{oc} and I_c respectively.

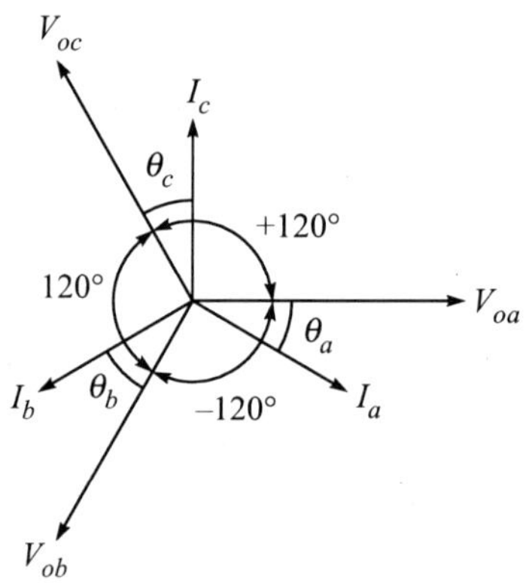

Fig. 1.30 Phasor representation of phase voltages and phase currents in *Y*-connected load (balanced).

The three-phase load can be connected in delta as shown in Fig. (1.31). The phasor diagram of the phase voltages and phase currents can be shown in a similar manner.

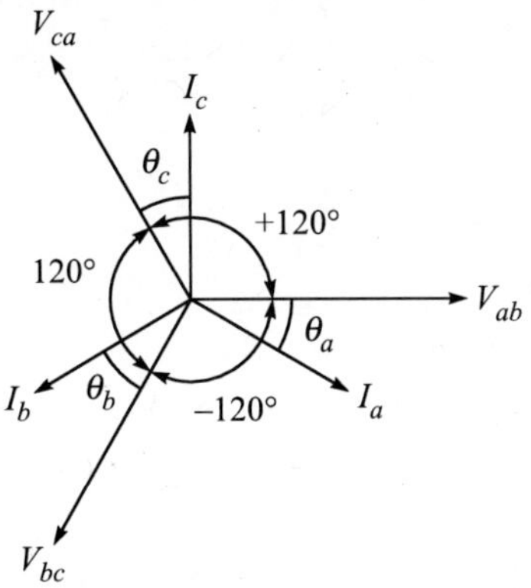

Fig. 1.31 Phasor representation of three-phase Δ connected system.

where, V_{ab}, V_{bc} and V_{ca} are the phase voltages.
I_a, I_b, and I_c are the phase currents.
θ_a, θ_b and θ_c are the phase angles of the impedances of each phase.

Both the phasor diagram as shown in Fig. (1.30) and Fig. (1.31) are drawn for an inductive load.

The phasor diagram for an unbalanced load is represented as shown in (Fig. 1.32).

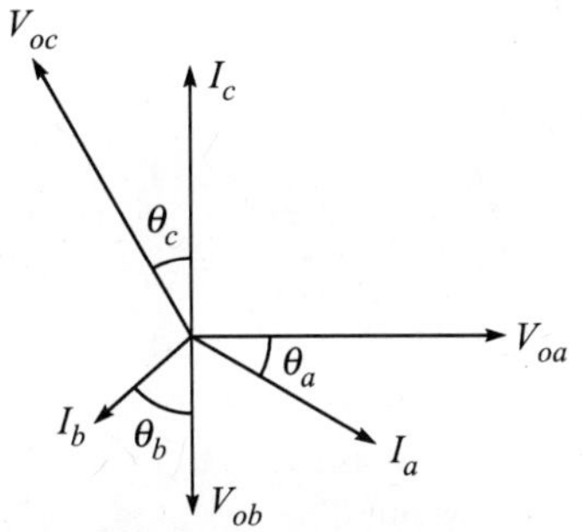

Fig. 1.32 Phasor diagram for a three-phase unbalanced load.

From the Fig. (1.32), we see that

$$I_a \neq I_b \neq I_c \tag{1.52}$$

$$V_{oa} \neq V_{ob} \neq V_{oc} \tag{1.53}$$

$$\theta_a \neq \theta_b \neq \theta_c \tag{1.54}$$

The unbalanced load affects the balanced three-phase source and will make it unbalanced. So it is necessary that the load should be balanced. But since with modernization, the use of non-linear devices is increasing day by day, these devices make the load unbalanced and so this unbalancing spreads in the grid based system. More researches are going on in the field for eliminating this non-linear effect on the power system network. The waveforms of the voltages and currents should be as sinusoidal as possible. Since more it deviates from sinusoidal nature, the more the generation of harmonics which is undesirable.

The balanced Δ connected load can be transformed into balanced Y connected load with phase impedance 'Z_Δ' and 'Z_Y' which is related by the relation

$$\boxed{Z_Y = \frac{Z_\Delta}{3}} \tag{1.55}$$

line currents remain same in both cases.

1.6 THREE-PHASE POWER

The three phases draws currents through it. Every phase has some resistance which consumes power. So there is power consumption by each phase. For balanced system, the power consumption is same for each phase. So the active on real power can be expressed as given below:

$$P_{3\phi} = 3V_p\ I_p\ \cos\theta, \text{ watts} \tag{1.56}$$

where, 'V_p' is the phase voltage, in volts.
'I_p' is the phase current, in amperes.
'$\cos\theta$' is the power factor of each phase and three phases respectively.

The above expression as given by equation (1.56) can be alternatively expressed as follows.

$$P = 3\frac{V_p^2}{R_p} = 3I_p^2 R_p \tag{1.57}$$

where, 'R_p' is the resistance of each phase in ohms.

The power expression remains same for Y and Δ connected system.

The total instantaneous power for a three-phase system is constant and is equal to three times the average power per phase. The total instantaneous power for any balanced polyphase system is constant. This advantage of three-phase system is clearly understood from the operation of three-phase induction motors. That is, the torque output remains constant and the motor is unaffected by the pulsations of the input supply. The reactive power for a three-phase balanced load is given by

$$Q_{3\phi} = 3\ V_p\ I_p\ \sin\theta \tag{1.58}$$

$$= 3\ I_p^2\ X_p,\ VA_r \tag{1.59}$$

where 'VA_r' → voltampere reactive

The reactive power is important for giving the ratings of a.c. machines. The reactive power is sometimes called the wattless power.

The total power supplied to the load (three-phase) is given in terms of apparent power and it is represented by '$S_{3\varphi}$'. Mathematically, it is given as follows:

$$S_{3\varphi} = 3 \cdot V_p I_p, \text{ Volt amperes} \tag{1.60}$$

Combining the three-power expressions as given by the expression eq. (1.59), eq. (1.59) and eq. (1.60), we have,

$$S_{3\varphi} = P_{3\varphi} + jQ_{3p} \tag{1.61}$$

So, the apparent power is sometimes called the complex power. The real part of the complex power is the real power and the imaginary part of the complex power is the reactive power. The phase angle 'θ' between the phase voltage 'V_p' and phase current 'I_p' is given by

$$\theta = \tan^{-1}\frac{X_p}{R_p} = \cos^{-1}\frac{R_p}{Z_p} = \sin^{-1}\frac{X_p}{Z_p} \tag{1.62}$$

Thus, the power factor of three-phase load is same as that of power factor of any one phase for a balanced three-phase system.

1.7 POWER MEASUREMENT IN SINGLE PHASE AND THREE PHASE A.C. CIRCUITS

Power in single phase circuits can be calculated using single phase wattmeter. It can be alternatively measured using voltmeter, ammeter and power factor meter. The accuracy in the measurement can be obtained using single phase wattmeter but as the number of meter increases the error in the measurement is increased.

The measurement using single phase dynamometer type wattmeter is shown in the (Fig. 1.33).

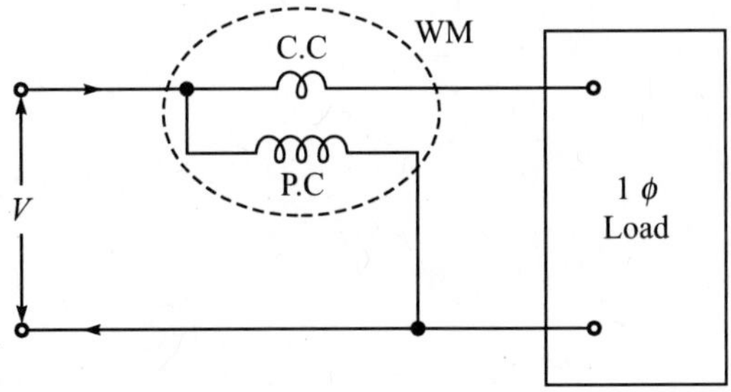

Fig. 1.33 Power measurement in single phase wattmeter.

Using single phase wattmeter it is possible to measure the three-phase power consumed by the load (three-phase type). By measuring power consumed by the impedance per phase and by multiplying with three. The total power consumed by the load can be determined as per the relation given by equation (1.63)

$$\boxed{P_{3\phi} = 3\, V_p I_p \cos\theta} \tag{1.63}$$

where, 'V_p' is the phase voltage, in volts
'I_p' is the phase current, in amperes
'cos θ' is power factor/phase.

The three-phase power measurement using single wattmeter method is applicable only for balanced loads. It is immaterial whether the load is connected in star or delta. For unbalanced loads, the error involved in the above method is large. So by employing two wattmeters and connecting the current coil (C.C.) and pressure coil (P.C.) of them in proper manner the three-phase power can be measured for balanced as well as for unbalanced loads.

Two-Wattmeter Method

The two-watt meter method connected diagram is shown in the Fig. (1.34) and Fig. (1.35) and for both types of loads it is applicable.

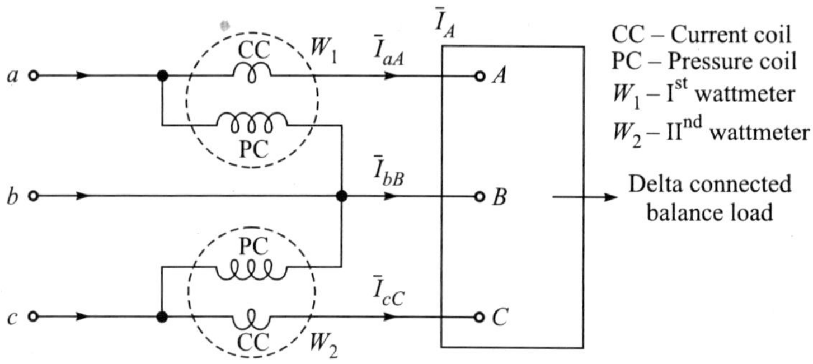

Fig. 1.34 Three-phase, three-wire system.

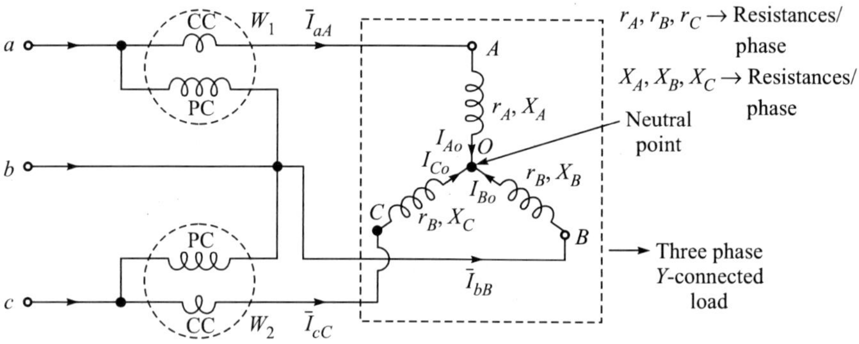

Fig. 1.35 Three-phase, three-wire system (*Y* connected Load).

The three-phase power is measured by two-wattmeter method using eq. (1.33)

$$\boxed{P_{3\phi} = W_1 + W_2} \tag{1.64}$$

The instantaneous three-phase power is given by

$$p = v_a i_a + v_b i_b + v_c i_c \tag{1.65}$$

The sum of the currents is equal to zero

$$i_a + i_b + i_c = 0 \tag{1.66}$$

$$i_b = -i_a - i_c \tag{1.67}$$

Substitute the value of 'i_b' in eq. (1.65), we get

$$p = v_a i_a + v_b(-i_a - i_c) + v_c i_c$$

$$p = i_a(v_a - v_b) + i_c(v_c - v_b)$$

$$p = i_a v_{ab} + i_c v_{cb} \tag{1.68}$$

The average power is given by integrating the above expression given by eq. (1.68) over full-time period and is given as below:

$$P_{3\phi} = W_1 + W_2 \tag{1.69}$$

where,

$$W_1 = V_{AB} I_{AO} \cos \theta'$$

$$W_2 = V_{CB} I_{CO} \cos \theta''$$

This could be explained using the phasor diagram shown in Fig. (1.36)

$$W_1 = V_{AB} I_{AO} \cos(30 + \theta) = V_{AB} I_{AO} \cos \theta' \tag{1.70}$$

$$W_2 = V_{CB} I_{CO} \cos (30 - \theta) = V_{CB} I_{CO} \cos \theta'' \tag{1.71}$$

Since $V_{AB} = V_{CB} = V_L$ and $I_{AO} = I_{CO} = I_P = I_L$

$\therefore$ $W_1 + W_2 = V_L I_L [\cos(30 + \theta) + \cos(30 - \theta)]$

$$\boxed{W_1 + W_2 = \sqrt{3} V_L I_L \theta} \tag{1.72}$$

$\therefore$ The total active power '$P_{3\varphi}$' consumed by the loads connected to the three phases are given by the eq. (1.73).

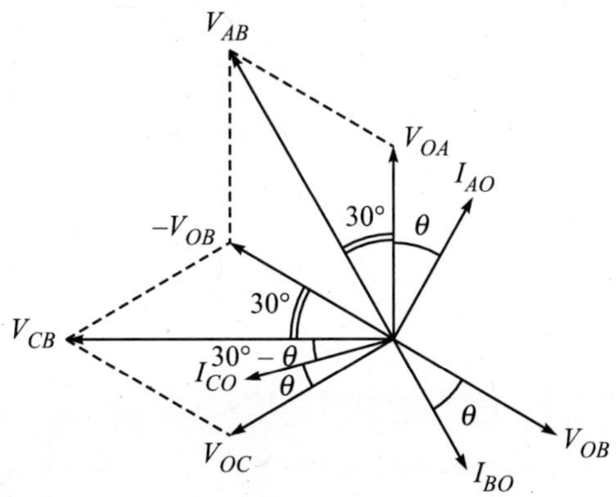

Fig. 1.36 Phasor diagram of two-wattmeter method.

$$P_{3\varphi} = W_1 + W_2 = \sqrt{3}\, V_L I_L \cos\theta \tag{1.73}$$

The above expression is valid for both the Y and Δ connected loads.

From the relations of 'W_1' and 'W_2' given in eqs. (1.72) and (1.73). The following important conclusions can be drawn.

(i) The readings of both wattmeters are equal i.e., $W_1 = W_2$ when $\theta = 0$ [upf loads i.e., resistive].

(ii) The reading of the first wattmeter is equal to zero i.e., $W_1 = 0$ if $\theta = +60°$ [i.e., $\cos\theta = 0.5$ (lead)] For all power factor < 0.5, 'W_1' is negative

(iii) The reading of second wattmeter is zero i.e., $W_2 = 0$ if $\theta = -60°$ [i.e., $\cos\theta = 0.5$ (lag)] For all power factor < 0.5, 'W_2' is negative.

(iv) For power factor >0.5 and $<$ unity, W_1 and W_2 readings are positive but are unequal.

For finding the expression of power factor of a three-phase load, we find total reactive power

$$Q = \sqrt{3}\,(W_2 - W_1) \tag{1.74}$$

and

$$\theta = \tan^{-1}\left(\frac{Q}{P}\right) \tag{1.75}$$

or,

$$\theta = \tan^{-1}\left[\frac{\sqrt{3}\,(W_2 - W_1)}{W_1 + W_2}\right] \tag{1.76}$$

Thus,

$$\boxed{\cos\theta = \cos\left[\tan^{-1}\frac{\sqrt{3}\,(W_2 - W_1)}{W_1 + W_2}\right]} \tag{1.77}$$

where, 'θ' is negative, for inductive loads

'θ' is positive, for leading loads.

The power triangle having V_A, V_{Ar} and W representing the three sides of it holds true for the three-phase and single phase respectively. So the expressions

$$P_{3\phi} = 3P_p = 3V_p I_p \cos\theta \tag{1.78}$$

$$Q_{3\phi} = 3Q_p = 3V_p I_p \sin\theta \tag{1.79}$$

$$S_{3\phi} = 3S_p = 3V_p I_p \tag{1.80}$$

holds for both Y and Δ connected loads. The power factor angle 'θ' is the angle between the phase voltage and phase current. This can be obtained by determining the angle of impedance per phase.

Single Line Diagram of Three-Phase a.c. Circuit

A balanced three-phase system can be reduced to a corresponding single phase system. Since the analysis of single phase system is easier, so after solving for the electrical quantities the single phase system is transformed back to three phase system.

The single line diagram of an interconnected system is shown below:

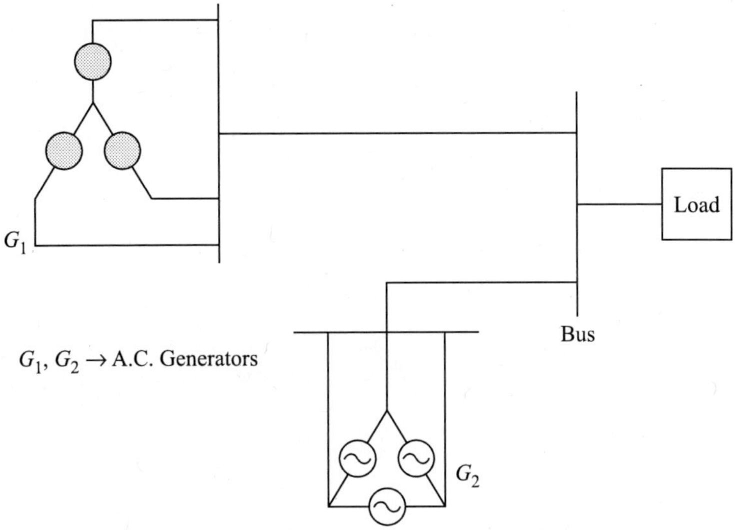

Fig. 1.37 Interconnected three-phase system.

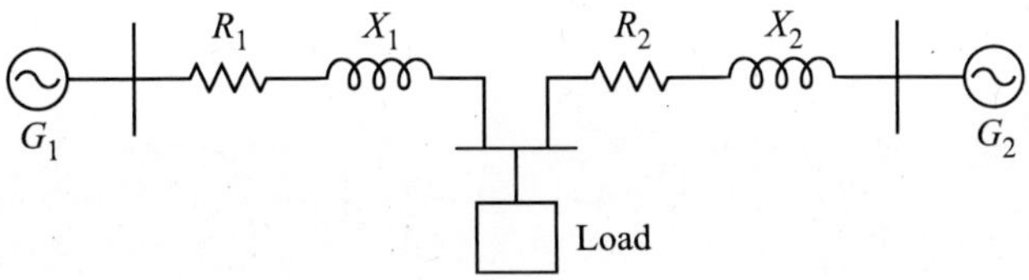

Fig. 1.38 Single line diagram of the three-phase interconnected system.

SOLVED QUESTIONS

Q.1 The power input to a 2000 V, 50 Hz, 3 phase motor running on full load at an efficiency of 90% is measured by two wattmeters which indicate 300 kW and 100 kW respectively. Calculate (i) input power (ii) power factor.

A.1

$$V_L = 2000 \text{ V}, f = 50 \text{ Hz}, \eta_{FL} = 90\% = 0.9$$

$$W_1 = 300 \text{ kW} = 300 \times 10^3 \text{ watt}$$

$$W_2 = 100 \text{ kW} = 100 \times 10^3 \text{ watt}$$

(i) Output power $= W_1 + W_2 = 400 \times 10^3$ watt

$$\eta_{FL} = \frac{\text{Output Power}}{\text{Input Power}}$$

$$\text{Input power} = \frac{\text{Output Power}}{\eta_{FL}}$$

$$= \frac{400 \times 10^3}{0.9} = 444.44 \text{ kW}$$

(ii) Power factor 'cos ϕ' $= \cos\left[\tan^{-1}\left\{\frac{(W_2 - W_1)\sqrt{3}}{W_1 + W_2}\right\}\right]$

$$\phi = \tan^{-1}\left[\frac{\sqrt{3}(W_2 - W_1)}{(W_1 + W_2)}\right]$$

$$\phi = \tan^{-1}\left[-\frac{\sqrt{3}}{2}\frac{200000}{400000}\right]$$

$$\phi = \tan^{-1}\left(\frac{\sqrt{3}}{2}\right) = -\tan^{-1}(0.866) = -0.7137 \text{ rad}$$

$$\phi = -40.913°$$

$$\cos\phi = 0.7559 \text{ (lag)}$$

Q.2 A balanced star connected load of (8 + j 6) Ω/ph is connected to a balanced three-phase, 400 V supply. Find the line current, power factor; power and total volt amperes?

A.2

$$Z_{ph} = 8 + j\,6,\ \Omega$$

$$V_L = 400 \text{ V}$$

$$I_L = ?,\ \cos\phi = ?\ P = ?\ ,\ S = ?$$

For Y-connected load

$$V_p = \frac{V_L}{\sqrt{3}}$$

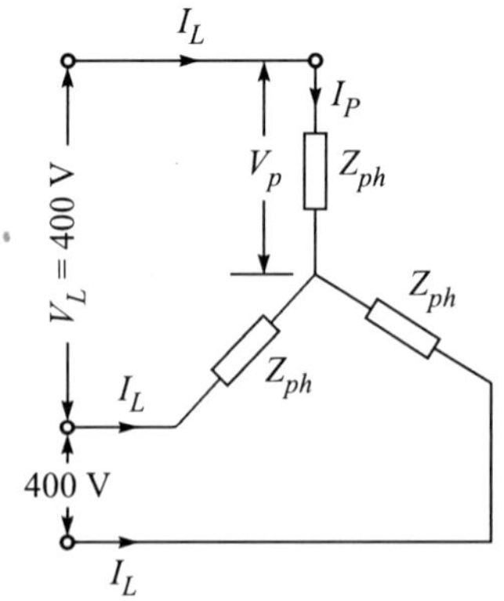

$$V_p = \frac{400}{\sqrt{3}} \text{ volts} = 230.9 \text{ V}$$

$$V_p = I_p \times Z_p, \; [\because \; I_P = I_L]$$

$$I_p = \frac{V_P}{|Z_p|} = \frac{230.9}{\sqrt{8^2 + 6^2}} = 23.09 \text{ A}$$

$\therefore$ Line current $I_L = I_P = 23.09$ A

(ii) Power factor $\cos \phi = \cos\left[\tan^{-1}\frac{6}{8}\right] = 0.8$

(iii) Power, $P = 3V_p I_p \cos \phi = \sqrt{3}\, V_L I_L \cos \phi = 12.79$ kW

(iv) Total $V_A = S = 3V_p\, I_p = 15.99$ KVA.

Q.3 Each phase of a star connected load consists of a non-reactive resistance of 100 ohm in parallel with capacitance of 31.8 μF. Calculate the line current, the power absorbed, the total KVA and the power factor when it is connected to a 400 V, 3 phase, 50 Hz supply?

A.3

$$X_c = \frac{1}{\omega C} = \frac{1}{2\pi f\, C}$$

$$X_c = \frac{1}{314 \times 31.8 \times 10^{-6}}$$

$$X_c = 100.782 \; \Omega$$

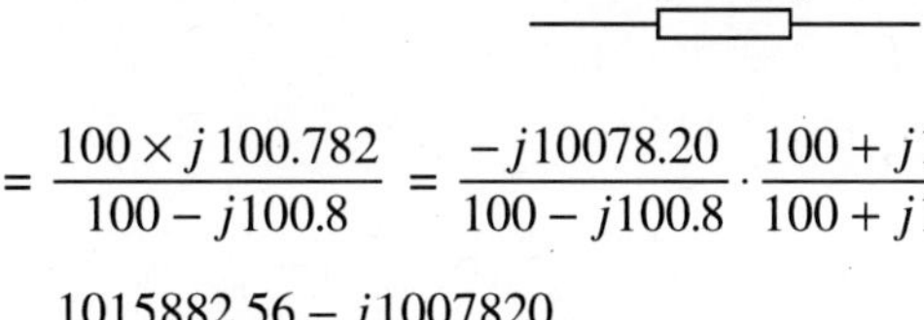

(i) $$\overline{Z}_{R-C} = \frac{100 \times j\,100.782}{100 - j100.8} = \frac{-j10078.20}{100 - j100.8} \cdot \frac{100 + j100.8}{100 + j100.8}$$

$$= \frac{1015882.56 - j1007820}{20160.64}$$

$$= 50.39 - j\,49.98 = \sqrt{2539 + 2499}\,\angle\tan^{-1}\left(\frac{49.98}{50.39}\right)$$

$$\overline{Z}_p = \overline{Z}_{R-C} = |\overline{Z}_{R-C}|\angle\theta = 70.98\,\angle{-44.8^\circ}$$

$$\therefore \qquad I_p = \frac{V_p}{|Z_p|} = \frac{400/\sqrt{3}}{70.98}$$

$$= 3.2537 \text{ Amp} = I_{\text{line}} \text{ (in } Y\text{-connected network)}$$

(ii) Power absorbed $= P = \sqrt{3}\, V_L\, I_L \cos\theta$

$$P = \sqrt{3} \times 400 \times 3.2537 \times 0.7096$$

$$P = 1599.5 \text{ W} = 1.6 \text{ kW}$$

(iii) Total KVA $= \sqrt{3}\, V_L\, I_L = 2{,}254.2$ W $= 2.2542$ kVA

(iv) Power factor $\cos\theta = \cos(-44.8°) = 0.7096$ (lag)

Q.4 The load connected to a three-phase supply comprises three-similar coils connected in star. The line currents are 25 A and the KVA and kW inputs are 20 and 11 respectively. Find the line and phase voltages and KVAR input.

A.4

$$\text{Input kVA} = 20$$

$$\text{Input kW} = 11$$

$$\text{Input kVA} = \sqrt{3}\, V_L\, I_L$$

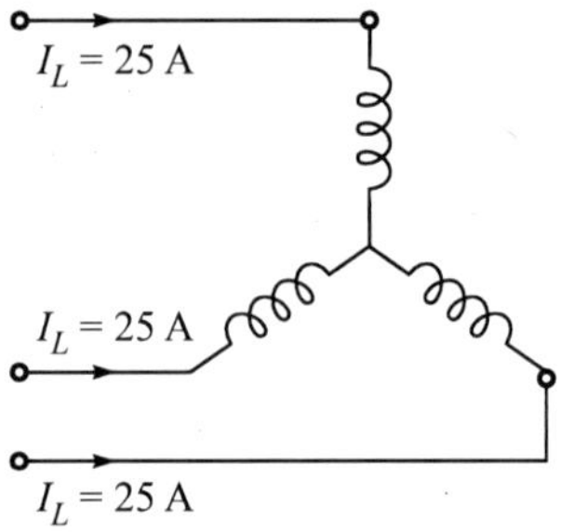

$$\text{Input} = \text{kW} = \left(\sqrt{3}\, V_L\, I_L\right) \cos\theta$$

$$\text{Input kVA} = \sqrt{3}\, V_L\, I_L = 20 \text{ kVA}$$

$$\text{Input kW} = 20 \times 10^3 \times \cos\theta = 11 \times 10^3$$

$$\cos\theta = \frac{11}{20} = 0.55 \text{ (lag)}$$

$$\sin\theta = \sqrt{1 - \cos^2\theta} = 0.8352$$

$$\text{Input KVAR} = \sqrt{3}\, V_L\, I_L \sin\theta = 16.7033 \text{ kVAR}$$

$$\sqrt{3}\, V_L \times I_L = 20 \times 10^3 \ VA$$

$$\sqrt{3}\, V_L \times 25 A = 20000 \ VA$$

$$V_L = \frac{20{,}000}{\sqrt{3} \times 25} V$$

Line voltage, $V_L = 461.9$ volt

Phase voltage, $V_p = \dfrac{V_L}{\sqrt{3}} = 266.7$ volt

Q.5 The 20 ohm non-inductive resistors are connected in star across a three-phase 400 V supply. Three other equal non-inductive resistors are connected in delta across the same supply, so as to take the same line current. What are the resistance values of these other resistors and what is the current flowing through each of them?

A.5

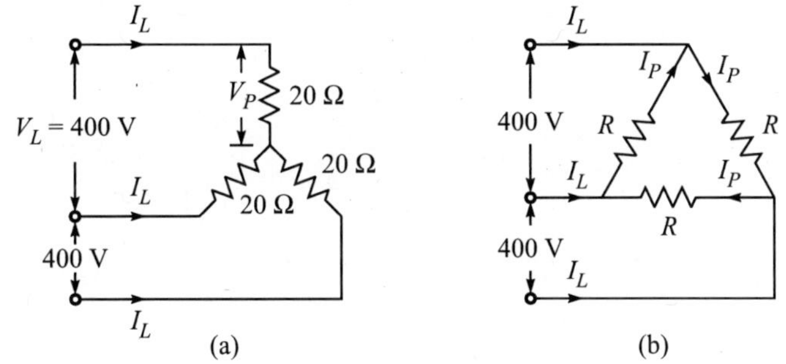

In Fig. (a) $\quad I_L = I_P,\ V_P = \dfrac{V_L}{\sqrt{3}} = \dfrac{400}{\sqrt{3}} = 230.9$ V

$$\therefore \quad I_P = \frac{V_p}{R_{20\Omega}} = \frac{230.9}{20}\text{ A} = 11.545 \text{ A}$$

$\therefore \quad I_L = I_P = 11.545$ A

In Fig. (b) $\quad I_L = 11.545$ A

$$I_P = \frac{I_L}{\sqrt{3}} = 6.6657 \text{ A}$$

$$\therefore \quad R = \frac{V_P}{I_p} = \frac{V_L}{I_p} = \frac{400}{6.6657} = 60.01\ \Omega$$

$\therefore$ Current through resistors 'R' = 6.67 Amp

Resistance 'R' $\approx$ 60 Ω

Q.6 A 400 V, three-phase voltage is applied to a balanced delta connected three-phase load of phase impedance (15 + j 20) Ω. Find the phase current in each line, power consumed per phase, phasor sum of three line currents.

A.6 $V_L = 400$ V, $Z_P = (15 + j\,20)\ \Omega$, $\bar{I}_p = ?$, $P_{ph} = ?$, $\sum_{i=1}^{3} \bar{I}_{L_i} = ?$

(i) $$\bar{I}_p = \frac{\bar{V}_p}{\bar{Z}_{ph}} = \frac{\bar{V}_p}{\bar{Z}_p}$$

$$\bar{Z}_p = \sqrt{15^2 + 20^2}\,\angle\tan^{-1}\frac{20}{15}$$

$$\bar{Z}_p = 25\angle 53.157°,\ \text{Ohms}$$

$$\therefore \quad |\bar{I}_p| = \frac{|\bar{V}_p|}{|\bar{Z}_p|} = \frac{400/\sqrt{3}}{25} = 9.2379\ \text{A}$$

$\therefore$ Phasor current in each line $\bar{I}_p = |\bar{I}_p|\angle-\theta = 9.2379\,\angle-53.157°$ Amp.

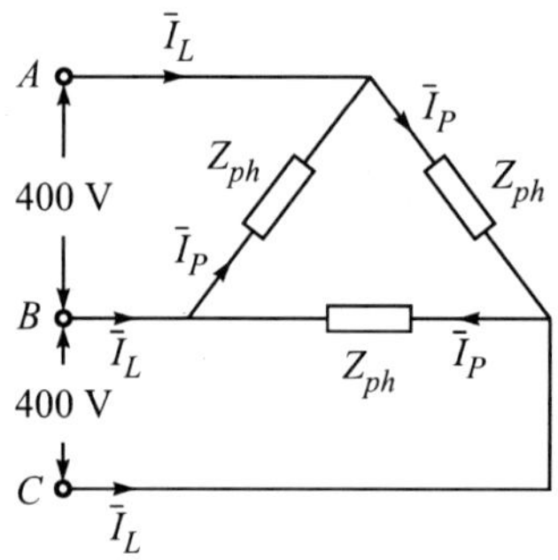

(ii) Power consumed per phase $= |\bar{I}_p|^2 \times R_p$

$= 1280.1$ W

$= 1.2801$ kW

(iii) Phasor sum of three line currents = 0

Q.7 If the phase voltage of a three-phase, Y connected system is 200 V, what will be the line voltages

(a) When the phases are correctly connected.

(b) When the connections to one of the phases are reversed.

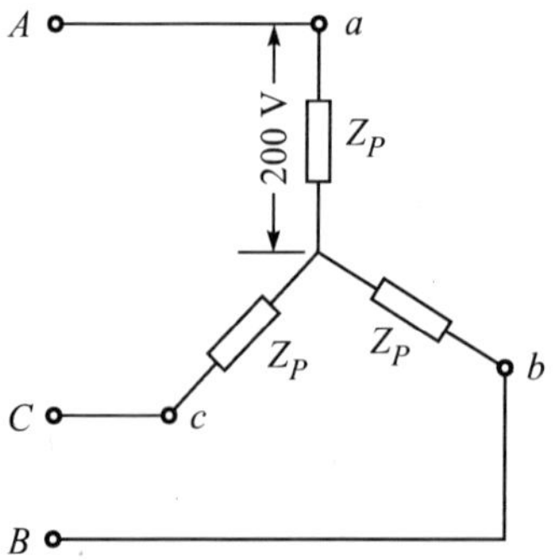

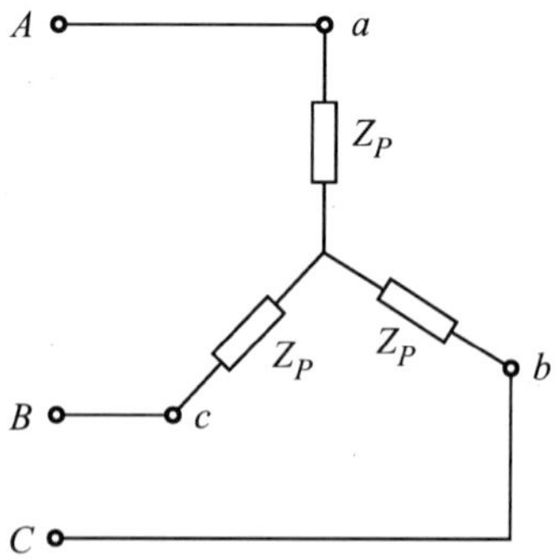

A.7 $$V_P = 200 \text{ V, in } Y\text{-connected system, } V_L = \sqrt{3}\, V_P$$

$$\therefore \quad V_L = \sqrt{3} \times 200 = 346.4 \text{ V}$$

On reversing the phase connections, the line voltage remains unaffected.

Q.8 Calculate the total power drawn by the three-phase load as shown in the figure below.

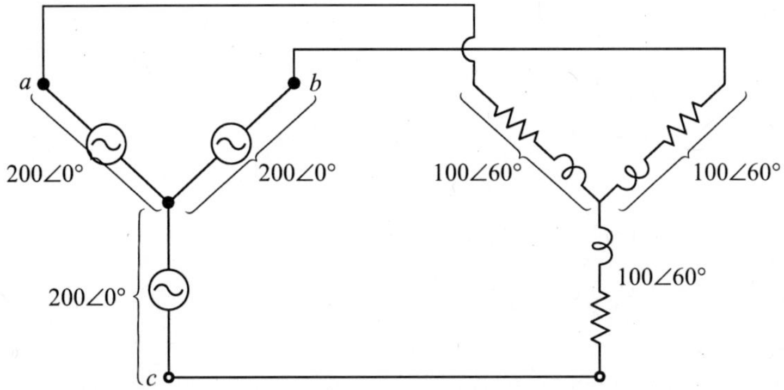

A.8 The power drawn by the load per phase

$$P_p = V_p I_p \cos\theta \qquad I_p = \frac{V_L/\sqrt{3}}{|\bar{Z}_p|} = \frac{V_p}{|\bar{Z}_p|} = \frac{200\angle 0^\circ}{100\angle 60^\circ}$$

$$= 200 \times 2 \times \cos 60$$

$$P_p = 200 \text{ W} \qquad I_p = 2\angle -60^\circ, \text{ Amp.}$$

$$\therefore \quad P_{3\phi} = 3P_p = 600 \text{ W}$$

Q.9 A three-phase 500 V motor has 0.4 power factor lagging. Two wattmeters are connected to measure the input, they show the total input to be 30 KW, find the reading of each wattmeter.

A.9 $$V_L = 500 \text{ V}, \cos\theta = 0.4 \text{ (lag)}, W_1 + W_2 = 30 \text{ kW}$$

$$\therefore \quad W_1 + W_2 = \sqrt{3}\, V_L I_L \cos\theta$$

$$I_L = \frac{W_1 + W_2}{\sqrt{3}\, V_L \cos\theta} = \frac{30000}{\sqrt{3} \times 500 \times 0.4} = 86.61 \text{ A}$$

$$\therefore \quad W_1 = V_L I_L \cos(30 + \theta)$$

$$W_2 = V_L I_L \cos(30 - \theta)$$

$$\cos\theta = 0.4 \Rightarrow \theta = \cos^{-1}(0.4) = 66.4^\circ$$

$$\therefore \quad W_1 = 500 \times 86.61 \times (-0.\ 1115) = -4.827 \text{ kW}$$

$$W_2 = 500 \times 86.61 \times 0.8049 = 34.856 \text{ kW}$$

Q.10 In a particular measurement the wattmeter readings are 5000 W and 1000 W. Calculate power and power factor if both the meters read positive values.

A.10 $$W_1 = 5 \text{ kW}, W_2 = 1\text{kW}$$

$\therefore$ Power measured $= W_1 + W_2 = (5 + 1)\text{kW} = 6 \text{ kW}$

$$\text{Power factor } \cos\theta = \cos\left[\tan^{-1}\left\{\frac{\sqrt{3}(W_2 - W_1)}{W_1 + W_2}\right\}\right]$$

$$\cos\theta = \cos\left[\tan^{-1}\left\{\frac{\sqrt{3}(1-5)\text{ kW}}{6\text{ kW}}\right\}\right]$$

$$= \cos\left[\tan^{-1}\sqrt{3}\cdot\left(-\frac{\not{4}^{2}}{\not{6}_{3}}\right)\right]$$

$$= \cos\left[\tan^{-1}\left(-\frac{2\sqrt{3}}{3}\right)\right] = 0.6547$$

$\therefore$ Power factor $= 0.6547$

Q.11 A three-phase 400 V, motor load has a power factor of 0.75 lagging. Two wattmeters connected to measure the power show the total input to be 26 kW. Find the reading of each instrument.

A.11 $$V_L = 400 \text{ V} \cos\theta = 0 \text{ (lag)}, W_1 + W_2 = 26 \text{ kW}$$

$$W_1 + W_2 = \sqrt{3}\, V_L I_L \cos\theta \Rightarrow I_L \frac{W_1 + W_2}{\sqrt{3}\, V_L \cos\theta} = 50.04 \text{ Amp}$$

$\therefore$ Readings of each wattmeter are

$$W_1 = V_L I_L \cos(30 + \theta)$$
$$W_2 = V_L I_L \cos(30 - \theta)$$
$$\theta = \cos^{-1} 0.75 = 41.43°$$

$\therefore$ $$W_1 = 500 \times 50.04 \times \cos(30° + 41.43°)$$
$$= -16.959 \text{ kW}$$
$$W_2 = 500 \times 50.04 \times \cos(-11.43°)$$
$$= 10.531 \text{ kW}$$

The readings of wattmeters are -16.96 kW and 10.53 kW respectively.

Q.12 In the two-wattmeter method of power measurement in a three-phase circuit, the readings of the wattmeters are 1200 W and 300 W. What is the power factor of the load?

A.12 $$W_1 = 1200 \text{ W}, W_2 = 300 \text{ W}$$

Power factor $= \cos\varphi$

$\therefore$ $$\cos\phi = \cos\left[\tan^{-1}\left\{\frac{\sqrt{3}(W_2 - W_1)}{(W_1 + W_2)}\right\}\right]$$

$$= \cos\left[\tan^{-1}\left\{\frac{\sqrt{3}(-900)}{(1500)}\right\}\right]$$

$$\boxed{\cos\varphi = 0.6934}$$

TUTORIAL QUESTIONS

Q.1 What do you understand by three-phase supply? Describe star and delta connections by showing the line voltages and phase voltages, line currents and phase currents.

Q.2 Obtain the line and phase relationships for voltage and currents for a delta connected three-phase power systems. Also obtain the expression of total power and per phase power.

Q.3 Explain three-wattmeter method for three-phase power measurement. Why it is not preferred?

Q.4 With the help of circuit diagram, explain two-wattmeter method for power measurement in a three-phase circuit.

Q.5 For power measurement of three-phase circuit by two-wattmeter method, when the value of power factor is less than 0.5 lagging. What will be the readings of both the wattmeters?

Q.6 In the two-wattmeter method of power measurement in a three-phase circuit, the readings of the wattmeters are 1000 W and 550 W. What is the power factor of the load?

(UPTU 2001-02)

Q.7 Each phase of a Δ connected load has a resistance of 25 Ω, an inductance of 0.15 H and a capacitance of 120 μF. The load is connected across a 400 V, 50 Hz, 3 phase supply. Determine the line current, active power and reactive volt ampere.

(UPTU 2001-02)

Q.8 With the aid of a phasor diagram show that the power and power factor of a balanced three-phase load can be measured by two wattmeters. For a certain load; one of the wattmeters read 20 kW and the other 5 kW after the voltage circuit of this wattmeter has been reversed. Calculate the power and the power factor of the load.

(UPTU 2001-02)

Q.9 Derive relationships of voltages and currents between their line and phase value in balanced star and delta connections. Three identical coils connected in delta across 400 V, 50 Hz, 3 phase, a.c supply, take a line current of 17.32 A at power factor of 0.8 lagging. Calculate

(i) The phase current

(ii) The resistance and inductance of each coil

(iii) The power drawn by each coil.

(UPTU-2001)

Q.10 A three-phase balanced load connected across a three-phase. 400 V a.c. apply draws a line current of 10 A. Two wattmeters are used to measure input power. The ratio of two wattmeters reading is 2:1. Find reading of the two wattmeter.

(UPTU-2001)

Q.11 A 1000 V, three-phase, star connected motor takes 60 kW at a power factor 0.85. Calculate the phase and line currents. If the motor reconnected in delta, calculate the current voltage of a three-phase supply suitable for this motor.

Q.12 A delta connected load draws a current of 20 amperes at a lagging power factor of 0.9 from a 400 V, 3 phase, 50 Hz supply. Find the resistance and inductance of each phase.

Q.13 Three similar coils each of inductance 0.1 H and resistance 30 Ω are connected to a 415 V, three-phase, 50 Hz supply (a) in star (b) in delta. Calculate the line current in each coil and the total power in the circuit in each case.

Q.14 Three coils, each with resistance of 50 Ω and with inductance 0.3 H are connected to a 415 V, three phase supply at 50 Hz in star. Calculate:

(i) Current in each coil

(ii) The total power in the circuit.

Q.15 A balanced three-phase, *Y*-connected load of 120 kW takes a leading current of 100 A when connected across a three-phase, 3.3 kV, 50 Hz supply. Determine the impedance, resistance, capacitance and power factor of the load.

Q.16 A three-phase star connected load is made up of three coils of equal impedances with the value of $R = 75\ \Omega$ and $L = 1.5$ H for each coil, while supply is of the rating 440 V, 50 Hz. Find (i) line current (ii) total power.

Q.17 The equal valued three branches of a delta-connected load has the branch impedances having resistance of 50 Ω and capacitance of 200 μF. If the power supply is of the value 220 V, 50 Hz then find (i) phase and line current (ii) active and reactive power.

Q.18 A balanced star connected load of $(3 + j\,4)\Omega$ per phase is connected to a balanced three-phase 230 V a.c. supply. Find the line current, power factor, power and volt ampere drawn by the load.

Q.19 During the measurement of power in a three-phase *Y*-connected load using two-wattmeter method, the load supplied was 50 kW at 0.707 lagging. Find the readings of the two wattmeters.

Q.20 A three-phase power supply supplies 400 V, 50 Hz to a Δ connected load where impedances per phase is $(10 + j\,10)\ \Omega$. Determine the line currents.

Q.21 A three-phase supply of 220 V, 50 Hz, supplies power to a balanced load of impedance of 10 Ω each per phase, inductive type with a power factor 0.866. Find the line currents.

Q.22 Three identical impedances are joined in delta connection. A three-phase, 400 V, 50 Hz supply is attached with the load. The value of $R = 10\ \Omega$ and $C = 150.0\ \mu F$. Find the phase currents, line currents and draw the phases.

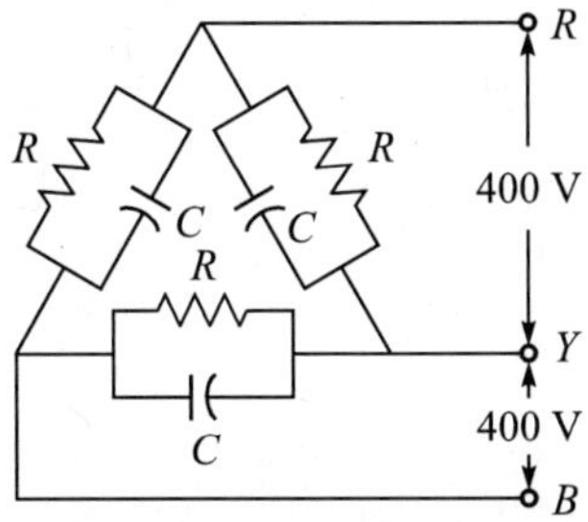

Q.23 The line currents of a delta connected voltage source is 10 A. A star connected load is also connected to the three-phase voltage source. The impedance of the load per branch is $(8 + j\,6)\Omega$, find the phase voltage of the source.

Q.24 A balanced *Y*-connected load has an impedance of $15\angle 60°$ (inductive) per phase. If phase voltage is $230\angle 0°$ volt, find the phase currents, line voltage.

Q.25 A three-phase, 50 Hz source supplies power to a balanced machine load having rated voltage 400 V, 20 A and 0.866 power factor. Find the resultant real and reactive power.

Q.26 The readings of two-wattmeter method are 20 kW and –5 kW. The three-phase supply voltage has 400 V, 50 Hz. Find power factor, line current and active power.

Q.27 A three-phase 400 V, 50 Hz supply is connected with two parallel loads, one of them in star with phase impedance $(8 + j\,6)\Omega$ and other in delta with impedance of each phase $(8 - j\,6)\Omega$. Find the total current drawn, power factor real and reactive power.

Q.28 The apparent power and true power are 150 kVA and 100 kW respectively; at 80 A in a three-phase star connected load. Find power factor phase and line voltage.

Q.29 Three equal impedances of $(R + j\,X_L)\Omega$ per phase are connected in '*Y*' and are supplied by 400 V, 50 Hz, three-phase supply. The two wattmeter readings are 5 kW and 1 kW. Determine '*R*' and '*L*' of the coil.

Q.30 A three-phase, 400 V, 50 Hz supplies power to a delta connected load having an impedance of $(8 + j6)\Omega$ per phase. Find the line current, power factor total power and readings of two wattmeters.

REVIEW QUESTIONS

Q.1 Illustrate the differences between linear and non-linear circuit.

Q.2 The current through a resistor '*R*' when connected to a 12 V battery is 1 A. Determine the resistance that must be inserted in series to reduce the current to half when connected across the same battery.

Q.3 If the current flowing through two parallel branches are 1 A and 2 A respectively. If the resistance of one is 10 Ω, what is the resistance of the other?

Q.4 If two resistances of 5 Ω and 10 Ω are connected across 1 volt source. What is the voltage that will appear across each resistance?

Q.5 A 1000 Ω electric heater operates on a 230 V, 50 Hz supply. What is the rms value of a.c. current through the heater? What is the maximum value of a.c. current? Determine the resistance of the heater.

Q.6 A series circuit consists of a resistance of 100 Ω, an inductance of 10 mH and a capacitance of 1000 μF. What is the impedance of the circuit at a frequency of 50 Hz? Discuss the nature of the circuit.

Q.7 A $230\angle 0°$ V a.c. voltage source supplies a current of $5\angle -30°$ A to a load. What is the power factor of the load? Compute the average power, reactive power and apparent power.

Q.8 An a.c. circuit can be represented by a voltage source of $24\angle 0°$ V in series with an impedance of $8 + j\,6$ Ω. In order to transfer maximum power to the load, what must be the load impedance? What is the current in the circuit? How much power is transferred to the load?

Q.9 Define apparent power, real power and reactive power.

Q.10 Show that the power factor of balanced three-phase load is

$$\cos\phi = \cos\left[\tan^{-1}\left\{\frac{\sqrt{3}\,(W_2 - W_1)}{W_1 + W_2}\right\}\right]$$

Q.11 Using superposition theorem find the current in the 10 Ω resistor of the circuit given as shown in figure below.

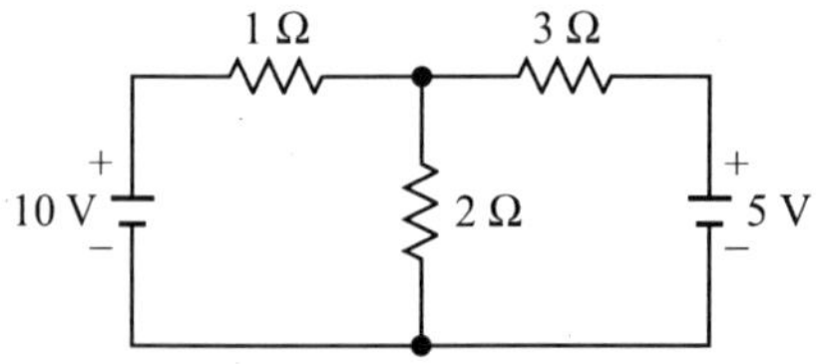

Q.12 A balanced three-phase load dissipates 15 kW at a power factor of 0.8 lagging from a balanced 230 V, 50 Hz, Δ connected system. For a Δ connected load obtain an equivalent circuit.

Q.13 Find the load current through 2 Ω resistance of the circuit shown below using Thevenin's theorem.

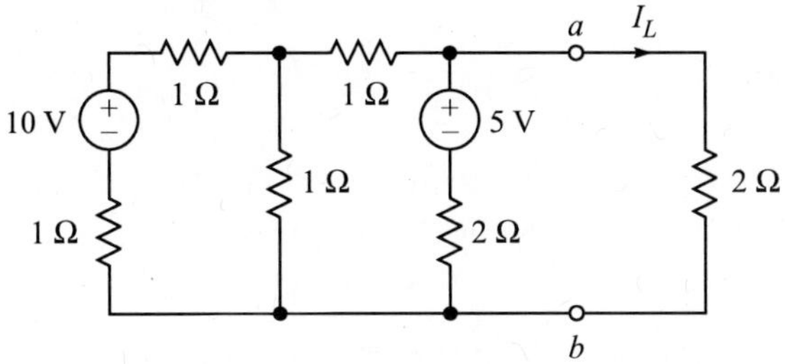

(JMI 2006)

Q.14 Determine the current supplied by the battery in the circuit shown below.

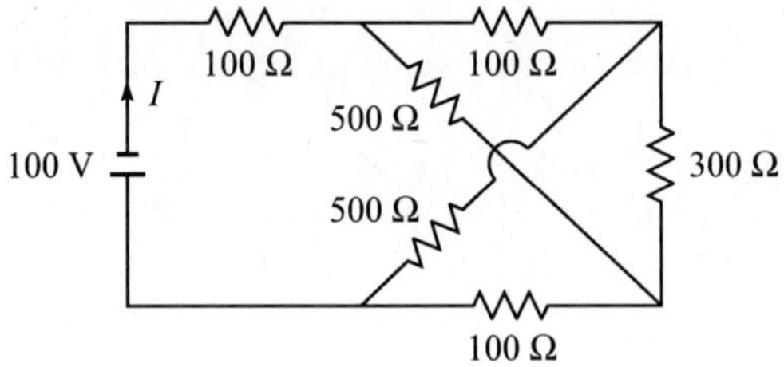

(JMI 2006)

Q.15 With the aid of phasor diagram obtain the relationship between line and phase values of voltages and currents in a three-phase, star connected system.

Q.16 Calculate the current in 20 Ω resistor in the circuit given in figure below using Thevenin's theorem.

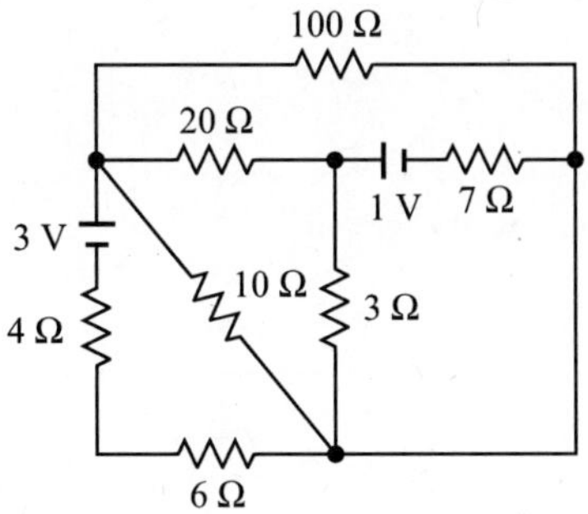

(JMI 2006)

Q.17 Find the Norton's equivalent circuit as seen by 'R_L' in the circuit shown below.

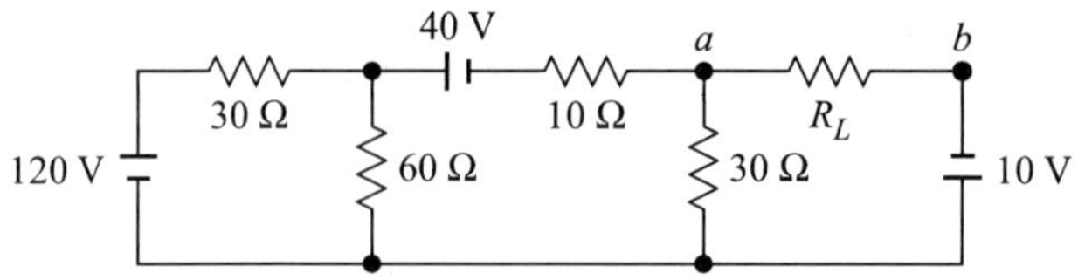

(JMI 2006)

Q.18 In the circuit shown in figure below, $V_b = 6\ V_a$ and the phase angle between V_a and V_b is 90°. Calculate the values of 'R' and 'X_c'.

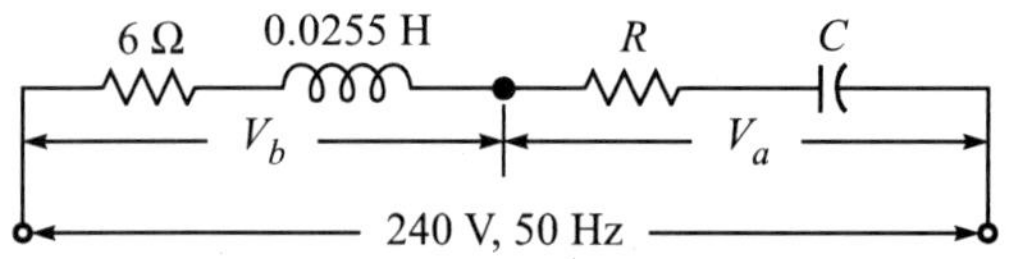

(JMI 2006)

Q.19 Calculate the average and rms value of voltage of a waveform shown in figure below.

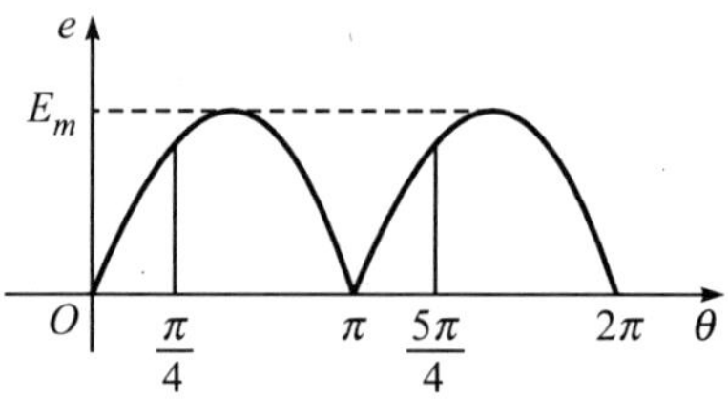

(JMI 2006)

Q.20 For the circuit shown below, compute

(i) Impedance of the circuit
(ii) Current I_1
(iii) Current I_2
(iv) Current I_3
(v) V_{BC}
(vi) Power delivered by the circuit

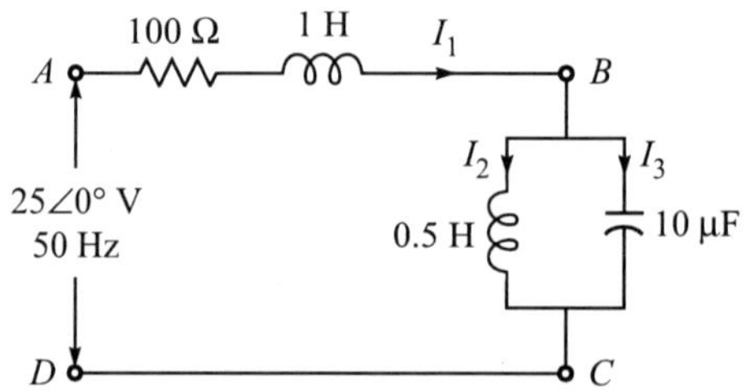

Q.21 Find 'I_o' in the circuit using source transformation

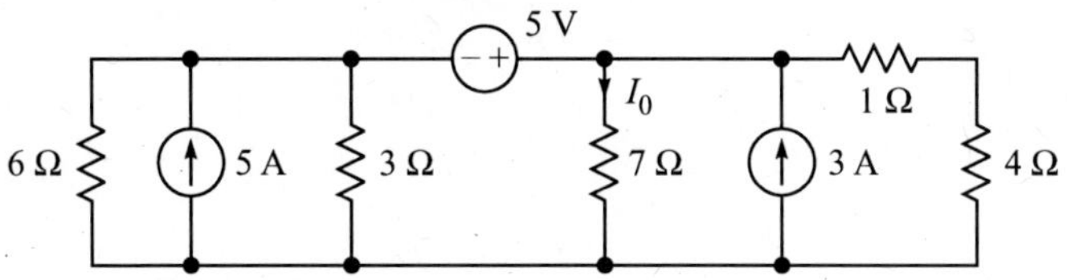

(JMI 06)

Q.22 Find the Thevenin's equivalent of the circuit of figure as shown at terminals XY:

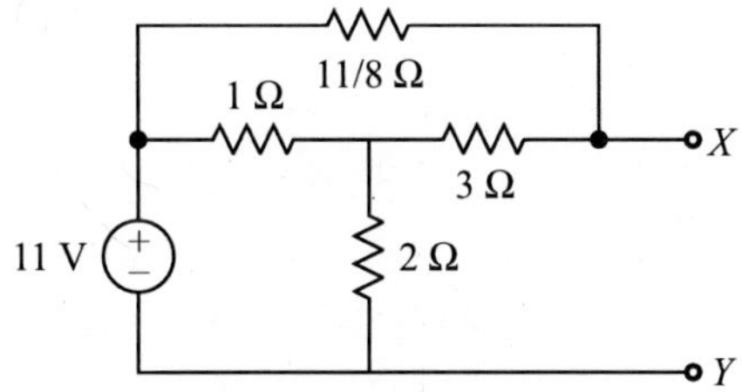

(JMI 06)

Q.23 Find the current and power supplied by the 100 V source in the circuit shown below. Also find the overall power factor.

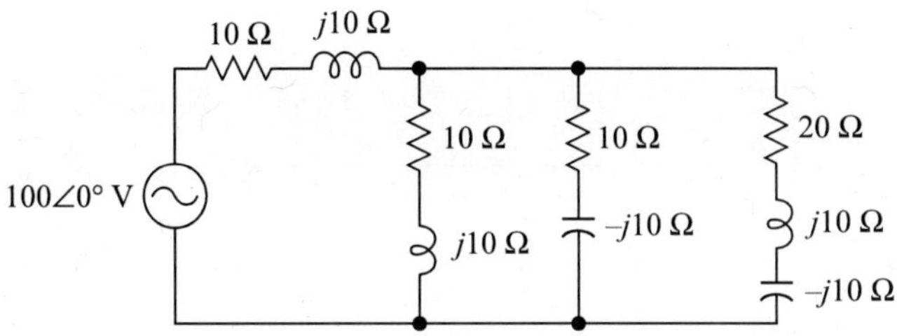

(JMI 06)

Q.24 A three-phase voltage source has a phase voltage of 120 V and supplies a star connected load having impedance $(36 + j\,48)\Omega$ per phase. Calculate

(i) line voltage
(ii) line current
(iii) total power factor
(iv) total three-phase power supplied to the load

(JMI 06)

Q.25 In the circuit shown below, find 'R' and 'V_s'. Also find the power output of the source 'V_s'.

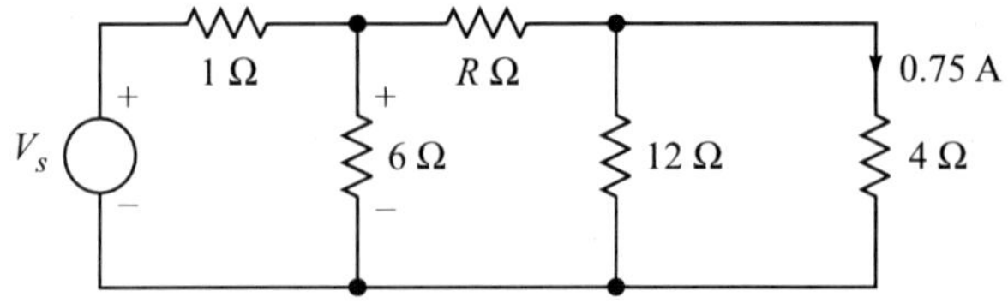

(JMI 06)

Q.26 For the circuit, find 'V' and 'I' using the principle of superposition.

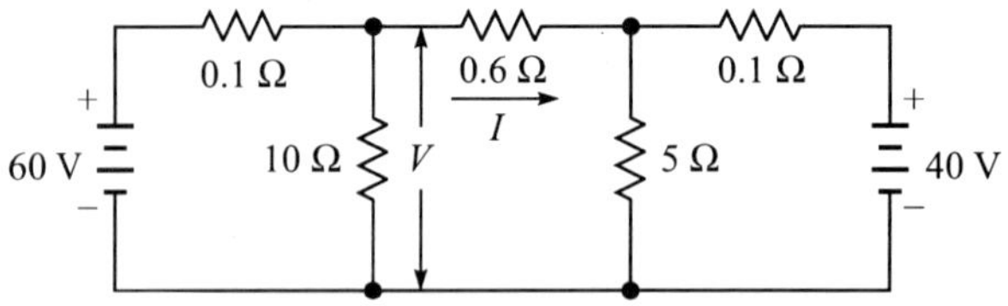

(JMI 06)

Q.27 In the circuit, the reactance of capacitor 'C_1' is 4 Ω, the reactance of 'C_2' is 8 Ω, and reactance of 'L' is 8 Ω A sinusoidal voltage having an effective value of 120 volts is applied to the circuit.

(i) Find the effective value of current delivered by the source.

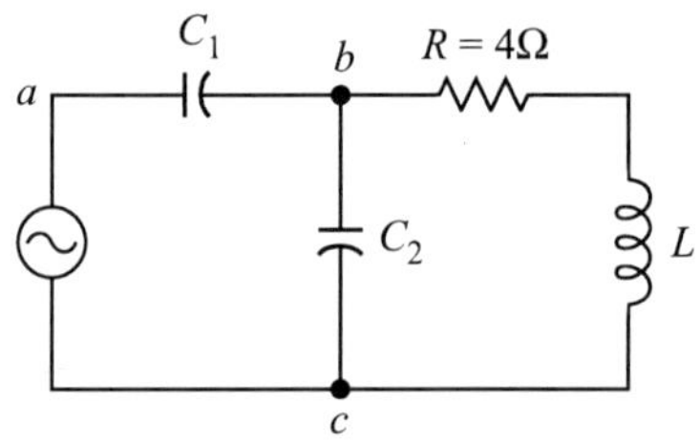

(ii) Draw the phasor diagram showing the source voltage, source current and voltages $\overline{V}_{ab}$ and $\overline{V}_{bc}$.

Chapter

2

Review of Basic Laws

2.1 Introduction
2.2 Maxwell's Equations
2.3 Self and Mutual Inductances
2.4 Transformer EMF
2.5 Motional EMF
2.6 Faraday's Law of Electromagnetic Induction
2.7 Lenz' s Law
2.8 Force on a Current Carrying Conductor in a Magnetic Field
2.9 Torque on a Current Carrying Coil in a Magnetic Field
2.10 Flemming's Right Hand Rule
2.11 Flemming's Left Hand Rule
2.12 Ampere's Law
2.13 Right Hand Rule
• Solved Questions
• Tutorial Questions

2.1 INTRODUCTION

The electrical machine irrespective of whether it is alternating or direct type involves the magnetic field as the reservoir where the electromechanical energy conversion takes place. So it is necessary to get a clear understanding of the magnetic field. The magnetic field is produced by a current carrying coil. So there may be a confusion as to what causes the electric field. Since the electric field is due to charge accumulation, so we can consider that the electric and magnetic fields exists simultaneously.

In the case of static machines as transformers the linkage of the two windings of transformer is due to magnetic fields or in other words flux linkages which is changeable in nature, that is, the flux changes with time.

In the case of direct current machines, the field which is produced due to field winding which is wound over the pole is steady whereas in the a.c. machines such as induction machines the flux is changeable so the field distribution under the poles also varies with time.

So without having concept of the basic laws it is difficult to analyze the machine, so the next coming section deals with the laws use in electromagnetics. It will help in the modelling of the machine.

2.2 MAXWELL'S EQUATIONS

The Maxwell equations are the basis for the operation of electromagnetic devices as electric motors, generators, etc.

The maxwell's equations are as follows :

(a) $\oint_s \overline{E} \cdot d\overline{A} = \frac{q_{\text{enclosed}}}{\varepsilon_0}$ (Gauss's law for electricity)

(b) $\oint_s \overline{B} \cdot d\overline{A} = 0$ (Gauss's law for magnetism)

(c) $\oint_c \overline{E} \cdot d\overline{S} = -\frac{d\phi_B}{dt}$ (Faraday's law)

(d) $\oint_c \overline{B} \cdot d\overline{S} = \mu_0 \varepsilon_0 \frac{d\phi_E}{dt} + \mu_0 \, i_{\text{enclosed}}$. (Ampere's Maxwell law).

Alternatively, the Maxwell's equations can be written as

$$\overline{\nabla} \times \overline{E} = -\frac{\partial \overline{B}}{\partial t} \tag{2.1}$$

$$\overline{\nabla} \times \overline{H} = \overline{J} + \frac{\partial \overline{D}}{\partial t} \tag{2.2}$$

$$\overline{\nabla} \cdot \overline{B} = 0 \tag{2.3}$$

$$\overline{\nabla} \cdot \overline{D} = P \tag{2.4}$$

where, $\overline{E}$ = Electric field intensity, in V/m

$\overline{H}$ = Magnetic field intensity, in A/m

$\overline{B}$ = Magnetic flux intensity, in Wb/m^2

$\overline{D}$ = Electric flux intensity, in C/m^2

$\overline{J}$ = Current density, in A/m^2

P = Volume charge density, in C/m^3

S = Surface

C = Contour is a closed loop surrounded by open surface's

V = Volume bounded by closed surface 'S'

Gauss's Law for Magnetic Fields

The simplest magnetic structures are magnetic dipoles which exist in the magnetic material. Gauss's law for magnetic fields

$$\phi_B = \oint_s \overline{B} \cdot d\overline{A} = 0 \tag{2.5}$$

The eqn. (2.5) suggests that the net magnetic flux through any (closed) guassian surface is zero. In other words, the magnetic monopoles do not exist.

The maxwell equations do not disobey the law of conservation of charge. The Lorentz force is often used to explain the principle of operation of motors. So we should consider the expression of it, as given by eq. (2.6)

$$\overline{F} = q(\overline{E} + \overline{v} \times \overline{B}) \tag{2.6}$$

The Lorentz force is experienced by a charge of magnitude 'q' and moving with velocity 'v' through an electric field of $\overline{E}$ and a magnetic field of $\overline{B}$.

The magnetic field is given by

$$\overline{B} = \mu\overline{H} = \mu_0\mu_r\overline{H} \tag{2.7}$$

where, μ_0 = permeability of free space = $4\pi \times 10^{-7}$H/m
μ_r = relative permeability

2.3 SELF AND MUTUAL INDUCTANCES

An inductor can be used to produce a magnetic field . If current 'i' passes through an inductor of 'N' turns, then flux of 'ϕ_B' is produced they are correlated as follows

$$L = \frac{N\phi_B}{i}, \text{ Henry}$$

where, 'L' is called the self inductance of coil. The flux linkage of the coil is given by 'λ' which is equal to the product of 'N' and 'ϕ_B' as given by eq. (2.8)

$$\lambda = N\phi_B \tag{2.8}$$

In the case of solenoid, the inductance per unit length for a long solenoid near its centre is

$$\frac{L}{l} = \mu_0 n^2 A \tag{2.9}$$

Thus, we see that the inductance 'L' depends on the geometry of device. The mutual inductance 'M' depends on other coil current which is discussed later.

2.4 TRANSFORMER EMF

The self-induction can be understood by considering the Fig. (2.1) as follows:

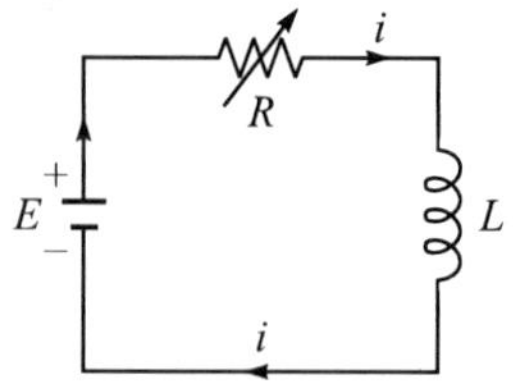

Fig. 2.1 Self-Induction.

When 'R' is changing then 'i' also changes so an induced emf appears in the coil 'L'. This process is called self-induction and the emf 'E_L' is called the self-induced emf. The inductance 'L' is called the self-inductance of the coil. The Faraday's law of electromagnetic induction gives the value of 'E_L'.

i.e., Self induced emf $$E_L = -\frac{d\lambda}{dt} \tag{2.10}$$

Flux linkage, $$\lambda = N\phi_B \tag{2.11}$$

with 'L' $$E_L = -\frac{d}{dt}(N\phi_B) \tag{2.12}$$

Since $$\lambda \propto i \tag{2.13}$$

For linear inductor, the above relation holds.

$$\lambda = Li \tag{2.14}$$

or, $$N\phi_B = Li \tag{2.15}$$

So, alternatively, 'E_L' can be expressed as follows

$$E_L = -L\frac{di}{dt} \tag{2.16}$$

'E_L' is called self-induced emf. So 'L' is called self inductance of the coil in Henry.

From the eqn. (2.16), we see that the self-induced emf exist only when the current 'i' through the coil is changing with time 't'. The magnitude of the current does not affect the value of 'E_L' only the rate of change of current affects 'E_L'. The negative sign of 'E_L' shows that the direction of this emf opposes the polarity of supply. So sometimes this emf is called the reaction or transformer emf. The direction of 'E_L' is given according to Lenz's law which states "that the effect will oppose the cause producing the effect".

Mutual Induction

In the Fig. (2.2) shown, two coils 1 and 2 are placed nearby. One coil carries current 'i_1' and a source of voltage 'v_1'. Due to Faraday's law of electromagnetic induction an emf is induced in the second coil. This is called mutual induction which is due to mutual interaction of the two coils. In the self-induction only one coil is involved.

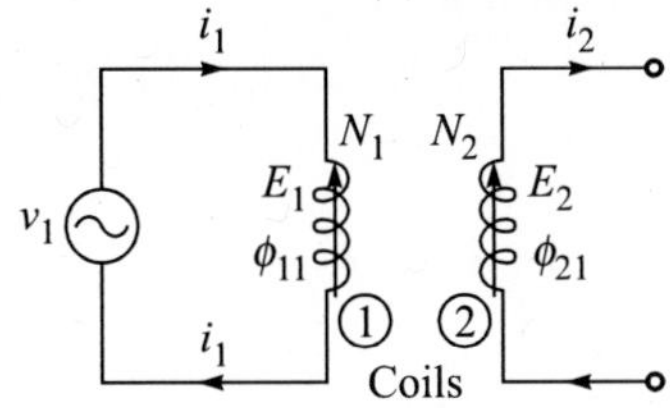

Fig. 2.2 Mutual Induction.

The flux linkage of second coil is '$N_2\phi_{21}$' (say)

then $$M_{21}\, i_1 = N_2\phi_{21} \tag{2.17}$$

If 'i_1' is varying with time then

$$M_{21}\frac{di_1}{dt} = N_2\frac{d\phi_{21}}{dt} \tag{2.18}$$

According to Faraday's law the R.H.S. of eq. (2.18) is the magnitude of E_2. Thus, the emf induced in second coil due to 'i_1' is given by

$$E_2 = -M_{21}\frac{di_1}{dt} \tag{2.19}$$

The negative sign indicates the direct of emf 'E_2' with respect to 'v_1' at any instant of time.

2.5 MOTIONAL EMF

The motional emf is related with the speed of conductor '$\bar{v}$' with respect to the magnetic field '$\bar{B}$'. In vector notation it may be given by eq. (2.20)

$$e_m = \oint_c (\bar{v} \times \bar{B}) \cdot d\bar{l} \tag{2.20}$$

where, e_m = motional emf, volts
$\bar{v}$ = velocity of conductor, m/s
$\bar{B}$ = magnetic flux intensity, tesla
$d\bar{l}$ = incremental length, m
c = contour

This emf originates due to flux cutting property of the moving conductor. The larger the cutting of flux the larger will be the magnitude of emf so as the cutting of flux can be increased with increase of speed of the conductor. So sometime this emf is referred as speed emf on motional emf for a closed loop moving in a magnetic field the total induced emf must be equal to the sum of both these emf as follows.

$$e = e_t + e_m \tag{2.21}$$

e_t = transformer emf, volts
e_m = motional emf, volts

The above expression can be written as

$$e = -\frac{d\phi}{dt} \text{ where } \phi = \text{B.A}$$

'A' is the area of conductor, m^2

In generators (dc) the emf induced in a armature conductor is given by

$$e = B\,l\,v, \text{ volts.}$$

2.6 FARADAY'S LAW OF ELECTROMAGNETIC INDUCTION

Faraday's First Law

A current appears only when there is relative motion between the loop and the magnet. This current in the loop is called induced current. The work done per unit charge to produce that current is called an induced emf and the process of producing the current and emf is called induction.

In short, the first law can be enunciated as follows

"An emf is induced in the loop when the number of magnetic flux lines passing through the loop is changing."

The actual number of flux lines passing through the loop does not matter; the value of the induced emf and the induced current are determined by the rate at which the number of flux line changes.

The amount of flux passing through the loop is given by

$$\phi_B = \int \overline{B} \cdot d\overline{A} = BA, \text{ weber} \tag{2.22}$$

If the loop lies in a plane and that the field $\overline{B}$ is perpendicular to the plane of the loop. Then $B \cdot dA \cos 0° = B \cdot dA$.

If 'B' is uniform and 'A' is the area of the loop, then we get the total flux 'ϕ_B' as given by the expression (2.22) as written before.

The unit of flux is weber, whereas the unit of magnetic flux intensity is tesla or Wb/m^2

Faraday's Second Law

Alternatively, the Faraday's law can be stated as 'The magnitude of the emf 'E' induced in a conducting loop is equal to the rate at which the magnetic flux 'ϕ_B' through that loop changes with time." The induced emf 'E' opposes the flux change. So, mathematically it can be stated as

$$E = -\frac{d\phi_B}{dt} \tag{2.23}$$

If the coil is tightly wound (closely packed), so that the same magnetic flux 'ϕ_B' passes through all the turns, the total emf induced in the coil is

$$E = -N\frac{d\phi_B}{dt} \tag{2.24}$$

'N' = number of coil turns

The flux change '$d\phi_B$' can be done by changing 'A' i.e., area of loop, 'B' magnetic field within the loop and the angle between the $\overline{B}$ and the plane of loop. The expression given by (2.24) is called the Faraday's second law.

2.7 LENZ'S LAW

The Lenz's law gave a rule for determining the direction of induced current in a loop.

Statement

It states that the direction of induced current opposes the change in the magnetic flux that induces the current. It may be noted that the direction of an induced emf is that of the induced current.

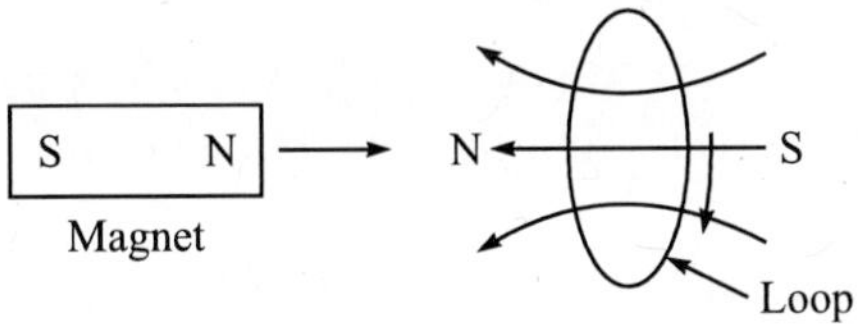

Fig. 2.3 Lenz's law.

The movement can be obtained in two ways i.e., the magnet may be moved towards the loop and also the loop can be moved towards the magnet with magnet kept stationary.

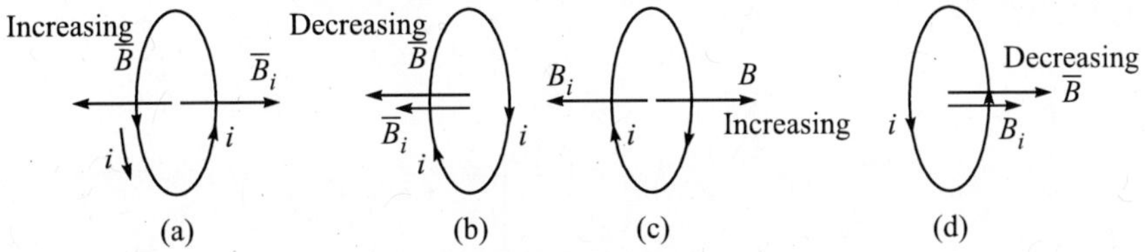

Fig. 2.4 Flux intensity due to induced current.

The Lenz's law can be realized by considering the Fig. (2.3) in which as the magnet's north pole is brought near the loop there is an increase of the magnetic flux in the loop and thereby induces a current in the loop. The loop then acts as a magnetic dipole with a north pole and a south pole. To oppose the magnetic flux increase being caused by the approaching magnet; the loop's north pole (and thus $\vec{\mu}$) must face toward the approaching north pole so as to repel it. Then the curled straight right hand rule for $\vec{\mu}$ tells us that the current induced in the loop must be clockwise.

If we next pull the magnet away from the loop a current will again be induced in the loop. If the loop's south pole facing the retreating north pole of the magnet, so as to oppose the retreat. Thus, the induced current will be counter-clockwise could be understood alternatively, with the magnet initially at a distant, no magnetic flux passes through the loop. As the north pole of the magnet then moves toward the loop with its magnetic field $\overline{B}$ directed toward the left; the flux through the loop increases. To oppose this increase in flux, the induced current 'i' must set up its own field $\overline{B_i}$ directed toward the right inside the loop as shown in Fig. (2.4); then the rightward flux of field $\overline{B_i}$ opposes the increasing leftward flux of field $\overline{B}$.The curled-straight right hand rule then tells us that 'i' must be counter-clockwise.

The flux of 'B_i' always opposes the change in the flux of $\overline{B}$, but that does not always mean that $\overline{B_i}$ points opposite $\overline{B}$. That is, if we move the magnet away from the loop, the flux 'ϕ_B' from the magnet is still directed to the left through the loop, but it is now decreasing. The flux of $\overline{B_i}$ must now be to the left inside the loop to oppose the decrease in 'ϕ_B' as shown in Fig. (2.4). Thus, $\overline{B_i}$ and $\overline{B}$ are now in the same direction.

In Fig. (2.3) and Fig. (2.4) shows the situations in which the south pole of the magnet approaches and retreats from the loop respectively.

2.8 FORCE ON A CURRENT CARRYING CONDUCTOR IN A MAGNETIC FIELD

As per Lenz's law, whether you move the magnet towards or away from the loop as shown in Fig. (2.5), a magnetic force resists the motion requiring the force applied to do positive work.

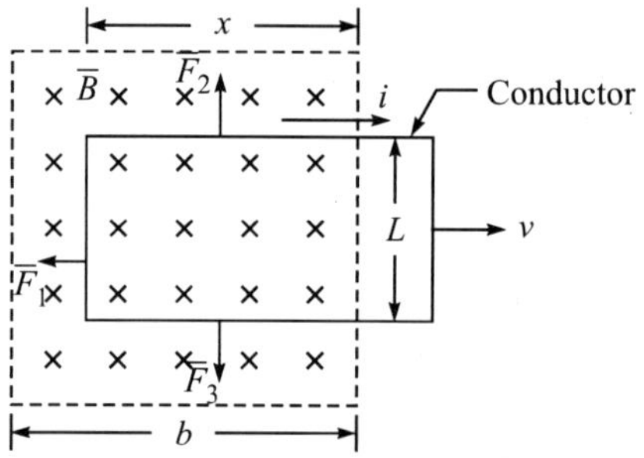

Fig. 2.5 Force on a current carrying inductor in a magnetic field.

At the same time, thermal energy is produced in the material of the loop i.e., conductor because of its resistance to the current that is induced by the motion. The energy transferred to the loop placed in the magnetic field due to applied force ends up in this thermal energy. The faster you move the magnet, the faster the force do work and greater the rate of energy transfer to meet thermal energy in the loop i.e., power transfer increases and thus its temperature increases accordingly.

Considering the Fig. (2.5) in which a loop of area 'A' in the field 'B' is pulled at a velocity 'v', placed in the magnetic field 'B'. As it is pulled the flux linked with the loop decreases. The area of the loop in field B is given by,

$$A = Lx \tag{2.25}$$

If the loop is moved at constant velocity $\overline{v}$, then a constant force $\overline{F}$ must be applied to it as a magnetic force of equal magnitude but opposite direction acts upon the loop to oppose $\overline{F}$. The rate of doing work is given by

$$\frac{dW}{dt} = |F|\cdot|\overline{v}| = F\cdot v \tag{2.26}$$

where F = magnitude of applied force, N
v = velocity, m/s

As the loop is moved towards right, the magnetic field decreases, so the flux through the loop also decreases and according to Lenz's law, a current is produced in the loop. It is the presence of this induced current that causes the force that opposes applied force '$\overline{F}$'.

To find the current, applying Faraday's law. If 'x' is the length of the loop still in the magnetic field, the area of the loop still in the field is given by equation (2.27)., the magnitude of the flux through the loop is

$$\phi_B = BA = BLx \tag{2.27}$$

As 'x' decreases, the flux decreases. Faraday's law tells us that with this flux decrease, an emf is induced in the loop given by equation (2.28)

$$e = -\frac{d\phi_B}{dt} = -\frac{d}{dt}(BLx) = -BL\frac{dx}{dt} \tag{2.28}$$

$$e = -BLv \tag{2.29}$$

where, 'e' is the induced emf, in volts.

In short the emf 'e' driving current 'i' is given by the circuit shown in Fig. (2.6).

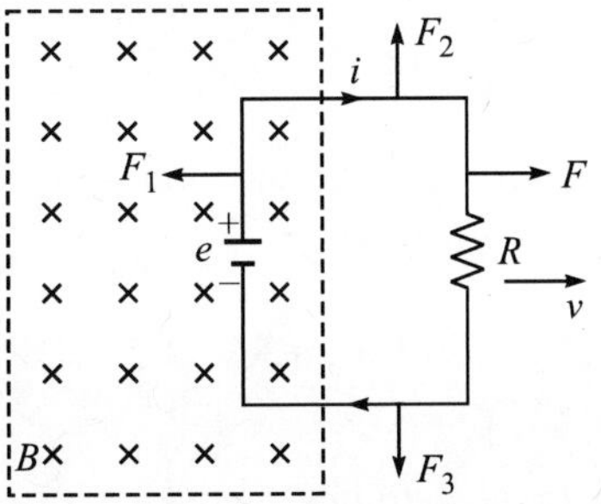

Fig. 2.6 Loop moving at velocity 'v'.

'R' is the resistance of loop, Ω
'i' is the induced current, A
'e' is the induced emf, V
The current 'i' is given by Ohm's law as per eq (2.30).

$$i = \frac{e}{R} \tag{2.30}$$

Substituting the values, we have

$$i = \frac{BLv}{R} \tag{2.30a}$$

Sideways, deflecting forces $\overline{F}_d$ act on the loop as given by $\overline{F}_2$ and $\overline{F}_3$. Due to equal magnitude of these forces, they cancel each other's effect. So the only force which acts on the loop is $\overline{F}_1$ which opposes $\overline{F}$.

So,
$$\overline{F}_1 = -\overline{F} \tag{2.31}$$

Considering angle between $\overline{B}$ and the length $\overline{L}$ for the part of loop inside the field is 90°, So

$$F = -\overline{F}_1 = i\overline{L} \times \overline{B} = i \cdot L \cdot B \sin 90°$$

or
$$\boxed{|\overline{F}_1| = F_1 = i\,L\,B} \tag{2.32}$$

So, we have

$$F = \frac{B^2 L^2 v}{R} \tag{2.33}$$

B, L, R = constants

v = velocity of loop is constant (for constant force F)
Substituting the values of 'F' in eq. (2.26), we have

$$P = Fv = \frac{B^2 L^2 v^2}{R} \tag{2.34}$$

The loop gets heated up due to thermal energy which can be derived as follows:

$$P = i^2 R \tag{2.35}$$

Substituting 'i' from eq. (2.30) in eq. (2.35), we get

$$P = \frac{B^2 L^2 v^2}{R} \tag{2.36}$$

Thus, we see that the thermal energy in the loop is equal to the rate of doing work on the loop.

2.9 TORQUE ON A CURRENT CARRYING COIL IN A MAGNETIC FIELD

The torque exerted on a current carrying conductor is given by eq (2.37) as follows

$$T = F_d \cdot r \tag{2.37}$$

where, 'F_d' is the deflecting force
and 'r' is the distance of the side of conductor from the central axis of loop.

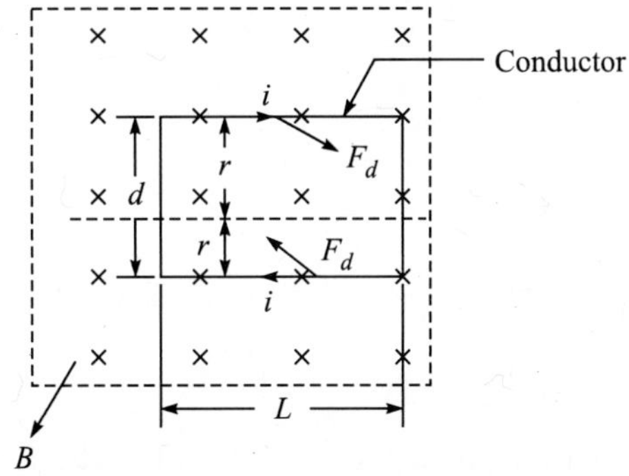

Fig. 2.7 Current carrying coil in field $\overline{B}$.

The general expression of $\overline{F}_d$ is given by Lorentz as

$$\overline{F}_d = i\,\overline{L} \times \overline{B} \tag{2.38}$$

Since the plane of field $\overline{B}$ is considered to be at 90° to the length vector $\overline{L}$, we have

$$|\overline{F}_d| = i \cdot L \cdot B \sin 90^\circ$$

$$F_d = i\,L\,B \tag{2.39}$$

Substituting the value of 'F_d' in eq. (2.39) we get torque as follows

$$T = i \cdot L \cdot B \cdot r$$

The single turn coil consists of two conductors so the total torque exerted on the coil carrying current is given by

$$T_t = 2\,T = 2\,i\,L\,B\,r \tag{2.40}$$

Since, $d = 2\,r$

So, $T_t = i\,L\,B\,d$

The area of the coil is A (say), then

$$A = L \cdot d$$

The torque exerted is given by

$$T_t = i \cdot B \cdot A, \text{ Nm} \tag{2.41}$$

If the coil is at an angle of 'θ' with respect to conductor or loop plane as shown in the Fig. (2.8) as shown below.

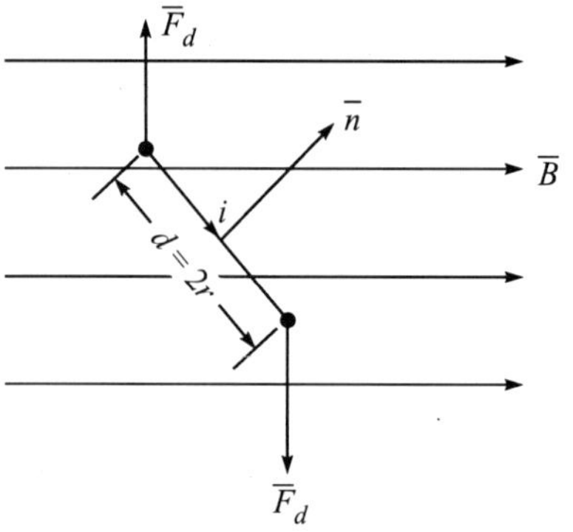

Fig. 2.8 Force exerted on a current carrying conductor.

2.10 FLEMMING'S RIGHT HAND (R.H.) RULE

When a conductor moves through an uniform magnetic field in such a way that it cuts the lines of magnetic force, an emf is induced in it, whose direction is given by Flemming's right hand rule. By holding the right hand as shown in Fig. (2.9) such that the thumb, forefinger and the middle finger are at right angles to each other.

If the *forefinger* points towards the *direction of the magnetic field* and the *thumb* points *towards the motion of the conductor*, then the *middle finger* points towards the *direction of the current.*

The Fig. (2.9) illustrates the Flemming's right hand rule.

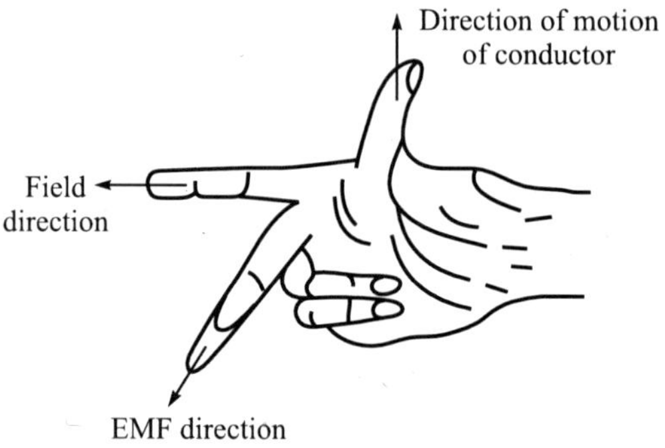

Fig. 2.9 Flemming's right hand rule.

The Flemming's right hand rule is used for determining the direction of emf. This rule is used for determining the direction of emf in the case of d.c. generator and other machines. The emf as derived in earlier section is given by

$$e = B\, l\, v \tag{2.42}$$

So, if the direction of 'B' and 'v' are known, then the direction of 'e' can be determined accordingly. If the direction of 'e' is known, then the direction of the current can be known determined.

2.11 FLEMMING'S LEFT HAND (L.H.) RULE

The magnetic of the force $\overline{F}_B$ due to field $\overline{B}$ on a current carrying conductor having current $\overline{I}$ is given by

$$|\overline{F}_B| = B\,I\,L \tag{2.43}$$

where,

'B' is the magnitude of magnetic field, in Tesla; 'I' is the current in the conductor, in ampere; 'l' is the length of the conductor, in m. Thus from the Flemming's L.H rule as shown in the Fig. (2.10).

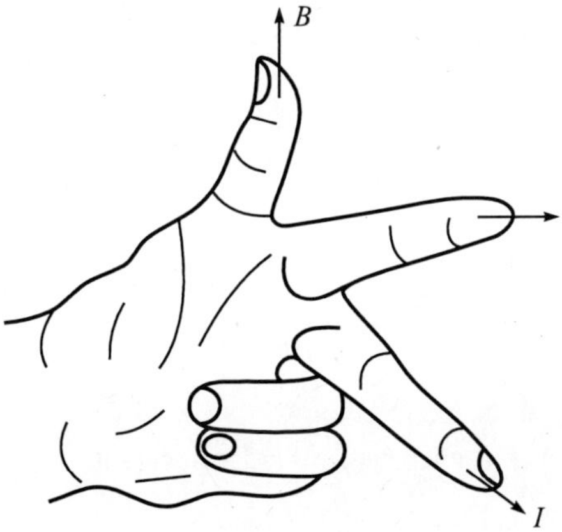

Fig. 2.10 Flemming's L.H. rule.

We see that if the thumb points in the direction of field, the forefinger points in the direction of the force then the middle or central finger points in the direction of current provided all the three fingers are kept at right angle to each other.

The Flemming's left hand rule is used to determine the direction of force exerted on a conductor carrying current as in the case of d.c. motor.

This could be understood by considering the Fig. (2.11).

The Fig. (2.11) shows that by applying the Flemming's L.H. rule on the upper conductor 'a' carrying current as shown by a cross. The direction of field 'B' is from north pole 'N' to south pole 'S' respectively. Then the direction of the force 'F_B' could be determined by using Flemming's L.H. rule which is towards right, while for the lower conductor 'a'' the direction of the force is towards left. So we see that a turning moment of force is exerted on the member carrying the conductor 'a' and 'a'' respectively. This concept is used for deriving the expression of torque in the case of d.c. motor.

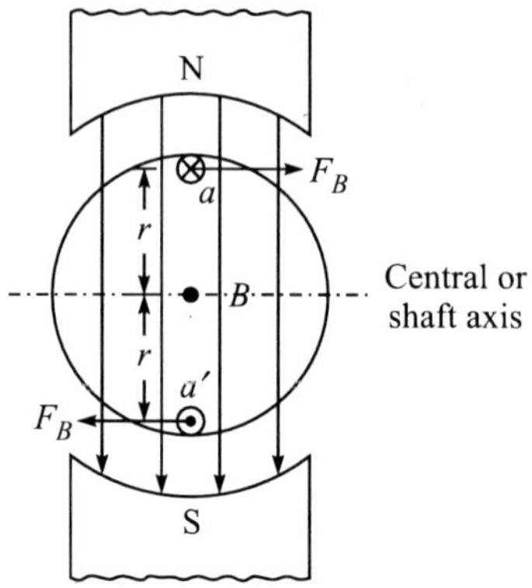

Fig. 2.11 Force on a conductor carrying current.

That is, the total turning moment of force is given by

$$T = 2\ F_B \cdot r \tag{2.44}$$

where 'r' is the distance between the centre of conductor and the central axis passing through the centre of the rotating member.

2.12 AMPERE'S LAW

The Ampere's law is mathematically expressed as

$$\oint B \cdot d\bar{s} = \mu_0\, i_{\text{enclosed}} \tag{2.45}$$

$d\bar{s} \rightarrow$ Incremental surface vector

$i_{\text{enclosed}} \rightarrow$ Net current encircled by amperian loop.

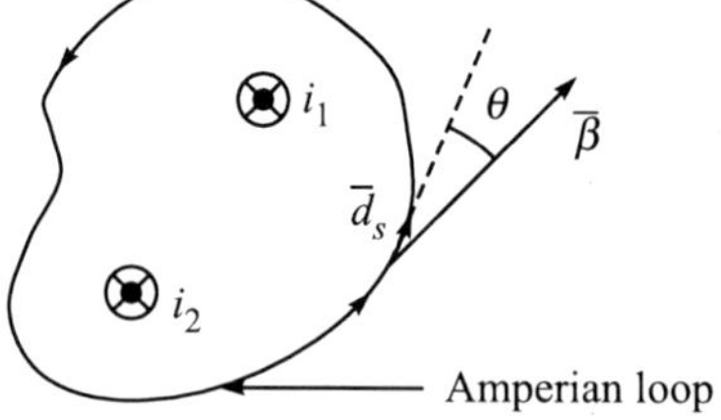

Fig. 2.12 Ampere's law applied to an amperian loop.

The direction of the current 'i_1' and 'i_2' can be cross or dot, the counter-clockwise direction marked on the loop indicates the arbitrarily chosen direction of integration. The direction of $d\bar{s}$ is tangential to the amperian loop at any instant.

$\bar{B}$ is the net magnetic field due to 'i_1' and 'i_2' the current carrying wires are considered perpendicular to the page so the magnetic field at $d\bar{s}$ due to each current is in the plane, thus the net magnetic field should also be in that plane. The direction of the field $\overline{B}$ can be arbitrary in that plane. In Fig. (2.12), it is directed at an angle of 'θ' with respect to the direction of $d\overline{s}$.

$$\oint B \cdot d\overline{s} = \oint B \cos\theta \cdot ds = \mu_0 \, i_{\text{enclosed}} \tag{2.46}$$

By taking the summation of the dot product of $\overline{B}$ and $\overline{ds}$ along the amperian loop. The direction of currents 'i_1' and 'i_2' is determined by right-hand palm rule.

Along the direction of integration if you placed the palm of right hand then the thumb points the direction of current if upward then it is considered positive if the direction of current is downward then it is considered negative according to the R.H. of palm rule. So if direction of 'i_1' is dot '.' and direction of 'i_2' is cross '×' then,

Net current in the amperian loop is equal to the difference of 'i_1' and 'i_2'. That is,

$$i_{\text{net}} = i_{\text{enclosed}} = i_1 - i_2 \tag{2.47}$$

$$\therefore \quad \oint B \cos\theta \, ds = \mu_0 (i_1 - i_2) \tag{2.48}$$

Similarly, the ampere's law can be used for determining the magnetic field produced by a current 'i' in a long straight wire as follows

$$B = \frac{\mu_0 i}{2\pi r} \tag{2.49}$$

Similarly, by applying Ampere's law we can find that a current 'i' produced inside a long straight wire of circular cross-section and by considering the current uniformly distributed over the cross-section of the wire then the magnetic field 'B' is given as

$$B = \left(\frac{\mu_0 i}{2\pi R^2} \right) r \tag{2.50}$$

where 'R' is the radius of circular sectional wire considering $R > r$
'r' is the radius of amperian loop.

For derivation of the exp (2.50), we consider that the amperian loop is inside the wire. Whereas the exp. (2.49) considers amperian loop outside the wire.

2.13 RIGHT HAND RULE

By grasping the current carrying conductor by the right hand and with the thumb along the direction of current then the direction of the curled fingers gives the direction of the flux 'ϕ' or field 'B'.

This rule is very much used in machines to determine the direction of current in a conductor placed in the static or rotating part. This enables to determine other important electrical and non-electrical parameters which is necessary in the analysis of the machine dynamics.

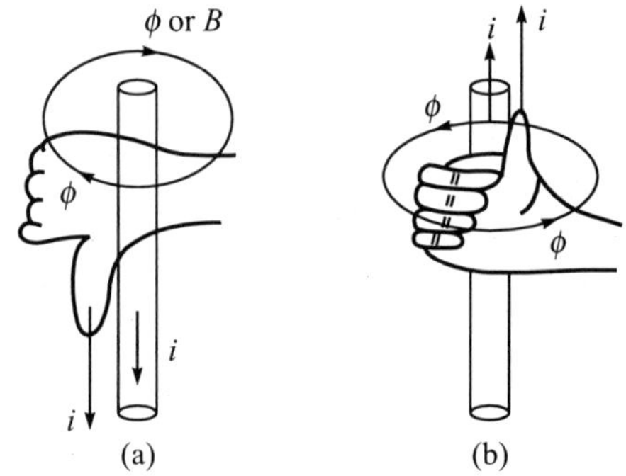

Fig. 2.13 R.H. rule.

The magnetic field produced by a current in a long conductor is shown in Fig. (2.14).

In which case we see that at points 'P_1' or 'P_2' on the magnetic flux line the field direction is perpendicular to the radial line connecting P_1, P_2 and the centre of the conductor.

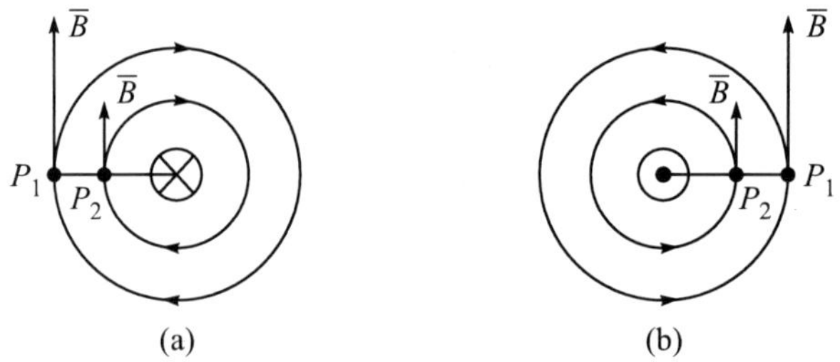

Fig. 2.14 Direction of field B due to current flowing in a conductor.

SOLVED QUESTIONS

Q.1. A long conducting cylinder with inner radius 1 cm and outer 2 cm. The cylinder carrying current as shown in Fig. 1 The current density 'J' is given by $J \propto r^2$. The constant of proportionality is 10^6 A/m^4 and 'r' is in metre. What is the magnetic field $\overline{B}$ at a point that is 1.5 cm from the central axis of cylinder?

A.1.

$$i_{\text{enclosed}} = \int J \cdot dA = \int_a^r c r^2 (2\pi r\, dr)$$

$$= 2\pi c \int_a^r r^3 dr = 2\pi c \left[\frac{r^4}{4}\right]_a^r$$

$$= \pi c \frac{(r^4 - a^4)}{2}$$

'i_{enclosed}' is the net current enclosed by an Amperian loop.

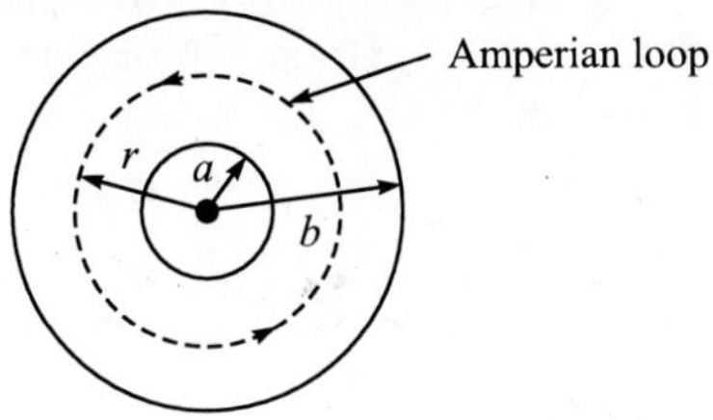

Fig. 1

where, 'r' $\rightarrow$ Amperian loop radius.

$$\oint B \cdot d\bar{s} = \mu_0 \, i_{\text{enclosed}}$$

$$B(2\pi r) = \frac{\mu_0 \pi c}{2}(r^4 - a^4)$$

$$B = \frac{\mu_0 c}{4r}(r^4 - a^4)$$

Solving for 'B', we have

$$B = \frac{4\pi \times 10^{-7}/m \times 10^6 \, A/m^2}{4(0.015\,m)} \times [(0.015)^4 - (0.01)^4]$$

$$B = 20.9 \, [5.0625 \times 10^{-8} - 10^{-8}]$$

$$B = 8.49 \times 10^{-7} \text{ Tesla}$$

Q.2. A coil of area 4 cm^2 and having 100 turns is placed in a field of 1 Tesla. If a current of 1mA produces an angular deflection of 30°. What is the spring constant considering $\theta \propto I$ for analog ammeter considering $\sin \phi$ as unity.

A.2.

$$A = 4 \text{ cm}^2 = 4 \times 10^{-4} \text{ m}^2$$

$$N = 100$$

$$B = 1 \text{ Tesla}$$

$$I = 1 \text{ mA} = 1 \times 10^{-3} \text{ A}$$

$$\theta = 30°$$

$$K = ?$$

We know that for analog ammeter

$$NIAB \sin \phi = K\,\theta$$

So,
$$K = \frac{NIAB \sin\phi}{\theta}$$

$$K = \frac{100 \times 10^{-3} \times 4 \times 10^{-4} \times 1}{3\phi} \text{Nm/degree}$$

$$K = 1.33 \times 10^{-6} \text{ Nm/degree}$$

Q.3. In a long solenoid, a current 1.0 A is passed. The solenoid has 200 turns/cm and its diameter is 3 cm. At its centre a 100 turn coil of diameter 2 cm is placed. The current through the solenoid is reduced to zero at a rate of 10 ms. What is the emf induced in the central coil?

A.3.

$$I = 1 \text{ A}$$

$$n = 200 \text{ turn/cm} = 20000 \text{ turn/m.}$$

$$A = \frac{1}{4}\pi d^2 = \frac{1}{4} \times \pi \times 9 \times 10^{-4} = 7.065 \times 10^{-4} \text{ m}^2$$

$$\therefore \quad \phi_{B,i} = B_A = (\mu_0 I n)A$$

$$= 4\pi \times 10^{-7} \times 1 \times 20000 \times 7.065 \times 10^{-4}$$

$$= 1.775 \times 10^{-5} \text{ Wb}$$

So, the rate of change of flux is

$$\frac{d\phi}{dt} = \frac{\phi_{B,t} - \phi_{B,i}}{t} = \frac{0 - 1.775 \times 10^{-5}}{10 \times 10^{-3}}$$

$$\frac{d\phi}{dt} = 1.775 \times 10^{-3} \text{ Wb/s}$$

$\therefore$ the induced emf $E = -N\frac{d\phi}{dt}$

$$E = 100 \times 1.775 \times 10^{-3}$$

$$E = 0.1775 \text{ V}$$

$$E = 177.5 \text{ mV}$$

Q.4. In a rectangular loop of wire placed in a non-uniform and varying field $\overline{B}$ that is perpendicular to the plane. The field magnitude is 4 t^2x^2 where 'B' is in tesla 't' is in seconds and 'x' in meters. The loop has width of 4 m and height 2 m. What are the magnitude and direction of the induced emf around the loop at $t = 0.1$ sec?

A.4.

$$\phi_B = \int B \cdot d\overline{A} = \int B \cdot dA = \int B \cdot H \cdot dx = \int 4\, t^2 x^3\, H dx$$

$$\phi_B = 4\, t^2 H \int_0^4 x^2 dx = 4t^2 H \left| \frac{x^3}{3} \right|_0^4 = 85.33\, t^2 H \cdot Wb$$

since $\quad H = 2 \text{ m}$

$$\therefore \quad \phi_B = 170.7\, t^2,\ Wb$$

From Faraday's law.

$$E = \frac{d\phi_B}{dt} = 170.7 \times 2 \times 1 = 341.33\, t, \text{ volt}$$

$\therefore$ At $\quad t = 0.1$ sec.

$$E = 341.33 \times 0.1, \text{ volt}$$

or, $\quad E = 34.133$ volt

Q.5. Two indentical 500 turn coils are wound on the same magnetic core. A current changing at the rate of 1000 A/s in first coil induces voltage of 10 V in the second coil. What is the mutual inductance of this arrangement? If the self-inductance of first coil is 10 mH, what per cent of flux set up by coil first links coil 2.

A.5. Mutual inductance between the coils is

$$M = \frac{10}{1000} = 0.01 \text{ H.}$$

Since for identical coils, self-inductances are equal. Then

$$M = K\sqrt{L_1 L_2} = K\sqrt{L^2} = KL$$

if $\quad L_1 = L_2 = L$

since; $\quad L_1 = 10 \text{ mH} = 0.01 \text{ H}$

$\therefore$ coefficient of coupling is 'K' $= \dfrac{M}{L}$

$$K = \frac{0.01}{0.01} = 100\% = 1$$

Hence, with this arrangement 100% or all the flux links with the coil 2.

TUTORIAL QUESTIONS

Q.1. A current $I = 1$ A circulates in a round thin wire loop of radius $R = 100$ mm. Find the magnetic induction (a) at the centre of the loop (b) at the point lying on the axis of the loop at a distance x = 100 mm from its centre.

Ans: (a) $B = \dfrac{\mu_0 I}{2R} = 6.3\ \mu\text{T}$

(b) $B = \mu_0 R^2 I/2(R_1^2 x^2)^{3/2} = 2.3\ \mu\text{T}$

Q.2. Find the magnetic induction of the field at the point 'O' of a loop with current 'I', whose shape is given in figure below.

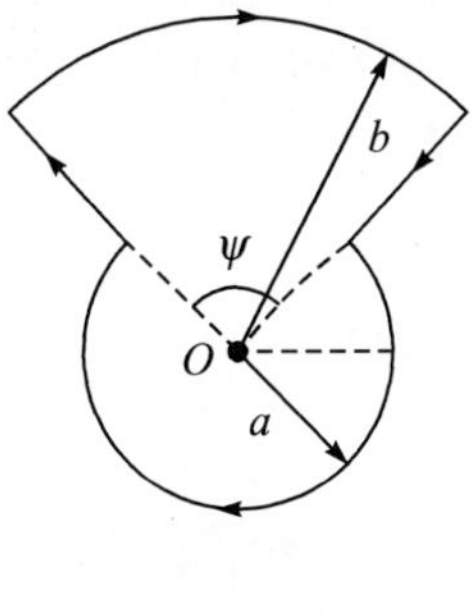

Ans: $B = \dfrac{\mu_0 I}{4\pi}\left[\dfrac{2\pi - \psi}{a} + \dfrac{\psi}{b}\right]$

Q.3. A small current carrying loop is located at a distance 'r' from a long straight conductor with current 'I'. The magnetic moment of the loop is equal to $\overline{M}_B$. Find the magnetic and direction of the force vector applied to the loop if the vector $\overline{M}_B$:

(a) is parallel to the straight conductor

(b) is oriented along the radius vector $\bar{r}$

(c) coincides in direction with the magnetic field produced by the current 'I' at the point where the loop is located.

Ans: (a) O (b) $F = \left(\dfrac{\mu_0}{4\pi}\right)\dfrac{2I \cdot M_B}{r^2}$

(c) $F = \left(\dfrac{\mu_0}{4\pi}\right) 2I \cdot M_B$

Q.4. A rectangular loop with a sliding connector of length 'l' is located in a uniform magnetic field perpendicular to the loop plane as shown in figure below. The magnetic induction is equal to 'B'. The connector has resistance 'R_1', the sides have resistances R_1 and R_2 respectively. Neglecting the self-inductance of the loop. Find the current flowing in the connector during its motion with a constant velocity 'v'.

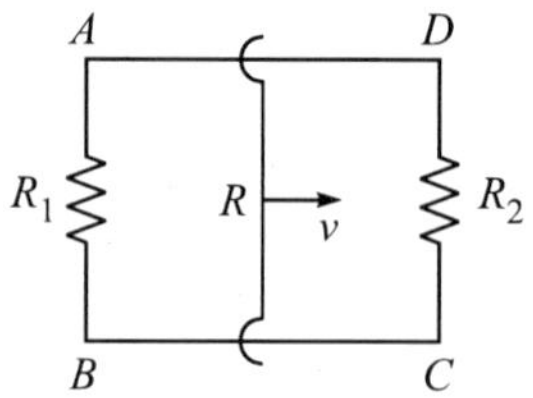

Ans: $I = \dfrac{Bvl}{R + \dfrac{R_1 R_2}{R_1 + R_2}}$

Q.5. There are two stationary loops with mutual inductance 'L_{12}' the current in one of the loops starts to be varied as $I_1 = \alpha t$, where 'α' is a constant 't' is time. Find the time dependence $I_2(t)$ of the current in the other loop whose inductance is 'L_2' and resistance R.

Ans: $I_2 = \dfrac{\alpha L_{12}}{R}(1 - e^{-tR/L2})$

Q.6 Two identical coils each of inductance 'L' are interconnected (a) in series (b) in parallel. Assuming the mutual inductance of the coils to be negligible, find the inductance of the system in both cases.

Ans: (a) $L_{\text{total}} = 2L$

(b) $L_{\text{total}} = L/2$

Q.7 Two solenoids of equal length and almost equal cross-sectional area are fully inserted into one another. Find their mutual inductance if their inductances are equal to 'L_1' and 'L_2'.

Ans: $L_{12} = \sqrt{L_1 L_2}$

Q.8 Using Maxwell's equations show that

(a) a time dependent magnetic field cannot exist without an electric field.

(b) a uniform electric field cannot exist in the presence of a time dependent magnetic field.

(c) inside an empty cavity a uniform electric (or magnetic) field can be time dependent.

Q.9 A wire shaped as a semicircle of radius 'a' rotates about an axis OO' with an angular velocity 'ω' in a uniform magnetic field of induction 'B' as shown in figure. The rotation axis is perpendicular to the field direction. The total resistance of the circuit is equal to 'R'. Neglecting the magnetic field of the induced current, find the mean amount of thermal power being generated in the loop during a rotation period.

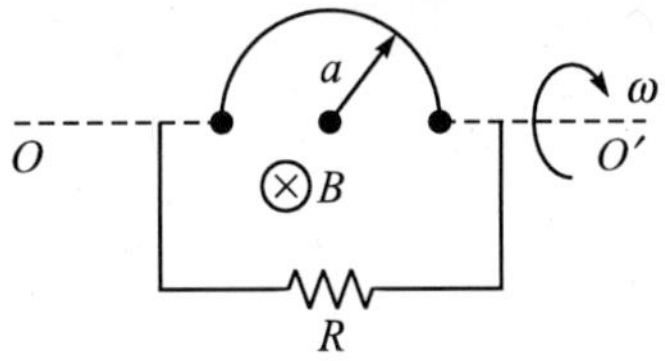

Ans: $P_t = \frac{1}{2}\left(\frac{(\pi\omega a^2 B)^2}{R}\right)$

Q.10 A small coil is introduced between the poles of an electromagnet so that its axis coincides with the magnetic field direction. The cross-sectional area of the coil is equal to $A = 3\ \text{mm}^2$, the number of turns is $N = 60$. When the coil turns through 180° about its diameter a ballistic galvanometer connected to the coil indicates a charge $q = 4.5\ \mu\text{C}$ flowing through it. Find the magnetic induction magnitude between the poles provided the total resistance of the electric circuit equals $R = 40\ \Omega$.

Ans: $B = \frac{1}{2}\frac{qR}{NA} = 0.5\ \text{T}$

Chapter

3

Magnetic Circuits

3.1 Introduction
3.2 Permanent Magnets
3.3 Magnetic Materials
3.4 Magnetic Circuit Terminology
3.5 Analogy of Magnetic Circuit with Electric Circuit
3.6 Series Magnetic Circuit
3.7 Parallel Magnetic Circuit
3.8 Fringing
3.9 Hysteresis
- Solved Questions
- Tutorial Questions

3.1 INTRODUCTION

This chapter describes the importance of magnetic materials in the practical electrical circuits. But before getting a clear understanding of the magnetic materials, one should be clear about the concept of magnets. In day-to-day life, we frequently come in contact with the magnets and magnetic materials which are found everywhere. So the magnetic materials play an important role in man's life. The effect of the magnets on the magnetic materials will appear strange if one do not have the concept of it.

The practical utility of these two materials are from small toys to huge alternators. One could have plenty of questions in the mind regarding the magnetic property as to what will happen to the poles or polarity of the magnet when the magnet gets broken or when the coil around the magnetic material is devoid of current. These are discussed one by one in the coming sections.

3.2 PERMANENT MAGNETS

The materials called magnets have a tendency to attract the iron pieces towards its poles. The direction of the iron pieces will be along the direction of field flux lines which can be imagined to originate from one pole and ending at the other pole. So we can imagine that a permanent magnet has a dipole. These dipoles cannot be separated to get monopole magnet. As soon as we break a magnet, automatically the opposite poles are generated as shown in the Fig. (3.1).

But the problem with the permanent magnet is that the field of the magnet weakens with ageing of the magnet. So the magnet operated device gets affected as a result of it. So, some other alternative is required to counter this effect. So the scientist and technologist are more interested in those magnets whose magnetic effect can be controlled by some external means. So with time there is the origin of electromagnets. Certain laws are enuntiated by the scientists which give a clear understanding of the behaviour of these magnets. Guass stated that the net magnetic flux through any closed Guassian surface is zero. Mathematically, the Guass law is given by eq. 3.1.

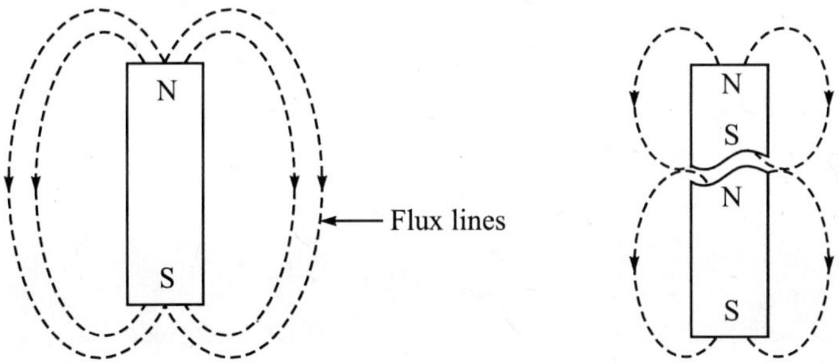

(a) Bar magnet (b) Broken magnet

Fig. 3.1

$$\phi_B = \oint_s \overline{B} \cdot \overline{dA} = 0 \tag{3.1}$$

The above expression is for magnetic fields. For electric fields, the Gauss law states that the net electric flux through the surface is proportional to the net electric charge 'q' enclosed by the surface.

Mathematically, it is given by eq. 3.2.

$$\phi_E = \oint_s \overline{E} \cdot \overline{dA} = \frac{q}{C_0} \tag{3.2}$$

where, 'S' is a Gaussian surface.

The simplest Guassian surface is a surface enclosing a dipole. It must be noted that earth behaves as a large magnet having two poles. If a bar magnet is suspended then the magnet will align in that direction in which the magnetic axis of the earth lies. The opposite poles of the magnets will get attracted by the opposite poles of earth's magnetic field. The direction of the magnetic axis is 11.5° with respect to the earth's rotational axis.

3.3 MAGNETIC MATERIALS

The use of the magnetic materials is enormous in the communication network, transportation, power plants, banks, audio headsets, etc. Magnetic materials are magnetic because of the presence of electrons in it. The flow of electrons will lead to current and hence produces magnetic field. The electrons present in a magnetic material are associated with two dipole moment. One is the spin magnetic dipole moment whereas the other is the orbital magnetic dipole moment. The derivation of the dipole moment expressions is out of the scope of this book. Only the expressions are given by equation (3.3) and equation (3.5) respectively.

$$\boxed{\bar{\mu}_S = -\frac{e}{m}\bar{S}} \qquad (3.3)$$

where, 'μ_s' is the spin magnetic dipole moment
'e' is the charge of electron (1.6×10^{-19} coulomb)
'm' is the mass of an electron (9.11×10^{-31} kg)
'$\bar{S}$' is spin angular momentum

$$\bar{S} = m_s \frac{h}{2\pi} \text{ for } m_s = \pm\frac{1}{2}$$

whereas, due to orbital motion of electron 'L_{orb}' originates given as,

$$\bar{L}_{orb} = m_l \cdot \frac{h}{2\pi} \text{ for } m_l = 0, \pm 1, \pm 2 \qquad (3.4)$$

where, 'L_{orb}' is the orbital angular momentum
'm_l' is the orbital magnetic quantum number
'h' is the Planck's constant (6.63×10^{-34} J.s)

From the above expressions given by (3.4) there is orbital magnetic dipole moment and are related by

$$\boxed{\bar{\mu}_{orb} = -\frac{e}{2m}\bar{L}_{orb}} \qquad (3.5)$$

The forces that act on magnetic materials when the materials are placed in a non-uniform field. Assume that the electrons orbit in the magnetic materials like tiny loops carrying current. We consider the north pole of the magnetic material as shown in Fig. (3.2).

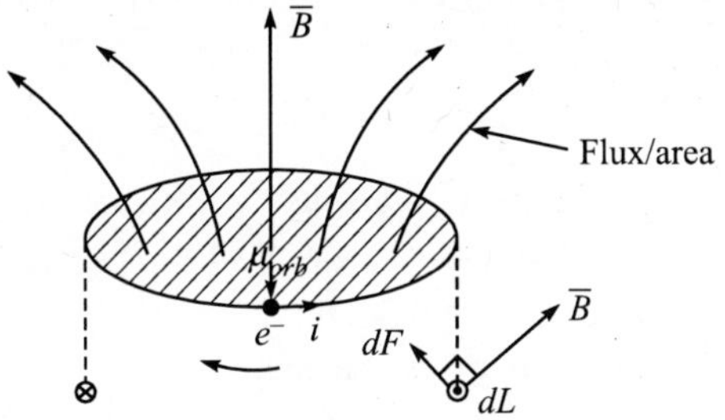

'dF' is an incremental magnetic force
'dL' is an incremental length of loop.

Fig. 3.2 Electron flowing in a loop.

Electrons can move in either direction. The direction of current will oppose the direction of electron. The field direction is given by Thumb's rule as shown in Fig. (3.3).

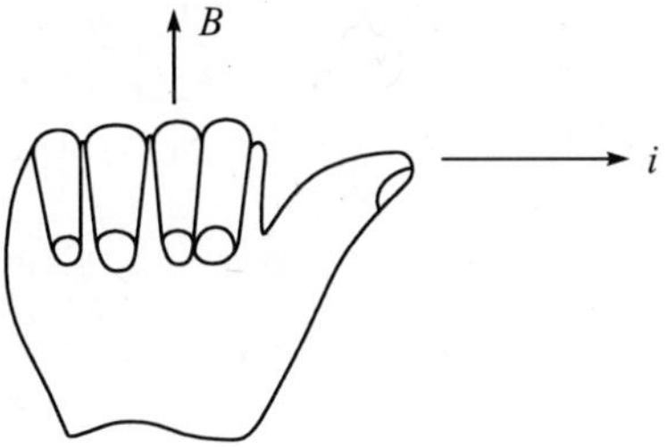

Fig. 3.3 Thumb Rule (Flemming's).

The current direction is seen with respect to the loop and is shown in Fig. (3.2). Thus, current in an incremental length 'dL' of loop in a field of $\overline{B}$ experiences a magnetic force $\overline{dF}$ given as follow.

$$\boxed{\overline{dF} = i\overline{dL} \times \overline{B}} \tag{3.6}$$

3.3.1 Types of Magnetic Materials

There are three types of magnetism and correspondingly three types of magnetic materials, namely diamagnetic, paramagnetic and ferromagnetic.

Diamagnetic Materials

Weak magnetic dipole moments are produced in the atoms of the material when the material is placed in an external magnetic field $\overline{B}$. The combination of all induced dipole moments gives the material as a whole only a feeble net magnetic field. The dipole moments and their net magnetic field disappear when $\overline{B}$ is removed. The diamagnetic material usually refers to materials that exhibit only diamagnetism. If the field is non-uniform, the diamagnetic material is repelled from a region of greater magnetic field towards a region of lesser field. Most of the living creatures are diamagnetic. Examples are bismuth, zinc, phosphorus, lead, etc.

Paramagnetic Materials

In the paramagnetic materials, the magnetic dipole moments do not cancel each other such that permanent magnetic dipole moment is present in it. If the paramagnetic material is placed in an external magnetic field $\overline{B}$, the magnetic dipole moments tend to line up with the field which gives the material a net dipole moment $\overline{\mu}$. This alignment with the external field is opposite to that of diamagnetic materials. Due to random collisions of its atomic dipoles due to thermal agitation transfer energy amongst them, destroying their alignment and thus reducing the materials net magnetic dipole moment. The magnetization $\overline{M}$ of the paramagnetic material sample is given by the equation (3.7) as follows.

$$\boxed{M = \frac{\text{Magnetic dipole moment}}{\text{Volume of sample}}}, \text{ A/m} \tag{3.7}$$

Complete alignment of the atomic magnetic dipole moments is called magnetic saturation of the sample.

According to Curie, the magnetization of the sample is directly proportional to the external magnetic field $\overline{B}$ and inversely proportional to the temperature 'T' in kelvins,

i.e,

$$\boxed{M = C\frac{B}{T}} \tag{3.8}$$

where 'C' is the curie constant
'T' is the temperature

The increase in 'B' increases the alignment of the dipoles whereas, 'T' decreases the alignment due to thermal agitation. The expression given by equation (3.8) is valid for high values of B/T. The plot between M/M_{max} and 'B/T' is called the magnetization curve.

where

$$M_{max} = \frac{N \cdot \mu}{V}$$

where 'N' is the total number of atoms.

Weak paramagnetic substances are known as non-magnetic substances. The diamagnetic substances are also considered in this category. They are not used where strong magnetic field is required. Paramagnetic materials get attracted by a magnet for e.g., manganese, aluminium, platinum, etc. super paramagnetic materials are made up of powdered iron or other magnetic materials such as permalloy.

Ferromagnetic Materials

Iron, cobalt, nickel and alloys such as Si steel, and M-5 steel contains these elements exhibit ferromagnetism because of exchange coupling in which spinning electron of one atom interacts with the neighbouring atoms. This result in alignment of the magnetic dipole moments of the atoms in spite of the random tendency of atomic collisions. Thus, the alignment persists so as to give permanent magnetism.

If the temperature exceeds curie temperature the ferromagnetic material becomes paramagnetic. For iron, curie temperature is 770°C.

The ferromagnetic materials have plenty of magnetic domains. These are regions of the crystal throughout in which the alignment of the atomic dipoles is essentially perfect. The magnetic domains are not all aligned as shown in the Fig. (3.4).

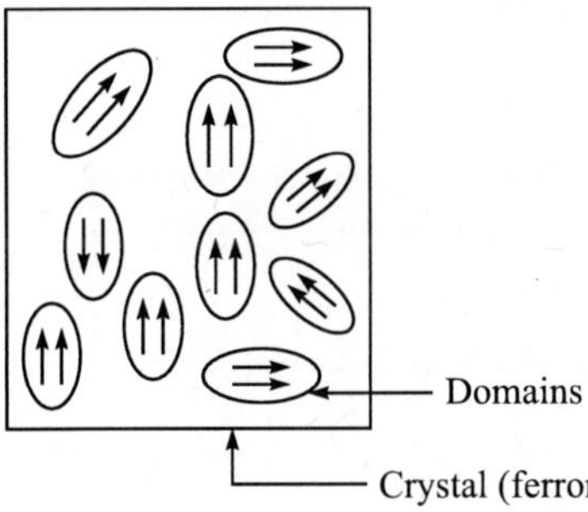

Fig. 3.4 Magnetic domains.

A ferromagnetic material placed in field $\overline{B}$ develops a strong magnetic dipole moment in the direction of $\overline{B}$. If the field is non-uniform, the ferromagnetic material is attracted towards a region of greater magnetic field from a region of lesser field. The examples of ferromagnetic materials are Si Steel, Cast Steel, Iron and its alloys with Ni, Co, Al, etc. The hard ones are permanent magnetic materials such as AlNiCo, chromium steels, etc. Soft ones are iron and its alloys.

There is another category of magnetic materials called ferrimagnetic materials which are oxides of iron or other magnetic materials such as Ni-Zn and Mn-Zn. The ferromagnetic and ferrimagnetic materials have high permeability even of the order of thousands.

3.4 MAGNETIC CIRCUIT TERMINOLOGY

(i) Magnetomotive force (M.M.F): The magnetomotive force may be defined as the work done in moving a unit magnetic pole from one point to another point. The mmf is represented by '$\mathcal{F}$' and the unit of mmf is ampere turns and represented by 'AT'.

The mmf can be mathematically expressed as

$$\boxed{\mathcal{F} = NI = \phi\mathcal{R} = Hl} \qquad (3.9)$$

where, N = Number of turns of coil wound over the magnetic material.
I = Coil current, in amperes.
ϕ = Magnetic flux, in webers
$\mathcal{R}$ = Reluctance of magnetic circuit, in AT/Wb
H = Magnetic field strength, in AT/m
l = Mean magnetic path length, in metres

(ii) **Magnetic flux 'ϕ':** The number of lines of magnetic field which is passing through the magnetic material is called magnetic flux and the unit of it is called weber.

For the flux prefers to pass through an area which has least reluctance, so most of the flux lines are passing through the magnetic material as it has least reluctance and the reluctance of the air is much more than the magnetic material.

(iii) **Magnetic flux intensity 'B':** The magnetic flux intensity may be defined as the ratio of the flux to the area through which the flux is passing. Mathematically, it is represented as

$$\boxed{B = \phi/A}, \text{ Wb/m}^2 \tag{3.10}$$

The unit of magnetic flux intensity is tesla or weber per meter square. The smaller units are gauss or maxwell.
The tesla is related to the flux lines as follows

$$1 \text{ Tesla} = 10^8 \text{ magnetic flux lines.}$$

(iv) **Permeability 'μ':** The permeability of the magnetic material is the conducting capacity of magnetic flux through a material. The permeability is mathematically expressed as

$$\boxed{\mu = B/H}, \text{ Henry/meter} \tag{3.11}$$

The higher the permeability, the better is the conductivity of the flux through the magnetic material. The air has less permeability with respect to that of magnetic material such as iron, nickel, etc.

(vi) **Relative Permeability 'μ_r':** The relative permeability of the material is the ratio of the permeability of the material to the permeability of free space. Mathematically, it is expressed as

$$\boxed{\mu_r = \frac{\mu}{\mu_0}}, \text{ where } \mu_0 = 4\pi \times 10^{-7} \text{ H/m} \tag{3.12}$$

The relative permeability of air is equal to one. The units of relative permeability is equal to weber per ampere turn.

(vi) **Magnetization force 'H':** The magnetization force is also known as magnetic field intensity and is expressed as ratio of the magnetomotive force to the mean length of magnetic flux path. Mathematically, it is expressed as

$$\boxed{H = \frac{F}{l} = \frac{NI}{l}}, \text{ AT/m.} \tag{3.13}$$

It is a measure of the intensity of magnetic flux lines per unit length. It denotes how much powerful is the magnet.

(vii) Reluctance '$\mathcal{R}$': The reluctance may be considered a major opposition to the magnetic flux through a magnetic circuit. Mathematically, it is represented as follows:

$$\boxed{\mathcal{R} = \frac{F}{\phi} = \frac{NI}{\phi}}, \text{ AT/Wb} \tag{3.14}$$

The reluctance is also expressed as

$$\mathcal{R} = \frac{l}{\mu_0 \mu_r A}, \text{ AT/Wb} \tag{3.15}$$

The higher the reluctance of magnetic circuit, lesser will be the magnetic flux lines passing through it and vice versa.

where, $\mu_0 = 4\pi \times 10^{-7}$ H/m

μ_r = Relative permeability of material, unitless

A = Area of cross-section, m^2

l = Mean length of material, 'm'

(viii) Permeance '$\mathcal{P}$': It may be expressed as the reciprocal of reluctance. It is a measure of acceptability of magnetic field flux. The permeance is expressed as follows:

$$\boxed{\mathcal{P} = \frac{1}{\mathcal{R}}}, \text{ Wb/AT} \tag{3.16}$$

The higher the permeance, the larger the flux lines will pass through the material.

(ix) Intensity of magnetization (I or J): It may be defined as the pole strength developed per unit area of the bar. Unit is Wb/m^2, mathematically given by

$$I = \frac{m}{A} = \frac{m \times l}{A \times l} = \frac{M}{V} \tag{3.17}$$

where,

m = Pole strength induced in the bar, in Wb

A = Pole face area of the bar,

M = Magnetic moment

V = Volume, mm^3

3.5 ANALOGY OF MAGNETIC CIRCUIT WITH THE ELECTRIC CIRCUIT

Magnetic Circuit	*Electric Circuit*
Similarities	
(1) The flux 'ϕ' behaves as current in electric circuit.	(1) Current 'I' flows through it
(2) The reluctance '$\mathcal{R}$' is offered to the flux.	(2) Resistance '$\mathcal{R}$' is offered to the current
(3) The mmf '$\mathcal{F}$' drives the flux 'ϕ' through the magnetic circuit.	(3) The emf 'E' drives the current 'I' through it.
(4) The magnetic ohm's law is $\mathcal{F} = \phi\mathcal{R}$.	(4) The Ohm's law is $E = IR$
(5) The reluctance is related as $\mathcal{R} = \frac{l}{\mu A}$.	(5) The resistance is related as $\mathcal{R} = \rho\frac{l}{A}$
(6) The permeance '$\mathcal{P}$' is given as $\mathcal{P} = \frac{1}{\mathcal{R}}$.	(6) The conductance 'G' is given as $G = \frac{1}{\mathcal{R}}$
(7) The flux 'ϕ' divides in parallel magnetic circuit.	(7) The current divides in parallel electric circuit.
Dissimilarities	
(1) No magnetic insulator exists.	(1) Electrical insulator exists called dielectric.
(2) No need of source affer production of flux in a magnetic material	(2) Need of source of emf for the current to flow.
(3) Flux can flow through open circuit	(3) Current cannot flow through open circuit
(4) Magnetic flux leaks	(4) Electric current cannot leak through air.
(5) Magnetic flux cannot completely flow through a material	(5) The current can flow completely through a conductor.

3.6 SERIES MAGNETIC CIRCUIT

The magnetic circuit in which the flux 'ϕ' produced by the mmf 'F' remains same throughout the magnetic circuit. The series magnetic circuit can have air gap or without air gap depending on the type of magnetic circuit. The practical application of the series magnetic circuit is armature attracted relay, single phase core type transformer, etc.

The series magnetic circuit is broadly divided into two types:

(i) Without air gap

(ii) With air gap.

The magnetic specimen may employ square or ring specimens as shown in the figure below.

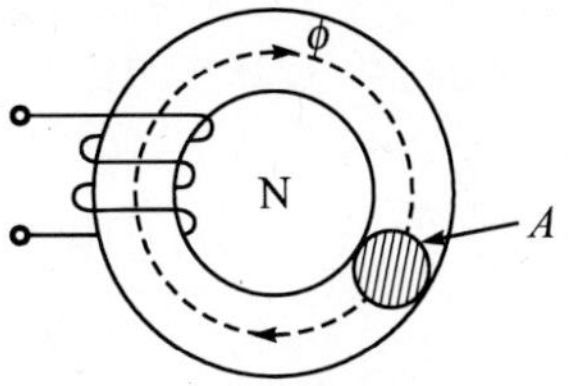

Fig. 3.5(a) Ring specimen (without air gap).

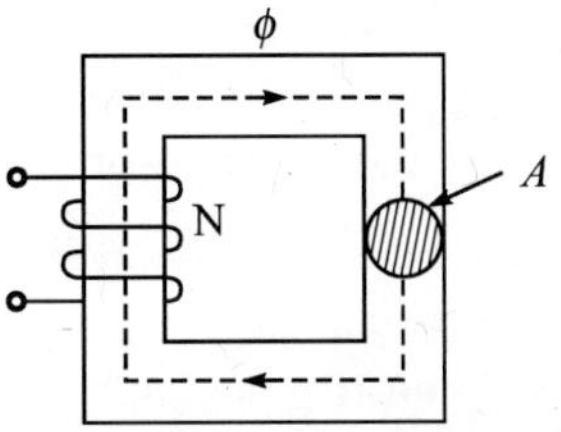

Fig. 3.5(b) Square specimen (without air gap).

(i) The magnetic equivalent circuit of the above magnetic circuit shown in the Fig. 3.5(a) and (b) can be represented as shown in Fig. (3.6).

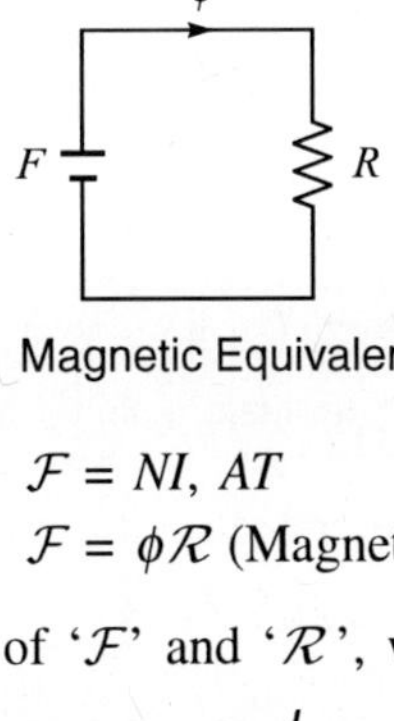

Fig. 3.6 Magnetic Equivalent Circuit.

where $\mathcal{F} = NI$, AT

$\mathcal{F} = \phi\mathcal{R}$ (Magnetic Ohm's Law)

Substituting the values of '$\mathcal{F}$' and '$\mathcal{R}$', we get

$$NI = \phi \cdot \frac{l}{\mu_0 \mu_r A}$$

Thus, the flux 'ϕ' is given as

$$\boxed{\phi = \frac{NI\mu_0\mu_r A}{l}}, \text{ weber} \tag{3.18}$$

The magnetic flux intensity 'B' is given

as $$\boxed{B = \frac{\phi}{A} = \frac{\mu_0\mu_r NI}{l}}, \text{ weber/m}^2 \tag{3.19}$$

(ii) Series magnetic circuit: (with air gap)

Considering ring specimen as shown in Fig. (3.7).

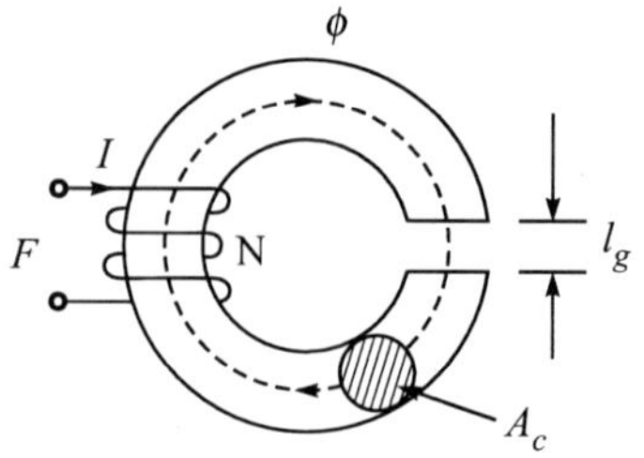

Fig. 3.7 Ring specimen (with air gap).

where, l_g = length of air gap, m
l_c = mean length of core, m

Neglecting fringing effect of magnetic flux in air gap. The area of the core 'A_c' is same as that of area of air gap 'A_g'.
The magnetic equivalent circuit is given as

where, $\mathcal{R}_c$ = core material reluctance
$\mathcal{R}_g$ = air gap reluctance.

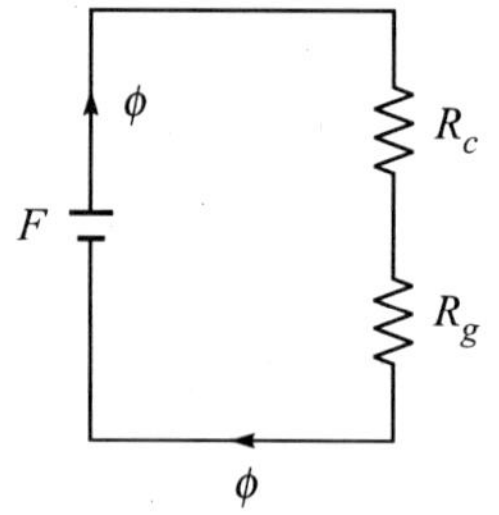

Fig. 3.8 Magnetic equivalent circuit of Fig. (3.7).

Applying, Ohm's law to the above circuit shown in Fig. (3.8).

$$\mathcal{F} = \phi(\mathcal{R}_c + \mathcal{R}_g) \quad (3.20)$$

where, $$\mathcal{R}_c = \frac{l_c}{\mu_0 \mu_r A_c} \quad \text{and} \quad \mathcal{R}_g = \frac{l_g}{\mu_0 A_g}$$

μ_r = relative permeability of material of core

Neglecting fringing effect, $A_g = A_c = A$(say)

$$\therefore \quad \mathcal{F} = \phi\left[\underbrace{\frac{l_c}{\mu_0 \mu_r A} + \frac{l_g}{\mu_0 A}}\right] \quad (3.21)$$

or, $$\mathcal{F} = \phi\, \mathcal{R}_{\text{total}} \quad (3.22)$$

$\mathcal{R}_{\text{total}}$ = Total reluctance of magnetic circuit

$$\mathcal{R}_{\text{total}} = \frac{F}{\phi} = \frac{NI}{\phi} \quad (3.23)$$

If 'ϕ' is unknown, then

$$\frac{\mathcal{F}}{\mathcal{R}_{\text{total}}} = \phi \tag{3.24}$$

From, expression

$$\mathcal{F} = \frac{\phi}{A}\left[\frac{l_c}{\mu_0 \mu_r} + \frac{l_g}{\mu_0}\right]$$

$$\therefore \qquad B = \frac{\phi}{A}$$

So, flux density is given as $\boxed{B = \dfrac{\mathcal{F}}{\left[\dfrac{l_c}{\mu_0 \mu_r} + \dfrac{l_g}{\mu_0}\right]}}$ (3.25)

(iii) Series magnetic circuit: (with different permeability materials used in core)

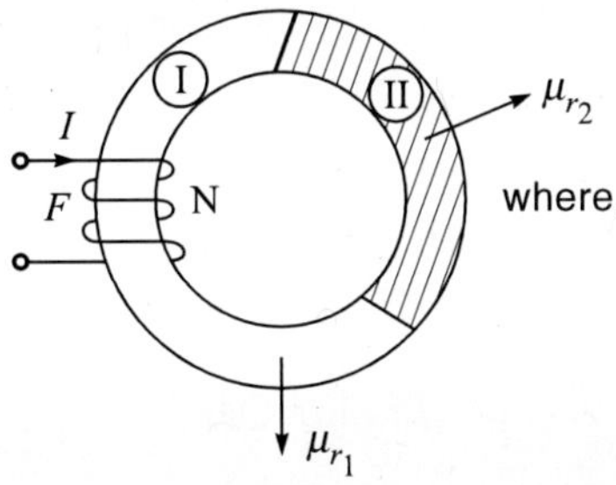

where, $\mu_{r_1} \neq \mu_{r_2}$
$A_1 = A_2 = A$(say)

Fig. 3.9 Core having different materials.

The magnetic equivalent circuit is given as follows:

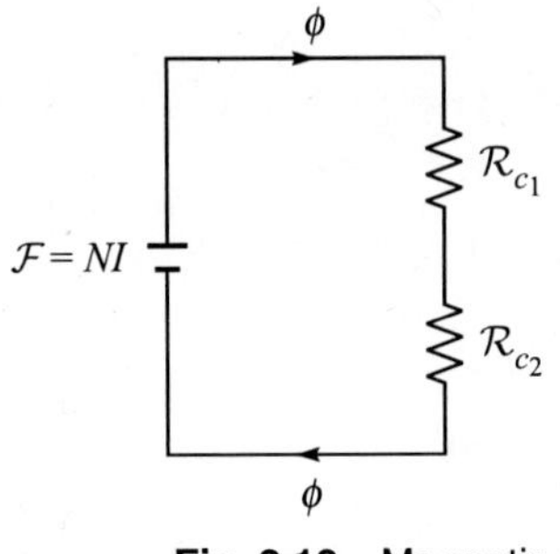

where, $\mathcal{R}_{c_1}$ and $\mathcal{R}_{c_2}$ are the reluctances of the part (I) and (II) of core.

Fig. 3.10 Magnetic equivalent circuit.

$\therefore$ Applying magnetic Ohm's law

$$\mathcal{F} = \phi(\mathcal{R}_{c_1} + \mathcal{R}_{c_2})$$

$$\mathcal{F} = \phi\left[\frac{l_1}{\mu_0 \mu_{r1} A} + \frac{l_2}{\mu_0 \mu_{r_2} A}\right] \quad (3.26)$$

$$\mathcal{F} = \phi\, \mathcal{R}_{\text{total}}$$

So, the flux and reluctance can be determined if the other parameters are known.

(iv) Series Magnetic Circuit: (with different cross-section and permeability of parts).

A_1, A_2 and A_3 are the areas of cross-sections

μ_{r1}, μ_{r2}, μ_{r3} are the relative permeability of the parts.

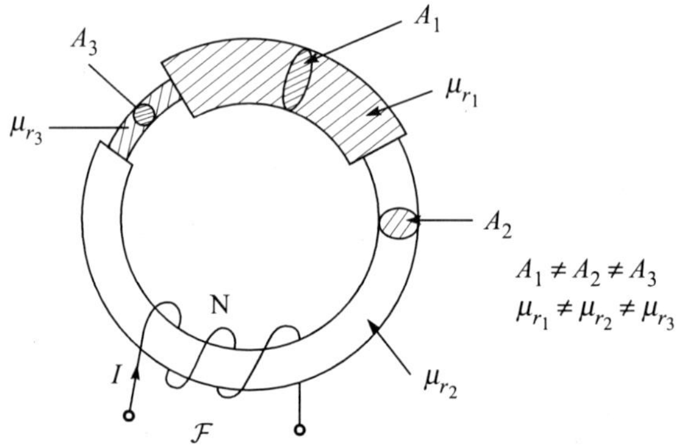

Fig. 3.11 Different permeability and cross-section areas in series magnetic circuit.

The magnetic equivalent circuit of the above Fig. (3.11) is given as

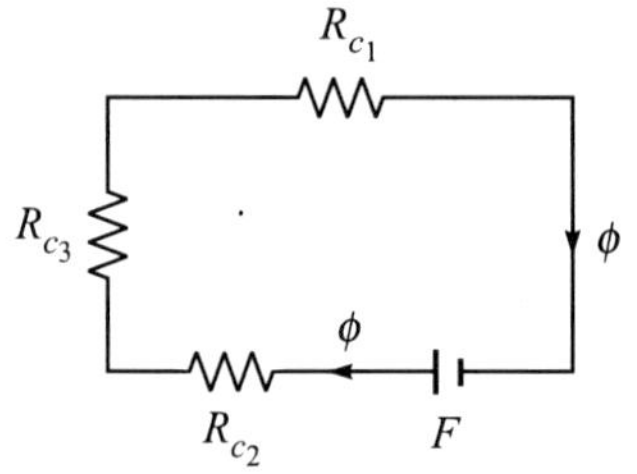

Fig. 3.12 Magnetic equivalent circuit of Fig. (3.11).

Applying magnetic Ohm's law, we have

$$\mathcal{F} = \phi(\mathcal{R}_{c_1} + \mathcal{R}_{c2} + \mathcal{R}_{c3}) \quad (3.27)$$

where, $\mathcal{R}_{c_1} = \dfrac{l_1}{\mu_0 \mu_{r_1} A_1}$; $\mathcal{R}_{c2} = \dfrac{l_2}{\mu_0 \mu_{r_2} A_2}$; $\mathcal{R}_{c3} = \dfrac{l_3}{\mu_0 \mu_{r_3} A_3}$

$$\mathcal{F} = \phi(\mathcal{R}_{\text{total}}).$$

(v) Series Magnetic Circuit: (with multiple excitation source)

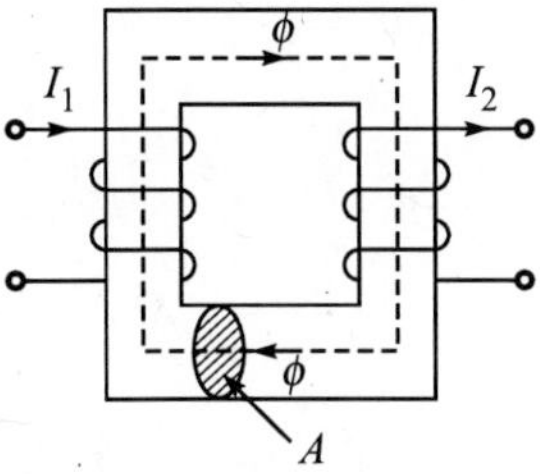

Fig. 3.13 Two excitations applied to a magnetic circuit.

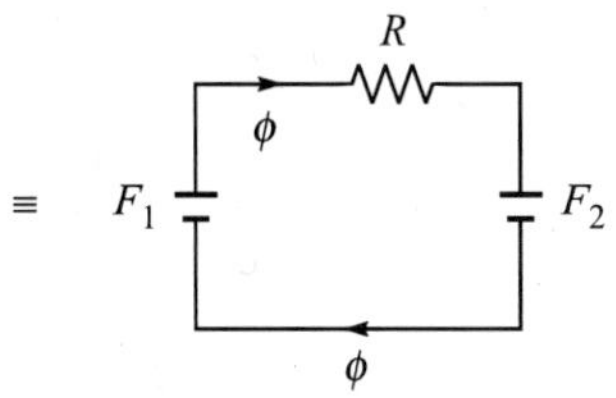

Fig. 3.14 Magnetic equivalent circuit.

If $\mathcal{F}_1 > \mathcal{F}_2$, then the flux '$\phi$' will flow in the direction as shown in the Fig. (3.14).

∴ Applying KVL to the magnetic circuit given in Fig. (3.14),
we have

$$\phi \mathcal{R} + \mathcal{F}_2 - \mathcal{F}_1 = 0 \tag{3.28}$$

or, $$\mathcal{F}_1 - \mathcal{F}_2 = \phi \mathcal{R}$$

or, $$\phi = \frac{\mathcal{F}_1 - \mathcal{F}_2}{R} \tag{3.29}$$

where, $$\mathcal{F}_1 = N_1 I_1, AT \tag{3.30}$$

and $$\mathcal{F}_2 = N_2 I_2, AT \tag{3.31}$$

The reluctance '$\mathcal{R}$' of the circuit is given as

$$\mathcal{R} = \frac{l}{\mu_0 \mu_r A}, \frac{\text{Amp} \cdot \text{Turn}}{\text{Weber}}$$

where, l = mean length of the magnetic circuit
A = Area of the core, m^2

3.7 PARALLEL MAGNETIC CIRCUIT

(i) Without Air gap: The parallel magnetic circuit involves parallel magnetic path so the main flux divides in it. As in the case of shell type transformers which involves three limbs and the excitation is given to the central limb. The magnetic circuit is given as

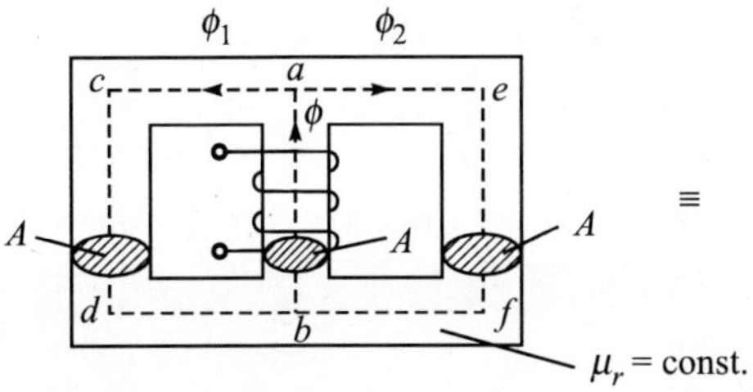

Fig. 3.15 Shell type transformer.

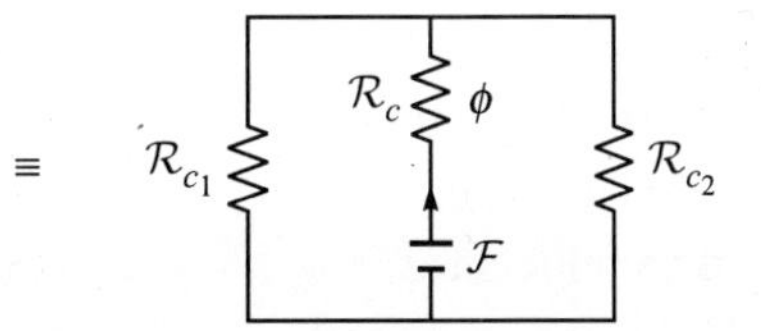

Fig. 3.16 Magnetic equivalent circuit (of shell type transformer).

Applying the Ohm's law to the above circuit, we get

$$\mathcal{R}_{\text{total}} = \mathcal{R}_c + \frac{\mathcal{R}_{c_1}\,\mathcal{R}_{c_2}}{\mathcal{R}_{c_1} + \mathcal{R}_{c2}} \tag{3.32}$$

$$\mathcal{F} = \phi \cdot \mathcal{R}_{\text{total}}$$

where, $l_c = l(ab),\ l_{c_1} = l(acbd),\ l_{c_2} = l(aefb)$

$$\therefore \quad \mathcal{R}_c = \frac{l_c}{\mu_0\,\mu_r\,A} = \mathcal{R}_{c_1} = \frac{l_{c1}}{\mu_0\,\mu_r\,A};\ \mathcal{R}_{c_2} = \frac{l_{c2}}{\mu_0\,\mu_r\,A}$$

The parallel magnetic circuit also has limbs of different cross sections i.e., $A_{c_1} \neq A_{c_2} \neq A_c$.

(ii) With Air gap: (with one excitation)

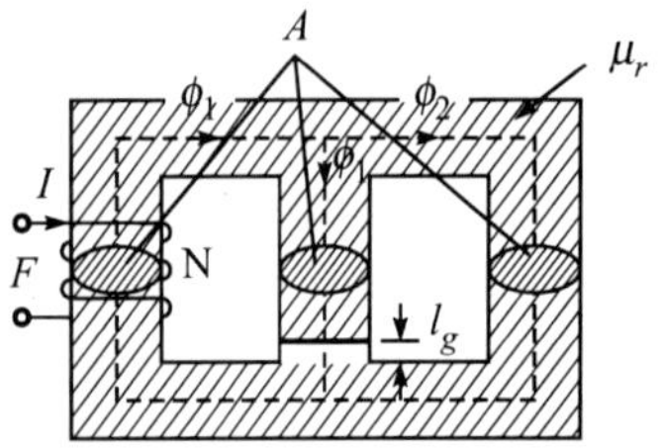

Fig. 3.17 Exact magnetic circuit.

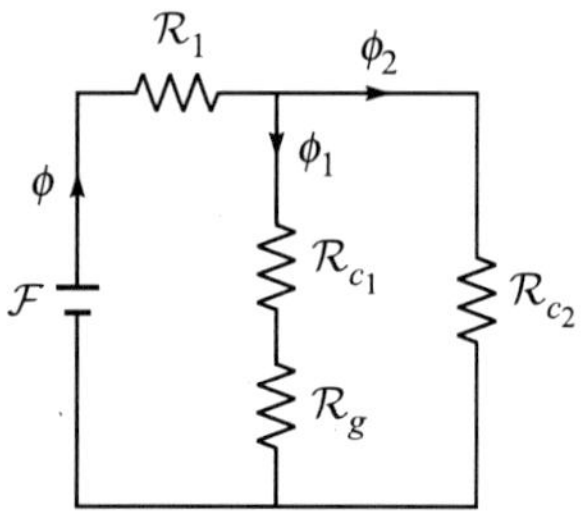

Fig. 3.18 Magnetic equivalent circuit.

The total reluctance of the magnetic circuit is

$$\mathcal{R}_{\text{total}} = \mathcal{R}_1 + \left[\frac{(\mathcal{R}_{c_1} + \mathcal{R}_g) \times \mathcal{R}_{c_2}}{\mathcal{R}_{c_1} + \mathcal{R}_g + \mathcal{R}_{c_2}}\right] \tag{3.33}$$

where, $\mathcal{R}_1 = \frac{l_1}{\mu_0\,\mu_r\,A};\ \mathcal{R}_g = \frac{l_g}{\mu_0\,A}$ (If no fringing effect)

$$\mathcal{R}_{c_1} = \frac{l_{c_1}}{\mu_0\,\mu_r\,A},\ \text{AT/Wb}$$

$$\mathcal{R}_{c_2} = \frac{l_{c_2}}{\mu_0\,\mu_r\,A},\ \text{AT/Wb}$$

$$\therefore \quad \mathcal{F} = \phi\ \mathcal{R}_{\text{total}}$$

$$\therefore \quad \phi = \frac{F}{\mathcal{R}_{\text{total}}} = \frac{NI}{\mathcal{R}_{\text{total}}},\ \text{weber}$$

(iii) **With three air gaps** (with multiple excitation)

Depending on the magnitudes of mmf's, we will have the flux direction. The reluctances can be determined as follows:

$$\mathcal{R}_{c_1} = \frac{l(gab)}{\mu_0\,\mu_r\,A}$$

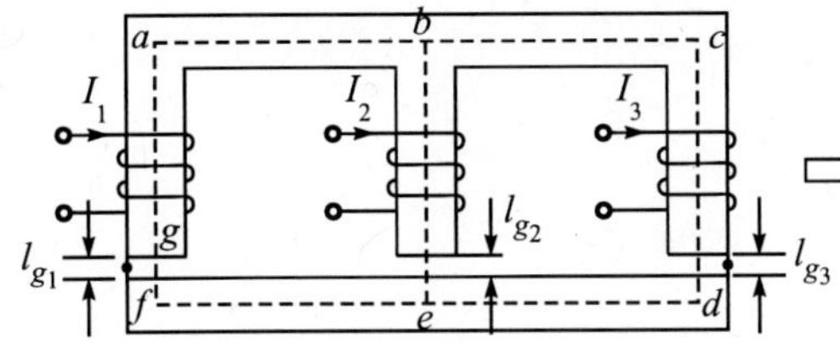

Fig. 3.19 Typical magnetic circuit.

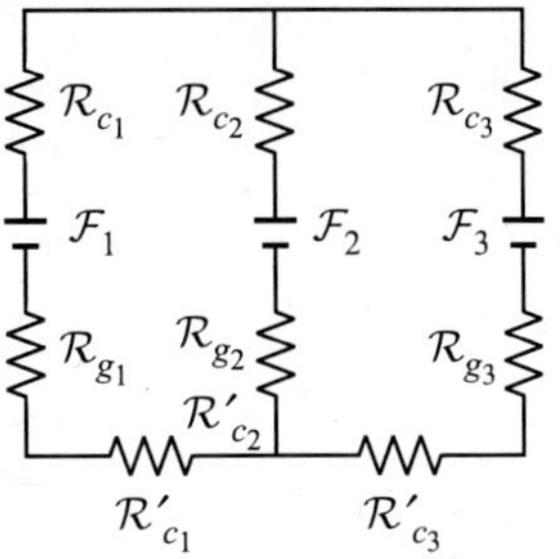

Fig. 3.20 Magnetic equivalent circuits.

Considering uniform cross-section area of the core. The core limb is made up of same material and hence have same permeability.

$$\mathcal{R}_{c2} = \frac{l(bh)}{\mu_0 \mu_r A}$$

$$\mathcal{R}_{c3} = \frac{l(bcl)}{\mu_0 \mu_r A}$$

$$\mathcal{R}_{g1} = \frac{l_{g_1}}{\mu_0 A} \text{ (Neglecting fringing effect)}$$

$$\mathcal{R}_{g2} = \frac{l_{g_2}}{\mu_0 A} \text{ (Neglecting fringing effect)}$$

$$\mathcal{R}_{g3} = \frac{l_{g_3}}{\mu_0 A} \text{ (Neglecting fringing effect)}$$

$$\mathcal{R}'_{c1} = \frac{l(ife)}{\mu_0 \mu_r A}$$

$$\mathcal{R}'_{c2} = \frac{l(je)}{\mu_0 \mu_r A}$$

$$\mathcal{R}'_{c3} = \frac{l(kde)}{\mu_0 \mu_r A}$$

By applying the Kirchoff law's and Ohm's law the flux through any branch could be determined.

3.8 FRINGING

The fringing may be defined as the bending of the flux lines when it crosses the air gap in a magnetic circuit. This bending of flux will increase the effective area of cross-section inside the gap as shown in the Fig. (3.21) below.

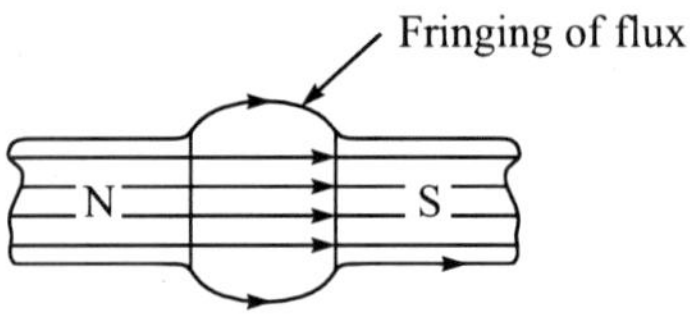

Fig. 3.21 Fringing effect.

The fringing effect lessens the flux under the poles and hence the flux inside the air gap decreases as a result of it.

The larger the length of air gap, the larger will be the fringing effect and hence further decrease in the value of flux.

In the analysis, normally the fringing effect is neglected so as to make the analysis simple. If there is no fringing, then $A_c = A_g$

where 'A_c' = area of cross-section of core
'A_g' = area of cross-section in the air gap having flux

With fringing, the effective area of cutting of flux in the air gap is greater than the actual area of cross-section of core.

3.9 HYSTERESIS

When the magnetic material carries a coil having a.c. current, then during positive cycle the magnetic material gets magnetized and during the negative cycle the magnetic material gets demagnetized. After demagnetization all the domains in the material cannot achieve its equilibrium position. At 50 Hz, the magnetic material goes through magnetization and demagnetization large number of times and gives specific hysteresis loop for a specific magnetic material. The hysteresis loop area is different for different magnetic material having varying permeability. The hysteresis loop for the cast iron and steel is shown in the Fig. (3.22).

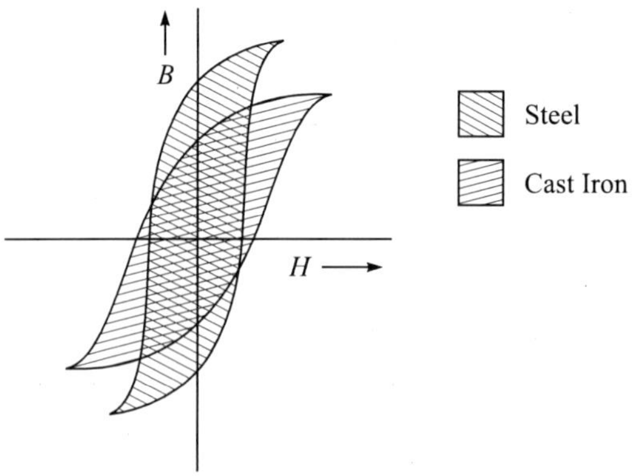

Fig. 3.22 Hysteresis loop.

The area of the hysteresis loop is given by

$$\boxed{W = [B][H]} \tag{3.34}$$

is called the energy product.

The hysteresis loss is given by

$$P_h = K_h \cdot B_m^x \cdot f \cdot V, \text{ watt} \tag{3.35}$$

where, $x = 1.6 - 2.6$

where 'K_h' is called Steinmetz constant
'B_m' is the maximum flux intensity, tesla
'f' is the supply frequency, Hz
'V' is the volume of material, m^3

SOLVED QUESTIONS

Q.1 For the a.c. excited magnetic circuit of Fig. (1), calculate the excitation current and induced emf of the coil to produce a core flux of 0.6 sin 314 t mWb.

UPTU (2001–02)

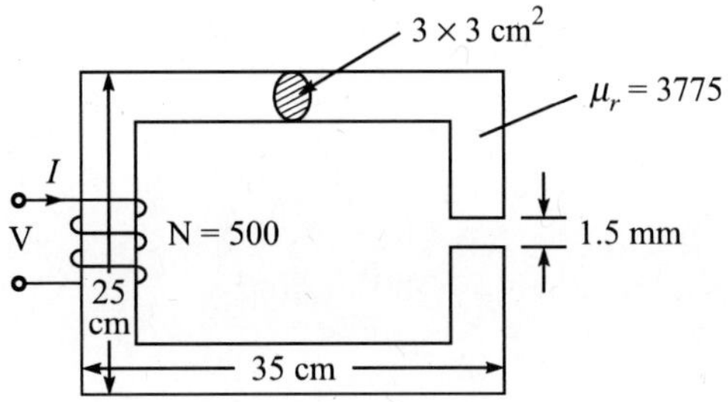

Fig. 1

A.1

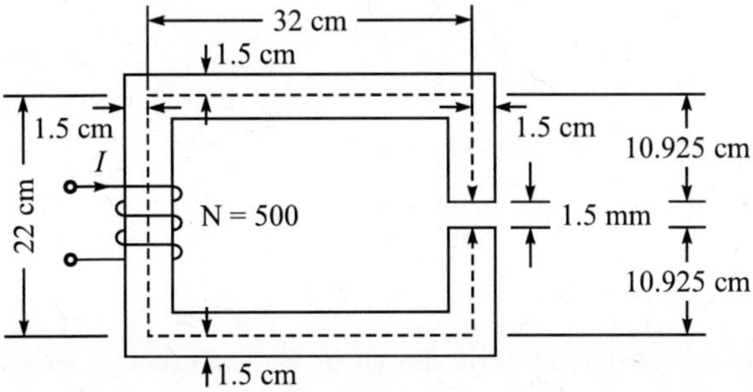

Fig. 1

(a) Mean lengths are

For air gap, l_g = 1.5 mm = 0.15 cm

For core l_c = 22 cm + 32 cm + 21.85 cm = 75.85 cm = 0.7585 m

$$\phi_c = 0.6 \sin 314t \times 10^{-3} \text{ Wb} \Rightarrow \phi_m = 0.6 \times 10^{-3} \text{ Wb}$$

$$A_c = 9 \text{ cm}^2 = 9 \times 10^{-4} \text{ cm}^2$$

$$\mathcal{R}_c = \frac{l_c}{\mu_r \mu_0 A_c} = \frac{0.7585}{3775 \times 4\pi \times 10^{-7} \times 9 \times 10^{-4}} = \frac{7585 \times 10^7}{426726}$$

$\mathcal{R}_c$ = 177, 748.722 AT/Wb

$$\mathcal{R}_g = \frac{l_c}{\mu_0 A_g} = \frac{0.15 \times 10^{-2}}{4\pi \times 10^{-7} \times 9 \times 10^{-4}} = \frac{15 \times 10^7}{113.04} = 1326963.9 \text{ AT/Wb}$$

$\mathcal{R}_{total} = \mathcal{R}_c + \mathcal{R}_g$ = 1504712.63 AT/Wb.

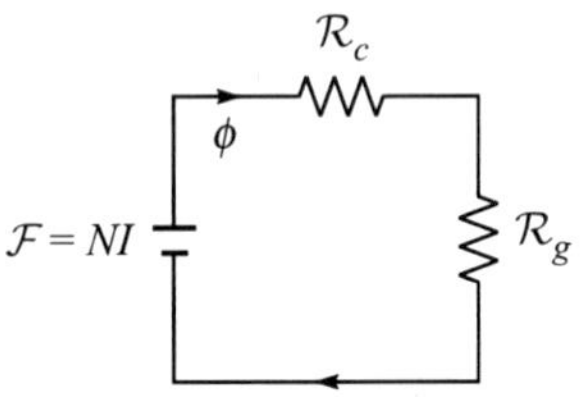

$$\mathcal{F} = \phi_m(\mathcal{R}_c + \mathcal{R}_g) = N\,I$$

$$I = \frac{\phi_m \mathcal{R}_{total}}{N} = \frac{0.6 \times 10^{-3} \times 1504712.63}{500} \text{ Amp}$$

$$I = 1.8057 \text{ Amperes}$$

∴ Excitation current = 1.8057 Amp.

(b) The induced emf $e = -N \dfrac{d\phi_c}{dt}$

$$e = -500 \times \frac{d}{dt}(0.6 \times 10^{-3} \sin 314\,t)$$

$$e = -500 \times 0.6 \times 10^{-3} \times 314 \cos 314\,t$$

$$e = 94.2 \sin(314\,t - \pi/2) \text{ V}$$

∴ E_m = 94.2 volts

So, rms value of induced voltage $E = \dfrac{E_m}{V_2}$

or, $\boxed{E = 66.62 \text{ Volt}}$

∴ Induced voltage = 66.62 volt.

Q.2 A rectangular magnetic core shown in Fig. (2) has square cross-section of area 16 cm^2. An air gap of 2 mm is cut across one of its limbs. Find the exciting current needed in a coil having 1000 turns wound on the core to create an air gap flux of 4 m Wb. The relative permeability of the core is 2000.

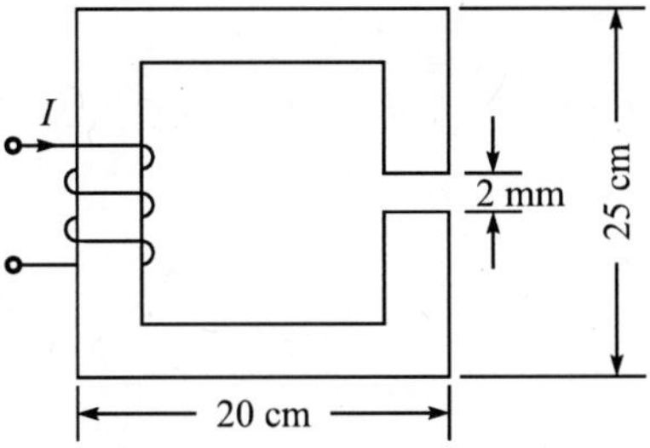

Fig. 2

UPTU (2001–02)

A.2

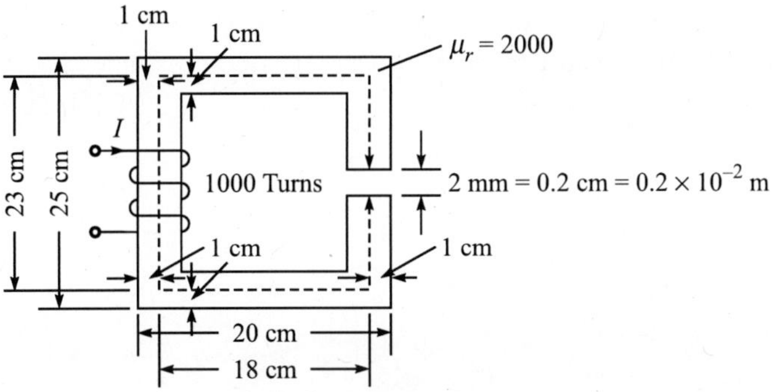

Fig. 2

Mean lengths are

$$l_c = 23 \text{ cm} + 18 \times 2 \text{ cm} + 23 \text{ cm} - 0.2 \text{ cm}$$
$$l_c = 81.8 \text{ cm} = 0.818 \text{ m}$$
$$l_g = 0.002 \text{ m}$$

Magnetic equivalent circuit is given below.

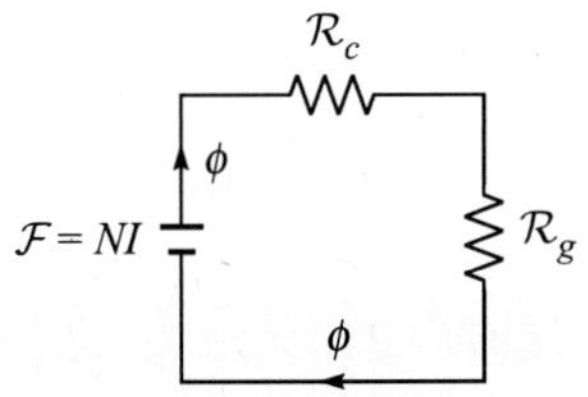

$$\phi = 4 \text{ m Wb} = 4 \times 10^{-3} \text{ Wb}; \; A_c = 16 \text{ cm}^2 = 16 \times 10^{-4} \text{ m}^2$$

$$\mathcal{F} = \mathcal{R} \cdot \phi \text{ where, } \mathcal{R} = \mathcal{R}_c + \mathcal{R}_g = \frac{l_c}{\mu_r \mu_0 A_c} + \frac{l_g}{\mu_0 A_g}$$

Let $A_c = A_g$ (No fringing effect)

$$\therefore \quad \mathcal{R} = \frac{1}{\mu_0 A_c}\left[\frac{l_c}{\mu_{r_c}} + l_g\right] = \frac{1}{4\pi \times 10^{-7} \times 16 \times 10^{-4}}\left[\frac{0.818}{2000} + 0.002\right]$$

$$\mathcal{R} = 1{,}198{,}746.02 \text{ AT/Wb}$$

$$\therefore \quad \mathcal{F} = NI = \phi\mathcal{R}$$

$$\text{So} \quad I = \frac{\phi\mathcal{R}}{N} = \frac{4 \times 10^{-3} \times 1{,}198{,}746.02}{1000} = 4.795 \text{ Amp.}$$

The exciting current = 4.795 A required for production of 4 m Wb flux inside the air gap.

Q.3 A wrought iron bar 30 cm long and 2 cm in diameter is bent into a circular shape as given in Fig. (3). It is then wound with 500 turns of wire. Calculate the current required to produce a flux of 0.5 m Wb in magnetic circuit with an air gap of 1 mm; μ_r (iron) = 4000 (assume constant).

UPTU (2004–05)

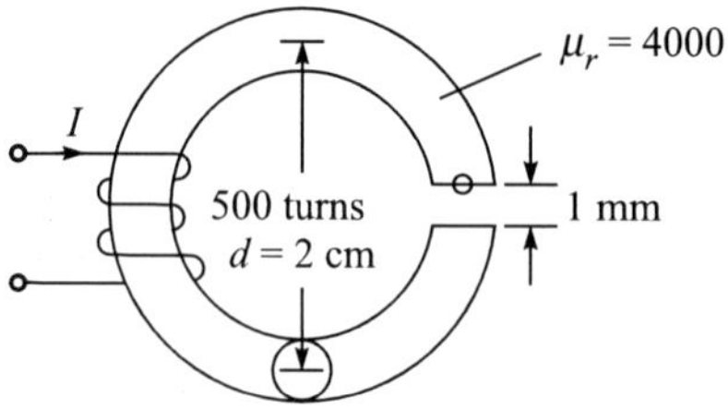

Fig. 3

A.3

$$l_c = 30 \text{ cm} = 0.3 \text{ m}$$

$$d = 2 \text{ cm} = 2 \times 10^{-2} \text{ m}$$

$$A = \frac{\pi d^2}{4} = \frac{3.14 \times 4 \times 10^{-4} \text{ m}^2}{4} = 3.14 \times 10^{-4} \text{ m}^2$$

$$\phi = \phi_g = 0.5 \text{ m Wb} = 0.5 \times 10^{-3} \text{ Wb (Assuming no fringing)}$$

$$l_g = 1 \text{ mm} = 0.001 \text{ m}$$

$$\therefore \quad \mathcal{R}_c = \frac{l_c}{\mu_r \mu_0 A}$$

$$\mathcal{R}_c = \frac{0.3}{4000 \times 4\pi \times 10^{-7} \times 3.14 \times 10^{-4}}$$

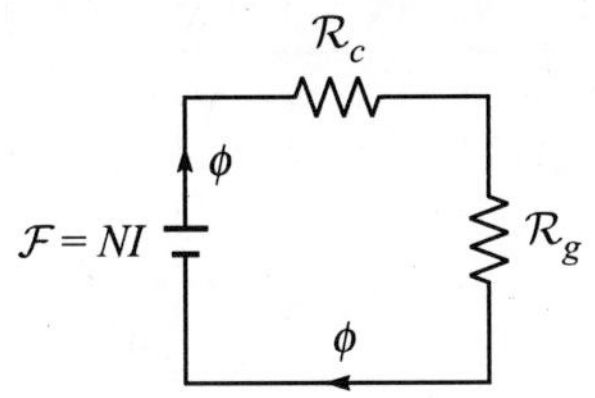

Magnetic equivalent circuit.

So, $\mathcal{R}_c = 190169.99$ AT/Wb

$$\mathcal{R}_g = \frac{l_g}{\mu_0 A} = \frac{0.001}{4\pi \times 10^{-7} \times 3.14 \times 10^{-4}}$$

$= 2{,}535{,}599.82$ AT/Wb

$\therefore$ $\mathcal{R} = \mathcal{R}_c + \mathcal{R}_g$ (In series magnetic circuit)

$\mathcal{R} = 2, 725, 769.811$ AT/Wb

Applying magnetic Ohm's law.

$$\mathcal{F} = NI = \phi\mathcal{R}$$

$$\therefore \quad I = \frac{\phi\mathcal{R}}{N} = \frac{0.5 \times 10^{-3} \times 2725769.811}{500} = 2.7258 \text{ A}$$

$$\boxed{I = 2.73 \text{ Amperes}}$$

Q.4 Calculate the mmf required to produce a flux of 0.015 Wb across an air gap of 2.5 mm long having an effective area of 200 cm^2.

UPTU (2005–06)

A.4

$\phi_g = 0.015$ Wb

$l_g = 2.5$ mm $= 0.0025$ m

$A_g = 200$ cm^2 $= 200 \times 10^{-4} = 0.02$ m^2

mmf required $= \mathcal{F} = \phi_g \mathcal{R}_g$

$$\mathcal{F} = \phi_g\left[\frac{l_g}{\mu_0 A_g}\right] \qquad \because \mathcal{R}_g = \frac{l_g}{\mu_0 A_g}$$

$$\mathcal{F} = 0.015\left[\frac{0.0025}{4\pi \times 10^{-7} \times 0.02}\right], \text{ AT}$$

$$\mathcal{F} = 99522.3 \text{ AT}$$

Q.5 The magnetic circuit frame shown in the following figure is built of iron of square cross-section of 3 cm. Each air gap is 2 mm wide and each of the coil is wound 1000 turns. The relative permeability of part *A* and *B* may be taken as 1000 and 1200 respectively. Calculate:

(a) Reluctance of part *A* (b) Reluctance of two air gaps

(b) Total reluctance of the magnetic path

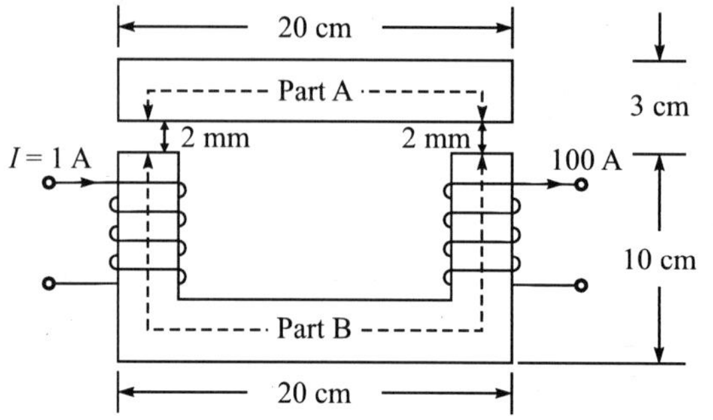

A.5 $l_A = 20\text{ cm} = 0.2\text{ m}$ $\quad A = 3\text{ cm} \times 3\text{ cm} = 9\text{ cm}^2 = 9 \times 10^{-4}\text{ m}^2$

$l_B = 34\text{ cm} = 0.34\text{ m}$ $\quad N = 1000$

$l_g = 2\text{ mm} = 0.002\text{ m}$ $\quad \mu_{r_A} = 1000,\ \mu_{r_B} = 1200$

(a) $$\mathcal{R}_A = \frac{l_A}{\mu_0 \mu_{r_A} A} = \frac{0.2}{4\pi \times 10^{-7} \times 1000 \times 9 \times 10^{-4}} = 176928.521\text{ AT/Wb}$$

$$\mathcal{R}_B = \frac{l_B}{\mu_{r_B} \mu_0 A} = \frac{0.34}{1200 \times 4\pi \times 10^{-7} \times 9 \times 10^{-4}}$$

$\mathcal{R} = 250648.738$ AT/Wb

(b) $$\mathcal{R}_g = \frac{l_g}{\mu_0 A} = \frac{0.002}{4\pi \times 10^{-7} \times 9 \times 10^{-4}} = 1769\ 285.209\text{ AT/Wb}$$

(c) Total reluctance $\mathcal{R}_{\text{total}} = \mathcal{R}' + \mathcal{R}''$

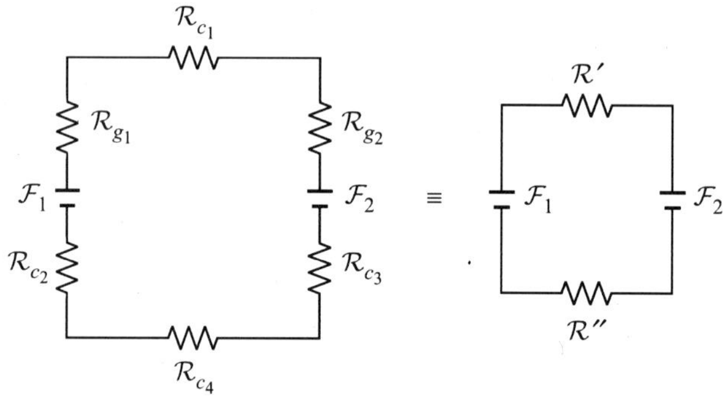

Fig. Magnetic Equivalent circuit.

$\mathcal{R}_{\text{total}} = \mathcal{R}_A + 2\mathcal{R}_g + \mathcal{R}_B$

$= 176928.521 + 2 \times 1769285.209 + 250648.738$

$= 3966147.677$ AT/Wb

Total reluctance of magnetic circuit $= 3.966 \times 10^6$ AT/Wb

Q.6 The value of permeability for a magnetic material is 0.13 H/m. Find the value of susceptibility for the material.

A.6

$$\mu = 0.13$$
$$\mu_r \mu_0 = \mu$$
$$\mu_r = \frac{\mu}{\mu_0} = \frac{0.13}{4\pi \times 10^{-7}} = \frac{13}{4\pi} \times 10^5 = 10{,}3{,}437.3$$

$\therefore$ Susceptibility $\chi = \mu_r - 1 = 1{,}03{,}436.3 = 1.034 \times 10^5$

Since χ is large, so the material is a ferromagnetic material.

Q.7 A coil of 307 turns of wire is wound on a circular ring of diameter 20 cm with a cross section area equal to 10 cm^2. Find the exciting current to establish flux density of 2 Wb/m^2 when μ_r = 650.

A.7

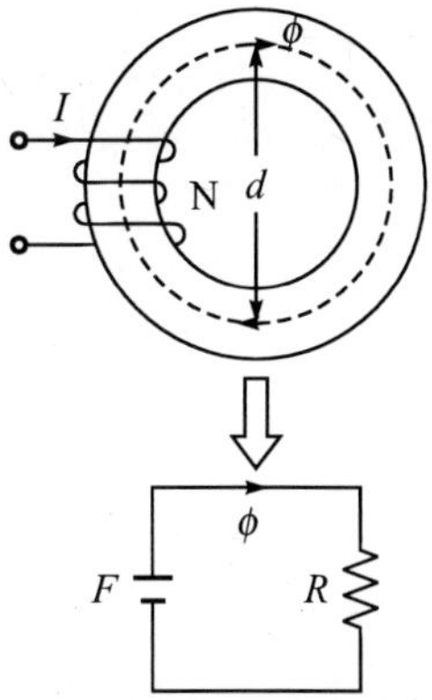

$$N = 307 \text{ turns.}$$
$$a = 10 \text{ cm}^2 = 10 \times 10^{-4} \text{ m}^2$$
$$d = 20 \text{ cm} = 0.2 \text{ m}$$
$$B = 2 \text{ tesla}$$
$$\mu_r = 650;\ \mu_0 = 4\pi \times 10^{-7} \text{ H/m}$$
$$\mu = \mu_0 \mu_r = 4\pi \times 10^{-7} \times 65$$
$$\mu = 8.164 \times 10^{-4}$$
$$\mathcal{F} = \phi\mathcal{R};\ \phi = Ba$$
$$\mathcal{F} = (B \cdot a) \cdot \mathcal{R};\ \mathcal{R} = \frac{l}{\mu a} = \frac{\pi d}{\mu a}$$
$$\mathcal{F} = 2 \times 10 \times 10^{+4} \times \frac{3.14 \times 0.2}{8.164 \times 10^{-4} \times 10 \times 10^{+4}}$$
$$\mathcal{F} = NI = 1538.5 \text{ AT}$$
$$I = \frac{1538.5}{N} = \frac{1538.5}{307} = 5.011 \text{ A}$$

Q.8 An iron ring with mean length of material 95 cm with an air gap of 5 mm has a winding of 450 turns which carries a current of 2.5 A. For a relative permeability of value 700; determine the flux density. The area of cross-section is 10 cm^2.

A.8

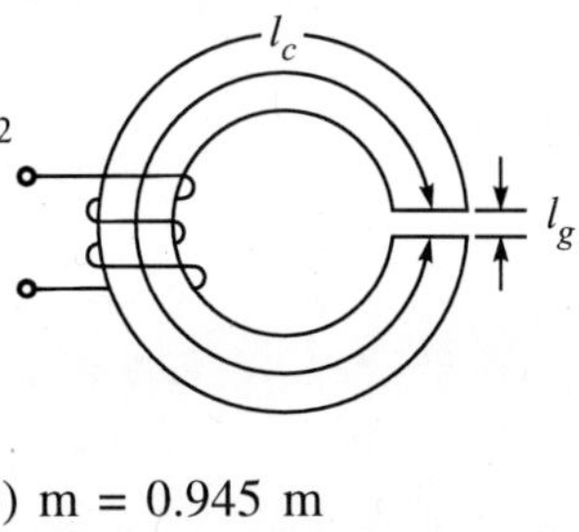

$$l = 95 \text{ cm} = 0.95 \text{ m};$$
$$a = 10 \text{ cm}^2 = 10 \times 10^{-4} \text{ m}^2$$
$$l_g = 5 \text{ mm} = 5 \times 10^{-3} \text{ m}$$
$$N = 450 \text{ turns}$$
$$I = 2.5 \text{ A}$$
$$\mu_r = 700$$
$$l_c = (l - l_g) = (0.95 - 0.005) \text{ m} = 0.945 \text{ m}$$

$$\mathcal{F} = NI = 450 \times 2.5 \text{ AT}$$
$$\mathcal{F} = 1125 \text{ AT}$$

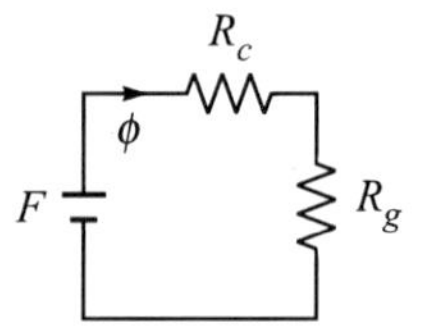

$$\mathcal{F} = \phi(\mathcal{R}_c + \mathcal{R}_g) = Ba\left(\frac{l_c}{\mu_0 \mu_r a} + \frac{l_g}{\mu_0 a}\right)$$

$$\mathcal{F} = \frac{B}{\mu_0}\left[\frac{l_c}{\mu_r} + l_g\right] \Rightarrow B = \frac{\mu_0 \mathcal{F}}{\left(\frac{l_c}{\mu_r} + l_g\right)} = \frac{4\pi \times 10^{-7} \times 1125}{\left[\frac{0.945}{700} + 0.005\right]}$$

$$B = 0.223 \text{ tesla.}$$

Q.9 A solenoid consisting of 100 turns and carrying 10 Amp current is 0.5 m long. Calculate (i) mmf (ii) total flux if area of cross-section is 0.0004 m^2. Consider $\mu_r = 1000$.

A.9 $N = 100$ turns ; $I = 10$ A ; $l = 0.5$ m

(i) $\mathcal{F} = NI = 100 \times 10 \text{ AT} = 1000 \text{ AT}$

(ii) $\mathcal{F} = \phi \mathcal{R}$

or, $$\phi = \frac{\mathcal{F}}{\mathcal{R}} = \frac{\mathcal{F}}{l/\mu a} = \frac{\mathcal{F}}{l/\mu_0 \mu_r a}$$

or, $$\phi = \frac{\mathcal{F} \cdot \mu_0 \mu_r a}{l} = \frac{1000 \times 4\pi \times 10^{-7} \times 1000 \times 0.0004}{0.5}$$

or, $$\phi = 1.0048 \text{ m Wb}$$

Q.10 An iron ring of mean radius 10 cm and area of cross-section 5 cm^2 has an air gap of 1 mm wide. The ring is wound uniformly with coil of 900 turns A flux of 10^{-4} Wb is required in the air gap. Find the strength of current required to produce the flux. Given $\mu_{\text{iron}} = 65 \times 10^{-4}$ H/m.

A.10

$$r_m = 10 \text{ cm} = 0.1 \text{ m}$$
$$a = 5 \text{ cm}^2 = 5 \times 10^{-4} \text{ m}^2$$
$$l_g = 1 \text{ mm} = 0.001 \text{ m}$$
$$N = 900$$
$$\phi_g = 10^{-4} \text{ Wb}$$
$$I = ?$$
$$\mu_{\text{iron}} = 65 \times 10^{-4} \text{ H/m}$$
$$l_m = 2\pi r_m = 2 \times 3.14 \times 0.1 = 0.628 \text{ m}$$

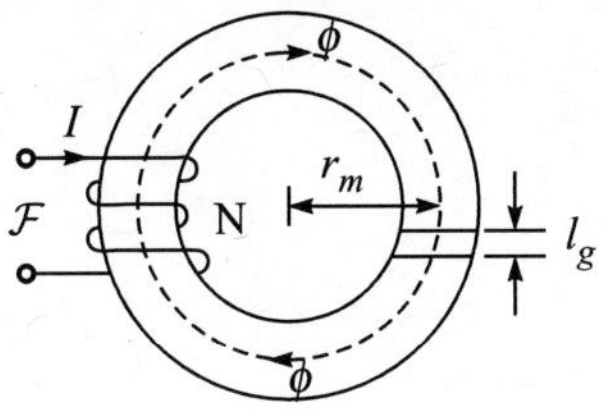

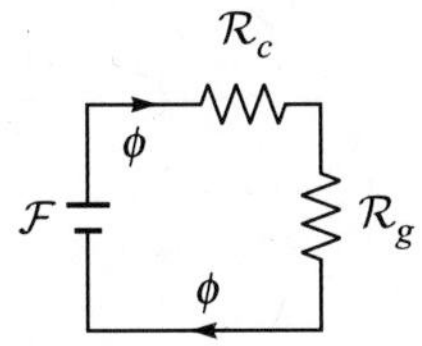

If no fringing effect, then $\phi_g = \phi$ and $a = a_g$

$$\mathcal{F} = \phi\,(\mathcal{R}_i + \mathcal{R}_g)$$

$$\mathcal{F} = \phi\left(\frac{l_i}{\mu_i \cdot a} + \frac{l_g}{\mu_0 a}\right)$$

$$\mathcal{F} = NI = 900\ I, \text{ Amp. turns}$$

$$900\ I = 10^{+4}\left[\frac{0.628 - 0.001}{65 \times 10^{-4} \times 5 \times 10^{+4}} + \frac{0.001}{4\pi \times 10^{-7} \times 5 \times 10^{+4}}\right]$$

$$900\ I = \left[\frac{0.627 \times 10^4}{325} + 159.24\right]$$

$$I = \frac{178.53}{900} = 0.1984 \text{ Amp.}$$

The current required to produce required flux = 0.2 Amp.

Q.11 An iron rod of length 50 cm and cross-section area 4 cm^2 is in the form of a closed circular ring. If the permeability of iron is 65×10^{-4} H/m, calculate the number of ampere turns required to produce a flux of 4×10^{-4} Wb.

A.11

$$l_c = 50 \text{ cm} = 0.5 \text{ m}$$

$$a = 4 \text{ cm}^2 = 4 \times 10^{-4} \text{ m}^2$$

$$\mu_{\text{iron}} = 65 \times 10^{-4} \text{ H/m}$$

$$\mathcal{F} = ?$$

$$\phi = 4 \times 10^{-4} \text{ Wb}$$

$$\mathcal{F} = \phi\,\mathcal{R}_i;\ \mathcal{R}_i \rightarrow \text{Reluctance of iron}$$

$$\mathcal{F} = \phi \cdot \frac{l_i}{\mu_i a}, \text{ AT}$$

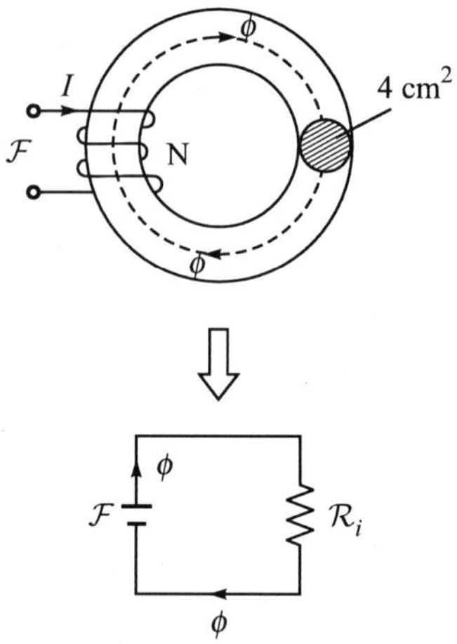

$$\mathcal{F} = 4 \times 10^{+4} \times \frac{0.5}{65 \times 10^{-4} \times 4 \times 10^{+4}}, \text{ AT}$$

$$\mathcal{F} = 76.923 \text{ Amp. turns}$$

Required AT's to produce flux of 4×10^{-4} Wb = 76.9 AT

Q.12 Estimate the number of ampere turns necessary to produce a flux of 0.001 Wb around an iron ring of 6 cm^2 cross-section and 20 cm mean diameter; having an air gap of 2 mm. The permeability of iron is 1200 μ_0.

A.12 $\mathcal{F} = ?$; $\phi = 0.001$ Wb; $a = 6 \text{ cm}^2 = 6 \times 10^{-4} \text{ m}^2$

$d_m = 20 \text{ cm} = 0.2 \text{ m}$ $\therefore l = \pi d_m = 0.2\pi$, m = 0.628 m

$l_g = 2 \text{ mm} = 0.002 \text{ m}$, $\mu_i = 1200 \mu_0 = 1200 \times 4\pi \times 10^{-7}$ H/m

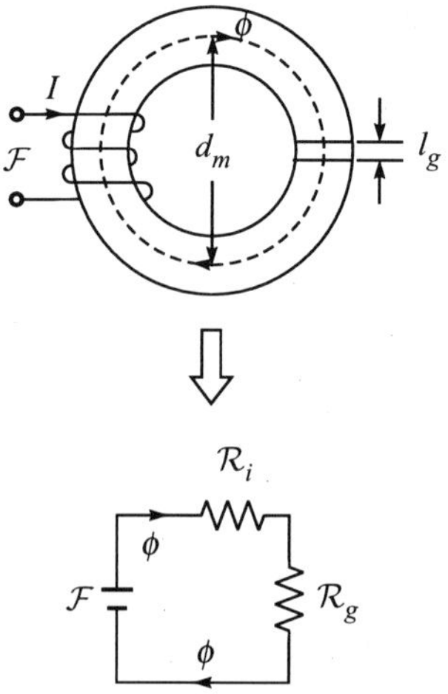

$$\mathcal{F} = \phi(\mathcal{R}_i + \mathcal{R}_g)$$

$$\mathcal{F} = \phi\left[\frac{l_i}{\mu_i a} + \frac{l_g}{\mu_0 a}\right]$$

$$\mathcal{F} = 0.001\left[\frac{(0.628 - 0.002)}{1200 \times 4\pi \times 10^{-7} \times 6 \times 10^{-4}} + \frac{0.002}{4\pi \times 10^{-7} \times 6 \times 10^{-4}}\right]$$

$$\mathcal{F} = \frac{10^4}{4\pi}\left[\frac{0.626}{0.72} + 3.333\right]$$

$$\mathcal{F} = 3345.9 \text{ Ampere turns}$$

Q.13 The air gap in a magnetic circuit is 1.5 mm long and 2500 mm^2 in cross-section. Calculate (a) the reluctance of the air gap (b) the mmf to send a flux of 800 μWb across the air gap.

A.13

$$l_g = 1.5 \text{ mm} = 1.5 \times 10^{-3} \text{ m}$$
$$a = 2500 \text{ mm}^2 = 2500 \times 10^{-6} \text{ m}^2 = 2.5 \times 10^{-3} \text{ m}^2$$
$$\mu_0 = 4\pi \times 10^{-7} \text{ H/m.}$$

(a) Reluctance in the air gap

$$\mathcal{R}_g = \frac{l_g}{\mu_0 a} = \frac{1.5 \times 10^{+3} \text{ m}}{4\pi \times 10^{-7} \times 2.5 \times 10^{+3} \text{ m}^2 \cdot \text{H/m}}$$

$$\mathcal{R}_g = 4.777 \times 10^5 \text{ H}^{-1} \text{ or AT/Wb}$$

(b)

$$\mathcal{F} = \phi_g \mathcal{R}_g = 800 \times 10^{-6} \times 4.777 \times 10^5 \text{ AT}$$
$$\mathcal{F} = 382.2 \text{ AT}$$

Q.14 A magnetic circuit made of transformer sheet steel is arranged as shown in Fig. below. The central limb has a cross-sectional area of 1000 mm^2 and each of the side limbs has a cross-sectional area of 600 mm^2. Calculate the mmf required to produce a flux of 1.2 m Wb in the centre limb. Assume the magnetic leakage to be negligible. Consider $\mu_r = 5000$.

A.14

$$a_0 = 600 \text{ mm}^2 = 600 \times 10^{-6} \text{ m}^2$$
$$a_c = 1000 \text{ mm}^2 = 1000 \times 10^{-6} \text{ m}^2$$
$$\phi_g = 1.2 \text{ m Wb} = 1.2 \times 10^{-3} \text{ Wb}$$

$$\mathcal{R} = (\mathcal{R}_c + \mathcal{R}_g) + \frac{\mathcal{R}_0}{2} \qquad \text{(I)}$$

$$\mathcal{R}_g = \frac{l_g}{\mu_0 a_c} = \frac{1.5 \times 10^{+3}}{4\pi \times 10^{-7} \times 1000 \times 10^{+6}}$$

$$\mathcal{R}_g = 1.194 \times 10^6 \text{ AT/Wb}$$

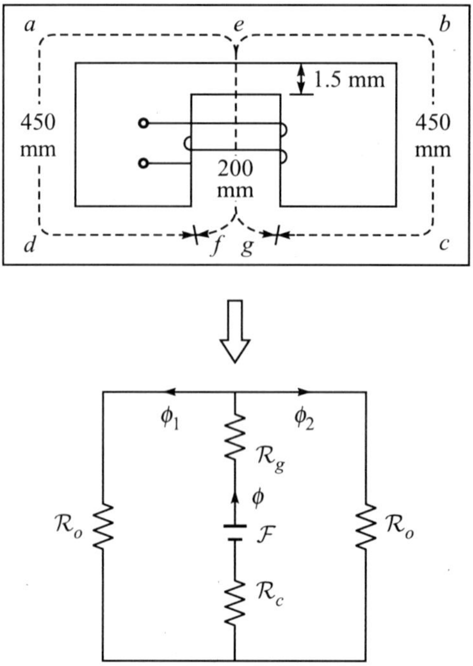

$$\mathcal{R}_c = \frac{l_c}{\mu_c a_c} = \frac{l_c}{\mu_0 \mu_r a_c}$$

$$= \frac{200 \times 10^{+3}}{\underset{2}{4\pi} \times 10^{-7} \times 5000 \times 10^{+3}} = 31847.1 \text{ AT/Wb}$$

$$\mathcal{R}_0 = \frac{l_0}{\mu_0 a_c} = \frac{400 \times 10^{+3}}{4\pi \times 10^{-7} \times 5000 \times 10^{+3}} = 63694.3 \text{ AT/Wb}$$

Substituting the values of $\mathcal{R}_c$, $\mathcal{R}_g$ and $\mathcal{R}_0$ in exp. (I)

$$\mathcal{R} = 1257694.\ 234 \text{ AT/Wb}$$

$$\therefore \quad \mathcal{F} = \phi_g\ \mathcal{R}, \text{ AT}$$

$$\mathcal{F} = 1509.2 \text{ AT}$$

Q.15 A magnetic core made of cast steel has the dimensions as shown below. There is an air gap of 1 mm long in one side limb and a coil of 400 turns is wound on the central limb. The cross-sectional area of the central limb is 1200 mm^2 and that of each side limb is 800 mm^2. Calculate the exciting current required to produce a flux of 600 μ Wb in the air gap. Neglect any magnetic leakage and fringing. Consider $\mu_r = 1500$.

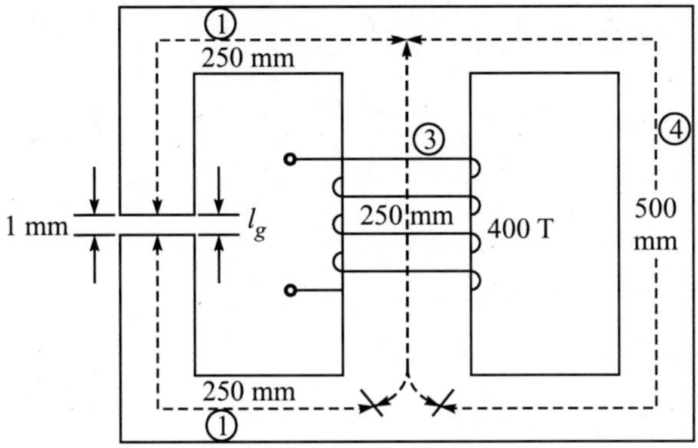

A.15

$$a_0 = 800 \text{ mm}^2 = 800 \times 10^{-6} \text{ m}^2 = 0.8 \times 10^{-3} \text{ m}^2$$
$$a_c = 1200 \text{ mm}^2 = 1200 \times 10^{-6} \text{ m}^2 = 1.2 \times 10^{-3} \text{ m}^2$$
$$\phi_g = 600 \ \mu \text{ Wb} = 600 \times 10^{-6} \text{ Wb} = 0.6 \times 10^{-3} \text{ Wb}$$

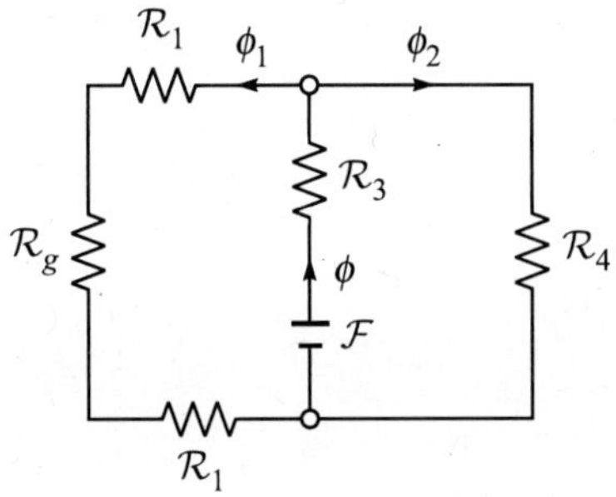

$$\mathcal{R}_1 = \frac{l_1}{\mu_1 a_0} = \frac{l_1}{\mu_0 \mu_r a_0} = \frac{250 \times 10^{-3}}{4\pi \times 10^{-7} \times 1500 \times 800 \times 10^{-6}}$$

$$\mathcal{R}_1 = 165870.5 \text{ AT/Wb}$$

$$\mathcal{R}_g = \frac{l_g}{\mu_0 a_0} = \frac{1 \times 10^{-3}}{4\pi \times 10^{-7} \times 800 \times 10^{-6}} = 995222.9 \text{ AT/Wb}$$

$$\mathcal{R}_4 = \frac{l_4}{\mu_0 \mu_r a_0} = \frac{500 \times 10^{-3}}{4\pi \times 10^{-7} \times 1500 \times 800 \times 10^{-6}} = 331741 \text{ AT/Wb}$$

$$\mathcal{R}_3 = \frac{l_3}{\mu_0 \mu_r a_c} = \frac{0.5}{4\pi \times 10^{-7} \times 1500 \times 1.2 \times 10^{-3}}$$

$$\mathcal{R}_3 = 221160.7 \text{ AT/Wb}$$

$$\mathcal{R} = \mathcal{R}_3 + \frac{(2\mathcal{R}_1 + \mathcal{R}_g)\cdot \mathcal{R}_4}{2\mathcal{R}_1 + \mathcal{R}_g + \mathcal{R}_4}$$

$$\mathcal{R} = 221160.7 + \frac{(331741 + 995222.9)\cdot 331741}{331741 + 995222.9 + 331741}$$

$$= 221160.7 + \frac{4.4 \times 10^{11}}{1658704.9}$$

$$= 221160.7 + 265267.2$$

$$= 486427.9 \text{ AT/Wb}$$

$$\therefore \qquad \mathcal{F} = \phi_g\ \mathcal{R} = 0.6 \times 10^{-3} \times 486427.9$$

$$\mathcal{F} = \phi = 291.9 \text{ AT}$$

Q.16 A 500 turn coil is wound on the central limb of the cast steel frame as shown below. A total flux of 1.8 m Wb is required in the gap. Find what current is required. Assume that the gap density is uniform and that all flux lines pass straight across the gap. Assume $\mu_r = 1500$.

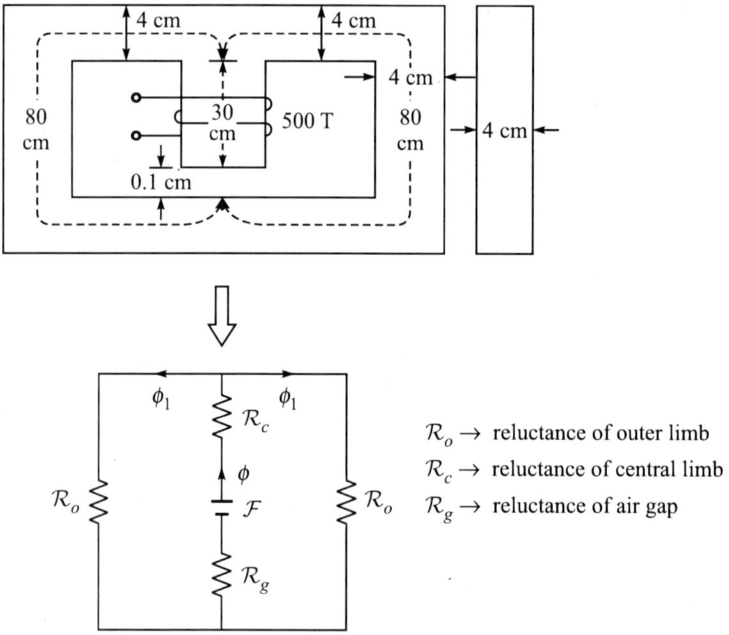

A.16

$$a = 4 \text{ cm} \times 4 \text{ cm} = 16 \text{ cm}^2 = 16 \times 10^{-4} \text{ m}^2$$

$$l_c = 30 \text{ cm} = 0.3 \text{ m}$$

$$l_0 = 80 \text{ cm} = 0.8 \text{ m}$$

$$\phi_g = \phi = 1.8 \text{ m Wb} = 1.8 \times 10^{-3} \text{ Wb}$$

$$\mu = \mu_0 \mu_r = 4\pi \times 10^{-7} \times 1500 = 18840 \times 10^{-7} \text{ H/m}$$

$$\mathcal{F} = NI = 500\, I = \phi \mathcal{R}$$

$$\mathcal{R} = \mathcal{R}_c + \mathcal{R}_g + \frac{\mathcal{R}_0 \mathcal{R}_0}{\mathcal{R}_0 + \mathcal{R}_0} = \mathcal{R}_c + \mathcal{R}_g + \frac{R_0}{2}$$

$$\mathcal{R}_c = \frac{l_c}{\mu a} = \frac{0.3 \times 10^7}{18840 \times 0.16 \times 10^{-2}} = 99522.3 \text{ AT/Wb}$$

$$\mathcal{R}_g = \frac{l_g}{\mu_0 a} = \frac{0.1 \times 10^{-2}}{4\pi \times 10^{-7} \times 16 \times 10^{-4}}$$

$$\mathcal{R}_g = \frac{10^{-3} \times 10^{11}}{64\pi} = \frac{10^8}{64 \times 3.14} = \frac{1000}{200.96} \times 10^5 \text{ AT/Wb}$$

$$\mathcal{R}_g = 4.976115 \times 10^5 \text{ AT/Wb} = 497611.5 \text{ AT/Wb}$$

$$\mathcal{R}_0 = \frac{l_0}{\mu \cdot a} = \frac{0.8}{18840 \times 10^{-7} \times 16 \times 10^{-4}}$$

$$\mathcal{R}_0 = 265392.8 \text{ AT/Wb}$$

$$\therefore \quad \mathcal{R} = 729830.2 \text{ AT/Wb}$$

$$\therefore \quad NI = \phi_g \mathcal{R}$$

$$I = \frac{\phi_g \cdot \mathcal{R}}{N}$$

$$I = \frac{1.8 \times 10^{-3} \times 729830.2}{500}$$

$$\boxed{I = 2.627 \text{ Amp}}$$

The current required for producing required flux is equal to 2.63 Amp (approx).

TUTORIAL QUESTIONS

Q.1 In a magnetic core with air gap, most of the field energy resides in the air gap. Justify.

Q.2 Explain the terms permeability, reluctance and permeance.

Q.3 Establish the relation between mmf, reluctance, flux in a magnetic circuit.

Q.4 The magnetic circuit frame shown in the following figure is built of iron of square cross of length 3 cm. Each air gap is 2 mm wide and each of the coil is wound 1000 turns. The relative permeability of part A and B are 1000 & 2000 respectively. Calculate.

(a) Reluctance of part A
(b) Reluctance of two air gaps
(c) Total reluctance of complete magnetic path.

[1.76×10^5 AT/Wb, 1.769×10^6 AT/Wb 3.95×10^6 AT/Wb]

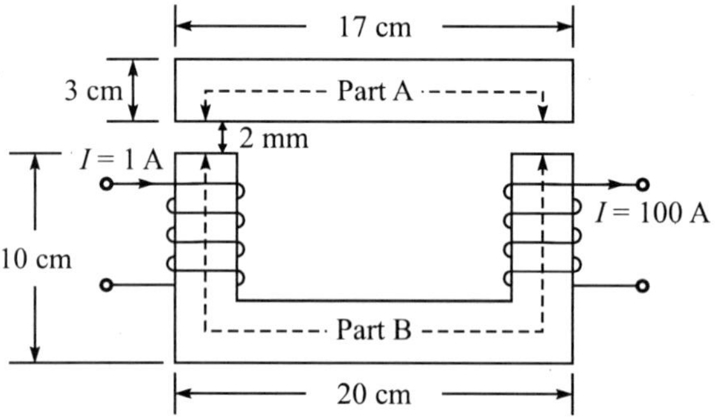

Q.5 For the magnetic circuit frame given above, calculate (a) mmf (b) total flux (c) flux density.

Q.6 A cast steel ring has a mean diameter of 200 mm and cross-sectional area of 250 mm^2. Calculate (a) the mmf to produce a flux of 0.3 m Wb and (b) the reluctance of the ring and the relative permeability.

[586 AT, 6.22×10^5 AT/Wb, 1024]

Q.7 A steel ring has a mean diameter of 20 cm, a cross-section of 25 cm^2 and a radial air gap of 0.5 mm cut in it. The ring is uniformly wound with 2000 turns of insulated wire and a magnetizing current of 1 A produces a flux of 1.25 μ Wb in the air gap. Neglect fringing and leakage. Calculate (a) Reluctance (b) Relative permeability of the steel.

[(a) 1.6×10^6 AT/Wb (b) 138]

Q.8 A mild steel ring has a mean circumference of 600 mm and a uniform cross-sectional area of 250 mm^2. Calculate the mmf required to produce a flux of 450 μ Wb. An air gap, 1.0 mm in length, is now cut in the ring. Determine the flux produced if the mmf remains constant. Assume the relative permeability of the mild steel to remain constant at 220.

[3906 AT, 329 μ Wb]

Q.9 The central limb of the magnetic circuit given below has an air gap of 2 mm. Assume $\mu_r = 800$. Determine the current in the 600 turns exciting coil to establish a flux of 100×10^{-3} m Wb in the air gap.

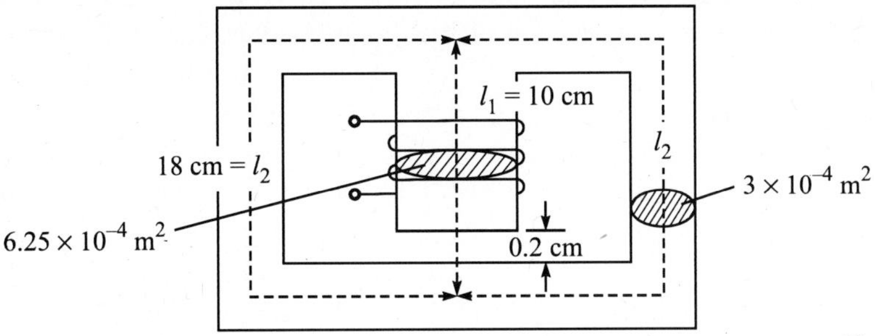

[0.5 A]

Q.10 A 680 turn coil is wound on the central limb of the cast steel frame as shown below. A total flux of 1.6 m Wb is required in the gap. Find the current required. Assume H = 900 AT/m for B = 1 Wb/m^2.

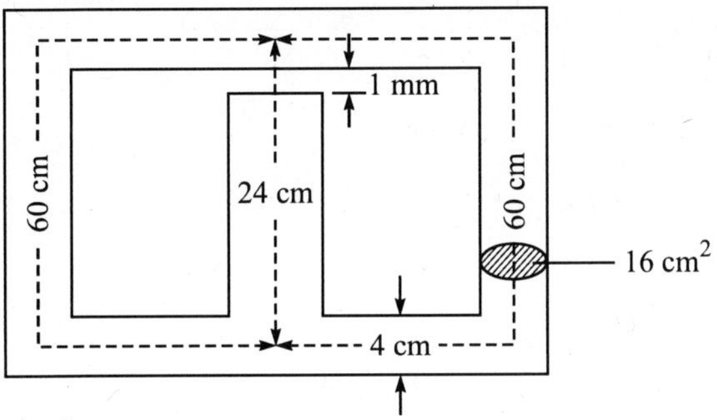

[1.9 A]

Q.11 An iron magnetic circuit has a uniform cross-sectional area of 5 cm^2 and a length of 25 cm. A coil of 120 turns is wound uniformly over the magnetic circuit. When the current in the coil is 1.5 A, the total flux is 0.3 m Wb and when the current is 5 A, the total flux is 0.6 m Wb. For each value of current calculate 'H' and 'μ_r'.

Q.12 An iron ring with a mean circumference of 140 cm and cross-section 12 cm^2 is wound with 500 turns of wire. When the exciting current is 2 A, the flux is found to be 1.2 m Wb. Determine relative permeability of iron.

Q.13 Explain the term: magnetic leakage and flux fringing.

Q.14 Discuss the similarities and dissimilarities between electric and magnetic circuits.

Q.15 Why is the reluctance of air gap greater than that of reluctance of ferromagnetic material?

Q.16 Why does the magnetic circuit have air gaps?

Q.17 A magnetic circuit has an area of 10 cm^2 and length of 0.2 m. If μ_r = 2000, find the reluctance.

[0.796 × 10^5 AT/Wb]

Q.18 An iron circuit has a reluctance of 2.8×10^5 AT/Wb. The coil has 1000 turns and an exciting current of 0.2. Find the inductance and flux.

[3.57 *H*, 0.714 m Wb]

Q.19 A magnetic circuit has an area of 10 cm^2 and length 0.1 m. If $\mu_r = 2000$, find the reluctance.

[0.4×10^5 AT/Wb]

Q.20 The core of a magnetic circuit has a mean length of 400 mm and a cross-sectional area of 4×10^{-4} m^2. An air gap 1 mm long is cut in the core. The coil has 1000 turns and $\mu_r = 1000$. Find the inductance of the coil.

[0.36 H]

Q.21 An air gap of 2 mm is cut in a magnetic circuit having mean circumference 40 cm. If $\mu_r = 600$ and the coil has 100 turns. Find the mmf and exciting current for a flux density of 0.8 tesla.

[1695.5 AT, 16.955 A]

Chapter

4

Principle of Electromechanical Energy Conversion

4.1 Introduction
4.2 Electric Field as a Medium
4.3 Magnetic Field as a Medium
4.4 Moving Coil in a Uniform Magnetic Field
4.5 Stationary Coil in a Time Varying Magnetic Field
4.6 Motoring Action
4.7 Generating Action
4.8 Singly Excited System
4.9 Doubly Excited System
4.10 Forces/Torques in Systems with Permanent Magnets
4.11 Dynamical Equations of Electromechanical Systems
- Solved Questions
- Review Questions
- Unsolved Questions

4.1 INTRODUCTION

It is seen in the real world that almost all electrical devices use energy in one form or other and convert it in other form. Depending on the principle of operation the devices are classified accordingly. The low power devices are called transducers which use the energy in electrical or non-electrical form. But as the power increases the rotating machines such as motors and generators are used. They are regarded as bulk power converting devices. As the power requirement is increasing day by day, so it is important to know about the power conversion devices specially motors and generators. Both motors and generators use the same principle. Moreover, they use the magnetic field as the medium where the power conversion takes place from electrical to mechanical and vice versa. Therefore, it can be thought as the medium where conversion of electromechanical energy takes place.

Thus, it can be thought that motoring operation and generating operation are opposite of each other. That is, the flow of energy is different for each case. In the case of motor the energy flows from electrical system to the mechanical system via the magnetic medium where electromechanical energy conversion takes place. Whereas, in the case of generator the energy flows from mechanical system to electrical system via the magnetic field where electromechanical energy conversion takes place. In some cases the electric field is used as electromechanical energy conversion device. Thus, it could be imagined that magnetic and electric fields act as reservoirs from which energy could be tapped out or energy could be stored. It could be compared with a tank full of water from which water could be taken out for some useful work or water could be stored inside it.

The energy which is stored inside the field is called stored field energy. So the electromechanical energy conversion takes place whenever there is any change in the stored field energy. If the change in the stored field energy is positive, then the generating action takes place as in the case of a.c. and d.c. generators. If the change in the field energy is negative, then motoring action takes place as in the case of a.c. and d.c. motors.

The electromechanical energy conversion could be easily visualized by the Fig. (4.1) as given below.

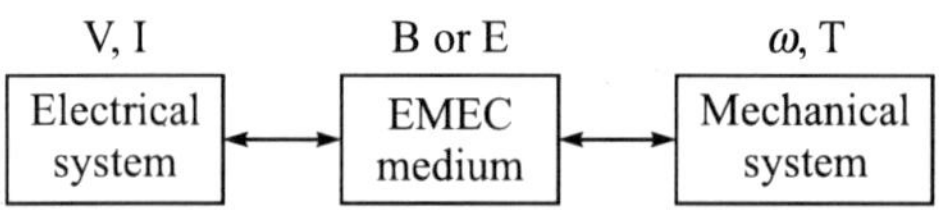

Fig. 4.1 Block diagram showing EMEC (Electromechanical energy conversion).

The general equation can be given as relation (4.1),

$$\boxed{W_i = W_0 + W_f} \tag{4.1}$$

The above relation is defined for lossless system. If the loss is also taken into consideration, then the relation given by expression (4.1) gets modified as per expression (4.2) as follows:

$$\boxed{W_i = W_0 + W_f + W_{\text{loss}}} \tag{4.2}$$

where, W_i = Input energy, in joules

W_0 = Output energy, in joules

W_f = Stored field energy, in joules

W_{loss} = Energy loss, in joules.

In the above expression as per (4.2), the energy loss 'W_{loss}' is irrecoverable and is lost in the form of heat. So all the input energy does not transform into equivalent ouput energy but it is less by the amount equal to the loss of energy as heat. So all the motors and generators cannot have 100% efficiency, but will be lesser than 100%. So by good design and by using good magnetic materials having higher permeability the loss parameter can be lessened by some value. So plenty of researches are going on the magnetic materials and as an outcome new better magnetic materials are coming. Thus, continuous efforts are done in this field to make highly efficient motors and generators.

4.2 ELECTRIC FIELD AS A MEDIUM

The electric field is present whenever two plates are charged oppositely and are separated by some distance 'd'.

Consider the case of a parallel plate capacitor whose plates are charged by +'Q' and –'Q' respectively. If the plates' area is 'A' (say), then the capacitance of the parallel plate capacitor is given by the expression (4.3), as follows:

$$C = \frac{\epsilon_0 A}{d} \tag{4.3}$$

The energy which is stored in the electrostatic field is given by expression (4.4) as given below.

$$W_f = \frac{1}{2} CV^2 \tag{4.4}$$

Thus, if one plate is fixed whereas other the plate is free to move as shown in Fig. (4.2) below.

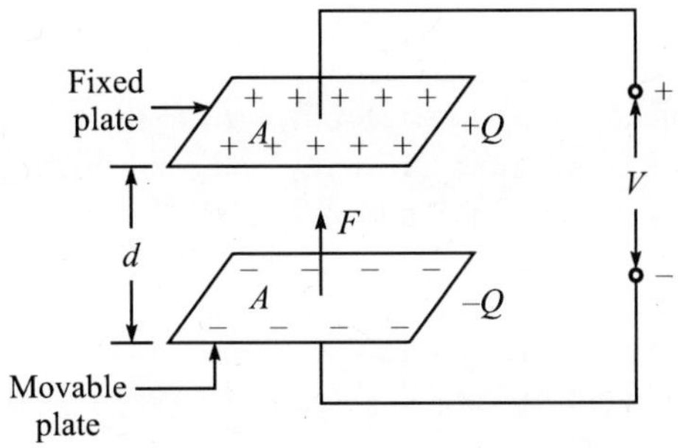

Fig. 4.2 Force due to electrostatic field.

As per the Fig. (4.2), there will be a force 'F' exerted on the movable plate of capacitor as a result the movable plate starts moving in the upward direction. This force 'F', as a result of Coulomb's law which states that the force of attraction 'F' between two charges 'q_1' and 'q_2' and is separated by distance 'r', is given by expression (4.5).

$$F = \frac{kq_1 q_2}{r^2} \tag{4.5}$$

Using the expression (4.1), we have

$$dW_f = dW_i - dW_0 \tag{4.6}$$

where, dW_f = Incremental change of stored energy, in joule
dW_i = Incremental change of input energy, in joule
dW_0 = Incremental change of output energy, in joules.

The input energy change is given as expression (4.7) as follows:

$$\boxed{dW_i = V\,dQ} \qquad (4.7)$$

Since, $$Q = CV \text{ so } I = \frac{dQ}{dt} = \frac{d(CV)}{dt} = C\frac{dV}{dt}$$

and $$dQ = Idt$$

So, $$VdQ = VIdt = P \cdot dt = dW_i$$

The power input consists of two electrical quantities, namely, 'V' and 'I' respectively. The product of power and time is nothing but the equivalent energy in joules. Since the power is the rate of doing work which is given by expression (4.8) as follows:

$$\boxed{P = \frac{W}{t}} \qquad (4.8)$$

Incremental power 'dP' is given by taking the derivative of the expression (4.8), that is,

$$dP = \frac{dW}{dt}$$

From expression (4.8), the work or energy is given by

$$\boxed{W = P \times t}$$

Taking derivative of both sides, we get $dW = Pdt + t \cdot dP$.

To get the expression of incremental energy 'dW' considering constant input power 'P' (equal to the product of V and I) is given by

$$\boxed{dW = P\,dt} \qquad \because \quad dP = 0 \text{ if } P = \text{const} \qquad (4.9)$$

Since power is the input to the capacitor, so the power could be assigned a subscript 'i' to specify input power 'P_i'. Since the energy also corresponds to the input power.

So the incremental energy is the incremental input energy 'dW_i' respectively. So with reference to the input the above expression (4.9) is given by (4.10) as follows:

$$\boxed{dW_i = P_i\,dt} \qquad (4.10)$$

Similarly, the incremental output energy 'dW_0' is given by (4.11),

$$\boxed{dW_0 = F \cdot dx} \qquad (4.11)$$

Since the work done is the product of the electrostatic force 'F' exerted on the bottom plate and the incremental distance 'dx' moved in the upward direction.

So, the incremental change in the field energy 'dW_f' is given by (4.12).

$$\boxed{dW_f = P_i\,dt - F \cdot dx} \tag{4.12}$$

Substituting, '$P_i dt$' equal to 'VdQ' in expression (4.12), we get

$$\boxed{dW_f = V\,dQ - F \cdot dx} \tag{4.13}$$

Since, 'dW_f' and 'dQ' are functions of 'V' and 'x' respectively. That is, $W_f = f(V, x)$ and $Q = f(V, x)$

So, the total derivative is given as

$$\boxed{dW_f = \frac{\partial W_f}{\partial x} \cdot dx + \frac{\partial W_f}{\partial V} \cdot dV} \tag{4.14}$$

$$\boxed{dQ = \frac{\partial Q}{\partial x} \cdot dx + \frac{\partial Q}{\partial V} \cdot dV} \tag{4.15}$$

Substituting the values of 'dW_f' and 'dQ' in expression (4.13), we have

$$F = \left[V \cdot \frac{\partial Q}{\partial x} - \frac{\partial W_f}{\partial x}\right] + \left[V \cdot \frac{\partial Q}{\partial V} - \frac{\partial W}{\partial V}\right]\frac{dV}{dx} \tag{4.16}$$

Considering $\frac{\text{'}dV\text{'}}{dx}$ negligible, we get

$$F = \left[V \cdot \frac{\partial Q}{\partial x} - \frac{\partial W_f}{\partial x}\right] \tag{4.17}$$

Since,

$$Q = CV$$

$$\therefore \quad \frac{\partial Q}{\partial x} = C\frac{\partial V}{\partial x} + V\frac{\partial C}{\partial x} \tag{4.18}$$

Similarly,

$$W_f = \frac{1}{2}CV^2$$

$$\therefore \quad \frac{\partial W_f}{\partial x} = \frac{1}{2}V^2\frac{\partial C}{\partial x} + CV\frac{\partial V}{\partial x} \tag{4.19}$$

Substituting the values of $\frac{\text{'}\partial Q\text{'}}{\partial x}$ and $\frac{\text{'}\partial W_f\text{'}}{\partial x}$ in exp. (4.19)

we get,

$$F = \left[V \cdot \left(C\frac{\partial V}{\partial x} + V\frac{\partial C}{\partial x}\right) - \frac{1}{2}V^2\frac{\partial C}{\partial x} - CV\frac{\partial V}{\partial x}\right]$$

$$F = \left[\cancel{CV\frac{\partial V}{\partial x}} + V^2\frac{\partial C}{\partial x} - \frac{1}{2}V^2\frac{\partial C}{\partial x} - \cancel{CV\frac{\partial V}{\partial x}}\right]$$

$$\therefore \qquad \boxed{F = +\frac{1}{2}V^2\frac{\partial C}{\partial x}} \tag{4.20}$$

Often negative sign is used in the exp. (4.20). The negative sign indicates the force is attractive in nature.

For isolated system (single conductor) $\boxed{F = -\frac{\partial W_f}{\partial x}}$ (4.21)

For system with fixed potential by external source (two conductors)

$$\boxed{F = +\frac{\partial W_f}{\partial x}} \tag{4.22}$$

4.3 MAGNETIC FIELD AS A MEDIUM

Whenever current carrying coil is wound over magnetic material then a magnetic field is produced, most of the magnetic field is concentrated in the magnetic material so the magnetic material behaves as a magnet. Consider a current carrying coil in an electromagnetic relay as shown in the Fig. (4.3) shown below.

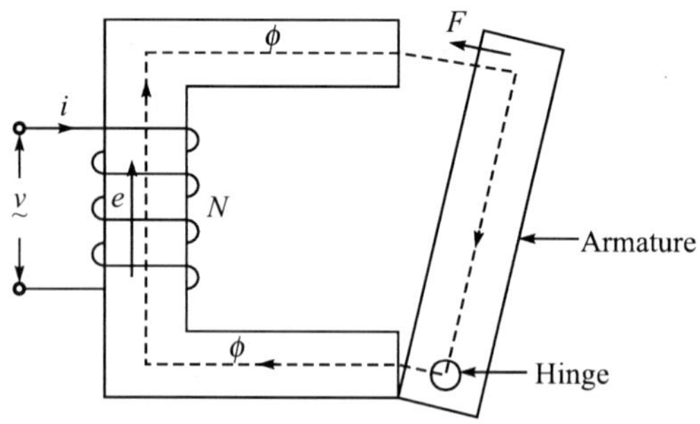

Fig. 4.3 Electromagnetic relay.

In Fig. (4.3), an attracted armature type electromagnetic relay is shown. The flux 'ϕ' flows through the magnetic circuit shown by the dotted lines. Due to the magnetization of the magnetic material it experiences a force 'F' on the armature which is attached by a hinge. The stored energy in the field experiences the force which is maximum in the air gap. Any change in the stored energy is responsible for the work done on the armature or the rotating member as in the case of rotating machines.

According to the conservation of energy, the input energy is equal to the output energy spent or retrieved. Mathematically, it could be expressed as

$$W_i = W_0 + W_f \tag{4.23}$$

where,

$W_i \rightarrow$ Input energy, in joules

$W_0 \rightarrow$ Output energy, in joules

$W_f \rightarrow$ Field energy, in joules
$W_i = P_i \cdot dt$, where 'P_i' is the input electrical power, in Watts.

$$W_i = V \cdot i\,dt, \text{ in joules} \tag{4.24}$$

The output energy is nothing but the mechanical work done on the armature or plunger

$$W_0 = F \cdot dx, \text{ in joules}$$

where, 'dx' is the incremental distance through which the armature is moved toward the magnetic member

$$v = iR + e \tag{4.25}$$

where,

$$e = \frac{d\lambda}{dt} = \frac{d}{dt}(N\phi) = N\frac{d\phi}{dt} \tag{4.26}$$

$$e\,dt = N\,d\phi \tag{4.27}$$

Multiplying 'i' with both sides, we get

$$ei\,dt = Ni\,d\phi = \mathcal{F}\,d\phi = dW_f = id\,(N\phi) = id\lambda \tag{4.28}$$

where,
$\mathcal{F} = Ni$ is the ampere turns of the coil which is called the m.m.f.

For ideal coil, $\mathcal{R} = 0$

then $dW_i = dW_f$ (4.29)

If the coil is purely inductive, then

$$v = -e \tag{4.30}$$

$$\boxed{W_f = \int_0^{\lambda} i(\lambda)\,d\lambda = \int_0^{\phi} \mathcal{F}(\phi)\,d\phi} \tag{4.31}$$

shaded portion $\rightarrow$ denotes the change in stored energy in the magnetic field.

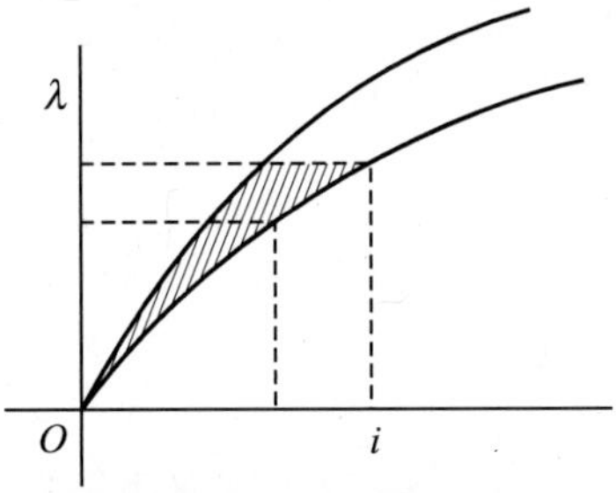

Fig. 4.4 $\lambda - i$ relationship.

The area under the curve corresponds to the field coenergy 'W_f' and above it is called the field energy 'W_f'.

4.4 MOVING COIL IN A UNIFORM MAGNETIC FIELD

When a coil rotates in a field '$\overline{B}$' then e.m.f. is induced in the coil due to Faraday's law. If the circuit is closed, then a current starts flowing through it. The coil ends are attached to split ring commutator segments which is in contact with the brushes which collect the d.c. current through it by reversing the coil current whenever the coil passes from one pole to another. The split ring commutator is acting as a switch. The split rings are fixed to the rotating member and so it also rotates at the same speed as that of the coil as shown.

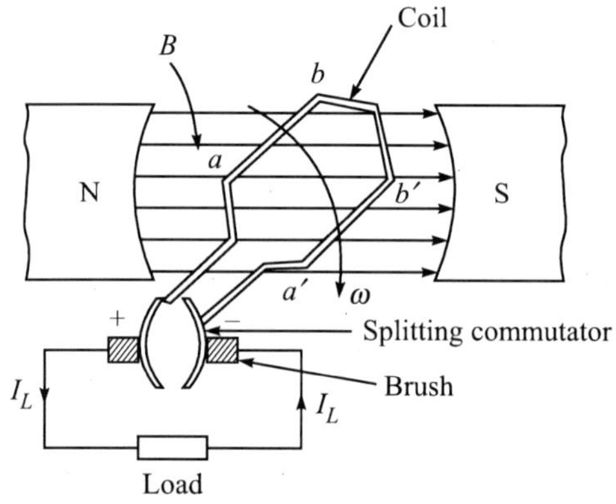

Fig. 4.5 Moving coil in field $\overline{B}$.

When the coil side 'ab' is under the north pole then the direction of voltage is induced and the current is in one direction and when it is under south pole it is in the opposite direction. That is, for the first half cycle i.e., when it is under the influence of north pole field the polarity is not changing. But as soon as it leaves the north pole and enters south pole then the polarity of the voltage reverses. The two poles are separated by one pole pitch corresponding to 180° electrical degrees. So the commutator reverses the connections with respect to brushes as soon as direction of the e.m.f. reverses or the coil sides leaves on pole and enters other pole field. The e.m.f. at brushes produced by coil rotating in a uniform magnetic field is shown in the Fig. (4.6).

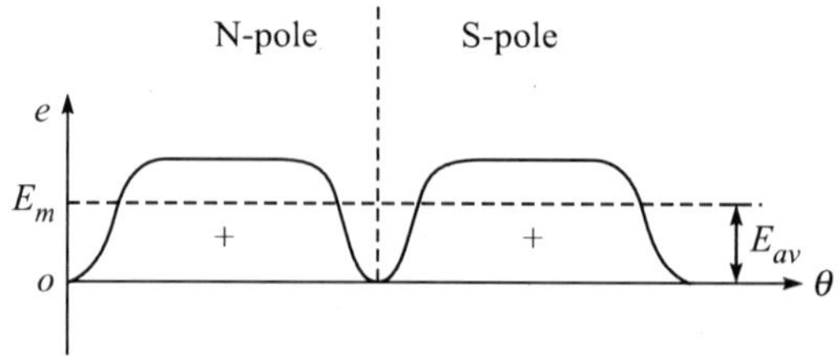

Fig. 4.6 E.M.F. across brushes of d.c. machine.

The brushes are normally placed in the inter-polar axis or region perpendicular to the magnetic axis passing through north-south pole axis. The pole faces are curved so that we get more or less constant flux per pole so that the electromotive force developed has more or less flat topped shape as shown in Fig. (4.6). The d.c. voltage and hence current has higher average value 'E_{av}' for larger period of time over the cycle.

The wave shown in Fig. (4.6) is a pulsating d.c. and contains larger harmonics. To reduce the harmonics and to reduce the pulsations in the output waveform a number of commutation segments are used thus we get more better d.c. The less harmonics corresponds to less heating of the machine cores. The open winding is not used practically because all the coil e.m.f.'s are not added as they are not in series, so the gramme ring winding are used.

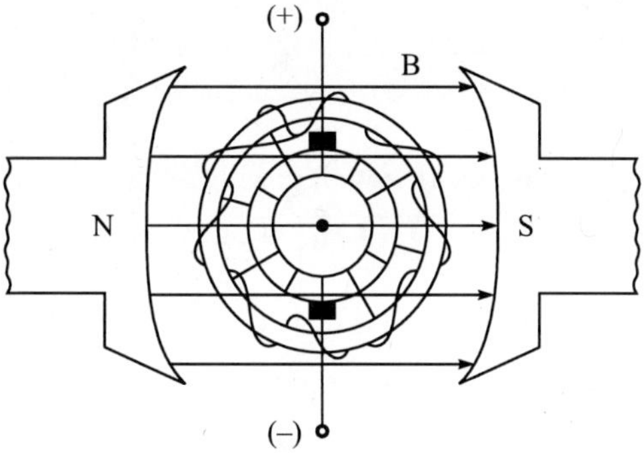

Fig. 4.7 Gramme ring winding.

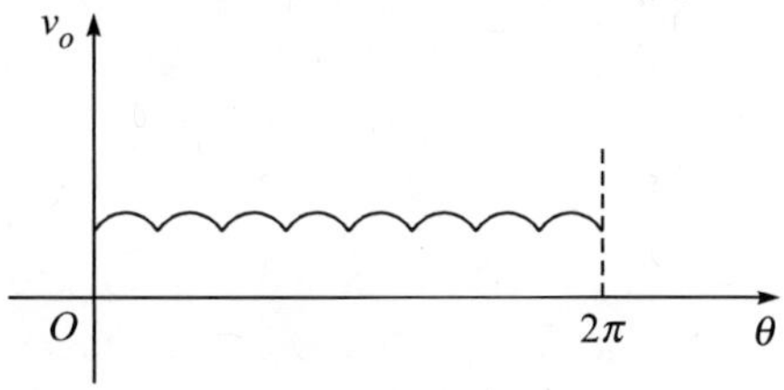

Fig. 4.8 Output voltage across brushes.

The gramme ring winding is a closed winding. The gramme ring windings are no longer used because the conductors inside the ring do not participate in e.m.f. generation as they are shielded magnetically by the armature magnetic iron care. In order to save copper and reduce the weight of the armature winding, drum wound armature are employed in which all the conductors lie on the outer surface of the armature slots such that the part of armature conductor which lie just under the pole shoe participate in the induction process and e.m.f. is produced in it.

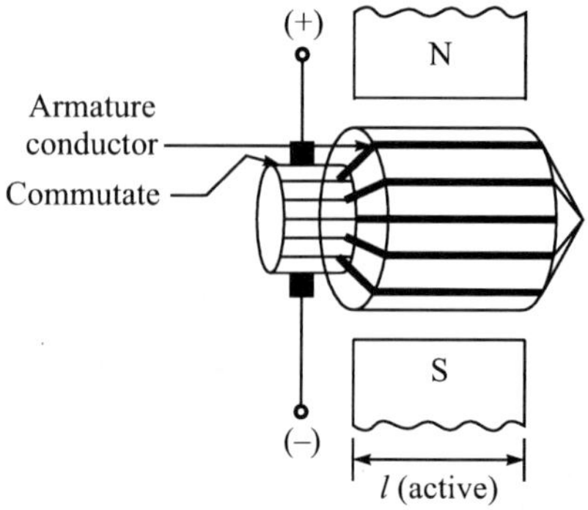

Fig. 4.9 Drum type armature winding.

In the case of gramme type winding the pole number can be increased hence is advantageous where large poles are needed.

The case of moving coil in an uniform magnetic field is observed in the case of d.c. machines, synchronous machine (with armature rotating) and dynamo's respectively.

The e.m.f. induced in a conductor rotating with a velocity 'v' is given by

$$\boxed{e = Blv} \tag{4.32}$$

where, 'B' is the average magnetic flux density, $\frac{\text{Wb}}{\text{m}^2}$

'l' is the active length of conductor, m

'v' is the conductor speed, m/s.

The e.m.f. 'e' given by exp. (4.32) is called speed e.m.f. or dynamically induced e.m.f.

4.5 STATIONARY COIL IN A TIME VARYING MAGNETIC FIELD

In many practical cases, a stationary coil in a time varying magnetic field are often implemented in the machines, such as, static devices like transformers to rotating devices such as synchronous machines (with rotating field) respectively. In many other applications such as relaying this is implemented. So we should see what happens in the stationary coil linked with or is in the influence of time varying field B and hence flux 'ϕ'.

We consider that a time varying flux $\underset{\sim}{\phi}$ represented as follows

$$\underset{\sim}{\phi} = \phi_m \sin \omega t \tag{4.33}$$

The flux $\underset{\sim}{\phi}$ is varying with different instant of time 't'.

Consider a stationary coil 'C_2' which is linked by a flux $\underset{\sim}{\phi}$ as shown in the Fig. (4.10) shown below.

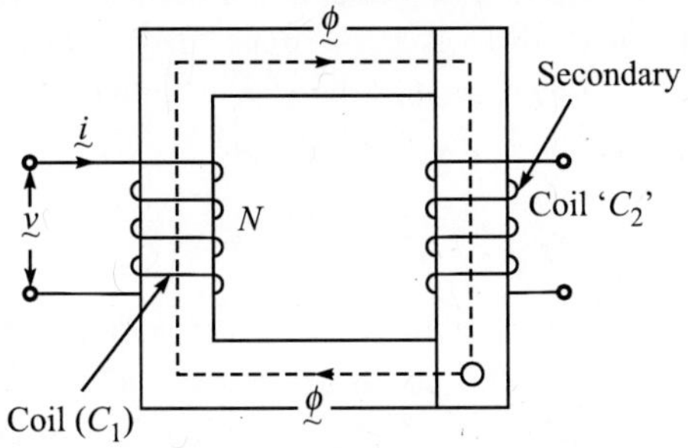

Fig. 4.10 Stationary coil in *B*.

If a coil 'C_2' is linked with $\underset{\sim}{\phi}$ then due to its changing nature with respect to 't' an e.m.f. is induced in the stationary coil 'C_2' given by the exp. (4.34) as given below

$$\boxed{e = -\frac{d\lambda}{dt}} \tag{4.34}$$

where, 'λ' is the flux linkage of coil

$\lambda = N\underset{\sim}{\phi}$, Weber turn

where, 'N' is the number of turns of coil. Substituting the value of 'λ' in expression of 'e' given by (4.34)

We get,

$$e = -\frac{d(N\underset{\sim}{\phi})}{dt} = -N\frac{d\underset{\sim}{\phi}}{dt} \tag{4.35}$$

Since, $\underset{\sim}{\phi} = \phi_m \sin \omega t$

So, $e = -N\dfrac{d(\phi_m \sin \omega t)}{dt}$

$$e = -N\phi_m\, \omega \cos \omega t$$

$$\boxed{e = N\phi_m\, \omega \sin\left(\omega t - \frac{\pi}{2}\right)} \tag{4.36}$$

Considering the r.m.s value of 'e', then

$$E_{\text{rms}} = \frac{N\phi_m \omega}{\sqrt{2}} = \frac{N\phi_m(2\pi f)}{\sqrt{2}}$$

$$\boxed{E_{\text{rms}} = 4.44\, N\phi_m f} \tag{4.37}$$

where, 'ϕ_m' is the maximum flux value, in Wb

'N' is the number of turns of coil 'C'

'f' is the frequency of variation of flux or current respectively, in Hz.

We see from the above expression that e.m.f. is induced in the coil linked with a time varying field. This e.m.f. is sometimes termed transformer e.m.f. or statically induced e.m.f. So, we see that the flux whether it is steady or varying is capable of producing e.m.f. in a coil placed in the vicinity of it.

4.6 MOTORING ACTION

The e.m.f. induction principle can be applied for understanding the motoring action in the a.c. machines and d.c. machines respectively. In the case of motor, we have two parts; one stationary and other rotating. The stationary part of the motor carrying the field current in the case of d.c. machines. The armature conductors are placed on the rotor respectively as shown in Fig. (4.11) below.

In the case of motor the electrical energy is given to the motor terminals so we can consider that a part of electrical energy is used for the production of flux or field in the air gap between the stator and rotor whereas the other part flows through the armature winding which is placed on the rotor respectively.

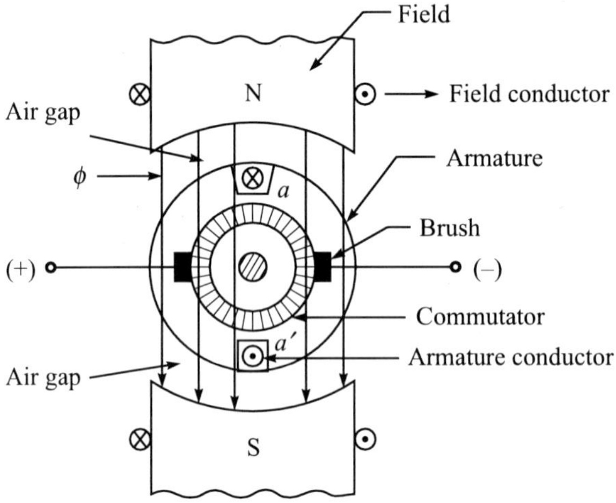

Fig. 4.11 D.C machine (motoring action).

So we can consider that whenever a current carrying coil is placed in a field is acted upon by a force 'F' given by

$$\boxed{F = BIl} \tag{4.38}$$

where 'B' is the average value of field, in Wb/m^2 or tesla
'I' is the current through armature coil, in ampere
'l' is the active length of armature conductor, in metre

The force acting on the armature conductor 'a' and 'a'' as shown in Fig. (4.12) acts in a direction given by Flemming's left hand rule as follows.

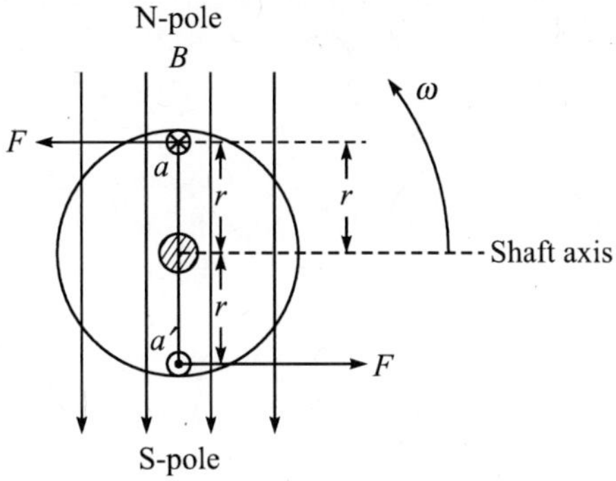

Fig. 4.12 Direction of force on the armature conductors '*a*' and '*a*′'.

From the centre of shaft axis, if the distance is equal to '*r*' upto the centre of conductor '*a*' and '*a*′' respectively. Then the torque which is acted upon the conductors also acts in the same direction as that of force and hence a turning moment is exerted on the armature in the anticlockwise direction and the rotor continues to move in this direction until some changes are done in the field energy stored in the air gap between the stator and rotor. The motor rotor continues to move till it gets current in the field and armature windings respectively.

Thus, the motor uses electrical energy and changes it into mechanical energy which can be used to perform some useful work such as lifting some load, pumping underground water from a certain depth, rolling sheets, winding and other plenty of applications due to which the motors are becoming popular day by day.

We can also visualize the motoring operation, by the electromechanical conversion which takes place in the air gap which is discussed at the start of this chapter.

4.7 GENERATING ACTION

The same motor which is designed for motoring operation could perform the work of generator by moving its rotor by some external means. The external means may be electrical or non-electrical means such as turbines, diesel, gas, etc.

In the case of generator the armature conductor is placed on the rotor in the case of d.c. machines and on the stator in the case of a.c. machines. The armature conductor can be rotated if placed on rotor as in d.c. generator or the field conductor can be rotated if it is placed on the rotor as in the case of synchronous generator or alternators. The generation of electricity takes place as a result of the induction which takes place in the armature conductor which cuts the flux produced by the field conductor. The energy which the generator gets is in the form of mechanical parameter, that is, angular speed 'ω' and torque 'T'. The product of which gives the mechanical power input to the rotor via shaft which is acting as a power transmitting device. This mechanical power can be obtained by non-electrical means such as steam, hydro, gas, etc.

The e.m.f. generated in the machine armature corresponds to the speed of the rotor and the flux value in which the armature is placed. Thus, the mechanical energy supplied to the rotor can be changed into electrical energy by doing some changes in the field flux in the air gap of the machine. The generated e.m.f. of the machine increases by increasing the flux of field or by increasing the speed of rotor or by both. But the field and speed cannot be increased above some value because field flux increase corresponds to the field current increase which could not be increased indefinitely as it would lead to the increase of diameter of the conductor and the slot size and depth should increase accordingly. This would increase the size, weight and cost of machine. Similarly, the increase of speed above its rated value with lead to deteriote the dynamic behaviour of machine as the centrifugal forces will be exerted on the machine rotor which will affect the dynamic balance of machine and shifting from the shaft axis respectively. So the speed as well as flux is specified for a particular machine. The increase in flux will also lead to saturation of core and ultimately overheating.

4.8 SINGLY EXCITED SYSTEM

Singly excited system has only one source of a.c. excitation such as in the case of relays, contactors, etc.

Consider the case of a hinged relay as shown in the Fig. (4.13) below.

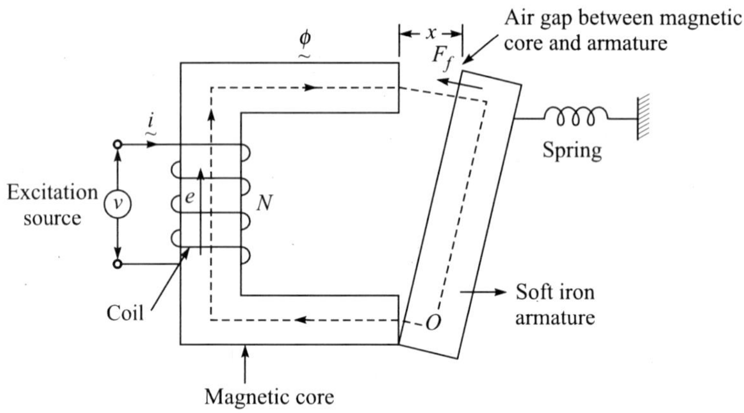

Fig. 4.13 Singly excited system (hinged relay).

The Fig. (4.13) shows an electromagnetic relay in which when current 'i' passes through the coil wound on the magnetic core. Then flux 'ϕ' flows through the magnetic circuit comprising core, air gap and armature then when the armature is moved toward the right, that is, when it is spaced apart and air gap is increased between the core and armature then due to electrical energy supplied to the coil, the energy is stored in the air gap as field energy. The mechanical spring gets compressed then a force 'F_f' acts on the armature towards left so as to decrease the magnetic circuit path. In the discussions, the flux is considered confined to the core and the coil is considered lossless

According to conservation of energy,

Input electrical energy = Mechanical energy output + Increase in field energy

$$id\lambda = F_f \cdot dx + dW_f \tag{4.39}$$

$\lambda = N\phi$ is called flux linkage of coil having 'N' turns

$$F_f dx = idA - dW_f \tag{4.40}$$

Neglecting, reluctance of iron

$$dW_f = \frac{1}{2} dL(x) i^2 \tag{4.41}$$

and
$$d\lambda = i \cdot dL(x) \tag{4.42}$$

Thus,
$$F_f \cdot dx = i^2 dL(x) - \frac{1}{2} i^2 dL(x) = \frac{1}{2} i^2 dL(x) \tag{4.43}$$

'$L(x)$' shows the self-inductance of coil as function of 'x'

Considering the coil current is constant and 'x' is small.

$$F_f = \frac{1}{2} i^2 \frac{dL(x)}{dx} = \frac{dW_f(i,x)}{dx} \tag{4.44}$$

Also,
$$L(x) = \frac{N^2}{\mathcal{R}(x)} \quad \text{where, } \mathcal{R}(x) = \frac{l(x)}{\mu A}$$

$$F_f = \frac{1}{2} i^2 \frac{d}{dx}\left[\frac{N^2}{\mathcal{R}(x)}\right] \tag{4.45}$$

$$= -\frac{1}{2}\frac{(Ni)^2}{\mathcal{R}^2(x)} \cdot \frac{d\mathcal{R}(x)}{dx} = \frac{1}{2}\left(\frac{\mathcal{F}}{\mathcal{R}(x)}\right)^2 \frac{d\mathcal{R}(x)}{dx} \tag{4.46}$$

$$= \frac{1}{2}\phi^2 \frac{d\mathcal{R}(x)}{dx}; \quad \because F(x) = \phi\, \mathcal{R}(x)$$

The reluctance and m.m.f. are functions of 'x'. From the expression (4.46), the negative sign implies that the force 'F_f' acts in a direction to decrease the air gap. In other words, we can say that the force acts such that the reluctance of the magnetic circuit is minimum and that is possible only when the air gap in the magnetic circuit flux path is minimum.

The singly excited system is also applicable to rotating systems such as machines. In that case, the torque replaces the force and the displacement 'x' is replaced by the angular displacement 'θ'.

So the torque which is exerted in the rotating part of the machines is given by

$$\boxed{T_f = \frac{dW_f(i,\theta)}{d\theta}} \tag{4.47}$$

The higher the change in the magnetic field energy 'W_f' in the air gap, the greater will be the torque which is exerted hence more will be the rotatory motion of the rotor. Since the electromechanical energy conversion is taking place in the air gap. Thus, we can consider that the air gap as a reservoir for stored magnetic energy 'W_f'.

4.9 DOUBLY EXCITED SYSTEM

In some of the practical applications often the doubly excited windings are used as in the case of synchronous machines; stepper motors, etc. The excitation to the windings may be both a.c. or may be a.c. and d.c. both as in the case of synchronous machine.

Where 'θ' is the angular displacement of rotor with respect to its reference axis. The motoring torque is in the clockwise direction (say), the field torque is in the opposite direction.

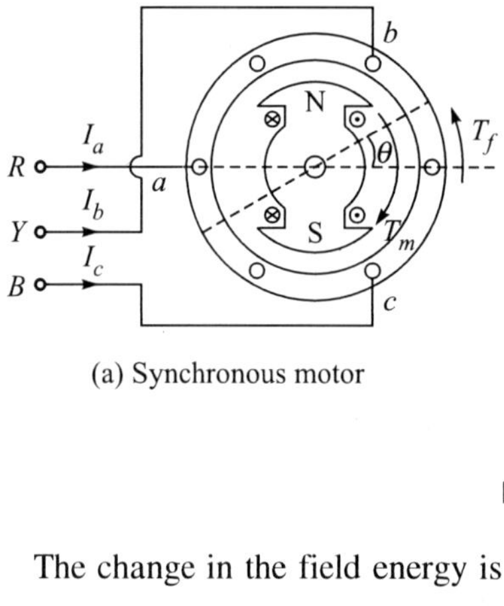

(a) Synchronous motor

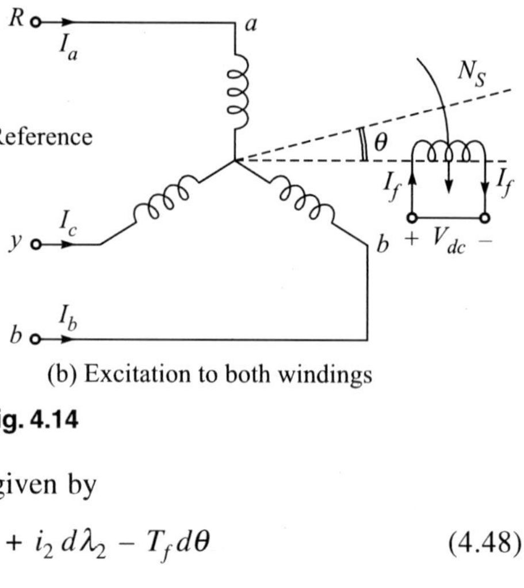

(b) Excitation to both windings

Fig. 4.14

The change in the field energy is given by

$$dW_f(\lambda_1, \lambda_2, \theta) = i_1\, d\lambda_1 + i_2\, d\lambda_2 - T_f\, d\theta \tag{4.48}$$

Since,

$$dW_m + dW_f = dW_e$$

$$dW_e = i_1 d\lambda_1 + i_2 d\lambda_2$$

$$dW_m = T_f\, d\theta$$

so,

$$i_1 = \left.\frac{\partial W_f}{\partial \lambda_1}\right|_{\lambda_2,\theta};\quad i_2 = \left.\frac{\partial W_f}{\partial \lambda_2}\right|_{\lambda_1,\theta};\quad T_f = \left.\frac{\partial W_f}{\partial \theta}\right|_{\lambda_1,\lambda_2}$$

$$W_f(\lambda_1, \lambda_2, \theta) = \frac{1}{2}\alpha_{11}\lambda_1^2 + \alpha_{12}\lambda_1\lambda_2 + \frac{1}{2}\alpha_{22}\lambda_2^2 \tag{4.49}$$

$$\beta_{11} = L_{22}/(L_{11} L_{22} - L_{12}^2);$$

$$\beta_{22} = L_{11}/(L_{11} L_{22} - L_{12}^2);$$

$$\beta_{12} = \beta_{21} = -L_{12}/(L_{11} L_{22} - L_{12}^2)$$

$$W_f'(i_1, i_2, \theta) = \frac{1}{2}L_{11}\,i_1^2 + L_{12}\,i_1\,i_2 + \frac{1}{2}L_{22}\,i_2^2 \tag{4.50}$$

4.10 FORCE/TORQUES IN SYSTEMS WITH PERMANENT MAGNETS

The magnetization curve of a permanent magnet is shown in the Fig. (4.15).

The magnetizing curve of the 'Permanent magnet' can be expressed as

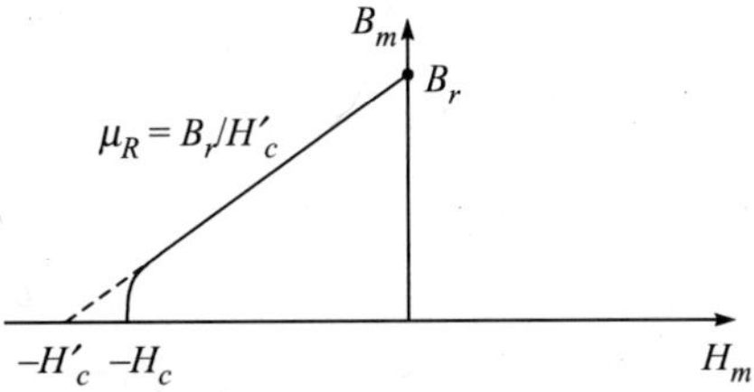

Fig. 4.15 BH curve of permanent magnet.

$$B_m = \mu_R\,(H_m - H_c') \tag{4.51}$$
$$B_m = \mu_R\,H_m + B_r \tag{4.52}$$

where, μ_R is the recoil permeability of a permanent magnet material

$$\mu_R = B_r/H_c'; \text{ assuming linear case}$$

Soft iron has very very high permeability and is selected for making plunger and armature of relays.

The expression for coenergy $W_f'(i, x)$ will reduce to $W_f'(x)$. Since in 'permanent magnet (PM)' the magnetic property is not dependent on the value of current 'i'.

Coenergy can be given by exp. (4.53) as written below

$$W_f' = \int_0^i \lambda\, di \tag{4.53}$$

The force on the armature is given by

$$F_f = \frac{dW_f'(x)}{dx} \tag{4.54}$$

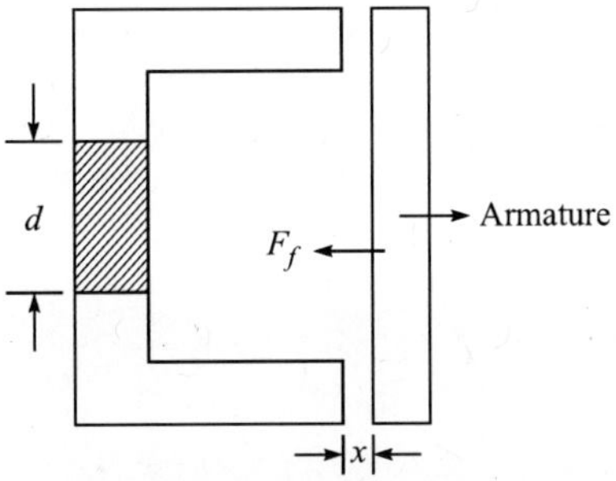

Fig. 4.16 PM attracting armature.

The coenergy is given by

$$W_f'(x) = \frac{\mu_R \cdot A \cdot H_c'^{\,} d^2}{2[d + 2(\mu_R/\mu_0)x]} \tag{4.55}$$

and

$$F_f = -\frac{A \cdot B_r^2}{\mu_0\left[1 + 2(\mu_R/\mu_0)\left(\frac{x}{d}\right)^2\right]} \tag{4.56}$$

The –ve sign indicates the force due to field will act so as to reduce 'x' that is spacing between the core and armature of relay or contactor.

4.11 DYNAMICAL EQUATIONS OF ELECTROMECHANICAL SYSTEMS

In the machines, we come across the following electrical and mechanical parameters such as frictional losses, electrical losses, forces and hence torques. To understand the operation of EMEC system, the dynamical changes occurring during operation of the machines is very important as it accounts for the EMEC mechanism and the effect it produces as a whole.

In Fig. (4.17) depicts a typical EMEC system with the consideration of the dynamical constants which accounts the electrical and mechanical changes taking place in the machine.

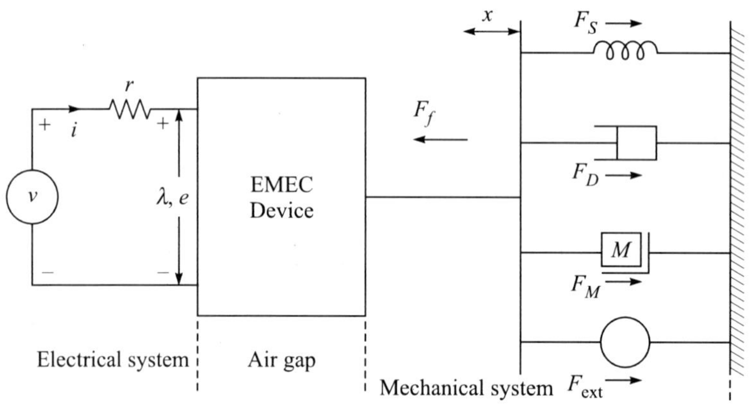

Fig. 4.17 Singly excited EMEC system.

The system consists of electrical system, mechanical system and the EMEC device or medium. The electrical system consists of a voltage source 'v' and resistance 'r'.

The KVL to the electrical system part gives

$$v = ir + e = ir + \frac{d\lambda}{dt};\ \lambda = L(x)i;\ L(x) = f(x)$$

$$v = ir + \frac{d}{dt}[L(x)i] = ir + L(x)\frac{di}{dt} + i\frac{dL(x)}{dx}, \frac{dx}{dt}$$

$$v = ir + L(x)\frac{di}{dt} + i\frac{dL(x)}{dt} \cdot \dot{x} \tag{4.57}$$

'$\dot{x}$' corresponds to speed

$$v = ir + v_{\text{induction}} + v_{\text{speed}} \tag{4.58}$$

where, $v_{\text{induction}}$ = induction voltage

v_{speed} = speed voltage

The speed voltage is required for EMEC. The 'f_k' is the restoring force of spring and is given by

$$f_k = kx;\ k = \text{spring constant} \tag{4.59}$$

where, 'f_D' is the damping force and is given by

$$f_D = Dv = D\dot{x} = D\frac{dx}{dt} \tag{4.60}$$

where, 'D' is called damping constant

'f_m' is the inertial force and is given by

$$f_m = M\alpha = M\ddot{x} = M\frac{d^2x}{dt^2} \tag{4.61}$$

where, M = mass of inertia.

The force due to magnetic field is required to balance all these force components. So

$$F_f = M\ddot{x} + D\dot{x} + kx + f \tag{4.62}$$

where, f = external mechanical force applied. Any changes in the EMEC medium is due to the change in any one component of force or due to the combined effect of two or more members.

SOLVED QUESTIONS

Q.1 A single conductor 10 cm long is moved by a force perpendicularly to a uniform field of 0.5 tesla, covering a distance of 1 m in a time of 2 sec. Calculate

(a) Induced e.m.f.

(b) Average e.m.f.

Ans: (a) $e = Blv = 0.5 \times \frac{100}{2} \times 10 \times 10^{-2}$

$= 2.5$ volt

(b) $$\phi = BA = 0.5 \times 10 \times 10^{-2} \times 1 = 0.05 \text{ Wb}$$

$$E = \frac{\phi}{t} = \frac{0.05}{2} = 0.025 \text{ V}$$

Q.2 A conductor of 20 cm is moved by a prime mover at a velocity of 10 cm/s but at an angle of 30° with respect to the same field. Calculate the average induced emf. If B is 1 Wb/ m^2.

Ans: $$E = B\, l\, v \sin \theta$$

$$E = 1 \times 0.2 \times 0.1 \times \sin 30 = 0.2 \times \frac{1}{2} = 0.1 \text{ volt}$$

$$E = 100 \text{ mV}$$

Q.3 The conductor of 10 cm is placed in between the pole faces having area of 50 cm^2 and the flux is 1 m Wb. If the current in the conductor is 5 A. Calculate the magnitude of the force and the direction of force.

Ans: (a) $$F = B\, I\, l = \frac{1 \times 10^{-3}}{\underset{10}{\cancel{50}} \times 10^{-4}} \times \cancel{5} \times 0.1 = 0.1 \text{ N}$$

The force will act in the downward direction.

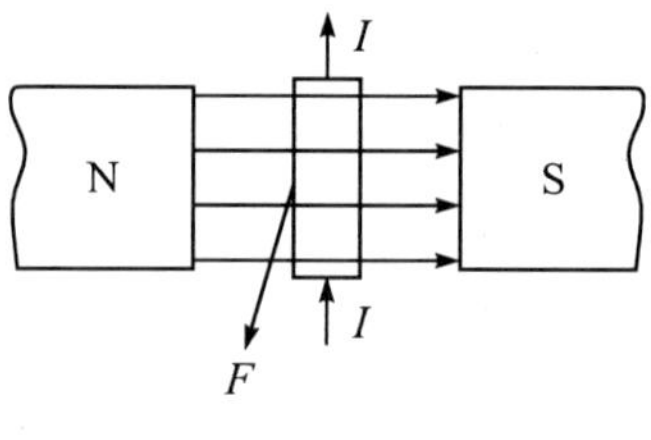

Q.4 Two parallel plates, each having dimensions of 20 cm by 20 cm, are held 2 mm apart in air. If the potential difference between the plates is 2 kV, determine the energy stored in the capacitor and the force acting on each plate.

Ans: $$W_e = \frac{1}{2} C V^2,\ C = \frac{\varepsilon_0 A}{d} = \frac{8.85 \times 10^{-12} \times \overset{200}{\cancel{400}} \times 10^{-4}}{2 \times 10^{-3}}$$

$$W_e = \frac{1}{2} \times \overset{8.85}{\cancel{17.70}} \times 10^{-11} \times (2 \times 10^3)$$

$$W_e = 8.85 \times 10^{-11} \times 4 \times 10^6 = 35.40 \times 10^{-5}$$

$$W_e = 3.54 \times 10^{-4} \text{ J} = 354\ \mu\text{J}$$

The force acting on each plate

$$F = \frac{1}{2} V^2 \cdot \frac{\partial C}{\partial x} = -\frac{1}{2} \varepsilon_0\, A \cdot \left(\frac{V}{x}\right)^2$$

$$F = -\frac{1}{2} \times 8.85 \times 10^{-12} \times \overset{200}{400} \times 10^{-4} \times \frac{\overset{(1000000)^2}{2000000}}{2}$$

$$F = -\,1770 \times 10^{-4}\ \text{N} = -0.177\ \text{N}$$

$$F = +(0.177)\ \text{N attractive force}$$

REVIEW QUESTIONS

Q.1 Explain the electromechanical energy conversion in rotating machines.

Q.2 Derive the expression of force in a singly excited system.

Q.3 Show that the torque developed in a doubly excited system is equal to the rate of increase of field energy with respect to displacement is constant.

Q.4 Explain the motoring and generating action in rotating machines.

Q.5 List the basic differences between the motoring operation and generating operation.

Q.6 Why the field rotates in three-phase machines?

Q.7 Discuss the transformer e.m.f. and motionally induced e.m.f.

Q.8 What is the difference between the electric field and magnetic field?

Q.9 How can the electric field be used for electromechanical energy conversion? Explain.

Q.10 What is the merit of electromagnets over the permanent magnets?

UNSOLVED QUESTIONS

Q.1 The energy stored in a magnetic circuit is given as $W_m = -5 \ln \lambda - \lambda$. What is the relationship between the applied current and the flux linkages? Calculate the inductance of the magnetic circuit when the current is 2 A.

Q.2 Determine the force acting on the movable part in Fig. 1 when the 100 turn coil carries a current of 20 A. The relative permeability of the magnetic material of 1000.

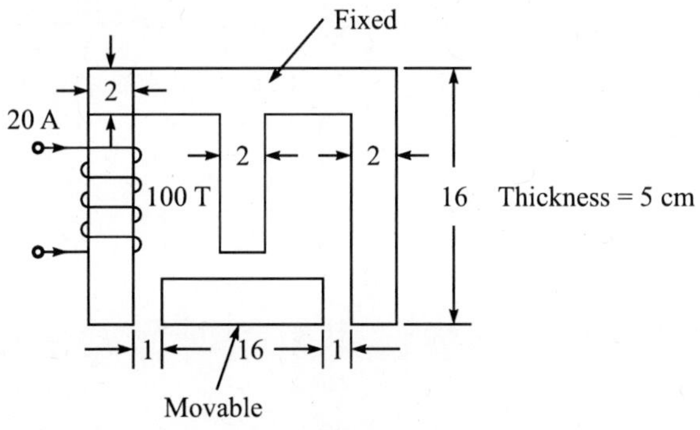

Fig. 1

Q.3 If the frequency of the induced e.m.f. in an 8 pole machine is 50 Hz, at what speed is the rotor rotating.

Q.4 A 120 turn coil on the rotor of a 2 pole motor carries a current of 10 A. The flux density per pole is 1.2 T. If the core diameter is 25 cm and its length is 10 cm, determine (a) the force per conductor (b) the force acting on the coil (c) the torque acting on the rotor.

Q.5 The inductance of a 2 pole motor is given as

$$L(\theta) = 5 + 2 \cos \theta, H$$

Determine the torque developed by the motor when the current in the 150 turn coil is 5 A and $\theta = 25°$.

Q.6 The coil inductance varies with rotor position of a motor is given by

$$L(\theta) = L_0 + L_2 \cos 2\theta$$

where, $L_0 = 10$ mH and $L_2 = 3$ mH

Find the torque as a function of 'θ' for a coil current of 2 A.

Q.7 A relay made up of high permeability has a movable plunger having high permeability. The plunger height 'h' is greater than air gap length 'g'. Calculate the magnetic stored energy 'W_f' as a function of plunger position 'x'. Write the assumptions made.

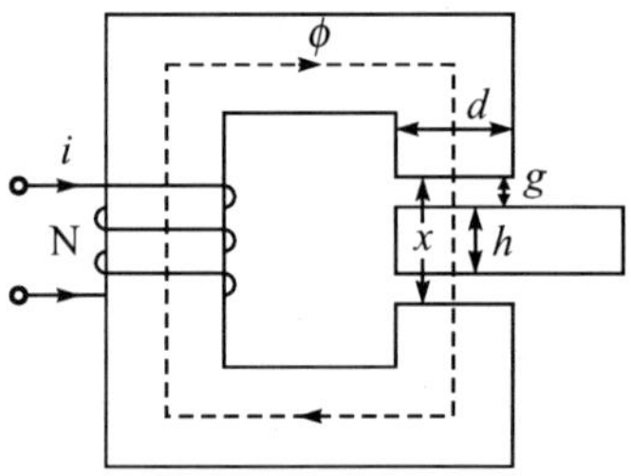

$N = 1200$, $g = 1.0$ mm, $d = 0.010$ m, $l = 0.05$ m, $i = 10$ A

Q.8 The self and mutual inductances of the two exciting coils of a doubly excited system are $L_{11} = L_{22} = \dfrac{1}{1+x}$, $L_{12} = L_{21} = \dfrac{2}{1-x}$

Calculate the time average force and coil currents as $x = 0.1$ m. When both coils are connected in series across $v = 100 \cos 314\, t$.

Chapter

5

Transformers

5.1 Introduction
5.2 Constructional Features of Transformers
5.3 Types of Transformers
5.4 Ideal Transformer
5.5 Phasor Diagram of Ideal Transformer
5.6 Practical Transformer
5.7 Phasor Diagram of Real Transformer
5.8 Equivalent Circuit of Transformer
5.9 Voltage Regulation of Transformer
5.10 Efficiency of Transformer
5.11 Transformer Testing
5.12 Auto Transformers
5.13 Three-Phase Transformer
5.14 Harmonics
5.15 Special Transformer
5.16 Per Unit System
• Solved Questions
• Unsolved Questions
• Exam Questions

5.1 INTRODUCTION

The electrical power system comprises an important electrical device called transformer which makes possible to transfer electrical energy from remote places to thousands of kilometres. So it is necessary to know about this important device. The transformers are available from very small sizes of less than an inch to large size transformers, in metres used in the power stations.

The transformers popularity is not only in the electrical field but also in the electronic industries. We can see transformers in the audio equipment, adapters of mobiles, computer CPU, stabilizers, telephones, etc. As the technology is developing day by day, transformers are also available in large number of varieties. Earlier, the power frequency transformers are developed but as the aviation and space shuttles are developed so there is a need of transformers which could operate at high frequencies. So the core design and material science is also explored and new materials which have a high value of permeability is coming in this decade.

Transformers are available in both single-phase and three-phase. The single-phase transformers are of smaller ratings and the three phase transformers are of larger ratings. Due to static components the transformer losses are less and thus this device has higher efficiency—greater than all rotating devices.

In India, the transformers which are used to transmit bulk power of the order of millions of voltamperes are sent via three-phase transformers kept at transmitting end substations whose primary voltage is lower of the order of 11 kV or 33 kV and secondary voltage is higher of the order of lakhs of volts (132 kV, 220 kV, 400 kV, 765 kV etc.) for large distance transmission.

Again at the receiving end via three phase transformers the voltage level is decreased to utility voltage of the order of 400 volts/230 volts for industrial and domestic uses. For large industries the voltage of 1.1 kV, 3.3 kV, 6.6 kV, 11 kV are also used. Now in India on the basis of consumer demand the transformer size is determined.

5.2 CONSTRUCTIONAL FEATURES OF TRANSFORMERS

Widely used transformers are of mainly two types: namely, core type or shell type transformers. The core type transformers are used for high power applications and the shell type transformer is used for low power applications. In the core type transformer, the winding bulges out of core outer limbs as shown in the Fig. 5.1(a) and in the shell type transformer the winding is bulges out of central limb, as shown in Fig. 5.1(b). In other words, in core type transformers the winding surrounds the core whereas in the shell type transformer, the core surrounds the winding.

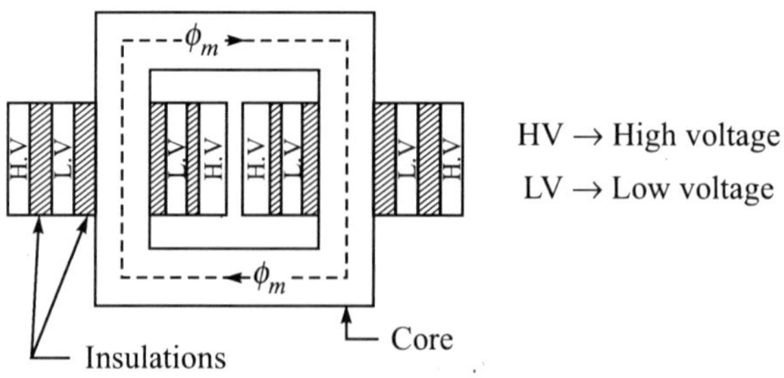

Fig. 5.1(a) Core type transformer.

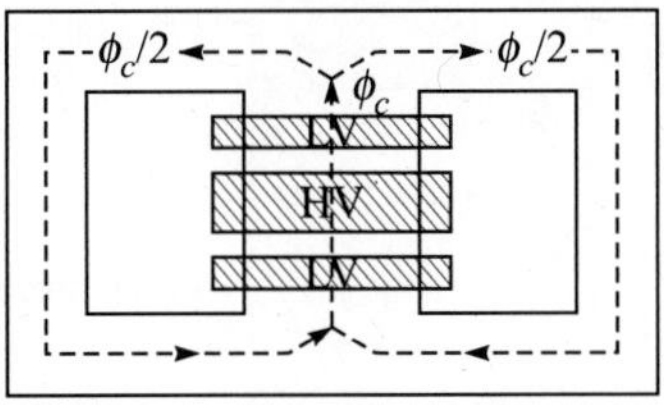

Fig. 5.1(b) Shell type transformer.

The laminations used for making the transformer core is given by the following English alphabets *L*, *I*, *U*, *E*, etc. as shown in Fig. (5.2),

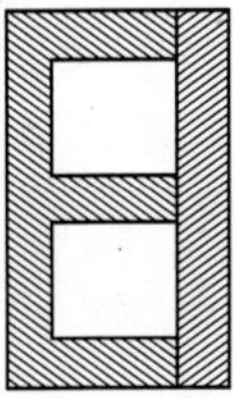

Fig. 5.2(a) Shell type transformer core made by *E*, *I* laminations.

Fig. 5.2(b) Core type transformer core made up of *L* laminations.

The above construction is for single phase transformer. The windings are wound on a former made up of cardboard, plastic, etc. The windings are, forewound using winding machine which may be hand driven, semi-automatic or automatic depending on the number of pieces needed. The windings number is determined before hand analytically, using the e.m.f. equation of the transformer as given by the exp. I

$$\left.\begin{aligned} E_1 &= 4.44\ B_m \cdot A \cdot f \cdot N_1, \text{ volts} \\ \text{and} \qquad E_2 &= 4.44\ B_m\ A \cdot f \cdot N_2, \text{ volts} \end{aligned}\right\} \qquad \text{(I)}$$

The windings on a plastic former is given by the Fig. (5.3)

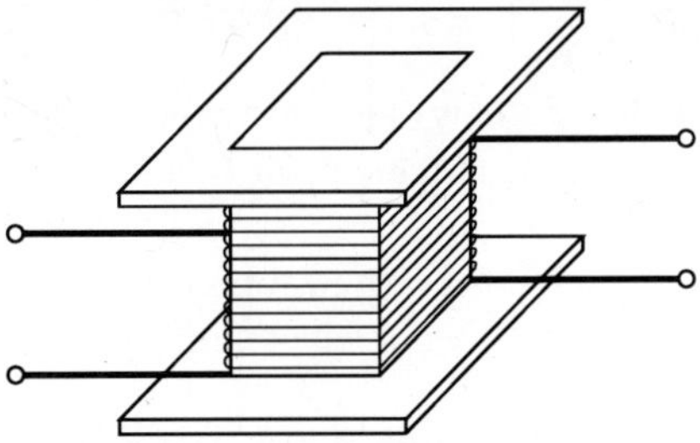

Fig. 5.3 Windings of shell type 1ϕ transformer on a plastic former or bobbin.

For very small transformers, the laminations are held by varnish, as the size of the transformer core is increased the bolts are necessary to held the insulated laminations as shown in Fig. (5.4).

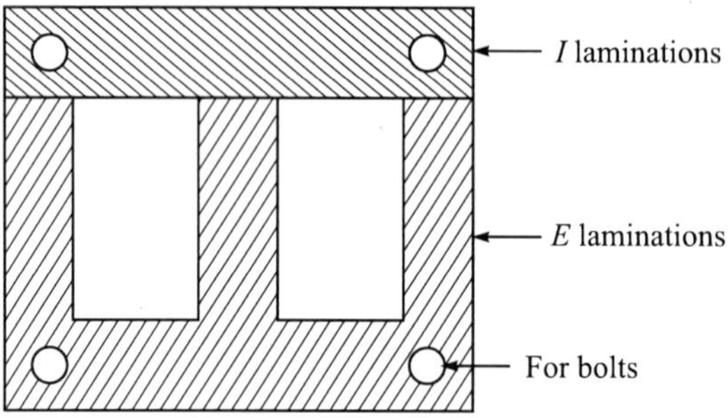

Fig. 5.4 Bolts positions on core of transformer.

For large transformers, there is a need of cooling of core. So oil filled transformers are used. A figure of oil filled three phase transformer is shown in the Fig. (5.5).

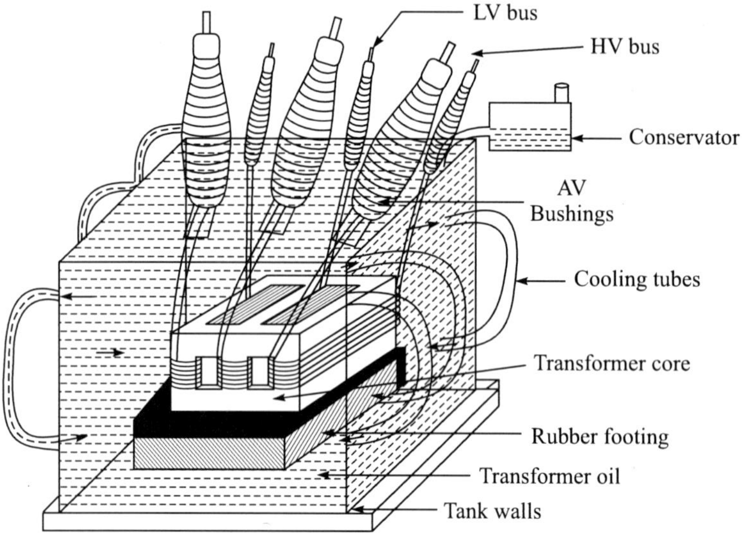

Fig. 5.5 Three-phase transformer.

In large transformers, radiators, fans of exhaust type, coolant, pumps are used. For alarms, buchholtz relay is used at the outlet of the transformer. The breather which is not shown is also attached to the conservator for limiting the moisture of air from contaminating the mineral oil (transformer oil) which is present in the transformer which is used as a coolant. For small transformers, natural air cooling is sufficient.

5.3 TYPES OF TRANSFORMERS

The transformers are classified on the basis of size, frequency, voltage, current, phase, etc.

On the basis of size, the transformers are divided as small size, medium size, large size transformer and very large size transformer. The small size transformer is of V A ratings, medium size transformer is from 1 kVA rating to 10 kVA ratings, large size transformer is from 10 kVA to 100's of kVA and very large transformer is larger than MVA ratings.

Transformers are classified on the basis of frequency, as power frequency transformer, audio frequency transformer and RF transformers.

On the basis of voltage, the transformers are divided as step up and step down transformer. In the step up transformer the secondary voltage is greater than the primary voltage whereas the step down transformer has lower value of secondary voltage as compared to primary side voltage.

On the basis of current, the transformers are classified as current transformers which steps down the current at secondary side to measurable values of 5 A, 10 A, etc. These types of transformers are used with the instruments so they are termed instrument transformers.

On the basis of phase, the single phase transformer and three phase transformers respectively. Many other types of transformers are available on the basis of the applications such as welding transformer, isolation transformer and many types which are discussed in the topic of special transformers.

Principle of Operation

All the transformers work on the basis of the mutual induction between the two windings which are electrically separate but magnetically interlinked. Consider an ideal transformer as shown in the Fig. (5.6) for gaining better insight of operation of transformer.

When interlinking between the two windings is maximum then maximum power can be transferred from one side to other side of the transformer. In the power transformers which operate at 50 Hz or 60 Hz this interlinking is maximum which is achieved by winding the primary and secondary on the same core of transformer which is made up of high permeability material so it allows maximum flux through it. But in the communication transformers the transformer core is made up of air cored transformers in which there is weak linking of flux as the core is made up of diamagnetic material. The main purpose of the transformer used in electronic circuits is to provide isolation of two circuits or for impedance matching.

5.4 IDEAL TRANSFORMER

In the Fig. (5.6), the parameters are defined as below.

$V_1(f)$: Primary voltage or excitation voltage, in volts at a frequency 'f' = 50 Hz or 60 Hz

$V_2(f)$: Secondary terminal voltage; in volts at the same frequency 'f'

$I_1(f)$: Primary current at frequency 'f', in amps.

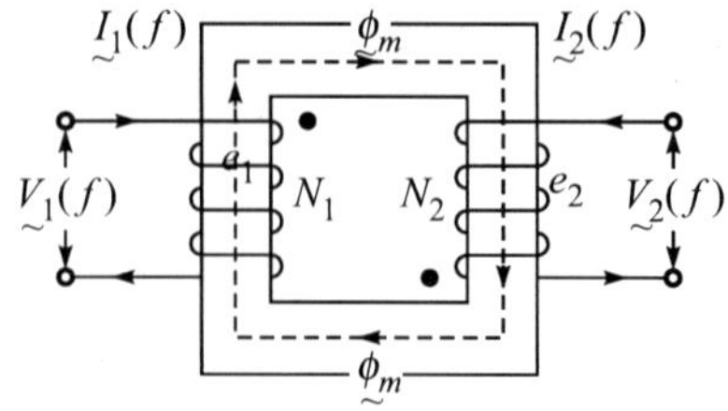

Fig. 5.6 Ideal Transformer.

$I_2(f)$: Secondary current at frequency 'f', in amps
N_1: Primary number of turns
N_2: Secondary number of turns
ϕ_m: Mutual flux, in weber
E_1, E_2: Reaction or back emf, in volts

In the case of an ideal transformer, the following assumptions are made as written below.

(i) The windings resistance is absent.
(ii) The windings leakage flux is absent.
(iii) The drops in the windings is zero
(iv) The permeability of core is infinite.
(v) Core is lossless.
(vi) The magnitudes of voltage 'V_1' and 'E_1' are equal. Similarly, the magnitudes of voltage 'V_2' and 'E_2' are equal.

The primary and secondary terminal voltage V_1 and V_2 are given as follows

$$V_1 = E_1 = 4.44\ N_1 \cdot \phi_m \cdot f \tag{5.1}$$

and

$$V_2 = E_2 = 4.44\ N_2 \cdot \phi_m \cdot f \tag{5.2}$$

Dividing the two expressions, we get

$$\frac{V_2}{V_1} = \frac{E_2}{E_1} = \frac{N_2}{N_1} = K \tag{5.3}$$

Also, the ampere turns of two sides are equal i.e.,

$$N_1\ I_1 = N_2\ I_2 \tag{5.4}$$

So,

$$\frac{I_1}{I_2} = \frac{N_2}{N_1} = K \tag{5.5}$$

Moreover, the voltampere of both sides are equal, i.e,

$$(VA)_1 = (VA)_2 \tag{5.6}$$

or,

$$V_1\, I_1 = V_2\, I_2 \tag{5.7}$$

or,

$$\boxed{\frac{I_1}{I_2} = \frac{V_2}{V_1} = K} \tag{5.8}$$

In the above expressions we see that a term 'K' appears which relates the various electrical parameters such as voltages and currents of either sides of transformer and thus can be transformed accordingly as per requirement. This factor 'K' is called transformation factor of transformer.

If $K > 1$, $V_2 > V_1$ and the transformer is known as step up transformer.

If $K < 1$, then $V_2 < V_1$ and the transformer is known as step down transformer.

The step up transformers are used in the power station for stepping up the alternator voltage of 11 kV to 132 kV, 220 kV, 400 kV, 765 kV and higher for sending the electrical power over large distances of the order of hundreds of kilometres. The step down transformer is used for stepping down the voltages from 132 kV, 220 kV, etc. to 6.6 kV, 1.1 kV, 400 V, etc. at the user end of power system network.

The transformer operation can be visualized indirectly by using the phasor concept. The transformer normally operates under two conditions, namely, no load condition and full load condition respectively.

5.5 PHASOR DIAGRAM OF IDEAL TRANSFORMER

No Load Condition ($I_2 = 0$)

For the phasor diagram of ideal transformer the following transformer equations are used.

For the primary side,

Primary voltage $\qquad \overline{V}_1 = -\overline{E}_1 \qquad$ (5.9)

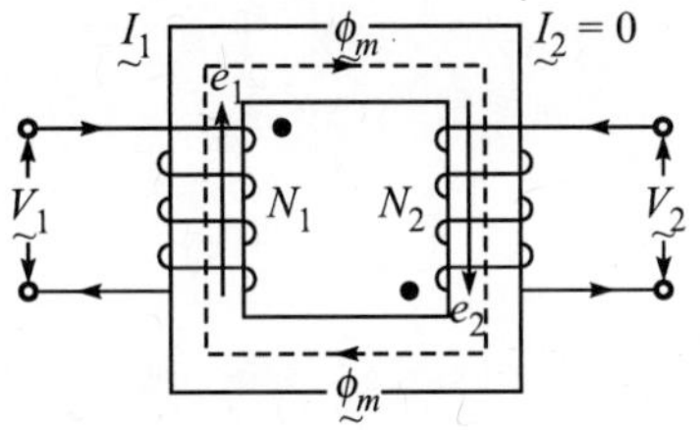

Fig. 5.7 Ideal Transformer (single phase type).

Since $\qquad j = \sqrt{-1}$, so $j^2 = -1$

or, the above equation can be represented as

$$\overline{V}_1 = j^2 \overline{E}_1 \qquad (5.10)$$

$$j = \angle +90°, \text{ so } j^2 = \angle +180°$$

That is, if the operator 'j' is multiplied with a phasor then it is rotated by 90° in the anticlockwise or positive direction.

The secondary terminal voltage $\overline{V}_2$ is represented by

$$\overline{V}_2 = \overline{E}_2 \qquad (5.11)$$

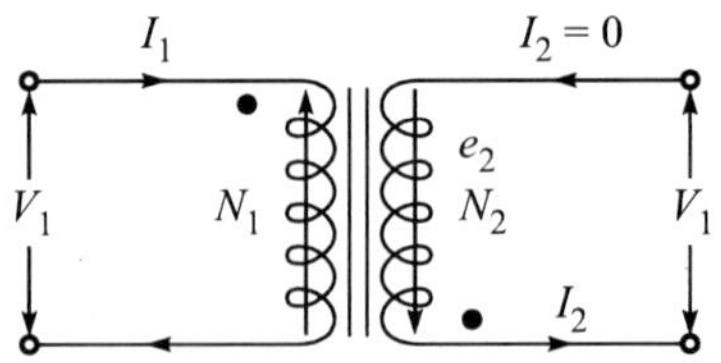

Fig. 5.8 Equivalent circuit of ideal transformer.

The primary current is represented by 'I_1' and expressed mathematically as follows

$$\bar{I}_1 = \bar{I}_0 = \bar{I}_m \tag{5.12}$$

where, $\bar{I}_0$ = No load primary current, in amp.

$\bar{I}_m$ = Magnetization component of primary current, in amp.

Since, '$\bar{I}_m$' produces the flux 'ϕ_m' in the core. So this current will be in phase with this mutual flux 'ϕ_m'. Deleting the subscript to avoid since $\overline{V}_1$ and $\bar{I}_1$ are sinusoidal quantities, so 'ϕ' is also sinusoidal say, $\phi = \phi_m \sin \omega t$

Since, $$\phi \propto \bar{I}_1 \text{ or } \bar{I}_m \text{ or } \bar{I}_0$$

So, the flux ϕ, I_1, I_m are in phase as shown in the phasor diagram below.

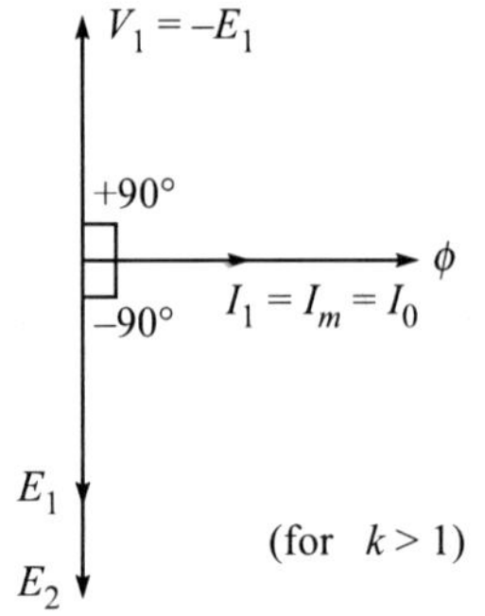

Fig. 5.9 Phasor diagram of 1 ϕ transformer under no load condition.

where, ϕ = Mutual flux in the core, in Wb

E_1 = Primary back e.m.f, in volts

V_1 = Primary voltage, in volts

I_1 = Primary current, in amp

I_m = Magnetization current, in amp

E_2 = Secondary back e.m.f., in volts

V_2 = Secondary voltage, in volts

But it is seen that with a load connected to the secondary. The phasor diagram no more remains same but it changes as shown in the Fig. (5.10) below.

Under Load Condition

Depending on the type of load attached across the secondary of transformer the phasors of voltage and current behaves differently. Firstly, we consider a pure resistive load connected at the secondary of transformer. Then the mathematical expressions of primary and secondary voltages are given by

$$\overline{V}_1 = -\overline{E}_1 = +j^2\ \overline{E}_1;\ j = \angle +90°$$

$$\overline{V}_1 = \overline{E}_1 \angle +180°$$

$\overline{V}_1$ leads $\overline{E}_1$ by 180° or primary voltage phasor $\overline{V}_1$ opposes the primary e.m.f. phasor $\overline{E}_1$

$$\overline{V}_2 = \overline{E}_2$$

$$\overline{I}_1 = \overline{I}_m + \overline{I}_2';\quad \overline{I}_m = \overline{I}_0 (\text{No load primary current}) \tag{5.13}$$

where, $\overline{I}_2'$ is the secondary current referred to or transferred to primary side and is equal to

$$\overline{I}_2' = K\,I_2 \tag{5.14}$$

where, $K = \dfrac{N_2}{N_1}$; N_1, N_2 are primary and secondary number of turns respectively.

$\overline{I}_2' \rightarrow$ load equivalent of primary current.

The phasor diagram of the transformer under resistive load condition is given below.

$\phi_1 \rightarrow$ Primary power factor angle

$(\cos\ \phi_1)_{\log} \rightarrow$ Primary power factor

$\cos\ \phi_2 = 1 \rightarrow$ Secondary power factor

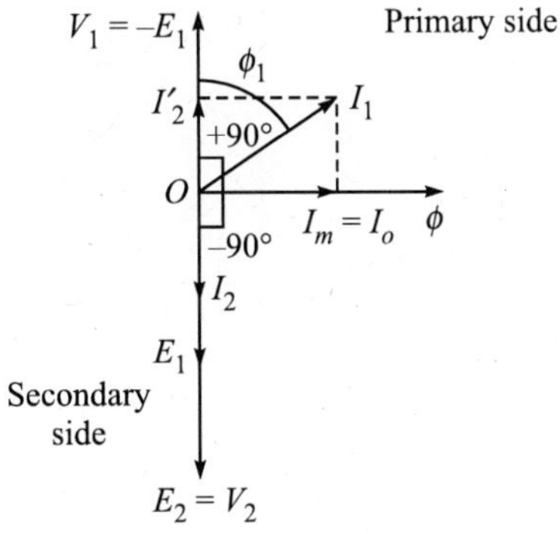

Fig. 5.10 Phasor diagram of 1 ϕ transformer (with resistance load).

The magnitude of no load current 'I_0' is 2–5% of the rated or full load value of transformer current which is marked over the name plate of transformer.

With the inductive load, the phasors of secondary current 'I_2' lags behind 'V_2' by 90° for purely inductive load i.e., ϕ_2 between V_2 and I_2 is equal to 90°. So the phasor diagram is given by

Primary power factor 'cos ϕ_1'

where, $\cos \phi_1 = 0$ (lagging)

Secondary power factor 'cos ϕ_2'

where, $\cos \phi_2 = 0$ (lagging)

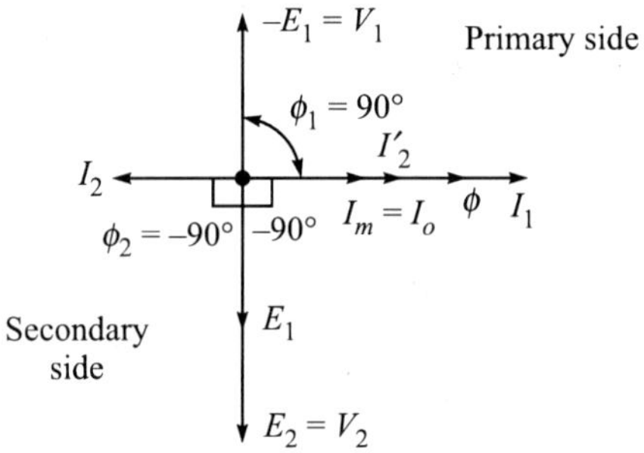

Fig. 5.11 Phasor diagram of 1 ϕ transformer with inductive load (pure '*L*').

Thus, we see that the power factor of the transformer is lagging in the case of both resistive and inductive load. But in the case of inductive load the power factor of the transformer is poorer.

With Capacitive Load

When capacitor is attached at the secondary of the transformer then the secondary terminal voltage 'V_2' will lag behind the secondary current 'I_2' by 90° if the load is purely capacitive as shown in the phasor diagram below.

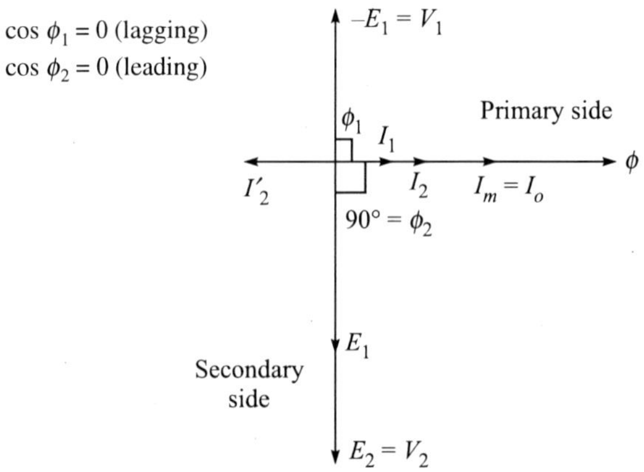

Fig. 5.12 Phasor diagram of 1 ϕ transformer (with capacitive load).

Thus, we conclude from the above three phasor diagrams that the power factor of the primary side of transformer keeps lagging irrespective of the power factor at the secondary side. Thus, we see that the transformer will draw reactive power and hence the transformer decreases the overall power factor of power system network or system with inductive load and not with capacitive load. The exciting current required also reduces with 'C' type load. In India, the power factor is taken as 0.8 (lagging) which is a standard value.

But it is seen that to get a lossless coil as in the case of ideal transformer and a lossless core it is not practically realizable. So we should consider the real parameters of a transformer so as to help a design engineer for coming to a feasible solution.

5.6 PRACTICAL TRANSFORMER

The real or practical transformer consists of the following electrical parameters in the primary and secondary side respectively.

(i) The resistance of windings are r_1 and r_2.
(ii) The leakage reactance of windings are x_1 and x_2.
(iii) The loss in the core which is equivalent to $I_e^2 R_0$.
(iv) The shunt branch comprises 'R_0' and 'X_0'.
(v) The permeability of core is finite i.e., $\mu \neq \infty$
(vi) There is leakage of flux across the core which passes through air.

So, considering the above conditions the transformer parameters are expressed as follows

$$\bar{V}_1 = -\bar{E}_1 + \bar{I}_1 r_1 + j\bar{I}_1 x_1 \tag{5.15}$$

$$\bar{V}_2 = \bar{E}_2 + \bar{I}_2 r_2 + j\bar{I}_2 x_2 \tag{5.16}$$

$$\bar{I}_1 = \bar{I}_0 + \bar{I}_2'$$

$$\bar{I}_0 = \bar{I}_m + \bar{I}_e \tag{5.17}$$

These relations are expressed graphically using the phasor diagrams as follows.

5.7 PHASOR DIAGRAM OF REAL TRANSFORMER

Under No Load $N_2 (I_2 = 0)$

If $\quad \bar{I}_2 = 0$ then, $\bar{I}_2' = k\,\bar{I}_2 = 0$

i.e., $\quad \bar{I}_2' = 0$

$$\bar{V}_2 = \bar{E}_2$$

$$\bar{V}_1 = -\bar{E}_1 + \bar{I}_1 r_1 + j\bar{I}_1 x_1$$

where, $\quad j = \angle +90^\circ$

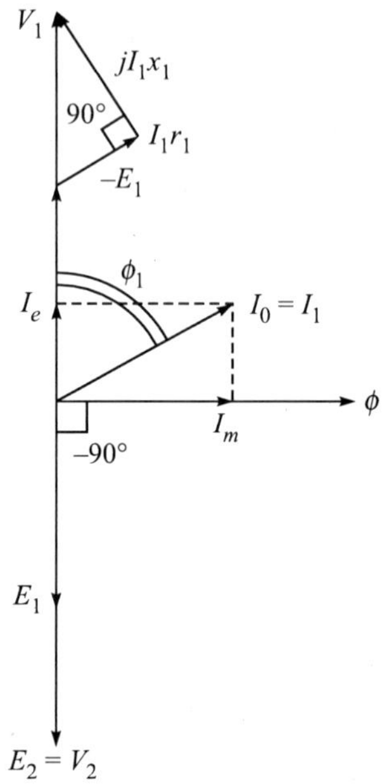

Fig. 5.13 Phasor diagram of real transformer (no load).

'*j*' is a operator which multiplied with a phasor quantity it rotates the phasor in the anticlockwise direction by 90°.

$$\phi_1 = \text{Phase angle of primary side}$$
$$\cos\,\phi_1 = \text{Power factor of primary side}$$

Since, 'I_1' lags 'V_1' by 'ϕ_1'. So the power factor 'cos ϕ_1' of the primary is lagging. As a whole the transformer acts as an inductive ($R - L$) device. It will draw reactive power from the supply.

But mostly, the transformer is operated under load conditions. So the type of load affects the power factor and the performance of the transformer. With loading the following things happens in the transformer.

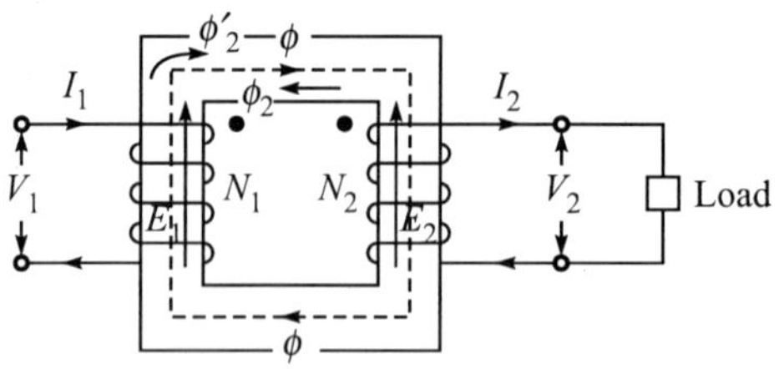

Fig. 5.14 Load applied to transformer secondary.

When a load is attached, it draws a current 'I_2' (say) as a result of which a flux 'ϕ_2' is produced due to m.m.f. $\mathcal{F}_2 = N_2 I_2$ of the secondary which opposes the flux 'ϕ' in the core as a result the net flux decreases in the core temporarily. Since the net flux affects the e.m.f. of primary 'E_1' which is equal to $N_1 \dfrac{d\phi}{dt}$. With the decrease of flux 'ϕ' by an amount 'ϕ_2' the emf 'E_1' also decreases accordingly. So the difference between the primary voltage 'V_1' and 'E_1' increases and hence larger primary current is drawn from the supply. Thus, this increase in primary current produces a flux 'ϕ_2'' (say) which will flow in the same direction as 'ϕ'. So this flux 'ϕ_2'' is equal and opposite to that of 'ϕ_2' so they cancel each other. Hence, the net flux 'ϕ' remains constant, so irrespective of the power factor of load the flux in the core remains invariant. So the transformer is rated in kVA and not in kW. So, the primary current '$\bar{I}_1$' can be considered to be sum of two components one is the no load primary current '$\bar{I}_0$' which flows through the transformer primary only when secondary is open circuited. This component is a very small value usually in the range of 2 – 5% of the rated value. The second component is '$\bar{I}_2$' which is called the load component of primary current which arises only when current '$\bar{I}_2$' is delivered to the load. So the primary current $\bar{I}_1$ (under loaded condition) of the transformer is expressed mathematically, as follows:

$$\bar{I}_1 = \bar{I}_0 + \bar{I}_2'$$

The type of load attached to the secondary of transformer affects the performance parameters such as voltage regulation of transformer. So, we consider different types of load one by one. There are primarily three types of loads, namely resistive (R only), inductive ($R - L$) and capacitive ($R - C$) loads.

With Resistive load

Through resistance 'R' if $\bar{I}_2$ current flows at the secondary and $\bar{V}_2$ is the voltage across the resistance 'R' then the secondary phase angle 'θ_2' between the voltage 'V_2' and '$\bar{I}_2$' is equal to zero. Or, the power factor of load is unity. Then the phasor diagram is given as follows.

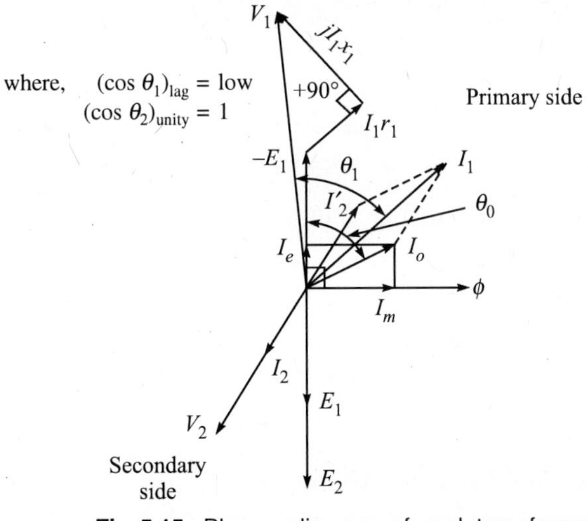

Fig. 5.15 Phasor diagram of real transformer (with purely resistive load).

In the figure, 'ϕ_0' is called hysteresis angle of advance which is of the order of 60° – 80°.

The overall power factor of the transformer is lagging.

For Inductive load (*R* – *L* type)

I_2 lags V_2 by 'θ_2' (say)
we see from the phasor diagram that

Primary power factor, $(\text{p.f.})_1 = \cos\theta_1$ is very low
Primary power factor, $(\text{p.f.})_2 = \cos\theta_2$ is high

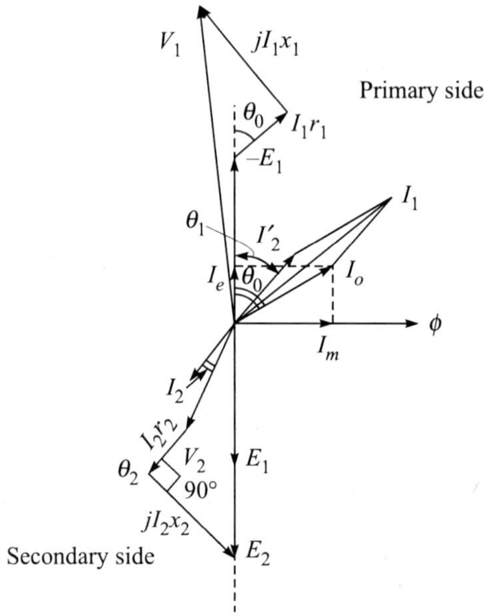

Fig. 5.16 Phasor diagram of real transformer with *R* – *L* load.

But both power factors at the primary and secondary are lagging. So the overall power factor of the transformer is poorer than the previous one.

With Capacitive Load

I_2 leads V_2 by 'θ_2' (say). From the phasor diagram we see that the power factor angle 'θ_1' is decreased with capacitive load. Hence, the power factor at the primary side is increased. So, the overall power factor of the transformer is improved. Hence, the transformer power factor is better with capacitive load with respect to the other types of load.

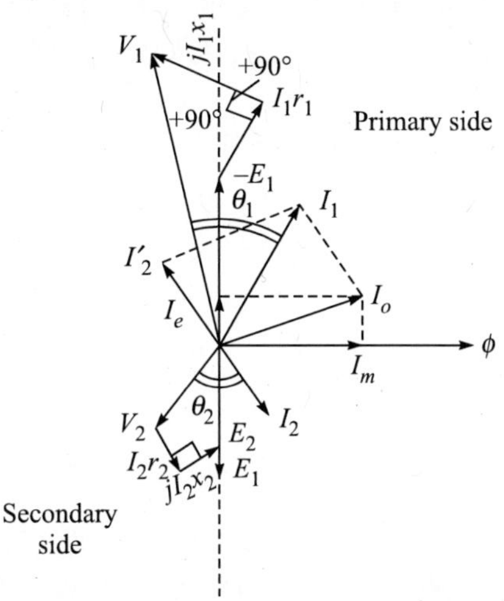

Fig. 5.17 Phasor diagram of real 1 ϕ transformer (for a capacitive load).

5.8 EQUIVALENT CIRCUIT OF A TRANSFORMER

The equivalent circuit of the transformer represents the electrical equivalent of the practical transformer. The equivalent circuit makes the analysis of the transformer easy. But mostly the approximate equivalent circuit is used for determining the performance parameters of the transformer. We can visualize the things which is occurring inside the transformer, also the effects on the electrical parameters with the loading of transformer.

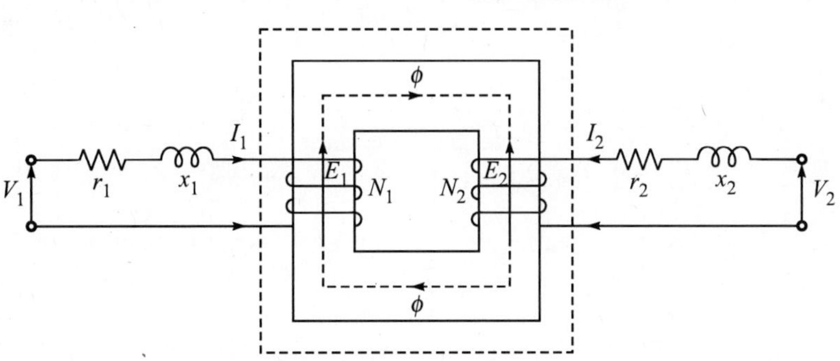

Fig. 5.18 Practical Transformer (1 ϕ).

where, V_1, V_2 = Primary and secondary terminal voltages, in volts.

I_1, I_2 = Primary and secondary currents, in amp.

r_1, r_2 = Primary and secondary resistances, in ohms

x_1, x_2 = Primary and secondary leakage reactances, in ohms.

ϕ = mutual flux, in webers.

N_1, N_2 = Primary and secondary number of turns.
E_1, E_2 = Induced e.m.f. in the primary and secondary in volts.

The leakage fluxes from core in the primary is shown by an inductive reactance (X_{L_1} or X_1) since the primary leakage flux 'ϕ_{l_1}' is proportional to the primary current 'I_1' i.e,

$$X_1 = 2\pi f\, L_1$$

Similarly, the secondary leakage flux 'ϕ_{l_2}' is shown by an inductive reactance X_{l_2} or X_2 which is dependent on the secondary winding inductive 'L_2'.

$$X_2 = 2\pi f L_2$$

The resistance and reactances of the windings are shown separately for better understanding the behaviour of the transformer windings.

The core of the transformer can be represented by a resistance 'R_0' and an inductance 'X_m'. The resistance 'R_0' represents the losses in the core which results in heating of core and 'X_m' represents an inductance which produces flux 'ϕ' in the core. Both are parallel to each other as the currents are not equal for the two effects i.e., heating and production of flux. So the primary and secondary circuit of the transformer can be represented by the equivalent circuit as shown below.

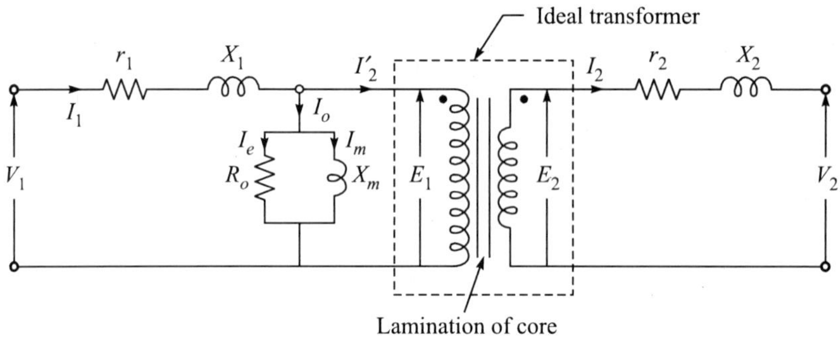

Fig. 5.19 Equivalent circuit of 1 ϕ transformer.

But for the analysis, it is often required to transfer or refer the quantities of the one side to the particular side. Consider the quantities of secondary referred to the primary side. This is possible only when the effects due to currents or voltages are producing the same effect as due to referred quantities.

The copper loss due to 'I_2' through 'R_2' produces the same copper loss due to 'I_2'' through 'R_2'' i.e.,

$$I_2^2 R_2 = I_2'^2 R_2' \tag{5.18}$$

or,

$$R_2' = \frac{I_2^2}{I_2'^2} \cdot R_2 = \left(\frac{I_2}{I_2'}\right)^2 R_2$$

Since $$I_1 = I_2' + I_0$$

'I_0' is about 2 – 5% of rated value of the transformer

So, $$I_0 << I_2'$$

or, $$I_1 \approx I_2' \tag{5.19}$$

Substituting the value of 'I_2'' in the exp. (5.18), we get,

$$R_2' \approx \left(\frac{I_2}{I_1}\right)^2 R_2$$

Since, $$\frac{I_2}{I_1} = \frac{1}{K}$$

so, $$\boxed{R_2' \approx \frac{R_2}{K^2}} \tag{5.20}$$

Similarly, we can determine the secondary leakage reactance referred to the primary side i.e., 'X_2', by considering the per unit reactive drop in both sides, as follows:

$$\frac{I_2' X_2'}{V_2'} = \frac{I_2 X_2}{V_2} \tag{5.21}$$

$$X_2' = \left(\frac{V_2'}{V_2}\right)\cdot\left(\frac{I_2}{I_2'}\right)X_2$$

$$V_2' \approx V_1 \; ; \; I_2' \approx I_1$$

$$X_2' \approx \left(\frac{V_1}{V_2}\right)\left(\frac{I_2}{I_1}\right)\cdot X_2 \qquad \left[\because \frac{V_1}{V_2} = \frac{1}{K} = \frac{I_2}{I_1}\right]$$

$$\approx \frac{1}{K}\cdot\frac{1}{K}\cdot X_2$$

$$\boxed{X_2' \approx \frac{X_2}{K^2}} \tag{5.22}$$

Similarly, we can find the value of 'V_2'' by considering the voltamperes delivered on both sides which should be equal.

$$V_2' I_2' = V_2 I_2 \tag{5.23}$$

or, $$V_2' = \left(\frac{I_2}{I_2'}\right)\cdot V_2$$

or, $$V_2' \approx \left(\frac{I_2}{I_1}\right)\cdot V_2$$

or, $$\boxed{V_2' \approx \frac{V_2}{K}} \tag{5.24}$$

Similarly, we can find the induced e.m.f. 'E_2'' in terms of 'E_2', we have

$$E_2' I_2' = E_2 I_2 \qquad \text{(VA Produced)} \tag{5.25}$$

$$E_2' = \left(\frac{I_2}{I_2'}\right) \cdot E_2$$

$$E_2' \approx \left(\frac{I_2}{I_1}\right) \cdot E_2 \qquad \left[\because \frac{I_2}{I_1} = K\right]$$

$$\boxed{E_2' \approx \frac{E_2}{K}} \tag{5.26}$$

Similarly, from the above exp.

$$I_2' = \left(\frac{V_2}{V_2'}\right) \cdot I_2 \tag{5.27}$$

$$\approx \frac{V_2}{V_1} \cdot I_2$$

$$\boxed{I_2' \approx K I_2} \qquad \left[\because \frac{V_2}{V_1} = \frac{N_2}{N_1} = K\right] \tag{5.28}$$

The equivalent circuit of the transformer is given by Fig. (5.20).

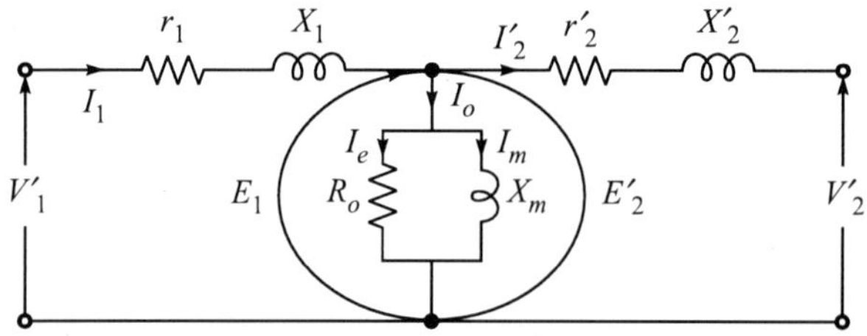

Fig. 5.20 Equivalent circuit (with referred values) to primary side of transformer.

Since 'I_0' is small as compared to 'I_2'', so by shifting the shunt branch to the supply end make not much error and the calculations become easy. But the equivalent circuit so obtained is the first approximation of the transformer equivalent circuit as shown in the Fig. (5.21).

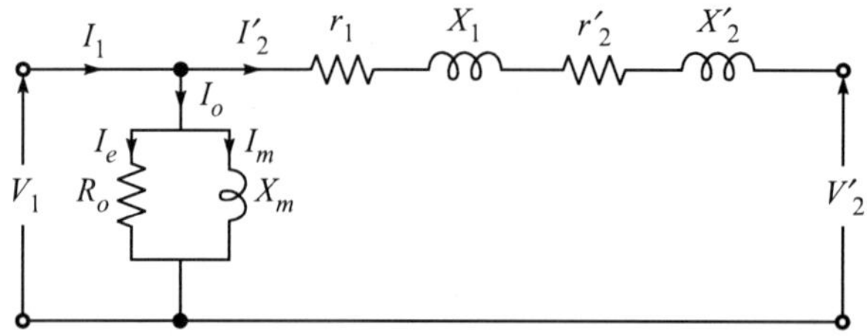

Fig. 5.21 First approximation of equivalent circuit of 1 ϕ transformer.

$$\text{Since, } r_1 + r_2' = R_{1eq} \quad \text{(Equivalent resistance referred to primary side)} \quad (5.29)$$

$$x_1 + x_2' = X_{1eq} \quad \text{(Equivalent reactance referred to primary side)} \quad (5.30)$$

So, the above equivalent circuit reduces to Fig. (5.22) as below.

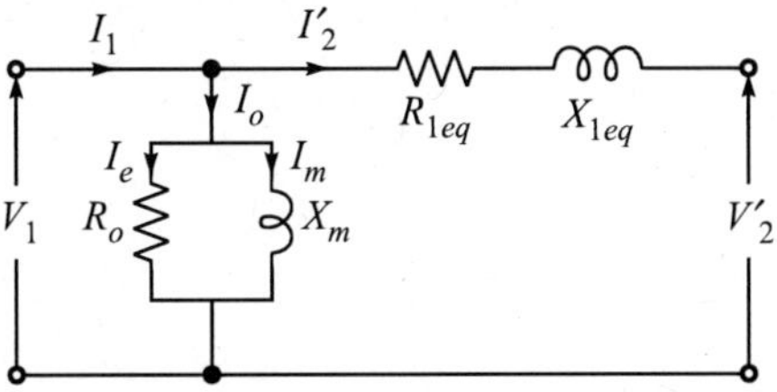

Fig. 5.22 Approximate equivalent circuit of 1 ϕ transformer.

Since 'I_0' is small, so for large transformers ~ 100 kVA the shunt branch is not considered. To avoid the analytical difficulties without much loss of accuracy.

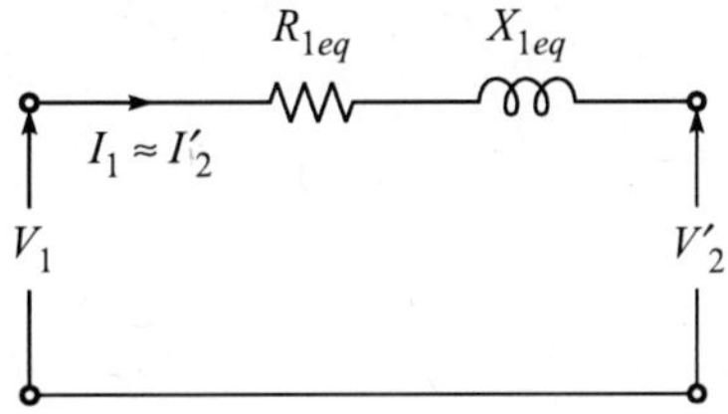

Fig. 5.23 Second approximation of equivalent circuit (of 1 ϕ transformers).

For very large transformers of the order of MVA or greater the resistance is also not considered only the inductive part is considered. Since the main motto is to realize the device practically and not the accuracy. So this leads to third approximation of the device equivalent circuit which is shown as in Fig. (5.24).

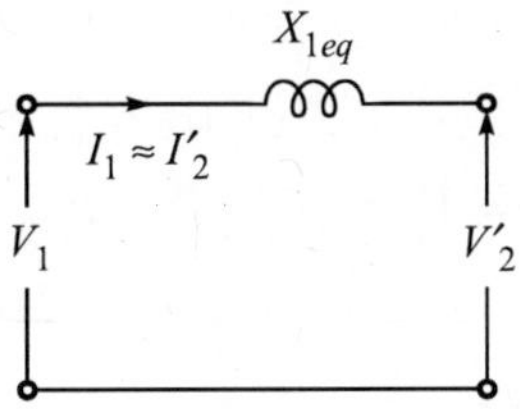

Fig. 5.24 Third approximate circuit of 1 ϕ transformer.

5.9 VOLTAGE REGULATION OF TRANSFORMER

The transformer performance is defined in terms of important performance parameter called voltage regulation. The voltage regulation is defined by the change in the transformer secondary terminal voltage 'V_2' when the load at a particular power factor is removed. Mathematically, it is expressed as follows:

$$\text{V.R.} = V_{20} - V_{2fl}, \text{ volts} \tag{5.31}$$

where, V.R. = Voltage regulation

If the value of V.R. is high, then the transformer is not good and vice versa.

The transformer's voltage regulation is normally expressed in terms of percentage, so it is known as percentage V.R., as given below.

$$\%\text{V.R.} = \frac{V_{20} - V_{2fl}}{V_{2fl}} \times 100\% \tag{5.32}$$

where, V_{20} = No load secondary voltage, in volts
V_{2fl} = Full load secondary voltage, in volts.

For different types of loads it has different values accordingly. The voltage regulation can have positive or negative values. For resistive and inductive loads it has positive values. But for the capacitive loads, it has negative value. It is preferred that the value of % V.R. as small as possible for better on load performance of the transformer. So care should be taken during design stage. The V.R. of the transformer can be easily analyzed using the phasor diagram corresponding to the approximate equivalent circuit of the transformer referred to load side.

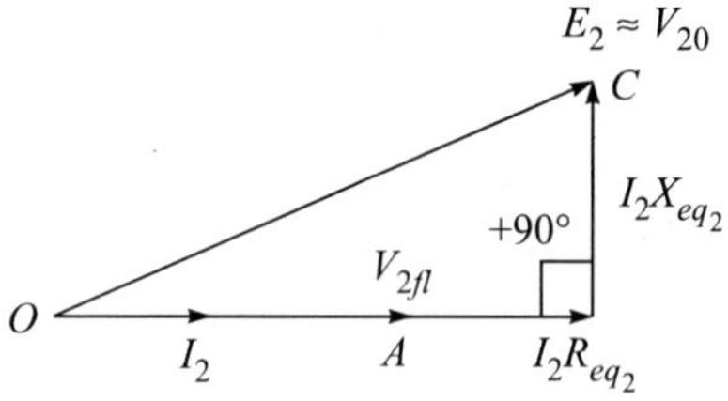

Fig. 5.25(a) Phasor diagram of secondary of 1 ϕ transformer (with resistive load).

From ΔOBC

$$OC^2 = OB^2 + BC^2$$

$$E_2^2 = (V_{2fl} + I_2R_{2\text{eq}})^2 + (I_2 X_{2\text{eq}})^2$$

$$V_{20} = E_2 = \sqrt{(V_{2fl} + I_2R_{1\text{eq}})^2 + (I_2X_{2\text{eq}})^2}$$

Using binomial expansion, the approximate value is given by

$$V_{20} \approx (V_{2fl} + I_2 R_{2\text{eq}}) + I_2 X_{2\text{eq}}$$

$$V_{20} - V_{2fl} \approx I_2 R_{2\text{eq}} + I_2 X_{2\text{eq}}$$

$$\%\text{V.R.} = \left(\frac{V_{20} - V_{2fl}}{V_{2fl}}\right) \times 100\% = \left(\frac{I_2 R_{2\text{eq}}}{V_{2\text{eq}}} + \frac{I_2 X_{2\text{eq}}}{V_{2\text{eq}}}\right) \times 100\%$$

$$\approx r_{\text{p.u}} + x_{\text{p.u}} \tag{5.33}$$

From ΔODC

We have,

$$OC^2 = OD^2 + DC^2$$
$$E_2^2 = (OE + ED)^2 + (DB + BC)^2$$
$$E_2 = V_{20} = \sqrt{(V_{2fl}\cos\theta_2 + I_2 R_{2\text{eq}})^2 + (V_{2fl}\sin\theta_2 + I_2 X_{2\text{eq}})^2}$$

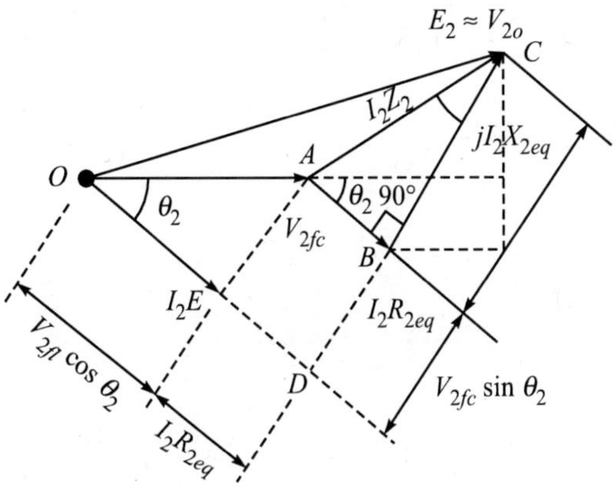

Fig. 5.25(b) Phasor diagram (RL load).

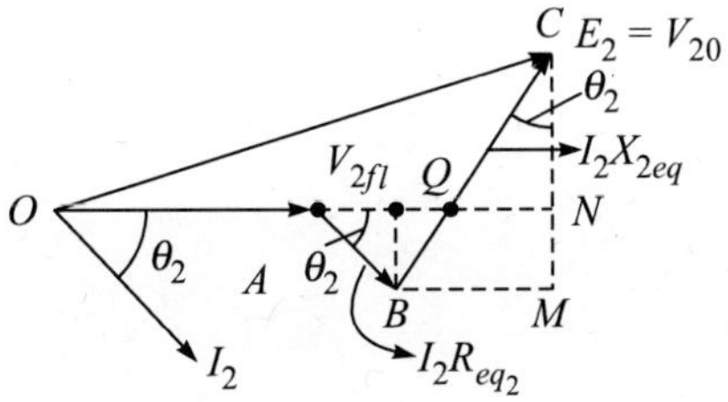

Fig. 5.25(c) Phasor diagram (RL load).

From ΔOCN

$$OC^2 = ON^2 + NC^2\,;\ QN = BM = I_2 X_{\text{eq}2}\sin\theta_2 \ \&\ QB = NM = I_2 R_{\text{eq}2}\sin\theta_2$$
$$E_2^2 = (OA + AQ + QN)^2 + (CM - NM)^2$$
$$= (V_{2fl} + I_2 R_{\text{eq}2}\cos\theta_2 + I_2 X_{\text{eq}2}\sin\theta_2)^2 + (I_2 X_{\text{eq}2}\cos\theta_2 - I_2 R_{\text{eq}2}\sin\theta_2)^2$$
$$E_2 = [(V_{2fl} + I_2 R_{\text{eq}2}\cos\theta_2 + I_2 X_{\text{eq}2}\sin\theta_2)^2 + (I_2 X_{\text{eq}2}\cos\theta_2 - I_2 R_{\text{eq}2}\sin\theta_2)^2]^{1/2}$$

So approximate value is given by

$$E_2 = V_{20} \approx V_{2fl} + I_2 R_{\text{eq}2}\cos\theta_2 + I_2 X_{\text{eq}2}\sin\theta_2$$

So, $V_{20} - V_{2fl} = I_2 R_{\text{eq}2}\cos\theta_2 + I_2 X_{\text{eq}2}\sin\theta_2$

So, $$\%\text{V.R.} = \frac{V_{20} - V_{2fl}}{V_{2fl}} \times 100\% = \frac{I_2 R_{eq2} \cos\theta_2 + I_2 X_{eq2} \sin\theta_2}{V_{2fl}} \times 100\%$$

$$= \left(\frac{I_2 R_{eq2}}{V_{2fl}} \times 100\%\right)\cos\theta_2 + \left(\frac{I_2 X_{eq2}}{V_{2fl}} \times 100\%\right)\sin\theta_2$$

$$\%\text{ V.R} = r_{p.u} \cos\theta_2 + x_{p.u} \cdot \sin\theta_2 \tag{5.34}$$

where, 'θ_2, is the load phase angle and '$\cos\theta_2$' is the load power factor; 'I_2' is the rated secondary current.

The $r_{p.u}$ and $x_{p.u}$ represents the per unit values of the resistance and reactance of transformer. In the above expressions, 'R_{eq2}' and 'X_{eq2}' are the equivalent resistance and reactance referred to the secondary side of transformer and is expressed as

$$R_{eq_2} = r_2 + r_1'$$

$$R_{eq_2} = r_2 + K^2 r_1;\ K = \frac{N_2}{N_1}$$

and

$$X_{eq_2} = x_2 + x_1'$$

$$X_{eq_2} = x_2 + K^2 x_1$$

The voltage regulation of the transformer for capacitive load can be determined using the phasor diagrams as shown in Fig. (5.26).

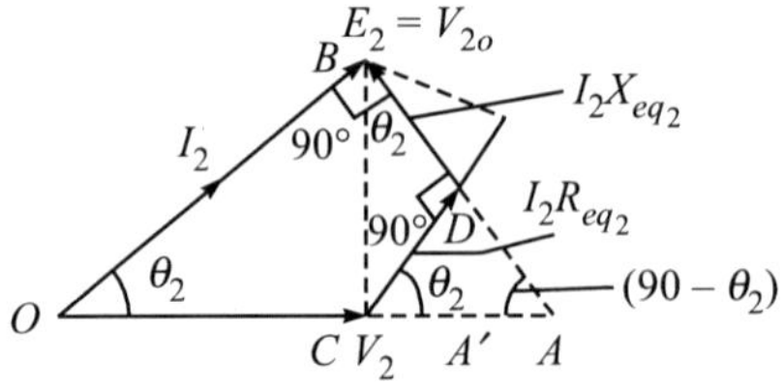

Fig. 5.26 Phasor diagram ($R - C$ laod).

On simplification, we get

The $$\%\text{V.R} = \frac{I_2 R_{eq2} \cos\theta_2 - I_2 X_{eq2} \sin\theta_2}{V_{2fl}} \times 100\%$$

In terms of per unit values, we get

$$\%\text{ V.R} = r_{p.u} \cos\theta_2 - X_{p.u} \sin\theta_2 \tag{5.35}$$

The % V.R. with capacitive load is always negative.

Zero Voltage Regulation

In the case of resistive and inductive loads the V.R can't be zero. So the zero voltage regulation is possible with capacitive loads only as it has a expression of difference of two quantities in the R.H.S. of eq. (5.35).

That is,

$$\%\text{V.R} = 0 = r_{pu} \cos\theta_2 - x_{pu} \sin\theta_2$$

so, $$x_{pu} \sin\theta_2 = r_{pu} \cos\theta_2$$

or, $$\frac{\sin\theta_2}{\cos\theta_2} = \frac{r_{pu}}{x_{pu}}$$

or, $$\tan\theta_2 = \frac{r_{pu}}{x_{pu}}$$

At $$\boxed{\theta_2 = \tan^{-1}\frac{r_{pu}}{x_{pu}}} \quad \text{for zero V.R.} \tag{5.36}$$

Zero V.R means no voltage drop with loading or unloading.

Maximum Voltage Regulation

The voltage regulation value will be higher in case of inductive loads since the voltage regulation value is the sum of two quantities

$$\%\text{V.R.} = r_{pu} \cos\theta_2 + x_{pu} \sin\theta_2$$

For maximum V.R.,

$$\frac{\partial(\%\text{V.R.})}{\partial\theta_2} = 0$$

$$r_{pu}(-\sin\theta_2) + x_{pu}(\cos\theta_2) = 0$$

or, $$x_{pu} \cos\theta_2 = r_{pu} \sin\theta_2$$

or, $$\tan\theta_2 = \frac{x_{pu}}{r_{pu}}$$

or, $$\boxed{\theta_2 = \tan^{-1}\frac{x_{pu}}{r_{pu}}} \tag{5.37}$$

That is, if 'θ_2' is equal to that given by eq. (5.37), the regulation of the transformer is poorest.

5.10 EFFICIENCY OF TRANSFORMER

The efficiency of the transformer is given by the ratio of output to input as given by the ratio of output to input as given by the expression (5.38)

At full load,

$$\eta = \frac{\text{Output power}}{\text{Input power}} \tag{5.38}$$

$$= \frac{\text{Output}}{\text{Output + losses}}$$

$$= \frac{(\text{Volt ampere})_{FL} \times \text{Power factor at full load}}{(\text{VA})_{FL} \times (\text{Power factor})_{FL} + P_i + (P_{cu})_{FL}}$$

$$\eta_{FL} = \frac{S_2 \cos\theta_2}{S_2 \cos\theta_2 + P_i + (P_{cu})_{FL}} \tag{5.39}$$

where, $S_2 = V_2 I_2$ is known as kVA rating of the transformer.

Generally, $V_2 I_2 = V_1 I_1$ (is considered) 'η_{FL}' is known as full load efficiency, which is determined at full load current or rated current value of the transformer.

At a current value less than rated value or full load value. Say $x\%$ of full load current the value of efficiency differs. Since the copper loss value changes with value of load current so it is called variable loss. The iron loss does not depend on the load current value so it is termed constant loss. The other losses are very negligible and not considered in the efficiency expression. At part load or fraction of full load say $x\%$. The efficiency is expressed as

$$\eta\Big]_{x\% \text{ of } FL} = \frac{x\, S \cos\theta_2}{x\, S \cos\theta_2 + P_i + x^2 (P_{cu})_{FL}} \tag{5.40}$$

The current at part load i.e., at $x\%$ (of I_{FL}) is given by

$$I_2 = x I_{2Fl} = x\, I_2 \text{ (rated)}$$

At half load, $x = 0.5$, so, $I_2 = I_2$ (rated)/2

At quarter load, $x = 0.25$, so, $I_2 = I_2$ (rated)/4

At full load, $x = 1.0$, so, $I_2 = I_2$ (rated)

The efficiency curve is given by

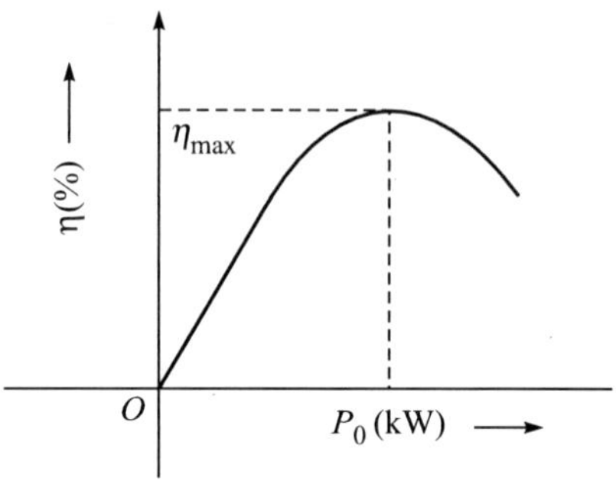

Fig. 5.27 Efficiency curve of 1 ϕ transformer.

So, we need in some places to determine the maximum efficiency of the transformer which is obtained by taking the first derivative of the exp. (5.41) with respect to current 'I_2' as follows:

$$\eta = \frac{V_2 I_2 \cos\theta_2}{V_2 I_2 \cos\theta_2 + P_i + I_2^2 R_{2eq}} \tag{5.41}$$

Put, $\frac{d\eta}{dI_2} = 0$, we get $\boxed{P_i = I_2^2 R_{2eq} = P_{cu}}$ (5.42)

The condition of maximum efficiency is that

$$\text{Iron loss} = \text{copper loss}$$

i.e.,

$$P_i = P_{cu}$$

At FL

$$I_2^2 \, R_{2\text{eq}} = P_i$$

$$\left. I_2 \right]_{\eta_{\max}} = \sqrt{\frac{P_i}{R_{\text{eq}2}}} \tag{5.43}$$

At part load 'x'

$$x^2 \, I_2^2 \, R_{2\text{eq}} = P_i = x^2 P_{cu}$$

$$x = \sqrt{\frac{P_i}{P_{pu}}} \tag{5.44}$$

∴

$$(\text{kVA}),_{\eta_{\max}} = x \, S \cos \theta_2 \tag{5.45}$$

∴

$$\left. x \right]_{\eta_{\max}} = \frac{(\text{kVA})\eta_{\max}}{(\text{kVA})_{\text{rated}}}; \text{ if } \cos \theta_2 = 1$$

or,

$$\left. x \right]_{\eta_{\max}} = \frac{\left. S \right]_{\eta_{mas}}}{S_{\text{rated}}} \tag{5.46}$$

The power transformers are designed at '$\eta_{\max}$' maximum efficiency at rated values of current of transformer. Since a power transformer has almost constant load at its secondary.

All Day Efficiency

The distribution transformer does not have constant load but it varies during the 24 hour period of day. So the efficiency of these transformers is designed for a high value at 0.75 I_{rated} value. The all day efficiency is expressed as follows

$$\eta_{\text{All day}} = \frac{(\text{kW hr})_{\text{output}}}{(\text{kW hr})_{\text{Input}}}$$

$$= \frac{(\text{kW hr})_{\text{o/p}}}{(\text{kW hr})_{\text{o/p}} \times \text{losses over 24 hrs}} \tag{5.47}$$

The input and output are determined over 24 hour period.

5.11 TRANSFORMER TESTING

The following tests are performed on a transformer

(a) Polarity test
(b) Open circuit test and short circuit test (OCT and SCT)
(c) Sumpner's test (back to back test)

Polarity Test

This test is generally performed on the transformer to determine the polarities of the terminals of transformer. This helps in connecting two transformers in parallel so that load can be shared between the two. Connecting two transformer terminals wrongly will cause maloperation and damage of either of transformer.

For the polarity test the transformers are connected in the fashion as shown.

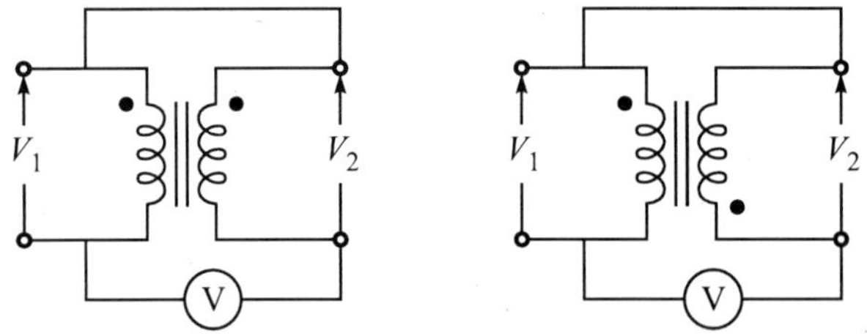

(a) Subtractive polarity (b) Additive polarity

Fig. 5.28 Polarity test on 1 ϕ transformer.

In the first case Fig. (a) the voltmeter 'V' will show the difference of 'V_1' and 'V_2' i.e., $V \approx V_1 \sim V_2$ whereas, in Fig. (b), the voltmeter will show sum of 'V' and 'V_2' as follows.

$$V \approx V_1 + V_2$$

So by using dot convention the higher potential at any particular instant can be measured.

Open Circuit Test or No Load Test (OCT)

This test is performed by keeping the secondary of the transformer open circuited as shown in Fig. There will be no drop in the secondary side as secondary current $I_2 = 0$. So $I_2' = 0$. Since the current 'I_1' of primary corresponds to the no load current 'I_0' so the current drawn during OCT is small about 2 – 5% of rated or full load current value. The voltage at which this test is conducted is the rated voltage at primary.

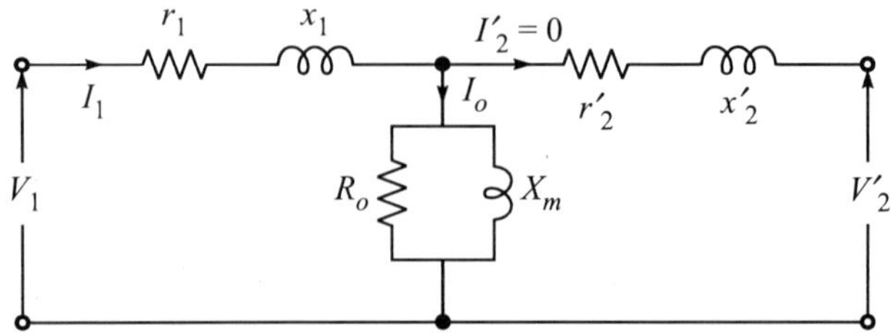

Fig. 5.29 Open circuit equivalent of 1 ϕ transformer.

If a wattmeter is attached in the circuit, then the measured power corresponds to the losses in the core of transformer if the copper loss of the primary is not considered.

The *OCT* is conducted connecting the measuring instruments i.e., ammeter, voltmeter and wattmeter as follows

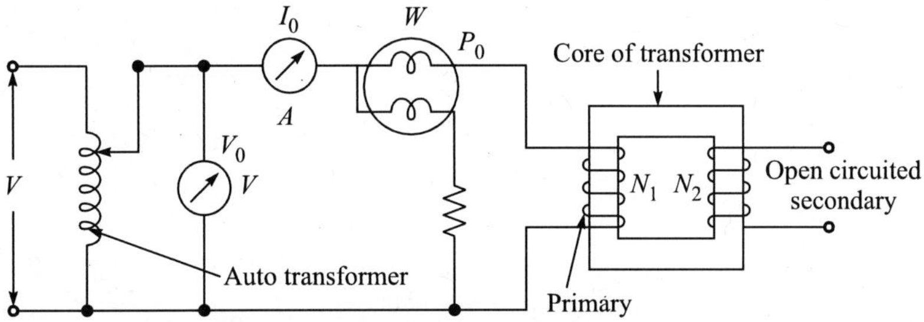

Fig. 5.30 OCT on 1ϕ transformer.

The readings of the *OCT* are

$$P_0 = V_0 I_0 \cos \theta_0 = P_{iron} \tag{5.48}$$

$$I_e = I_0 \cos \theta_0$$

$$I_m = I_0 \sin \theta_0$$

$$R_0 = \frac{V_0}{I_e}; \qquad X_m = \frac{V_0}{I_m} \tag{5.49}$$

So, from this test the shunt branch parameters 'R_0' and 'X_m' are determined. Moreover, the iron or constant losses can also be determined by this method.

Short Circuit Test

The secondary of the transformer if short circuited with full load current passed through the transformer s.c. winding then the equivalent circuit of transformer is given by

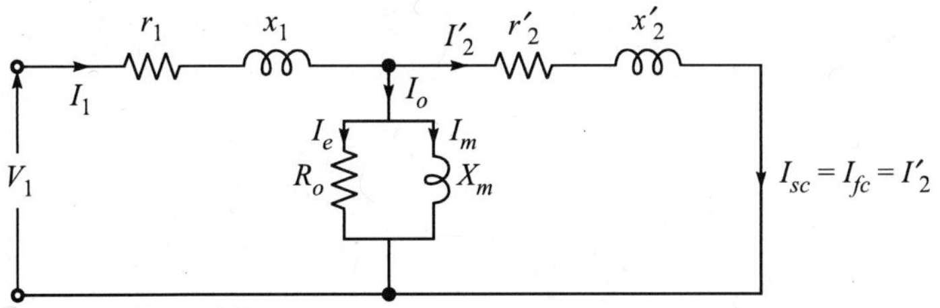

Fig. 5.31 Short circuit equivalent of 1 ϕ transformer.

Since the load primary current 'I_0' is very small so the shunt branch can be regarded as open circuit. With such approximations, the equivalent circuit under short circuit conditions can be replaced by the equivalent as shown in Fig. (5.32).

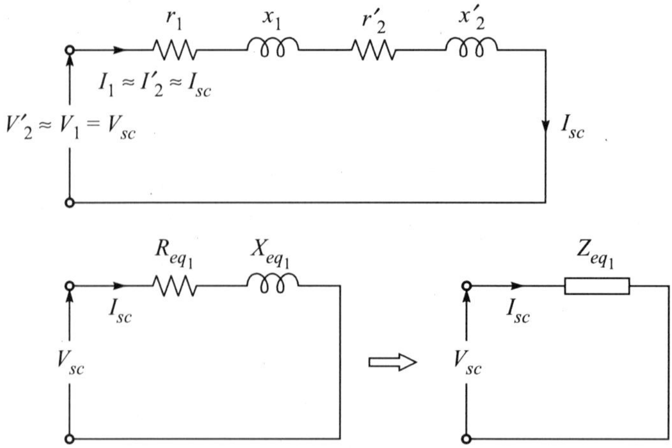

Fig. 5.32 Short circuit equivalent.

So, the readings of the meters connected for the short circuit test is given as 'P_{sc}', 'V_{sc}' and 'I_{sc}' respectively.

The connections of the meters are as shown in the Fig. (5.33) below.

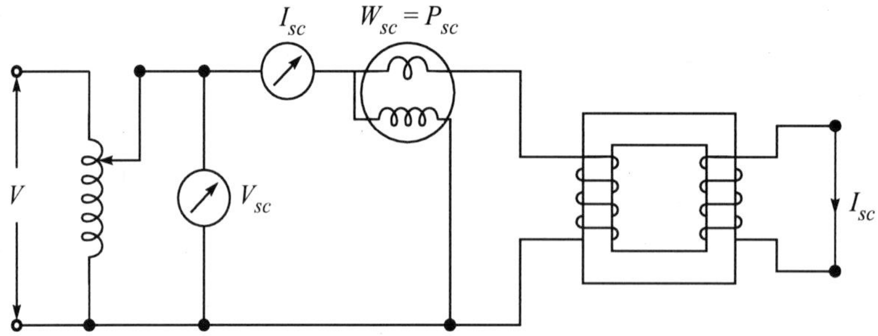

Fig. 5.33 SCT on 1 ϕ transformer.

From the Fig. (5.32), we have

$$Z_{eq_2} = \frac{V_{sc}}{I_{sc}} = Z_{eq1} \cdot K^2 \; ; \; K = \frac{N_2}{N_1}$$

$$P_{sc} = I_{sc}^2 \, R_{2eq} = P_{cu} \tag{5.50}$$

$$R_{2eq} = \frac{P_{sc}}{I_{sc}^2} \tag{5.51}$$

The equivalent reactance $$X_{2eq} = \sqrt{Z_{2eq}^2 - R_{2eq}^2} \tag{5.52}$$

So by SCT, the series branch parameters are determined. Moreover, the copper loss in the windings can also be determined under full load conditions or part load condition.

Sumpner's Test (Back to Back Test)

The back to back test is performed on two identical transformers. The iron and copper losses can be determined simultaneously without loading the transformers at the secondary. The connections of the test is shown below in Fig. (5.34).

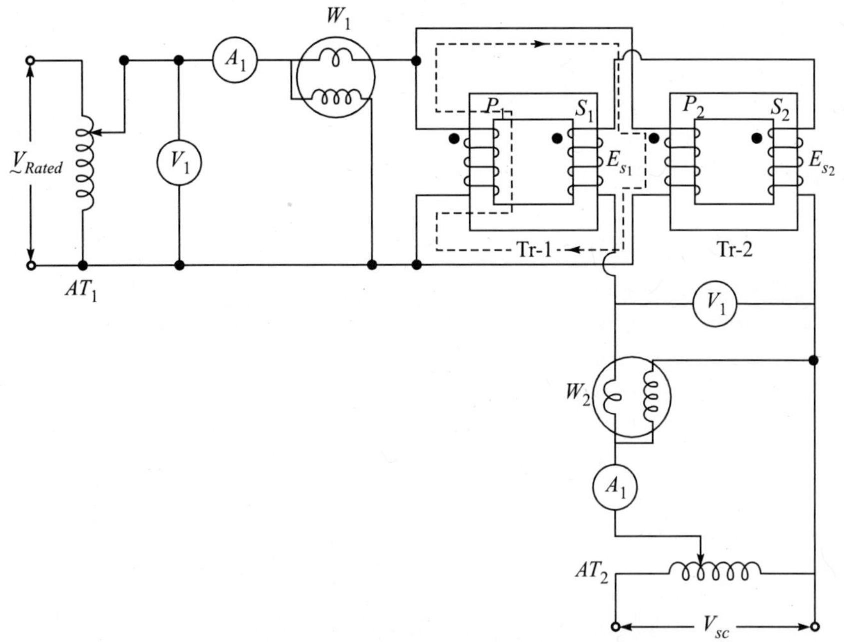

Fig. 5.34 Sumpner's test on two 1 ϕ transformer.

The primaries of two transformers are connected in parallel across the excitation voltage which is equal to the rated voltage of primary of transformer. The secondaries are connected in series and in phase opposition such that the voltage across the two windings is equal to zero i.e., $V_2 = E_{S_1} \sim E_{S_2} = 0$

With the full load current drawn by the secondary winding using auto transformer 'AT_2'. The wattmeter 'W_2' corresponds to the copper losses. There is a regeneration of current in primary windings as shown by the dotted lines. The wattmeter reading 'W_1' corresponds to the iron losses incurred in the transformers with no current in the secondaries i.e., open circuit condition. The wattmeter readings corresponds to twice the loss incurred in each transformer.

If 'P_i' and 'P_{cu}' are the iron and copper losses of each transformer, then the wattmeter reading is given by

$$W_1 = 2P_i$$
$$W_2 = 2P_{cu}$$

$$\therefore \quad \frac{W_1 + W_2}{2} = P_i + P_{cu} = P_{\text{total loss}} \tag{5.53}$$

So the efficiency of the transformer is determined using the relation

$$\eta = \frac{S\cos\theta_2}{S\cos\theta_2 + P_i + P_{cu}}$$

$$\eta = \frac{S\cos\theta_2}{S\cos\theta_2 + \left(\dfrac{W_1 + W_2}{2}\right)} \tag{5.54}$$

where, S = kVA rating of transformer

$\cos\theta_2$ = Power factor of load.

This test is also called regenerative test and can be used to determine the heat rise of transformer due to losses.

5.12 AUTO TRANSFORMERS

It is often required to regulate the input a.c. voltage at fixed frequency. So for that some regulating device is necessary. This is employed by the auto transformer which regulate the a.c. voltage continuously from zero volts to the maximum voltage required by the equipment. So the need of steps on the secondary winding of two winding transformers is eliminated. The auto transformers have high efficiency as compared to the two winding transformers.

The auto transformers are available for both single-phase and three-phase applications. The single phase auto transformer is employed for low power electrical apparatus and the three-phase auto transformer used for high power applications such as induction motor starting, interconnection of high power lines, etc. The single-phase auto transformers are only considered in this chapter. The auto transformers are available in two types, namely, step up auto transformer and step down auto transformers. The auto transformer used for single phase applications have single winding with part of winding connected to the input supply, it becomes primary winding and part of winding is connected to the load and is known as secondary winding.

Figure 5.35 depicts the step down auto transformer.

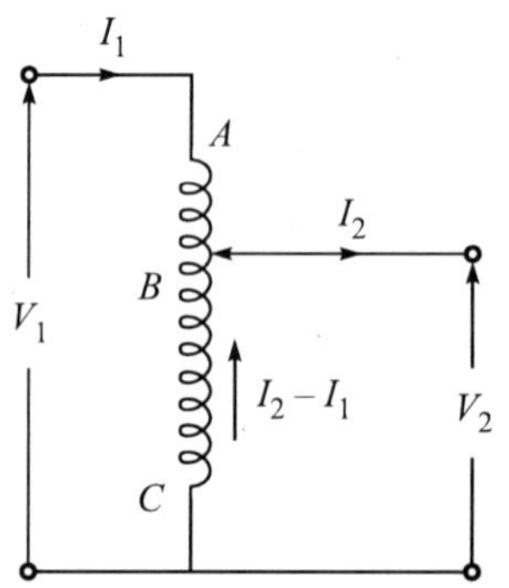

Fig. 5.35 Step down auto transformer (single phase).

The part 'AB' of the winding is insulated heavily due to the stress over the insulation of winding due to high voltage of primary. The power from the primary to the secondary side is due to conductive coupling of primary and secondary (AB) and partly by inductive coupling of primary and secondary (BC). The winding is on a core of toroidal form.

The volt ampere of the auto transformer is given by

$$\begin{aligned} V_2 I_2 &= V_2 I_2 + V_2 I_1 - V_2 I_1 \\ &= \underset{\text{(I)}}{V_2(I_2 - I_1)} + \underset{\text{(II)}}{V_2 I_1} \end{aligned} \tag{5.55}$$

The (I) term implies the power due to magnetic induction or transformer action and the second term (II) corresponds to the conduction of 'I_1' from primary to secondary. If we consider the two winding equivalent transformers corresponding to the auto transformer then we have to see the advantages offered by the auto transformer. Firstly, we consider the copper required for windings in both types of same ratings.

$$\begin{aligned} \frac{\text{Conductor in auto transformer material}}{\text{Conductor in two winding transformer material}} &= \frac{I_1(N_1 - N_2) + (I_2 - I_1)N_2}{I_1 N_1 + I_2 N_2} \\ &= \frac{2(I_2 - I_1)N_2}{2 I_2 N_2} = 1 - \frac{I_1}{I_2} \\ &= 1 - \frac{V_2}{V_1} = \frac{a-1}{a} = 1 - K \end{aligned} \tag{5.56}$$

where, $$a = \frac{N_1}{N_2} = \frac{1}{K},\ N_1 > N_2$$

we see that as $V_2 \to V_1$, the copper required for auto transformer is less. So the auto transformer is preferred for the voltage ratio 2:1 respectively About 50% saving will be there and cost will reduce by 30% approximately.

The kVA rating of equivalent auto transformer is given by

$$\begin{aligned} (\text{kVA})_{\text{two winding tr.}} &= V_1 I_1 = V_2 I_2 \\ (\text{kVA})_{\text{auto tr.}} &= (V_1 + V_2)I_1 \\ &= V_1 I_1 + V_2 I_1 \\ \frac{(\text{kVA})_{\text{auto}}}{(\text{kVA})_{\text{Tw}}} &= \frac{(V_1 + V_2)I_1}{V_1 I_1} \\ &= 1 + \frac{V_2}{V_1} = 1 + K \end{aligned} \tag{5.57}$$

$$\frac{(\text{KA})_{\text{auto tr.}}}{(\text{VA})_{\text{Tw}}} = 1 + K = 1 + \frac{1}{a} = \frac{a+1}{a}$$

Since $a = \dfrac{N_1}{N_2}$ and for step down auto transformer

$$N_1 > N_2$$

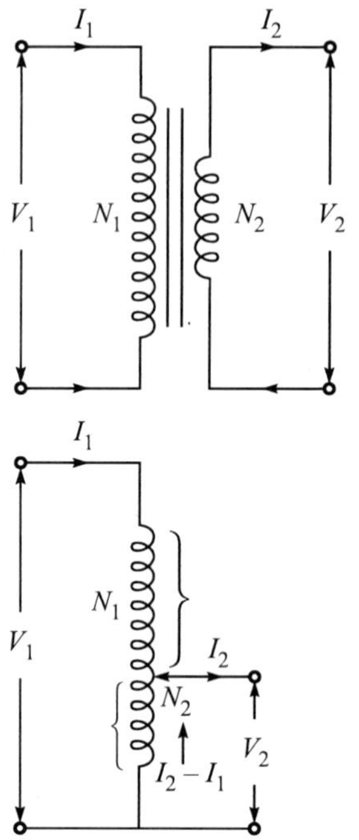

Fig. 5.36 Equivalent auto transformer of two winding transformer.

For 2:1, auto transformer

$$\frac{V_1}{V_2} = \frac{2}{1} = \frac{N_1}{N_2} = a$$

of $\quad a = 2$, substituting this value in exp.

we get $$\frac{(\text{VA})_{\text{auto transr}}}{(\text{VA})_{\text{Two winding tr.}}} = 1 + 2 = 3$$

$$\therefore \quad (\text{VA})_{\text{auto transformer}} = 3 \times (\text{VA})_{\text{Two winding tr.}}$$

Therefore, if a two winding transformer is made to work as auto transformer then the rating of auto transformer will increase.

5.13 THREE-PHASE TRANSFORMER

Three-phase transformers are normally employed for high power applications in the power industries, heavy industries, mining, etc. The three-phase transformers have various advantages over the single phase transformers. It is more compact and comparatively cheap with respect to the total cost of bank of three single phase transformers. The coverage area is also small with respect to bank of three transformers. The primary and secondary of transformers can be connected in star or delta depending on the requirement. The three-phase connections are available as *Y/Y*, Δ/Δ, Δ/*Y*, *Y*/Δ *V/V*, zig-zag connections respectively. Every connection has its own merits and demerits. The *Y/Y* connection is used at the utility end of the power system network. It is advantageous having a neutral point so 3 phase, 4 wire system is possible.

The connections of the Y/Y connection is shown in the Fig. (5.37).

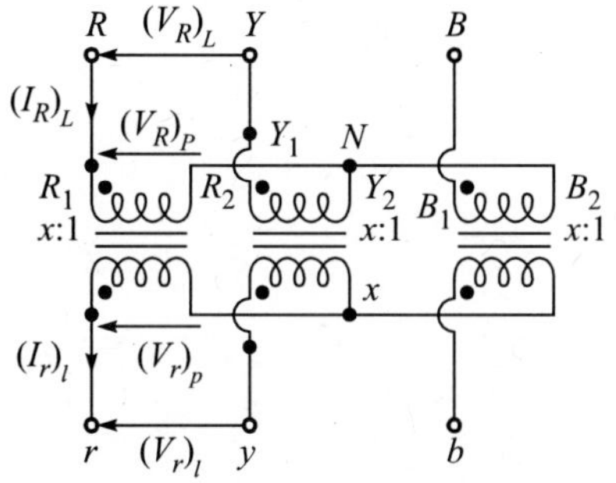

Fig. 5.37 Y_{y0} connection of 3 ϕ transformer.

In the Fig. (5.37)

$(V_R)_L$ = Line voltage of primary, *V*
$(I_R)_L$ = Line current in primary, *A*
$(V_r)_P$ = Phase voltage of secondary, *V*
$(I_r)_l$ = Line current in secondary, *A*
N, *n* = Neutral point of primary and secondary

Since the connections of primary and secondary are in *Y* (star). So the line and phase quantities are related by the relation.

$$V_L = \sqrt{3}\, V_P$$

and

$$I_L = I_P$$

So, the phase voltages at the primary of transformers is given by

$$(V_R)_P = \frac{(V_R)_L}{\sqrt{3}}$$

$$(I_R)_P = (I_R)_L$$

where, subscript '*P*' and '*L*' represents phase and line respectively.

Similarly, the phase voltage and phase current at the secondary side is given by

$$(V_r)_P = \frac{(V_r)_L}{\sqrt{3}}$$

$$(I_r)_P = (I_r)_l$$

The turns ratio $\quad$ 'a' $= \dfrac{N_1}{N_2}$ is given as x:1

So, using the relation

$$\frac{V_1}{V_2} = a = \frac{I_2}{I_1}$$

$$\frac{(V_R)_P}{(V_r)_p} = \frac{x}{1};\ \frac{(I_r)_P}{(I_R)_p} = \frac{x}{1} \tag{5.58}$$

So, the phase transformations ratio P.T.R for Y/Y connected is x:1
The ratio of line voltages are given by

$$\frac{(V_R)_L}{(V_r)_l} = \frac{x}{1} = \frac{\sqrt{3}\,(V_R)_P}{\sqrt{3}\,(V_r)_P} \tag{5.59}$$

Similarly,
$$\frac{(I_r)_l}{(I_R)_L} = \frac{x}{1} \tag{5.60}$$

So, the line transformation ratio is equal to phase transformation ratio. The phasor diagram to primary and secondary induced e.m.f.'s is given by

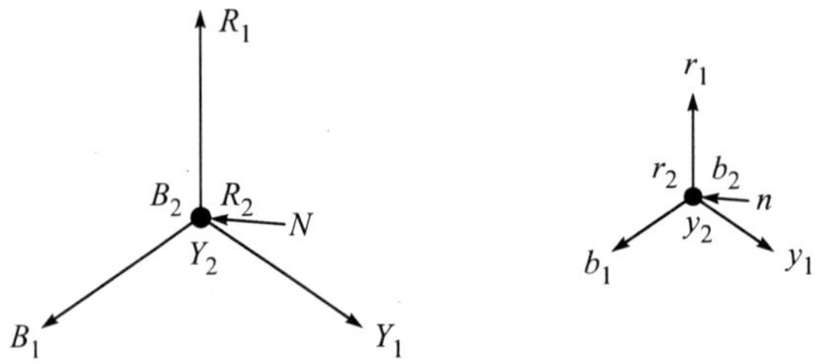

Fig. 5.38 Phasor diagram of Y_{y0} connection of 3 ϕ transformer.

In this connection, the phase difference between the two side electrical quatities such as voltages and currents are equal to zero. By changing secondary connections the Y_{y_6} connection or 180° connections is possible.

Δ/Δ Connection: In this connection the primary and secondary terminals of the transformer are connected in delta. This type of connection is preferred for power transformers which are used in transmission purpose. The connections are shown in Fig. (5.39).

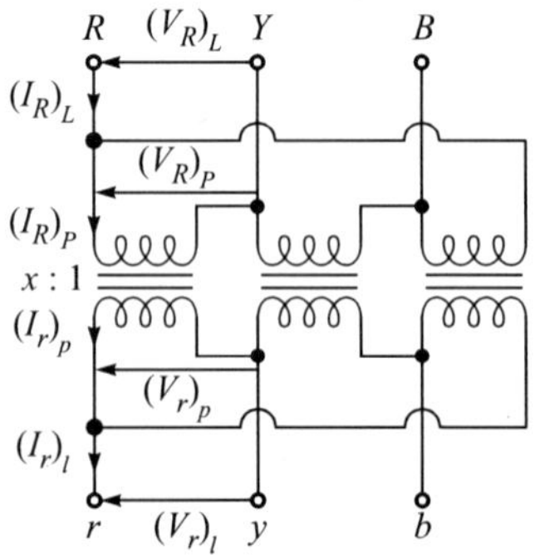

Fig. 5.39 Δ/Δ connections of 3 ϕ transformer.

For the delta connected system.

$$V_L = V_P$$
$$I_L = \sqrt{3} I_P$$

Suppose, the turn ratio of transformer each phase is 'x'. Then, the line transformation ratio and phase transformation ratio are given by

$$\frac{(V_R)_P}{(V_r)_p} = \frac{x}{1} = \frac{(V_R)_L}{(V_r)_l} = \text{P.T.R.} = \text{L.T.R} \qquad (5.61)$$

In the delta connected system there is a problem of cyclic current which flows in the delta connected system.

The phasor diagram of the two sides is shown is Fig. (5.40).

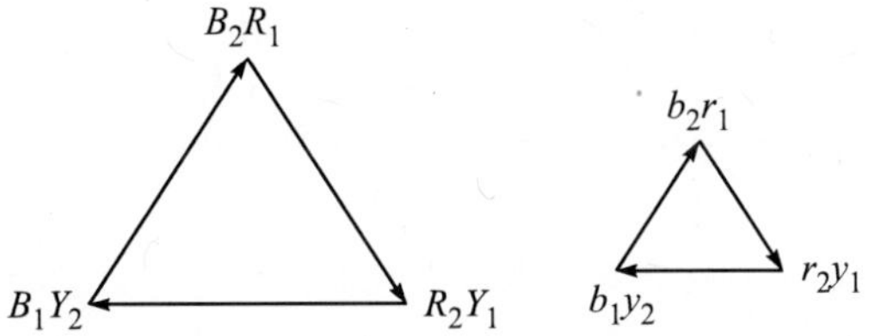

Fig. 5.40 Phasor diagram of Δ/Δ connected system. By changing the connections of secondary, we will have Dd6 or 180° connections.

Δ/*Y* connections

This type of connections are used at the distribution substation for stepping down voltage from a very high level (220 kV, 400 kV) to low voltage level of the order of 1100 V, 400 V or lesser. This connection suffers from the demerits of having a phase difference of 30° between the primary and secondary side quantities. So this will lead to unbalancing of phases and hence incursions of harmonics. The connection diagram of Δ/*Y* connection is shown below.

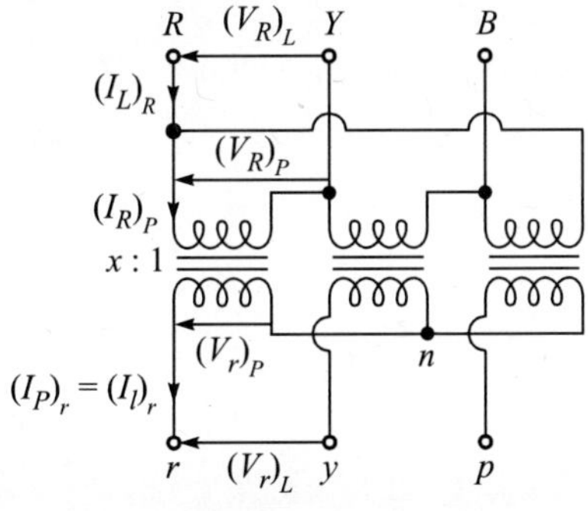

Fig. 5.41 Δ/Y connections of 3 φ transformer.

The phase transformation ratio (PTR) of Δ/Y connected transformer is given as follows:

$$(V_R)_P = V_{RL} \ (\because \ \Delta \text{ connected})$$

$$(I_R)_P = \frac{(I_R)_L}{\sqrt{3}} \ (\because \ \Delta \text{ connected})$$

$$(V_r)_P = (V_r)_L/\sqrt{3} \ (\because \ Y \text{ connected})$$

$$\frac{(V_R)_P}{(V_r)_p} = \frac{x}{1} = \frac{(V_R)_L}{(V_R)_L\sqrt{3}}$$

So, the P.T.R of transformer $= x:1$

and L.T.R of transformer $= x:\sqrt{3} = 1:\dfrac{\sqrt{3}}{x}$ (5.62)

There will be 30° phase shifting between the secondary and primary quantities respectively.

The Δ/Y connection is preferred at the primary distribution substations transformers.

Y/Δ Connections

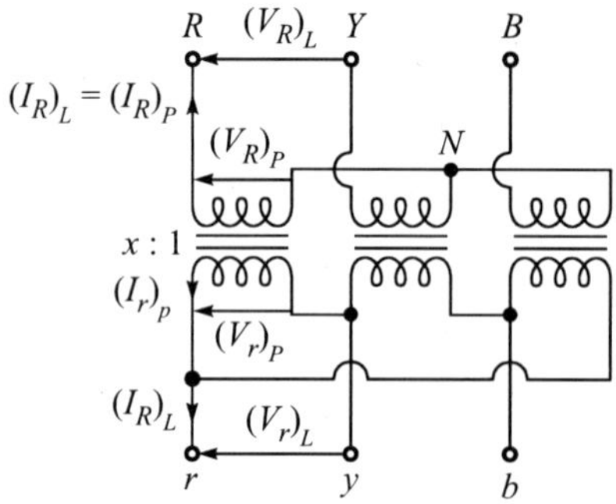

Fig. 5.42 *Y*/Δ connection of 3 ϕ transformer.

The relations between the phase and line quantities of the primary and secondary sides are

$$\left.\begin{aligned}(V_R)_L &= \sqrt{3}(V_R)_P\\ (I_R)_L &= (I_R)_P\end{aligned}\right\} \text{ since 'Y' connected}$$

$$\left.\begin{aligned}(V_r)_L &= (V_r)_P\\ (I_r)_L &= \sqrt{3}(I_r)_P\end{aligned}\right\} \text{ since '}\Delta\text{' connected}$$

$$\frac{(V_R)_P}{(V_r)_P} = \frac{x}{1} = \frac{(V_R)_L/\sqrt{3}}{(V_r)_L}$$

So, $$\text{P.T.R} = x:1$$

$$\text{L.T.R} = x:\sqrt{3} = \frac{x}{\sqrt{3}}:1 \qquad (5.63)$$

The Y/Δ connected transformer connections are done for the transformer (3 ϕ) connected to the alternators phase conductors at generating side.

In the Y/Δ connection there is a 30° phase shift between the primary and secondary side quantities (voltages and currents). The secondary sides connections could be changed so the phase angles could be changed accordingly.

But due to this phase shift when two or three phase transformers are needed to satisfy some conditions such as ratings, phase sequence, phase shift, polarity, etc. So in that case the phase shift between the two transformers should be equal.

In such cases, the zigzag connection is used in place of star connection as shown in Fig. (5.43).

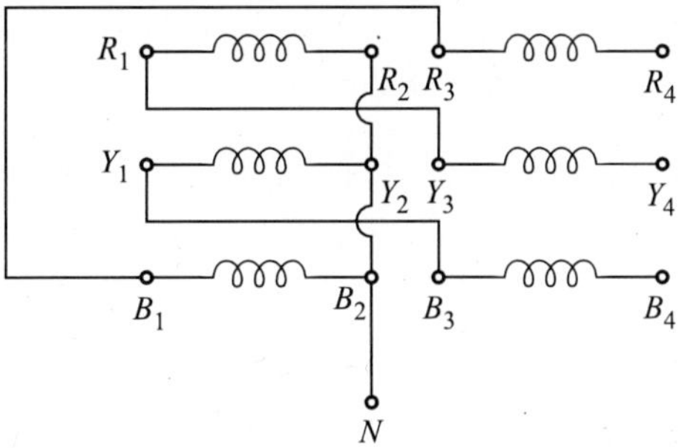

Fig. 5.43 Zigzag connections.

The phasor diagram of zigzag connection is given by

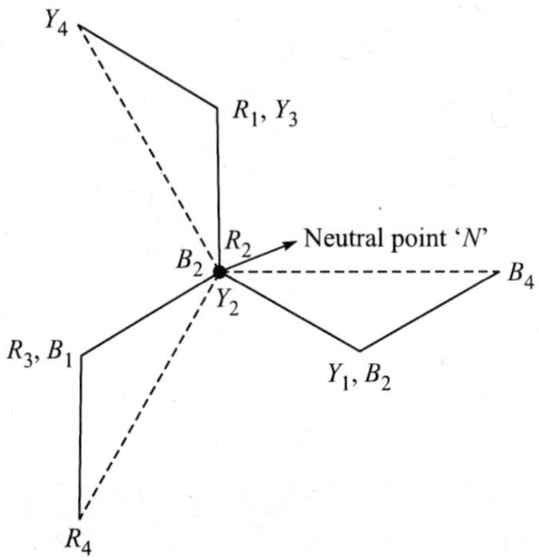

Fig. 5.44 Phase diagram of zigzag connected 3 ϕ transformer.

V/V Connection

In the case of three-phase transformer, which is made up using three single-phase transformers, one transformer may not work due to burning of one of its windings or for some maintenance work carried out on it. In such case the continuity of supply is the main concern though at a less rating. The connection of three-phase winding (Δ) when one winding gets open is called *V* connection or open delta connection. The *V*/*V* connected transformer is given as follows.

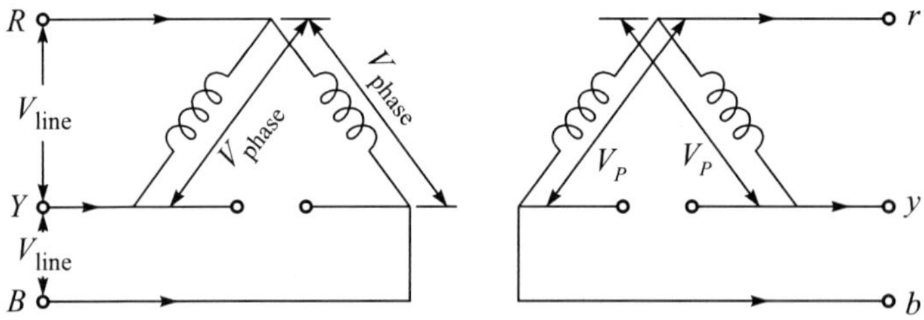

Fig. 5.45 *V*/*V* connection.

The power rating of the '*V*' connected transformer is 2 $V_p I_p$ whereas the power rating of 'Δ' connected transformer is 3 $V_p I_p$, using *V*/*V* connections both three-phase and single-phase supply could be obtained. But higher power is dealt with delta connections.

5.14 HARMONICS

It is seen that the single-phase transformers produce harmonics when excited at the primary with rated voltage. This is as a result of the magnetic material used for the core. The magnetization current 'I_m' which produces flux 'ϕ' in the core gets distorted due to saturation of the magnetic material. In the case of single-phase transformer the distortion of flux does not create much problem since the harmonics cause small effect on the secondary voltage. But in the case of three-phase transformers the harmonics are high because of the triplen harmonics along with other harmonics of higher order. These harmonics circulate in the case of 'Δ' connected transformer but in the case of '*Y*' connected transformer the harmonic will lead to oscillating neutral which implies the neutral to phase voltage will not be equal and vary with time and thus it will appear as if the neutral is oscillating.

The *Y*/Δ and Δ/*Y* connected transformers are needed to be provided with a grounded neutral connection. This will suppress the harmonics from spreading to the other side of the transformer. Thus, improving the efficiency of the transformer. Moreover, the life of the transformer will also be increased.

5.15 SPECIAL TRANSFORMER

Three-Winding Transformers

The three winding transformer provides two voltage levels at the secondary side. This is obtained by providing two secondary windings having different number of turns corresponding to two different levels say 1.1 kV and 440 V respectively as shown below.

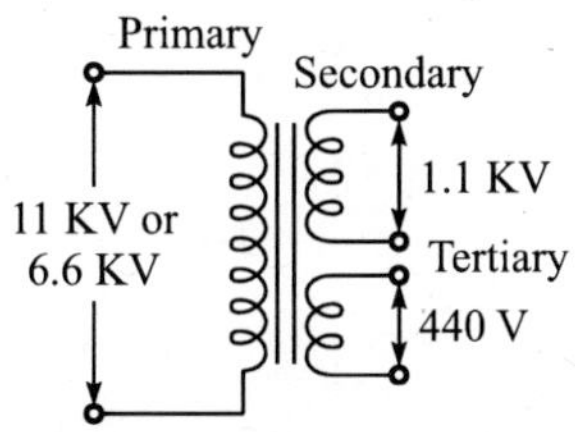

Fig. 5.46 Three-winding transformer.

The three-winding transformer is used for a number of applications such as supplying loads and static condenser at different voltage levels, testing transformer, etc.

Instrument Transformers

The instrument transformers are classified as C.T. and P.T. respectively, the C.T. or current transformer is used in conjuction with the ammeter of normal operating range of 5 A and 10 A so the currents of the order of 100's of amperes could be measured which is passing through the primary. There is no need of breaking the primary circuit from on condition to off position. By using split core type C.T., the measurement of high current at the primary could be made from a large safe distance. The C.T. is shown in the Fig. (5.47) below.

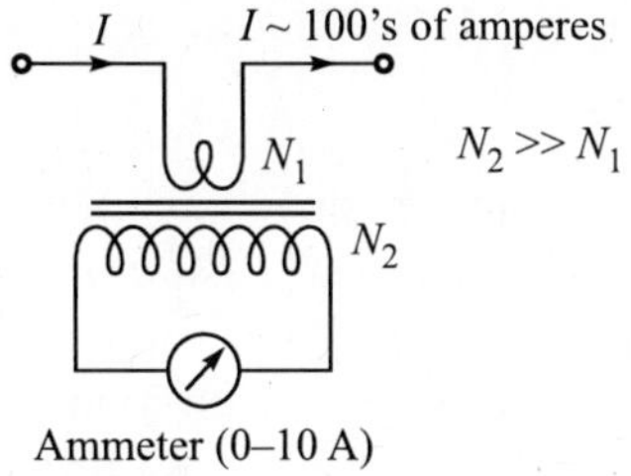

Fig. 5.47 Current transformer.

Care should be taken that there should be a small load at the secondary of transformer even when the secondary of the transformer is open circuited. Since the open circuited secondary will lead to damage of transformer and also it could provide a fatal shock to the personnel operating at the secondary.

Potential Transformer

The potential transformer operates in conjunction with low voltage range voltmeter to measure very high voltages. The core is not same as just that of the ordinary transformer. The core gets saturated later as compared to ordinary transformer. Both the potential and current transformer have errors in the magnitude of quantities measured and the phase of the quantities. But it is a very small percentage of the measured quantities so could be neglected for most practical purposes.

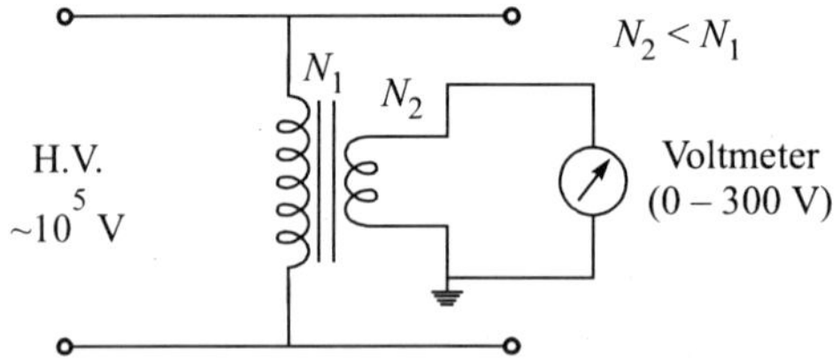

Fig. 5.48 Potential transformer.

Wide Band Transformer

The transformer at 50 Hz such as power and distribution transformers used in the power system network the main concern is the performance of the transformers i.e., efficiency and voltage regulation respectively.

But the transformers used for electronic devices and communications the main concern is the shape of transmitted signal respectively. Any change in the output signal shape of transformer will affect the operation of the communication system performance. The wide band transformers are classified on the basis of frequency function employed and the application performed.

The core material, number of turns, thickness of both windings also differ from the normal transformer. In this category the transformers are pulse transformer, audio transformers, etc. The applications are in aircraft, space, etc.

Grounding Transformer

The grounding transformer is meant for suppressing the harmonics in the transformer connected in Y/Δ and Δ/Y connected transformer. The ratings of grounding transformer also are available in large number of ratings. It also takes care of L-G (line to ground) faults in the 3 ϕ system which commonly occurs in the three-phase power system network.

High Impedance Transformers

The transformers discussed so far have low leakage of flux from core. But in some purposes such as welding, arc furnace, reactive power regulators, etc. there is a need of high impedance transformers because it will cause the arc to be stable for a longer period. The open circuit secondary voltage is also small. The primary and secondary windings are not coupled as in ordinary power frequency transformers.

The secondary windings are coupled to the controller which is electronically controlled and causes leakage flux change. The change in leakage flux causes change in the power factor for the transformer secondary.

The Fig. (5.49) connections can be employed for improving the power factor of line. The tertiary winding suppresses the harmonics to pass to the load.

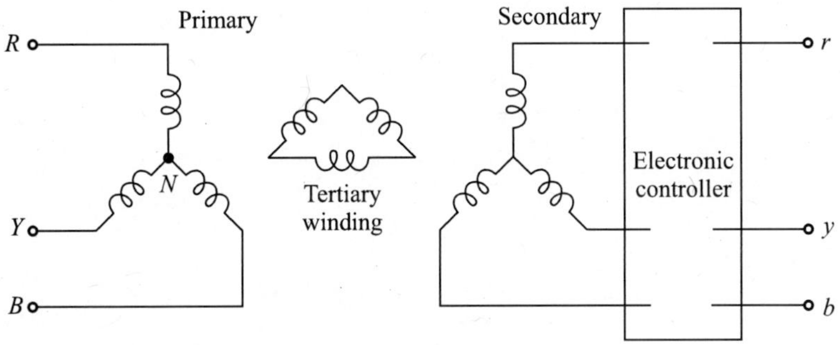

Fig. 5.49 Variation of leakage flux of secondary using electronic controller.

Induction Heating Transformers

The induction principle is used to melt the metal. The ratings of these transformers are from 15 kVA to 40,000 kVA. The 'μ' of the core material is very high. The secondary coil is loosely coupled to primary so leakage in the secondary is very high. The power factor is high (0.6–0.8). The applications is in aluminium, copper and steel industries.

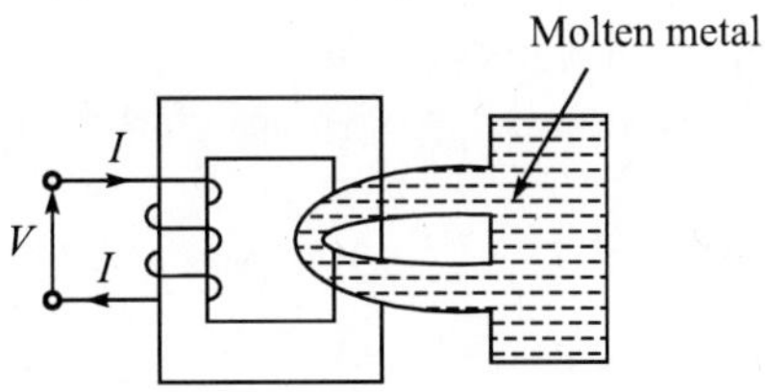

Fig. 5.50 Induction furnace.

5.16 PER UNIT SYSTEM

Per unit system makes the calculations easy. The per unit quantity is expressed as follows:

$$\text{Quantity (in p.u.)} = \frac{\text{Actual quantity}}{\text{Base value of quantity}}$$

Base value is generally the rated value of that particular quantity considered. Such as base values of power and voltage are given by

$$P_{base} = V_{base}\, I_{base} = (VA)_{base} = (S)_{base}$$
$$V_{base} = V_{rated} \rightarrow \text{(Marked over name plate of transformer)}$$
$$I_{base} = \frac{P_{base}}{V_{base}}$$
$$Z_{base} = \frac{V_{base}}{I_{base}} = \frac{V_{base}^2}{P_{base}}$$

If the p.u. values are considered then its value is same on both sides.

$$Z_{eq2,\,pu} = Z_{eq1,\,pu}$$
$$R_{eq2,\,pu} = R_{eq1,\,pu}$$
$$X_{eq2,\,pu} = X_{eq1,\,pu}$$

The primary side p.u. value of impedance is given by

$$Z_{eq1,\,pu} = \frac{Z_{eq_1}}{Z_{B_1}} \quad \text{where } Z_{B1} = \frac{V_{B_1}}{I_{B_1}}$$

where, V_{B1}, I_{B1} are rated voltage and current of primary.

Similarly, $$Z_{eq2,\,pu} = \frac{Z_{eq_2}}{Z_{B_2}};\ Z_{B2} = \frac{V_{B_2}}{I_{B_2}}$$

for secondary side.

$$I_{1,\,pu} = \frac{I_1}{I_{B_1}} = \frac{I_1}{I_{R_1}} = \frac{I_2 \cdot K}{I_{R_2} \cdot K} = \frac{I_2}{I_{R_2}} = \frac{I_2}{I_{B_2}} = I_{2,\,pu}$$

where, I_{R_1}, I_{R_2} are rated current of primary and secondary side. 'K' is the transformation ratio.

Similarly, $$V_{1,\,pu} = \frac{V_1}{V_{B_1}} = \frac{V_1}{V_{R_1}} = \frac{V_2/K}{V_{R_2}/K} = \frac{V_2}{V_{R_2}} = \frac{V_2}{V_{B_2}} = V_{2,\,pu}$$

So, we see that in terms of p.u. values it is immaterial whether we are referring to primary or secondary sides. Since the p.u. values of each quantity remain same on both sides.

The merits of p.u. system are:

(1) The value of quantities in terms of p.u. value falls in a small range.
(2) Referring to any side not necessary.
(3) Computation is easy.
(4) Verification of results easily done.

SOLVED QUESTIONS

Q.1 A step down transformer operates on a 50 Hz a.c. supply with a primary voltage of 230 V. The cross-sectional area of the core in 50 cm^2. Calculate

(i) Maximum flux

(ii) Maximum flux density

(iii) Voltage induced on the secondary side

Assume the primary and secondary turns to be 500 and 200 respectively.

Ans: $E_1 = V_1 = 230\ V$, $A_c = 50\ \text{cm}^2 = 50 \times 10^{-4}\ \text{m}^2$, $N_1 = 500$, $N_2 = 250$

$$\text{(i) } \phi_m = \frac{E_1}{4.44\ N_1 f} = \frac{230}{4.44 \times 500 \times 50} = 2.07 \text{ m Wb}$$

$$\text{(ii) } B_m = \frac{\phi_m}{A_c} = \frac{2.07 \times 10^{-3}}{50 \times 10^{-4}} = 0.4144\ T$$

$$\text{(iii) } E_2 = 4.44\ \phi_m \cdot N_2 f = 4.44 \times 2.07 \times 10^{-3} \times 250 \times 50 = 114.88 \text{ V.}$$

Q.2 A 3300 V/200 V, 50 Hz, 100 kVA transformer has its LV winding with 80 turns. Calculate

(i) The currents in both the windings

(ii) Number of turns of HV winding

(iii) Maximum value of the flux. Given that the transformer is loaded to its full capacity.

Ans: (i) $$I_1 = \frac{S}{V_1} = \frac{100 \times 10^3 \text{VA}}{3300 \text{ V}} = 30.3 \text{ A}$$

$$I_2 = \frac{S}{V_2} = \frac{100 \times 10^3 \text{ VA}}{200 \text{ V}} = 500 \text{ A}$$

(ii) $$\frac{V_1}{V_2} = \frac{N_1}{N_2}$$

So, $$N_1 = \frac{V_1}{V_2} \cdot N_2 = \frac{3300}{200} \times 80$$

$$N_1 = 1320$$

$$\text{(iii) } \phi_m = \frac{E_1}{4.44\ f\ N_1} = \frac{3300}{4.44 \times 50 \times 1320} = 11.26 \text{ m Wb}$$

Q.3 A 400 V/100 V transformer takes a no load current of 5 A at 0.2 lagging p.f. Secondary winding supplies a load of 100 A at a p.f. of 0.8 lagging. Find the primary input current.

Ans:

$$\cos\ \theta_0 = 0.2 \text{ (lag)}$$
$$\cos\ \theta_2 = 0.8 \text{ (lag)}$$
$$I_2 = 100$$
$$V_1 = 400 \text{ V}$$
$$V_2 = 100 \text{ V}$$

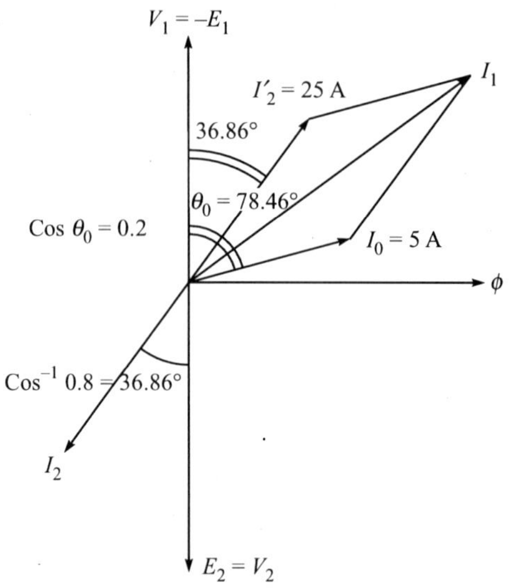

$$I_0 = 5 \text{ A}$$

$$\frac{E_2}{E_1} = \frac{V_2}{V_1} = K = \frac{100}{400} = \frac{1}{4}$$

$$I_2' = K\, I_2 = \frac{1}{4} \times 100 = 25 \text{ A}$$

$$\therefore \quad I_1 = \sqrt{I_0^2 + I_2'^2 + 2\, I_0\, I_2' \cos(78.5° - 36.9°)}$$

$$\boxed{I_1 = 28.9 \text{ A}}$$

Q.4 A 25 kVA, 4 kV/200 V, 50 Hz transformer has $R_1 = 3.45\ \Omega$, $R_2 = 0.009\ \Omega$, $X_1 = 5.2\ \Omega$ and $X_2 = 0.051\ \Omega$. Calculate the equivalent resistance and equivalent reactance referred to (i) primary (ii) secondary. Also calculate the net power lost due to winding resistance.

Ans: $$S = 25 \text{ kVA}, \quad K = \frac{V_2}{V_1} = \frac{200}{4000} = \frac{1}{20}, \quad f = 50 \text{ Hz}$$

Equivalent resistance referred to primary

$$R_{eq_1} = R_1 + \frac{R_2}{K^2} = 3.45 + \frac{0.009}{(1/20)^2} = 7.05\ \Omega$$

Equivalent resistance referred to primary

$$X_{eq_1} = X_1 + \frac{X_2}{K_2} = 5.2 + \frac{0.051}{(1/20)^2} = 25.6\ \Omega$$

Equivalent resistance referred to secondary

$$R_{eq_2} = R_2 + K^2 \cdot R_1 = (0.009) + \left(\frac{1}{20}\right)^2 \cdot (3.45)$$

$$= 0.0176\ \Omega$$

$$X_{eq_2} = X_2 + K^2 X_1 = (0.051) + \left(\frac{1}{20}\right)^2 \cdot (5.2)$$

$$= 0.064\ \Omega$$

Power loss in winding resistance R_1 and R_2

$$= I_1^2 \cdot R_{eq_1} = \left(\frac{S}{V_1}\right)^2 \cdot R_{eq_1} = (6.25)^2 \times 7.05$$

$$= 275.4 \text{ watt}$$

Q.5 The efficiency of 400 kVA, single-phase transformer is 98.77% when delivering full load of 0.8 pf and it is 99.13% at half load unity p.f. Calculate (i) Iron loss (ii) Full load copper loss.

Ans: The efficiency 'η' is given by

$$\eta\Big]_{FL} = \frac{S \cos\theta_2}{S \times \cos\theta_2 + P_i + P_{cu}}$$

$$\therefore \quad 0.9877 = \frac{400 \times 10^3 \times 0.8}{400 \times 10^3 \times 0.8 + P_i + P_{cu}}$$

At half load,

$$\eta\Big]_{HL} = 0.9913 = \frac{\frac{1}{2} \times 400 \times 10^3 \times 1}{\frac{1}{2} \times 400 \times 10^3 \times 1 + P_i + \left(\frac{1}{2}\right)^2 P_{cu}}$$

$$\therefore \quad P_i + P_{cu} = 3985$$

$$P_i + \frac{P_{cu}}{4} = 1755$$

Solving the above equations, we get

$$P_i = 1012 \text{ W} \quad \text{and} \quad P_{cu} = 2973 \text{ W}.$$

Q.6 The iron loss and full load copper loss of a 100 kVA, 6600/400 V, single phase transformer are 600 W and 900 W. Calculate the efficiency at full load and half load, at 0.8 p.f. lag. Calculate the load at which maximum efficiency is obtained and its magnitude at the same power factor.

Ans:

$$\eta_{FL} = \frac{S \cos\theta_2}{S \cos\theta_2 + P_i + P_{cu}}$$

$$= \frac{100 \times 10^3}{100 \times 10^3 \times 0.8 + 600 + 900}$$

$$= 98.159\%$$

$$\eta\Big]_{HL} = \frac{\frac{1}{2} S\cos\theta_2}{\frac{1}{2} S\cos\theta_2 + P_i + \left(\frac{1}{2}\right)^2 P_{cu}}$$

$$= \frac{\frac{1}{2}\times 100\times 10^3\times 0.8}{\frac{1}{2}\times 100\times 10^3\times 0.8 + 600 + 255}$$

$$= 97.98\%$$

For η_{max} $\quad P_i = P_{cu}$

$$\therefore \quad \text{kVA}\Big]_{\eta_{max}} = \text{kVA}\times\sqrt{\frac{P_i}{(P_{cu})_{fl}}}$$

$$= 100\times\sqrt{\frac{600}{900}} = 81.65 \text{ kVA}$$

$$\therefore \quad \%\eta_{max} = \frac{\text{kVA}\Big]_{\eta_{max}}\cos\theta_2}{\text{kVA}\Big]_{\eta_{max}}\cos\theta_2 + P_i + P_{cu}\Big]_{\eta_{max}}}$$

$$= \frac{81.65\times 10^3\times 0.8}{81.65\times 10^3\times 0.8 + 600 + 600}$$

$$= 98.196\%$$

Q.7 A single phase transformer delivers 10 A, 220 V to a resistance load while the primary draws 6 A at 0.9 lagging p.f. from 450 V, 50 Hz supply. The turn ratio of the transformer is 2. Calculate efficiency and regulation under this condition.

Ans:

$$\eta = \frac{\text{Output}}{\text{Input}} = \frac{V_2 I_2\cos\theta_2}{V_1 I_1\cos\theta_1} = \frac{220\times 10\times 1}{450\times 6\times 0.9} = 90.53\%$$

No load secondary voltage $E_2 = \dfrac{N_2}{N_1} E_1$

$$E_2 = \frac{1}{2}\times 450 = 225 \text{ V}$$

$$\therefore \quad \text{Regulation} = \frac{E_2 - V_2}{E_2}\times 100 = \frac{225-220}{225}\times 100$$

$$= 2.22\%$$

Q.8 A 10 kVA, 440 V/100 V, 50 Hz transformer gave the following test results.

O.C.T: 440 V, 1.1 A, 80 W
S.C.T: 10 V, 22.7 A, 120 W

Both the tests are conducted from H.V side. Determine the values of the transformer parameters; i.e., R_0, X_0, R_{eq1} and X_{eq1}. Also calculate the efficiency and regulation of transformer at (i) full load p.f. being 0.8 lag (ii) half load p.f. being 0.9 lag.

Ans:

$$\cos\theta_0 = \frac{W_0}{V_1 I_0} = \frac{80}{440 \times 1.1} = 0.1653$$

$$\theta_0 = 80.49^\circ$$

$$I_e = I_0 \cos\theta_0 = 1.1 \times 0.1653 = 0.1818 \text{ A}$$

$$I_m = I_0 \sin\theta_0 = 1.1 \times 0.9863 = 1.0849 \text{ A}$$

$$X_0 = \frac{V_1}{I_m} = \frac{440}{1.0849} = 405.57\ \Omega$$

$$R_0 = \frac{V_1}{I_e} = \frac{440}{0.1818} = 2420.24\ \Omega$$

$$R_{eq1} = \frac{W_{sc}}{I_{sc}^2} = \frac{120}{(22.7)^2} = 0.2323\ \Omega = R_1 + R_2'$$

$$Z_{eq1} = \frac{V_1}{I_{sc}} = \frac{10}{22.7} = 0.44\ \Omega$$

$$\therefore \quad X_{eq1} = 0.374\ \Omega$$

$$\text{Full load current} = \frac{S}{V_1}$$

$$= \frac{10000}{440} = 22.7 \text{ A}$$

SCT is conducted at FL current

So
$$W_{sc} = P_{cu} = 120 \text{ W}$$
$$W_0 = P_i = 80 \text{ W}$$

$$\text{Copper loss at half load} = \left(\frac{1}{2}\right)^2 P_{cu} = 30 \text{ W}$$

$$\therefore \quad \eta_{FL} = \frac{S\cos\theta_2}{S\cos\theta_2 + P_i + (P_{cu})_{FL}} = \frac{10 \times 10^3 \times 0.8}{10 \times 10^3 \times 0.8 + 80 + 120}$$

$$= 97.56\%$$

$$\text{Regulation}]_{FL} = \frac{I_1 R_{eq_1}\cos\theta_2 + I_1 \times X_{eq_1}\sin\theta_2}{V_1}$$

$$= \frac{22.7 \times 0.2323 \times 0.8 + 22.7 \times 0.37 \times 0.6}{440}$$

$$= 2.585\%$$

$$\eta_{HL} = \frac{\frac{1}{2} S \cos\theta_2}{\frac{1}{2} S\cos\theta_2 + P_i + \left(\frac{1}{2}\right)^2 P_{cu}}$$

$$= \frac{\frac{1}{2}\times 10\times 10^3\times 0.9}{\frac{1}{2}\times 10\times 10^3\times 0.9 + 80 + \left(\frac{1}{2}\right)^2\times 120}$$

$$= 97.61\%$$

$$\text{Regulation}]_{HL} = \frac{\frac{1}{2}(I_1)[R_{eq_1}\cos\theta_2 + X_{eq_1}\sin\theta_2]}{V_1}$$

$$= 0.961\%$$

Q.9 A transformer has copper loss of 1.5% and reactance drop of 3.5%. When tested at full load. Calculate the full load regulation at (i) 0.8 p.f. lagging (ii) 0.8 p.f. leading.

[B.Tech UPTU, 03]

Ans:

$$\%\text{V.R.}]_{0.8\text{ lag}} = R_{p.u}\cos\theta_2 + X_{pu}\sin\theta_2$$
$$= 1.5\times 0.8 + 3.5\times 0.6$$
$$= 1.2 + 2.1 = 3.3\%$$
$$\%\text{V.R.}]_{0.8\text{ lead}} = R_{pu}\cos\theta_2 - X_{pu}\sin\theta_2$$
$$= 1.5\times 0.8 - 3.5\times 0.6$$
$$\%\text{V.R} = -0.83$$

Q.10 A 100 kVA, single-phase transformer has $N_1 = 400$, $N_2 = 600$ $R_1 = 0.3\ \Omega$ $X_1 = 1.1\ \Omega$, $R_2 = 0.01\ \Omega$, $X_2 = 0.035\ \Omega$. The applied voltage is 2.2 kV. Find Z_{eq_1}, V.R, $V_2]_{0.8\text{ p.f.}, FL}$.

Ans:

$$R_{eq_1} = R_1 + \frac{R_2}{K_2} \qquad K = \frac{V_2}{V_1} = \frac{N_2}{N_1} = \frac{80}{400} = \frac{1}{5}$$

$$R_{eq_1} = 0.3 + \frac{0.01}{(1/5)^2} = 0.55\ \Omega$$

$$X_{eq_1} = X_1 + \frac{X_2}{K^2} = 1.1 + \frac{0.035}{(1/5)^2} = 1.975\ \Omega$$

$$Z_{eq_1} = \sqrt{R_{eq_1}^2 + X_{eq_1}^2} = 2.05\ \Omega$$

$$I_1 = \frac{S}{V_1} = \frac{100\times 10^3}{2{,}200} = 45.45\text{ A}$$

$$\%\text{V.R.} = \frac{I_1[R_{eq_1} \cos\theta_2 + X_{eq_1} \sin\theta_2]}{V_1} \times 100$$

$$= 45.45\ [0.55 \times 0.8 + 1.975 \times 0.6]/2200$$

$$= 3.357\%$$

$$V_2 = K\ V_1$$

$$= \frac{1}{5} \times 2200 = 440 \text{ Volt.}$$

UNSOLVED QUESTIONS

Q.1 A 3.3 kV/240 V, 1 phase transformer draws no load current of 0.7 A and absorbs 650 W on no load. Find the magnetizing and iron loss currents.

(B.Tech II, UPTU, 05)

Q.2 Calculate the regulation of a single phase transformer in which the copper loss is 1% of the output at full load unity power factor and the reactance drop 5% of the rated voltage at full load when the power factor is 0.8 leading.

(B.Tech II, UPTU, 05)

Q.3 A 6600 V, three-phase transformer takes 10 A. Calculate the secondary line voltage, line current and output for the following connections:

(i) Δ/Δ

(ii) Y/Y

(iii) Δ/Y

(iv) Y/Δ

The ratio of turns per phase is 12, neglect losses.

[B.Tech III, UPTU, 2002]

Q.4 Derive the EMF equation of single phase transformer. The no load current of a transformer is 4.0 A at 0.25 p.f when supplied at 250 V, 50 Hz. The number of turns on the primary winding is 200. Calculate

(i) Flux in the core

(ii) The core loss and

(iii) Magnetising current

[B.Tech II, UPTU, 03]

Q.5 Define voltage regulation of a transformer. A transformer has copper loss of 1.5% and reactance drop of 3.5% when tested at full load. Calculate its full load regulation at

(i) 0.8 p.f lagging

(ii) 0.8 p.f leading

[B.Tech II, UPTU, 03]

Q.6 A 600 kVA, single-phase transformer when working at unity power factor, has an efficiency of 92% at full load and also at half load. Determine its efficiency when it operates at unity p.f and 60% of full load.

[B.Tech II, UPTU, 03]

Q.7 Describe Sumpner's test (back to back test) and explain how losses and efficiency are determined. A 10 kVA, 1 phase, 500/250 V, transformer gave the following test results:

OCT: 250 V, 3 A, 20 W
SCT: 15 V, 30 A, 300 W

Calculate efficiency and regulation at full load, 0.8 p.f lagging.

[B.Tech II, UPTU, 03]

Q.8 A three-phase, delta/star connected 11 kV/400 V, 50 Hz transformer takes a line current of 5 amp, when secondary load of 0.8 lagging p.f. is connected. Determine each coil current and output of the transformer.

[B.Tech II, UPTU, 03]

Q.9 Define voltage regulation of a transformer. The following test data were taken on 25 kVA, 2400 V/240 V, 50 Hz transformer with low voltage winding shorted and measurements made on h.v. side

$$V = 41.5 \text{ V}, I = 21.7 \text{ A}, P = 350 \text{ W}$$

Calculate the voltage regulation of this transformer when it supplies full load of 0.8 p.f. lagging. Neglect the magnetising current.

[B.Tech II, UPTU, 2002]

Q.10 A three-phase transformer bank consisting of three single-phase transformers is used to step down the voltage of a three-phase, 6600 V transmission line. If the primary line current is 10 A, calculate the secondary line voltage, line current and output kVA for the following connections

(i) Star/delta
(ii) Delta/star

The turn ratio is 12.

[B.Tech II, AMU, 92]

Q.11 Explain why an open delta connection has a VA rating of 0.58 of the rating of the normal Δ/Δ connections.

[B.Tech II, AMU, 92]

Q.12 A 50 kVA, 2200 V/110 V transformer when tested, gave the following result.

O.C.T (LV side): 400 W, 10 A, 110 V
S.C.T (HV side): 808 W, 20.5 A, 90 V

If the transformer is operating at its maximum efficiency and supplying u.p.f load at rated secondary voltage. Calculate the primary voltage and current.

[B.Tech II, AMU, 93]

Q.13 Two similar 250 kVA, single phase transformers when tested by back to back method, gave the following results:

Main Wattmeter, $W_1 = 5$ kW

Primary series circuit wattmeter $W_2 = 7.5$ kW (at full load). Find out the individual transformer efficiencies at 75% full and 0.8 p.f. (leading).

[BSc. Engg II, AMU, 93]

Q.14 A small industrial unit draws an average load of 100 A at 0.8 p.f (lagging) from the secondaries of its 2000/200 V, 60 kVA, star/delta transformer bank. Find

(i) the power consumed by the unit, in kW.
(ii) the total kVA used.
(iii) rated current available from the transformer bank.
(iv) the rated transformer phase currents of the delta secondaries.
(v) the primary line and phase currents.

[B.E. II, AMU, 93]

Q.15 Two single-phase transformers each rated as 250 kVA, 11/2 kV and 50 Hz are connected in open delta on both primary and secondary sides. Find the load kVA that can be supplied from this transformer combination.

[B.E. II, AMU, 92]

Q.16 A delta connected three-phase load of 250 kVA, 0.8 p.f., 2 kV is connected to the low voltage terminals of this open delta transformer. Determine the currents on 11 kV side of this connection.

[B.E. II, AMU, 92]

Q.17 An ideal three-phase step down transformer delivers power to a balanced three-phase load of 120 kVA at 0.8 power factor. The input line voltage is 11 kV and the turn ratio of the transformer phase to phase is 10. Determine the line voltages, line currents, phase voltage and phase currents on both the primary and secondary sides for the following connections
(i) Star-delta (ii) Delta-star.

[B.E. II, AMU, 92]

Q.18 A 4 kVA, 200 V/400 V, 50 Hz single-phase transformer gave the following test results.

No load test (LV side): 200 V, 0.7 A, 60 W
Short circuit test (HV side): 9 V, 8 A, 21.6 W.

Calculate:

(a) The magnetizing current and component corresponding to iron loss at normal voltage and frequency.

(b) The efficiency on full load and unity power factor.

(c) The voltage regulation at 0.8 power factor lagging.

[B.E. II, AMU, 94]

Q.19 A 200/400 V, 20 kVA, 50 Hz single phase transformer is connected as an auto transformer to transform 600 V to 200 V.

(i) Determine the auto transformer turn ratio

(ii) Determine the kVA rating of the auto transformer.

(iii) With a load of 20 kVA, 0.8 power factor connected to 200 V terminals, determine the currents in the load and the two winding transformer windings.

[B.E. II, AMU, 1994]

Q.20 In a 25 kVA, 2000/200 V transformer, the iron and copper losses are 350 W and 400 W respectively.

(i) Calculate the efficiency on unity power factor at full load and half load.

(ii) Determine the load for maximum efficiency and iron and copper loss in this case.

[B.Tech II, JMI, 2006]

Q.21 A 200/400 V, 20 kVA and 50 Hz transformer is connected as an auto transformer to transform 600 V to 200 V.

(i) Find the auto transformer ratio.

(ii) Find kVA rating of auto transformer.

(iii) With a load of 20 kVA, 0.8 p.f. lagging connected to 200 V terminals, find the currents in the load and the two transformer windings.

[B.Tech II, JMI, 2006]

Q.22 Three 1100/110 V transformers connected Δ/Δ supply a lighting load of 100 kW. One of these transformers is damaged and hence removed for repair. What currents will be flowing in each transformer when

(i) the three transformers were in service.

(ii) the two transformers are in service.

[B.Tech II, JMI, 2006]

Q.23 A 1-phase 2 kVA, 250/500 V, 50 Hz transformer gave the following results:

O.C.T: 250 V, 1 Amp, 80 W on l.v. side

S.C.T: 20 V, 12 Amp, 100 W on h.v. side

Calculate the circuit constants. Also determine its efficiency and regulation at full load at 0.8 p.f. lagging.

[B.Tech II, JMI, 05]

Q.24 An auto transformer supplies a load of 5 kW at 115 V, at unity power factor. If the primary voltage is 230 V, determine:

(i) transformation ratio
(ii) secondary current
(iii) primary current
(iv) power transformed
(v) power conducted directly from the supply mains to the load.

[B.Tech II, JMI, 03]

Q.25 Two similar 40 kVA, single-phase transformers gave the following result when tested by the Sumpner's test. Wattmeter reading in the parallel connected circuit = 800 W; Wattmeter reading in the series connected circuit (at I_{rated}) = 1000 W. Determine the efficiency of each transformer at

(a) full load unity power factor, and
(b) half load 0.8 power factor (lagging).

[B.E. II, AMU, 2006]

Q.26 A 50 kVA, 2200/110 V, transformer when tested gave the following a test results:

Open circuit test (HV side open); 400 W, 10 A, 110 V

Short circuit test (LV side shorted); 808 W, 20.5 A, 90 V

Compute all the parameters of the equivalent circuit referred to the HV and LV sides of the transformer.

[B.E. II, AMU, 2006]

Q.27 A 400 V/100 V, 10 kVA, 2 winding, single phase transformer is to be employed as an auto transformer to supply 400 V circuit from a 500 V source. When tested as a 2 winding, transformer at rated, load, 0.85 power factor (lagging) its efficiency is 97%.

(i) Determine its kVA rating as an auto transformer.
(ii) Find its efficiency as an auto transformer.

[B.E. II, AMU, 2006]

Q.28 A 50 kVA, 2200/110 V transformer when tested gives following results:

O.C.T (LV): 400 W, 10 A, 110 V

S.C.T (HV): 808 W, 20.8 A, 85 V

Compute all the parameters of the equivalent circuit and referred them to the H.V and L.V sides. Also obtain per unit values.

[B.E. Tech II, AMU, 96]

EXAM QUESTIONS

Q.1 Derive the e.m.f equation of a single phase transformer.

[B.Tech II, UPTU, 03]

Q.2 Draw the equivalent circuit of a single phase transformer. Explain the tests performed to obtain the parameters of the equivalent circuit.

[B.Tech II, UPTU, 03]

Q.3 Define voltage regulation of a transformer.

[B.Tech II, UPTU, 03]

Q.4 Describe Sumpner's test and explain how losses and efficiency are determined.

[B.Tech II, UPTU, 03]

Q.5 Define transformer. Explain its construction and working.

[B.Tech II, UPTU, 04]

Q.6 Draw no load phasor diagram of transformer and explain the significance of each phasor. Also draw corresponding equivalent circuit.

[B.Tech II, UPTU, 04]

Q.7 Discuss the effects of load and load power factor on the efficiency of transformer. Obtain expression for fraction of load at which maximum efficiency occurs.

[B.Tech II, UPTU, 04]

Q.8 Define voltage regulation and transformer and explain its singificance. With the help of approximate equivalent circuit and phasor diagram obtain an expression for it.

[B.Tech II, UPTU, 04]

Q.9 Give circuit diagrams and explain the procedures of open circuit and short circuit tests on a two-winding transformer. How can parameters of equivalent circuit be determined from these tests?

[B.Tech II, UPTU, 05]

Q.10 Describe the back to back test for determining the regulation and efficiency of a pair of similar trasformer. What are the limitations of this test?

[B.Tech II, UPTU, 05]

Q.11 Write technical notes on polarity test and Sumpner's test on transformers. Make suitable illustrations and mathematical derivations.

[B.Tech II, UPTU, 2005]

Q.12 Describe with a neat diagram, the working principle of instrument transformer.

[B.Tech II, UPTU, 2003]

Q.13 What is an auto transformer? Also show that for the same capacity and voltage ratio, the auto transformer requires less copper than a two-winding transformer.

[B.Tech II, UPTU, 2004]

Q.14 Discuss the advantages, disadvantages and applications of auto transformers. Compare the conductor savings of auto transformer with a two-winding transformer.

[B.Tech II, UPTU, 2005]

Q.15 Describe the working principle of current transformer and potential transformer.

[B.Tech II, UPTU, 2005]

Q.16 Why the transformer is rated in kVA and not in kW? Explain.

[B.Tech II, UPTU, 2005]

Q.17 Explain the back to back method of testing of two identical single phase transformers.

[B.Tech II, UPTU, 2005]

Q.18 Draw the different three-phase transformer connections. Also derive the phase and line transformation ratio expression for each connection.

[B.Tech II, UPTU, 2005]

Q.19 Explain the construction and possible methods of connecting a three-phase transformer.

[B.Tech II, UPTU, 2000]

Q.20 What are the distinguishing features of $Y - Y$, $Y - \Delta$, $\Delta - Y$ and $\Delta - \Delta$, three-phase connections?

[B.Tech II, UPTU, 2000]

Q.21 What is meant by tertiary winding? Why is it needed? Explain.

[B.Tech II, UPTU, 2000]

Q.22 Describe various losses in a transformer. Explain how each loss varies with load current, supply voltage and frequency.

[B.Tech II, UPTU, 2002]

Q.23 Discuss relative advantages and disadvantages of employed three one phase transformers for use in three-phase operation over employing a single unit three-phase transformer.

[B.Tech II, UPTU, 2002]

Q.24 Explain leakage flux and leakage reactance in reference to a transformer and obtain its equivalent circuit with secondary quantities referred to the primary side.

[B.Tech II, UPTU, 2001]

Q.25 Why are the iron losses in a transformer substantially independent of the load current? Explain.

[B.Tech II, UPTU, 2001]

Q.26 Explain the constructional difference between a current transformer and a potential transformer.

[B.Tech II, UPTU, 2001]

Q.27 Explain how a short circuit test is performed on a single-phase transformer and why it gives the copper losses only and not the iron losses.

[B.Tech II, UPTU, 2001]

Q.28 What are main features of $Y - Y$, $\Delta - \Delta$, $\Delta - Y$, $Y - \Delta$ three-phase connections?

[B.Tech II, JMI, 2005]

Q.29 For what purposes are tertiary windings used on a three-winding transformer. Explain how they can assist in unbalanced loading condition if suitably connected.

[B.Tech II, JMI, 2005]

Q.30 Define voltage regulation of a transformer and derive its expression for a leading power factor load. Find out the condition for which the voltage regulation is zero.

[B.E. Tech II, AMU, 2006]

Q.31 Explain why copper loss in a transformer can be neglected in open circuit test and iron loss can be neglected in short circuit test.

[B.E. II, AMU, 2006]

Q.32 Draw the connection and phasor diagrams of a star-delta transformer.

[B.E II, AMU, 2006]

Q.33 Define all day efficiency and discuss its significance.

[B.E. II, AMU, 1995]

Q.34 Draw connections and the phasor diagram of the following 3 ϕ transformers (i) *Yd* 1 (ii) *Yd* 11 (iii) *Dd* 6.

[B.E II, AMU, 1992]

Q.35 An iron cored transformer is working at a maximum flux-density of 0.8 tesla. Its core is replaced by Si steel core, working at a maximum flux density of 1.2 tesla. If the total flux is to remain the same, what is the reduction in volume expressed as percentage of original volume? The frequency and voltage per turn are the same in both cases.

(**Ans:** 33%)

[IES-2007]

Q.36 The core of a two-winding transformer is subjected to a magnetic flux variation as indicated in the figure.

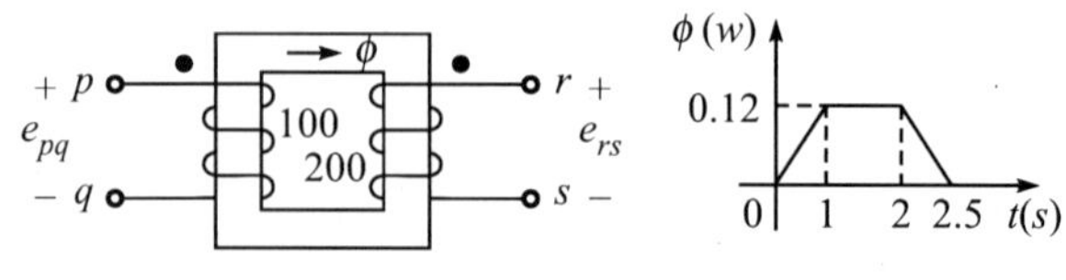

The induced e.m.f. (e_{rs}) in the secondary winding as a function of time will be of the form.

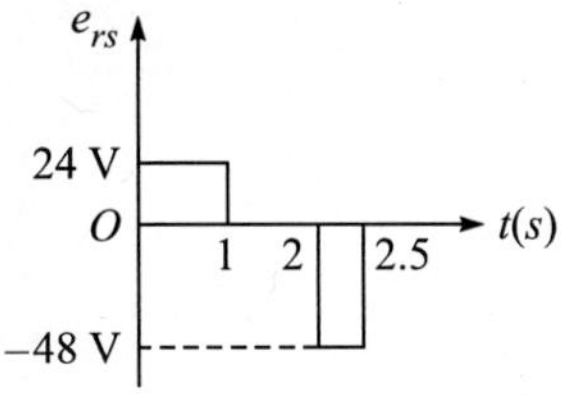

(a)

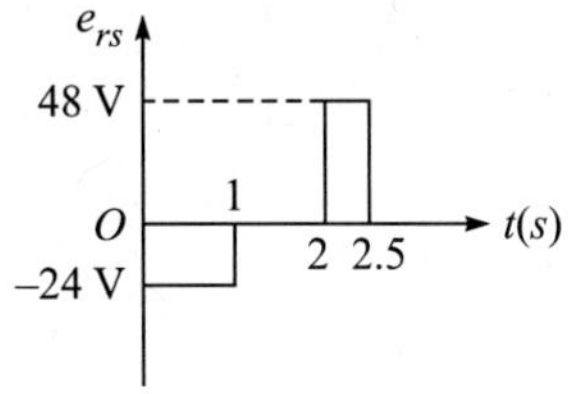

(b)

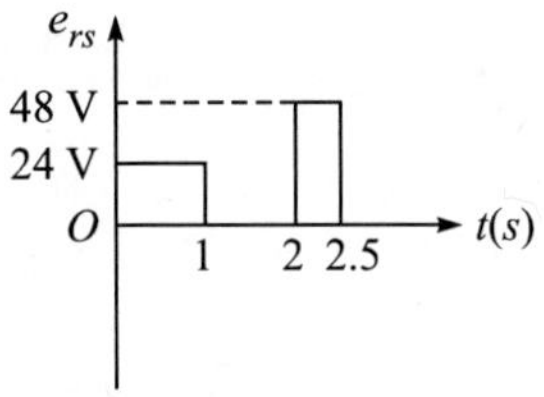

(c)

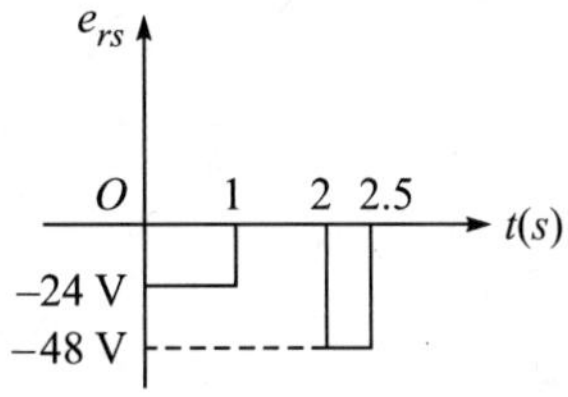

(d)

(**Ans:** (b))
[GATE-2008]

Q.37 Three single-phase transformers are connected to form a 3 phase transformer bank. The transformers are connected in the following manner. The transformer connection will be represented by

(a) *Yd* 0
(b) *Yd* 1
(c) *Yd* 6
(d) *Yd* 11

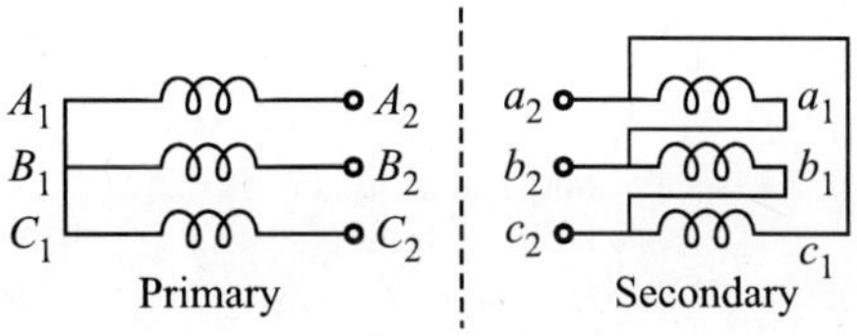

(**Ans:** (b))
[GATE-2008]

Chapter

6

D.C. Generator

6.1 Introduction
6.2 Constructional Features of D.C. Machines
6.3 Types of D.C. Machines
6.4 Principle of Operation
6.5 Symbolic Representation
6.6 Characteristics of D.C. Generators
6.7 E.M.F. Equation of D.C. Generator
6.8 Commutation
6.9 Armature Reaction
6.10 Losses in D.C. Machines
6.11 Voltage Regulation of D.C. Generator
6.12 Efficiency of D.C. Generator
6.13 Application of D.C. Generator
Solved Questions (Exam)
Tutorial Questions
Review Questions

6.1 INTRODUCTION

The importance of the d.c. machines can be understood by its applications in the fractional horsepower range to thousands of horsepower in the steel industries. The d.c. machines were developed before the advent of a.c. machines. But because of the reliability of the a.c. supply the a.c. machines superseded the d.c. machines. In spite of that the d.c. machines of higher ratings are still used in the traction purpose.

Now the applications of the d.c. machines are seen mostly in the FHP ratings. The electronic appliances employ the FHP motors. So we cannot neglect these d.c. machines.

So to get a clear understanding of the d.c. machines, the internal structure and the operating principle of the machines should be clear in our mind. In the coming sections the different types of machines are discussed in detail.

The permanent magnet d.c. machine are used in the electronic circuits as it needs no extra circuit required for creating the field poles. The brushless d.c. machines are also preferred. The modern d.c. motors of small ratings uses the transistorized, switching circuits made on the printed circuit boards and are used for switching the motor armature circuit when the rotor passes from one pole to another.

The large rating d.c. motors are restricted to a few industries. The high efficient robust induction motor is fast replacing the d.c. motors of larger ratings due to many advantages over the d.c. equivalent rating motor.

6.2 CONSTRUCTION FEATURES OF D.C. MACHINES

The d.c. machine basically consists of two main parts, namely, stator and rotor. The stator is the static part, whereas the rotor is the rotating part of the machines. The stator consists of field pole assembly, yoke or frame of machines, brush assembly. The rotor consists of shaft, commutator, armature, respectively. These are represented in the Fig. (6.1) as given below.

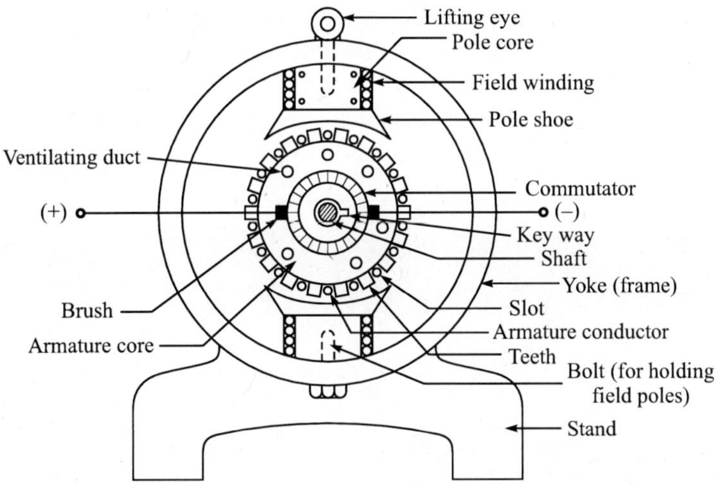

Fig. 6.1 Sectional view of D.C. machine.

The stator parts can be taken one by one so as to get a clear picture of machine in the mind. The first important part is the yoke or frame of the machine.

Yoke

The yoke of the machines serves as a protective covering for the machine internal parts i.e., rotor and other stator parts. The yoke in the smaller machines is made up of cast iron but for medium and large machines it is made up of cast steel. The saturation of the cast iron is at 0.8 tesla whereas the saturation of cast steel is at 1.5 tesla. So the steel frame is better than cast iron frame. The other important function of the yoke is to provide a path for the magnetic flux to pass through it. That is, it is a part of magnetic circuit of the machine. Figure 6.2 depicts the yoke.

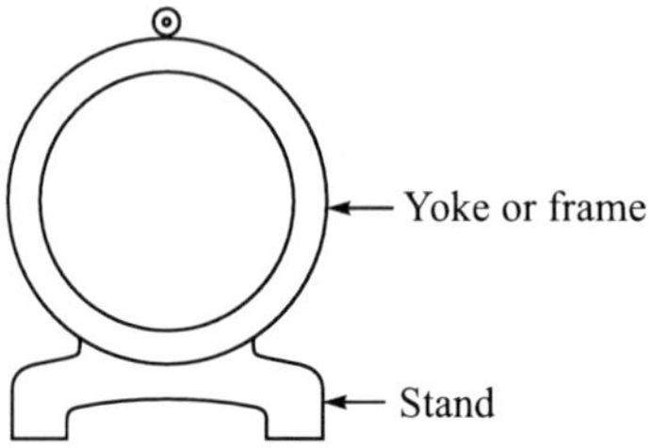

Fig. 6.2 Yoke of DC machine.

Field Assembly

The field core is made up of magnetic material having higher permeability so as to increase the flux through the material and to enable larger flux per pole of the machine. The field core is solid assembly for smaller machines but it is laminated for larger rating machines so as to decrease the heating of core. The field core laminations are rivetted together so as to make it a complete assembly so that the field winding can be placed around it. The field winding which is made up of copper and is coated with varnish is prewound over a former and it is placed over the core as shown in Fig. (6.3).

Fig. 6.3 Prewound field winding on former.

The field assembly basically consists of the following parts, namely, field core, field winding and pole shoe.

The field core is made up of high permeable magnetic material such as iron, steel etc. with addition of material such as nickel, molybdenum, etc. This will concentrate the magnetic flux lines in the core. The field core is attached to the pole shoe using bolts. The pole shoe is also made up of magnetic material of high permeability. The field winding which is former wound also gets a support by the pole shoe. The pole shoe is also laminated in case of larger rating machines.

The field winding carries current which magnetizes the field core and hence due to electromagnetism effect, the magnetic material gets magnetized as a result it behaves as a magnet and produces flux due to the magnetomotive force produced by the field coils. This flux passes through the magnetic path which comprises the field core, pole shoe, air gap, armature and the yoke. The intensity of the flux can be increased by increasing the current through the field winding. The higher the field current the more powerful field is created by the field poles. The flux under the pole shoe is called flux per pole or useful flux. To increase the useful flux of the machine the field pole area is increased by chamfering of the poles.

The chamfering of the poles is done on the pole shoe by lengthening the pole shoe ends such that the distance between the pole shoe and the armature is lesser at the centre of pole shoes and larger at the end of pole shoes. This will increase the flux lines under the poles such that larger number of the armature conductors cut these flux. The useful flux per pole is intentionally increased during the designing of machines. This will increase the electromotive force generated in the armature in the case of d.c. generators. The Fig. (6.4) shows the chamfering of poles of the field core.

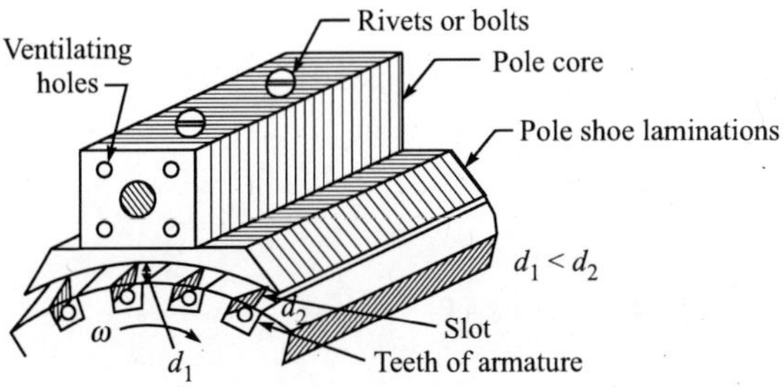

Fig. 6.4 Field core showing chamfering.

Brush Assembly

The brush assembly comprises brush holder, rocker, spring adjustment lever and brush. Figure 6.5 shows the arrangement of brush in the brush holder (box type).

The brush is placed in a box type holder. The brush is pressed by a spring whose tension can be adjusted with a tension adjustment lever. The brush is kept in contact with the commutator segments. The contact should be such that there should not exist any air gap between them. If there is any air gap present between them then there will be arcing and the arcing will lead to temperature rise which will cause pitting of the commutator segments. The pitting continues and the copper commutator segment gets damaged as time elapses. There is a copper pig tail which is firmly connected to the other end of brush which is free for external

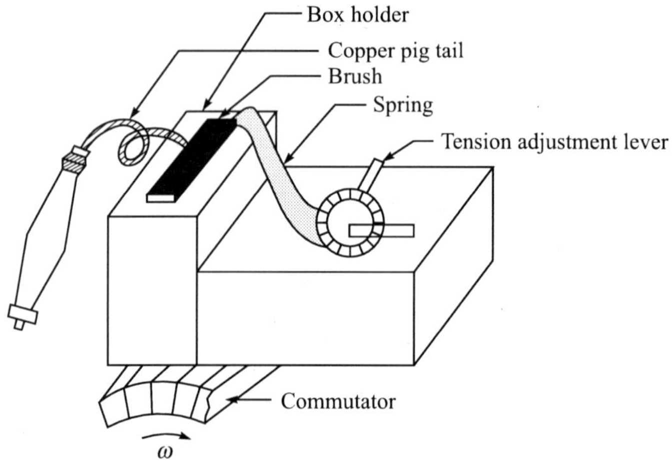

Fig. 6.5 Brush assembly of D.C. machine.

circuit connections. This copper pig tail connects the machines rotor armature conductor to the external electrical circuit through brush. When the brush gets wears or tears then it is replaced with a new brush.

The brush rocker is firmly attached to the frame of the machine. If the brush is not placed properly with respect to the main poles then the arcing persist. So the position of the brush is adjusted so as to minimize or eliminate the arcing near the brush due to the armature reaction.

End Shields

The end shields are the side coverings of the machine. This is also an important part of machine which consist of the set of bearings on which the shaft rests. Its alignment is necessary from point of view of dynamic balance of machine. The end shields have voids or openings for letting the cold air to pass through the machines thus it maintains temperature of the rotor and stator. The end shields are also made up of low permeable magnetic material and they are two in number placed diametrically opposite to each other.

The other important part of machine is the rotor which comprises armature, armature windings, commutator, shaft, etc.

The armature is made up of hard steel and is usually laminated to prevent the eddy losses. The laminations are of 0.35 mm to 0.5 mm thickness depending on the size of machine. One lamination of the armature is shown in Fig. (6.6).

The laminations are insulated from each other and consist of a key way which is fixed over the shaft. So that proper alignment of slots and teeth of the armature laminations is obtained. This will help in the placement of armature conductors or coil sides in the slots of the armature. The slot and teeth number is same for all the rest of the lamination. They are rivetted or held by bolt or welded so as to make

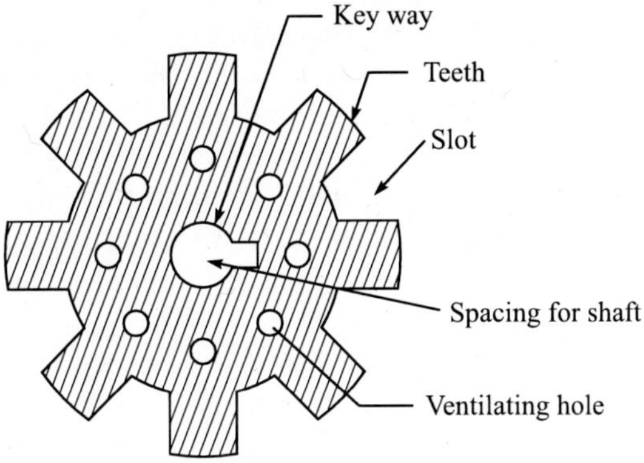

Fig. 6.6 One lamination of armature core.

it a single armature assembly. The complete armature core assembly is depicted in the Fig. (6.7).

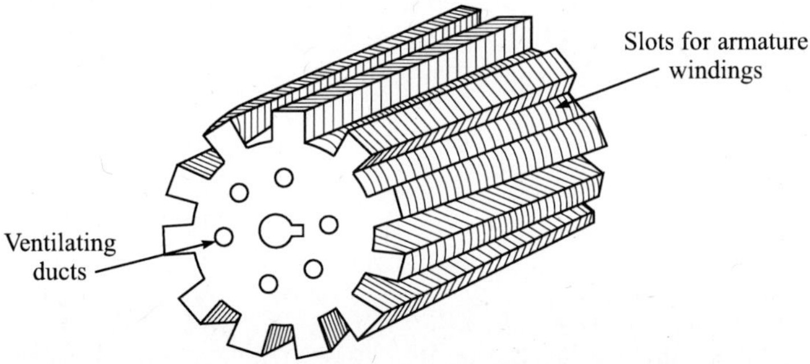

Fig. 6.7 Armature core assembly.

The ventilating ducts enable the cool air to pass through the armature core radially or longitudinally depending on the design of machine. In small rating machines the longitudinal ducts are present but for larger ratings machines both radial and longitudinal ducts are present. The slots of armature core can be semi-closed type or open type. In Fig. (6.7), open type slot is shown. But practically the semi-closed slots is preferred which is shown in the Fig. (6.8) below.

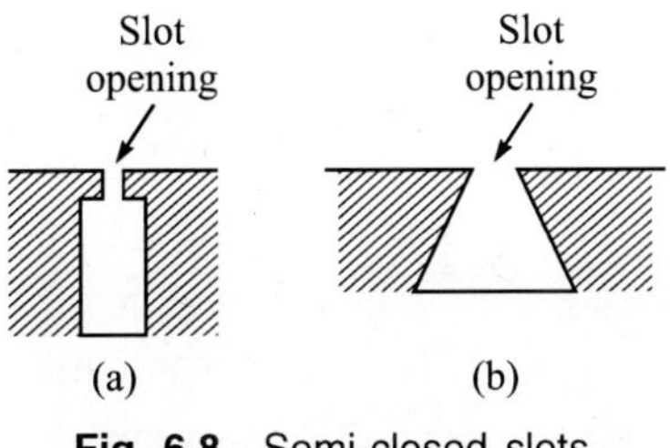

Fig. 6.8 Semi-closed slots.

The slot design is varied for different machines. In the semi-closed type slots, the armature conductor is restricted in the slot and during motion it does not come out. The slot opening is closed by wooden wedge. The slot can have single layer or double layer depending on the rating of the machines. In double layer winding the two layers are separated by insulation. The lower and upper conductors are insulated from the sides of slot. Usually, double layer winding is preferred in larger rating machines. The steel bands are used to keep the conductor in the slots and the steel bands is wound around the core. Thick bands are used whose number vary depending on the length of armature core.

The armature windings are made up of copper and is insulated with enamel. The armature windings are of closed type or open type. If the windings are closed in itself, then it is called closed winding. If the windings end does not close or the ends remain open, then it is called open winding. In the case of d.c. machines, the close type windings are used. The armature windings are placed in slot as shown in Fig. (6.9) and Fig. (6.10). The Fig. (6.9) shows single layer winding whereas Fig. (6.10) shows double layer winding.

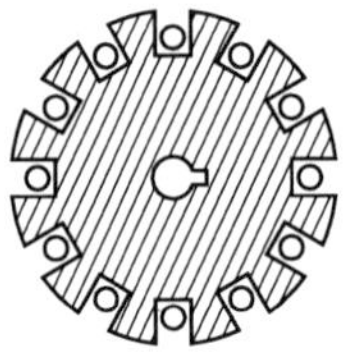

Fig. 6.9 Single layer winding.

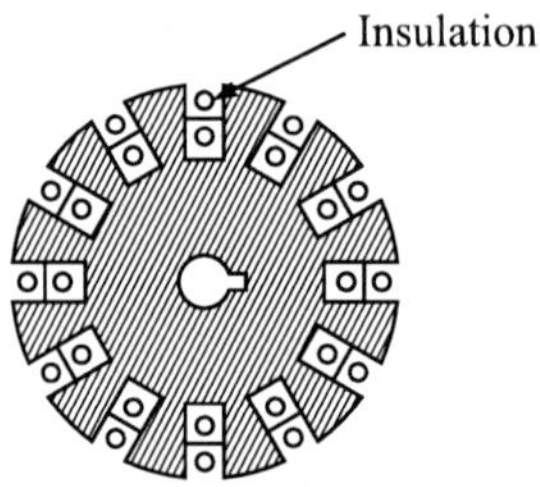

Fig. 6.10 Double layer winding.

The armature windings are further divided into two main types i.e., lap winding and wave winding. Lap winding is used when the current rating of machine is high. Since the lap winding comprises large number of parallel paths depending on the number of poles of machines. The lap winding can be progressive or retrogressive which is shown in Fig. (6.11).

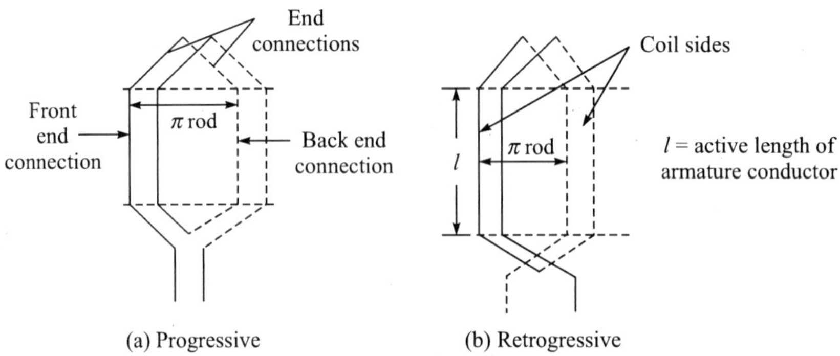

Fig. 6.11 Lap winding (two turn).

In the progressive lap winding, the winding progresses ahead on the armature core in the forward direction. In the retrogressive winding, the winding is done against the forward direction.

The other armature winding connection which is used for high voltage, low current machine is the wave winding. In this winding the number of parallel path is fixed and that is only two. The wave winding is shown in Fig. (6.12).

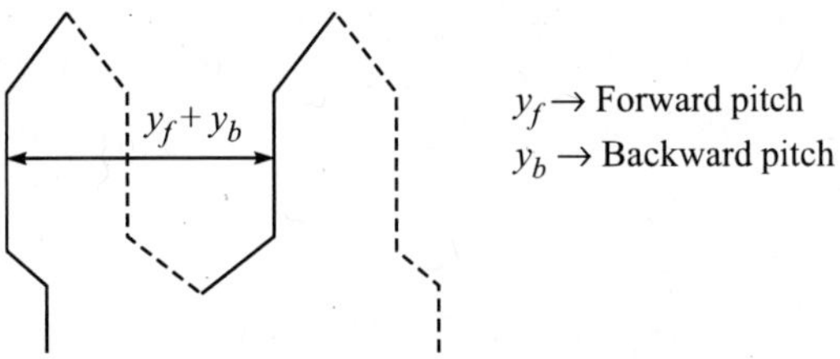

Fig. 6.12 Two-turn wave winding.

The armature conductors are connected to the commutator segments which is also made up of hard drawn copper. The armature conductor is connected to the commutator segments through lugs or riser as shown in the Fig. (6.13).

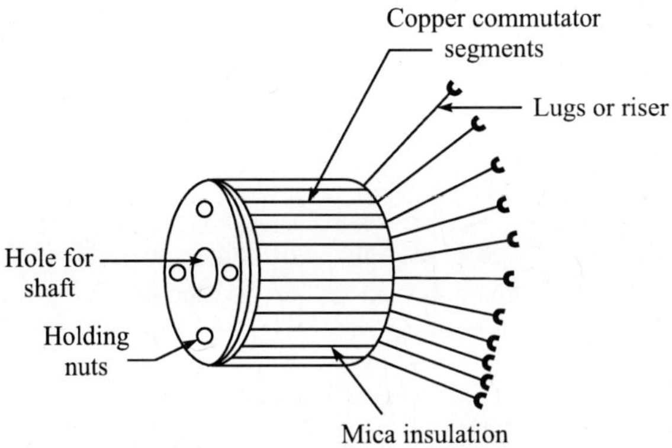

Fig. 6.13 Commutator assembly.

The commutator segments collects the current from the armature windings in the case of d.c. generator and sends it to the external circuit through brush. The current in the armature winding is a.c. which is converted to d.c. by the brush commutator combination. In short, we can say that the brush commutator combination acts as a mechanical rectifier. The brush thickness is usually 1.5 times that of commutator segment thickness. The commutator segments are separated by micanite layer to prevent shorting of armature conductors. To understand the rectification process the commutation concept of d.c. machine should be clear. It is discussed later in this chapter.

The side view of the armature along with the commutator and shaft is shown in Fig. (6.14). It will give a better understanding of the rotor.

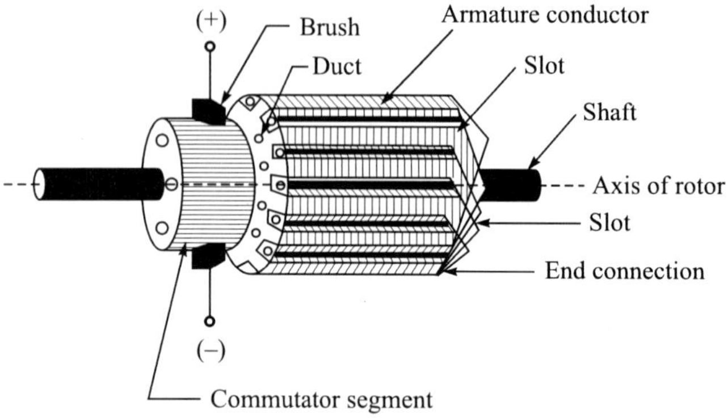

Fig. 6.14 Side view of rotor assembly.

The end connections of the armature windings are taken together and are held by tape. The end connections do not take part in the induction process nor they contribute to the electromotive force generated. The shaft which is made up of steel passes through the rotor axially. The mechanical member is attached to the shaft of the machine. The shaft is placed over the bearings which is fixed to the end shields. The bearings are of different types depending on the machine size. It may be roller bearings or pedestal bearings depending on the size of machine.

6.3 TYPES OF D.C. MACHINES

The d.c. machines are broadly divided into two main types, namely, d.c. motor and d.c. generator. The d.c. generator is further divided into two types on the basis of excitation. The classification of d.c. machines is given below

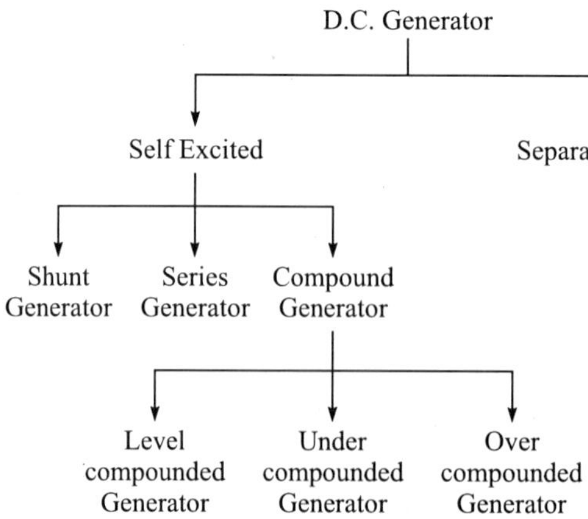

The separately excited generator have a separate d.c. source which is used for d.c. field excitation. The Fig. (6.15) shows the line diagram of separately excited d.c generators.

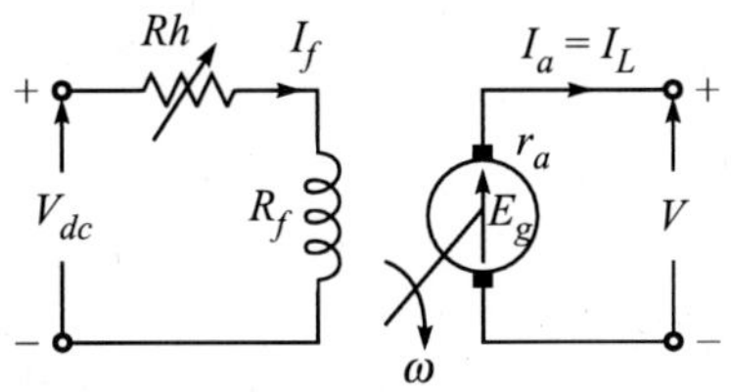

Fig. 6.15 Separately excited D.C. generator.

In the above figure, we see that the field winding is excited by 'V_{dc}' separately. The rheostat '*Rh*' controls the exciting field current 'I_f' and hence controls the field produced by the field poles. Thus, it controls the terminal voltage '*V*' of the generator.

The generated voltage 'E_g' is related to the terminal voltage '*V*' by the eq. (6.1) as follows:

$$\boxed{E_g = V + I_a r_a + V_b} \tag{6.1}$$

where, E_g = Generated voltage, *V*
V = Terminal voltage, *V*
I_a = Armature current, *A*
r_a = Armature winding resistance, Ω
V_b = Brush voltage drop, *V*

Neglecting the brush drop, we have

$$\boxed{E_g = V + I_a r_a} \tag{6.2}$$

The brush drop per brush = 1 volt (approx)
In the field circuit,

$$\phi_f \propto I_f \tag{6.3}$$

ϕ_f = Field flux, *Wb*
I_f = Field current, *A*

If the rheostatic resistance is increased, then 'I_f' is decreased and hence 'ϕ_f' is decreased simultaneously.

$$E_g \propto \phi_f \text{ (If } \omega = \text{constt)} \tag{6.4}$$

where, 'ω' = Prime mover angular speed, rpm

As 'ϕ_f' is decreased, 'E_g' is decreased accordingly. For constant '$I_a r_a$' drop, the terminal voltage 'V' decreases.

As the separately excited generator requires a separate d.c. source, so it adds an extra cost. So there is a popularity of the self excited generator which is discussed in the next topic.

Self-Excited Generator

The self-excited generator uses its own generated current for exciting the field winding. So the cost of the generator is decreased for the same capacity as that of separately excited generator. Thus depending on the connections of the field winding with respect to that of armature winding it is further divided into three main types: namely, shunt, series and compound d.c. generators.

Shunt Generator

The shunt generators have the field winding placed in parallel with respect to that of armature winding as shown in Fig. (6.16) given below.

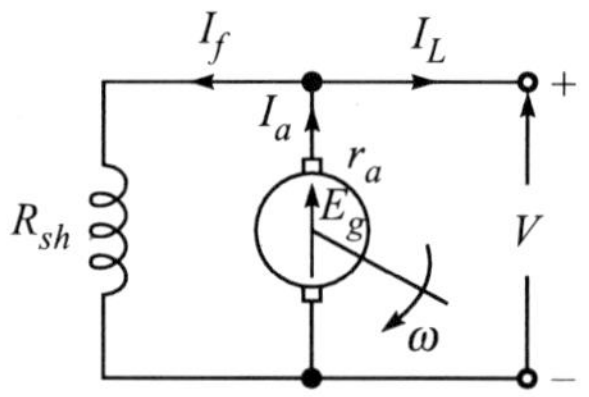

Fig. 6.16 D.C shunt generator.

The voltage equations of the d.c. generator is given by KVL as given below.

$$E_g = V + I_a r_a + V_b \quad (6.5)$$

Neglecting 'V_b', since it is very small

$$E_g \cong V + I_a r_a \quad (6.6)$$

Using KCL,

$$I_a = I_L + I_f \quad (6.7)$$

$$V = I_f \cdot R_{sh} \quad (6.8)$$

So,

$$I_f = \frac{V}{R_{sh}} \quad (6.9)$$

If 'R_{sh}' is constant, then I_f = constant

So, the field flux 'ϕ_f' is also constant in the case of shunt generator.

Mathematically,

$$\phi_f = \text{constt} \quad (6.10)$$

$\therefore$

$$\phi_f \propto I_f$$

Voltage Build up in D.C. Shunt Generator

In the case of d.c. shunt generator for voltage build-up, the following conditions are necessary:

(a) Presence of residual field and hence residual flux under the field poles.
(b) The machine rotor speed should be greater than critical speed of machine.
(c) The machine field resistance should be less than that of critical field resistance.

The machine field pole core is magnetized and then does not demagnetized completely. So the field pole core have some magnetic domains which are aligned in the same direction. Thus, it provides small magnetic field under the pole shoe called residual field 'B_r'. The flux produced by this field is called residual flux 'ϕ_r'. The rotor armature conductors cuts this flux 'ϕ_r' at a rate depending on the frequency of rotation. Hence, electromagnetic induction takes place according to the eq. (6.11),

$$e = B_r lv \tag{6.11}$$

where, 'e' = emf induced in an armature conductor, V
'B_r' = residual field, tesla
'l' = active length of armature conductor, m
'v' = linear velocity of armature, m/s.

There are several armature conductors per parallel path. So the emf gets added per parallel path armature conductor to give a net emf 'E_r' which will drive an armature current 'I_{a_1}' which will excite the field winding, thus increasing the net field flux to a value equal to ϕ_{f_1} (say) which will increase the generated emf 'E_g' to a value equal to 'E_{g_1}' (say). Thus, again increasing the armature current to a value 'I_{a_2}' which will again increase the field flux to a value equal to ϕ_{f_2} (say). Thus, further increasing to a value of emf to 'E_{g_2}'. This process continues till the saturation of the magnetic material occurs. After the magnetic material of pole assembly reaches saturation. The field flux cannot be further increased and so the generated emf cannot be increased. The voltage builded up of the d.c. shunt generator is shown in Fig. (6.17).

Thus, for voltage build-up, the residual flux 'ϕ_r' is needed. The voltage will build up till the machine rated voltage is reached.

The d.c. shunt generator under loaded condition had only small drop in the terminal voltage so it is considered constant voltage generator. The drop in the terminal voltage is small about 5% from the rated voltage. The d.c. shunt generator are used in offices, hotels etc.

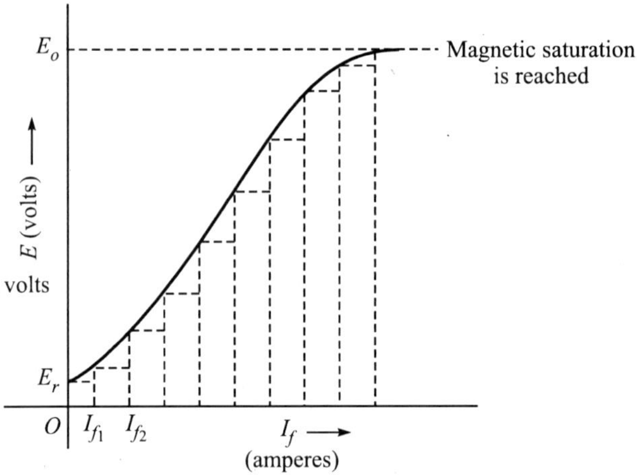

where,

E_r = residual emf, volts

Fig. 6.17 Voltage build-up of d.c. shunt generator.

Series Generator

The series generator has the field winding placed in series with respect to the armature winding. So the armature current is equal to that of the field current. In the case of shunt generator we have seen that the field flux is almost constant. But in the case of series generator the field flux 'ϕ_f' depends on the armature or field current.

Mathematically,
$$\phi_f \propto I_a \tag{6.12}$$

The line diagram of the series generator is given as

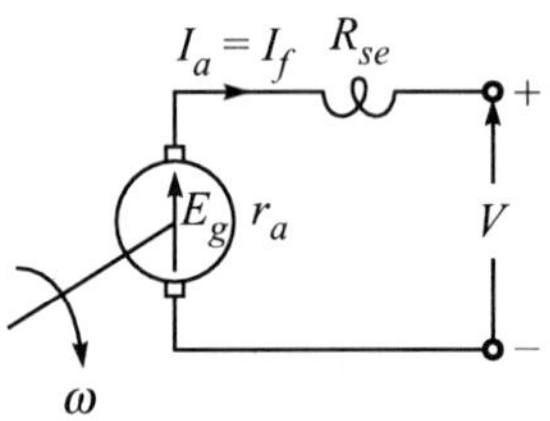

Fig. 6.18 D.C. series generator.

The voltage expression of d.c. series generator is given as in eq. (6.13).

$$E_g = V + I_a(r_a + R_{se}) \tag{6.13}$$

where,
$$R_{se} = \text{Series field resistance, } \Omega$$

$$I_a = I_f = I_L \tag{6.14}$$

We see from eq. (7.12) that the field flux is directly proportional to the armature current. So as the armature current and hence field current increases the field flux 'ϕ_f' increases accordingly. Since the generated emf 'E_g' is directly related to the field flux 'ϕ_f', so the generated emf 'E_g' also increases as field flux 'ϕ_f' increases. The expression as given by eq. (6.13) does not take the effect of brush drop. Including the brush drop 'V_b' the generated emf 'E_g' is given by eq. (6.15),

$$E_g = V + I_a(r_a + R_{se}) + V_b \tag{6.15}$$

As $V_b << V$, so, 'V_b' is neglected. The series generator is used as boosters to increase the voltage at the receiving end of a long feeder.

Compound Generator

The compound generator consists of both series and shunt field winding. Depending on the series and shunt field winding connections the compound generator is further divided into two main types as follows:

Cummulative Compounded Generator: The field produced by the shunt field winding if it aids the series field winding then the net flux inside the air gap under pole shoe is increased. This could be understood by considering the Fig. (6.19).

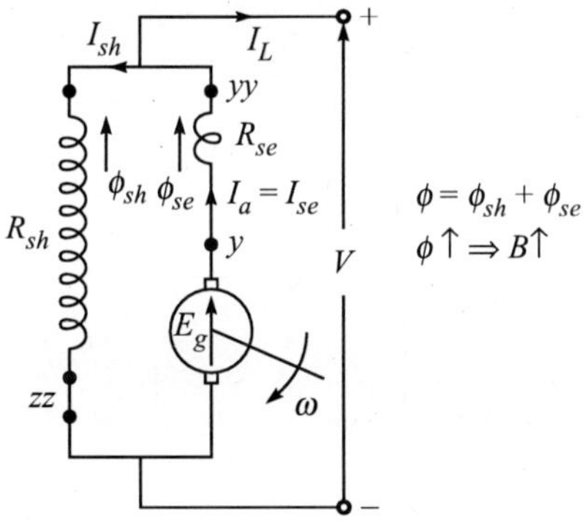

Fig. 6.19 Cummulative compound generator.

The cummulative compound generator is again subdivided into three types : under compound, level compound and over compound generator. Depending on the magnitude of the field fluxes (shunt and series) magnitudes the above classification is done.

Under Compound Generator: If the shunt flux 'ϕ_{sh}' is less than that of series flux 'ϕ_{se}'. This is done by placing a field rheostat in series with the shunt field winding and a diverter resistance in parallel with the series field winding as shown in Fig. (6.20).

where, $R_h \rightarrow$ Field rheostat, Ω
$R_s \rightarrow$ Divertor resistance, Ω

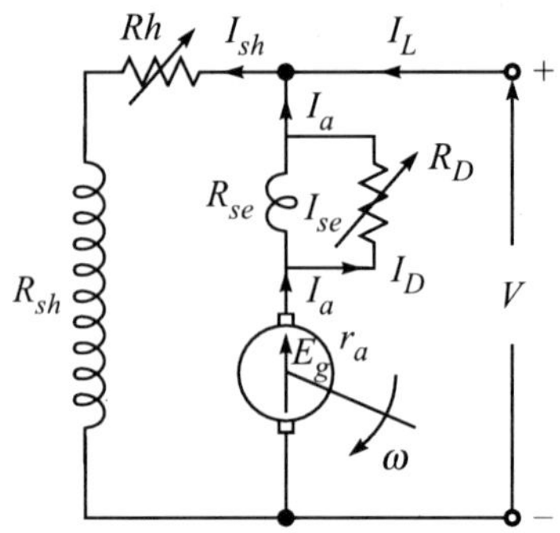

Fig. 6.20 Compound generator (with R_h & R_s).

By decreasing 'R_D', 'I_{se}' can be decreased hence, 'ϕ_{se}' can be decreased. By decreasing 'Rh', 'I_{sh}' can be increased hence, 'ϕ_{sh}' can be increased. Thus, 'ϕ_{sh}' can be increased to a value greater than that of 'ϕ_{se}'. Thus, the generator so obtained is called under compounded cummulative generator.

The voltage equations are

$$V = I_{sh}(R_{sh} + R) \tag{6.16}$$

where, 'R' is the rheostatic resistance, Ω

$$E_g = V + I_{se} R_{se} + I_a r_a \tag{6.17}$$

$$I_a = I_{sh} + I_L \tag{6.18}$$

$$I_{se} = I_a \cdot \left(\frac{R_D}{R_D + R_{se}}\right) \tag{6.19}$$

I_{se}, I_{sh} = series and shunt field winding current, A

Over Compound Generator

In the over compounded generator, the series field winding flux exceeds that of shunt field winding flux. This is done by varying the shunt field and diverter resistances. The over compounded generator boost the terminal voltage so it is preferred for such cases where there is a need of increasing the terminal voltage such as short feeder.

Mathematically, $\phi_{se} > \phi_{sh}$ (For over compound generator) (6.20)

The fluxes can be changed by changing the number of turns of the field winding. But it is seldom used.

Level Compound Generator

It is often required that the voltage at the terminals should not vary from no load to full load conditions. For that it is necessary that the magnitude of shunt and series fluxes should be equal as per the eq. (6.21).

$$\phi_{se} = \phi_{sh} \tag{6.21}$$

The cummulative generator can be short shunt or long shunt depending on the connections of windings. If the shunt field winding is connected across the armature winding and series field winding as shown in Fig. (6.20), then such a generator is called long shunt compound generator.

If the shunt field winding is connected across the armature, then it is called short shunt compound generator as shown in Fig. (6.21).

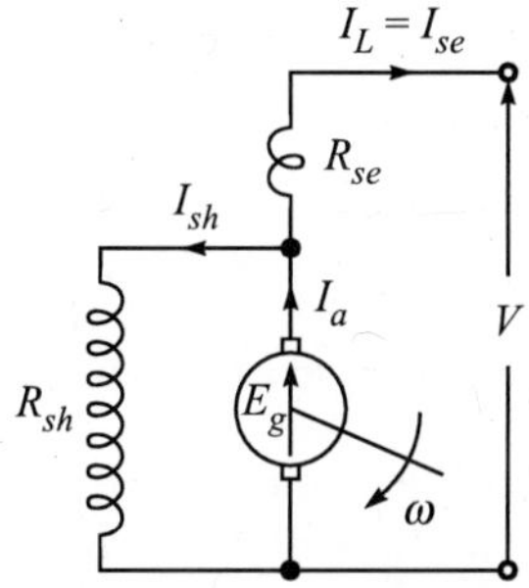

Fig. 6.21 Short shunt compound generator.

The voltage and current equations for short shunt generator are given as

$$E_g = V + I_a r_a + I_L R_{se} \tag{6.22}$$

$$V - I_{se} R_{se} = I_{sh} R_{sh} \tag{6.23}$$

$$I_a = I_{sh} + I_L \tag{6.24}$$

$$I_L = I_{se} \tag{6.25}$$

In the above voltage equations, the brush drop is neglected which is usually very small with respect to the rated voltage.

6.4 PRINCIPLE OF OPERATION

When a current carrying conductor placed in a field is rotated at an angular velocity 'ω' (say). Then due to cutting of fluxes by the conductor an emf is induced in the conductor coil sides according to Faraday's law of e.m.f induction. The direction of the emf is given by Flemming's right hand rule.

The principle can be alternatively explained as follows: with the conductor placed on rotor which is rotating at a fixed frequency. The flux linkage with the conductor varies for different positions of the rotor. We can imagine that there is a change in flux linkage with the conductor with change in time.

Mathematically, it can be expressed as

$$e = -\frac{d\lambda}{dt} \tag{6.26}$$

where, e = Induced emf, V
$d\lambda$ = Incremental change in flux linkage
dt = Incremental change in time, sec

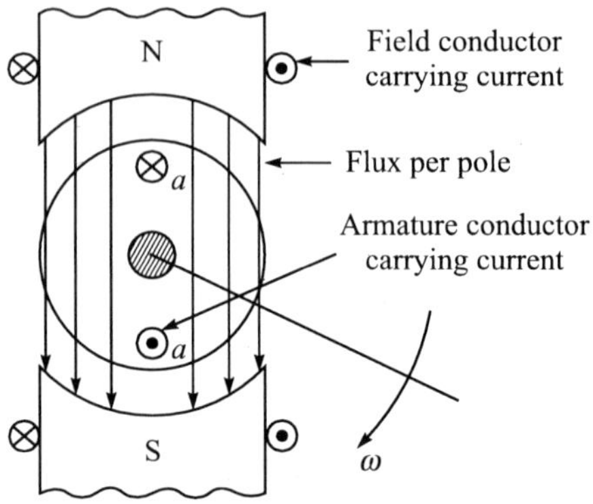

Fig. 6.22 Emf generation in D.C. Generator.

The negative sign in eq. (6.26) gives the direction of emf which is given by Lenz's law. That is, it will oppose the very cause producing the emf.

From Fig. (6.22), we conclude that the emf is induced only in the active length of armature conductor coil sides and not in the overhang of coil.

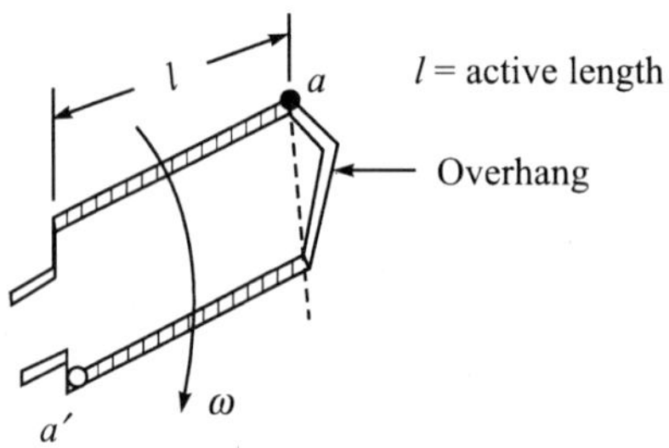

Fig. 6.23 One turn of an armature coil.

In a d.c. machine there are large number of armature conductor per parallel path between the brushes. So, the emf's are added for the armature conductors which are in series, that is, which carries the same current. Thus, these small magnitude emf's are added up to give a large emf magnitude across the brushes. The e.m.f which appears across the brushes remains constant irrespective of the number of parallel paths between the brushes.

6.5 SYMBOLIC REPRESENTATION

The d.c. machine can be represented symbolically by a battery of magnitude 'E_g' having an internal resistance 'r_a' in series with it as shown in Fig. (6.24).

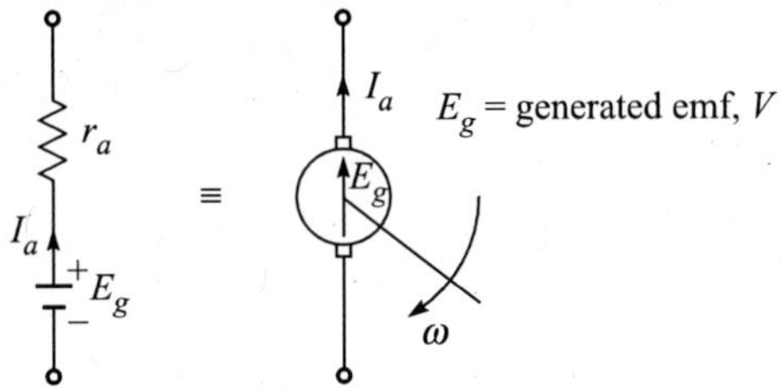

Fig. 6.24 Symbol of d.c. machine.

The symbolic representation of d.c. motor is given in Fig. (6.25).

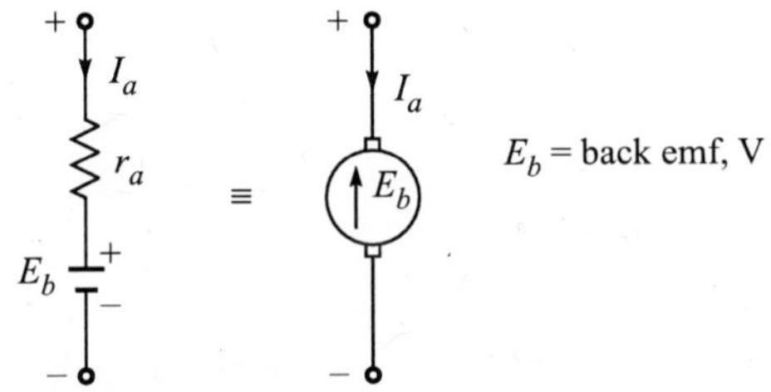

Fig. 6.25 Symbolic representation of d.c. motor.

The symbolic representation of the d.c. machine simplifies the mathematical calculation and hence simplifies the analysis of the d.c. machine. This enables us to correlate the machine with electrical parameters such that the Kirchhoff's law and Ohm's law can be applied effectively.

6.6 CHARACTERISTICS OF D.C. GENERATORS

The characteristics of the generators enable us to get the understanding of machine performance and to determine relevant performance parameters. Two machines can be compared. Basically, three characteristics are obtained in the case of generators, namely, internal characteristic, external characteristic and the magnetization characteristic.

Internal Characteristic: The internal characteristic of the machine is drawn between 'E' and I_L where 'E' is the induced emf in the armature, in volts and 'I_L' is the load current, in amperes.

External Characteristic: The external characteristic is obtained between 'V' and 'I_L'.

$$V = \text{Terminal voltage, volts.}$$
$$I_L = \text{Load current, amps.}$$

The external characteristic is often called the load characteristic or performance characteristic of the machine.

Magnetization Characteristic: This characteristic is obtained between 'E_o' and 'I_f' where 'E_o' is the no load emf, in volts and 'I_f' is the field current, in amperes. The magnetization characteristic is also called no load or OCC (open circuit characteristic) or no load saturation characteristic. Using magnetization characteristic, we get the information of the magnetic material used in the machine.

Magnetization Characteristic of Separately Excited D.C. Generator

The magnetization characteristic of the separately excited generator has the same shape as BH curve as shown in the Fig. (6.26).

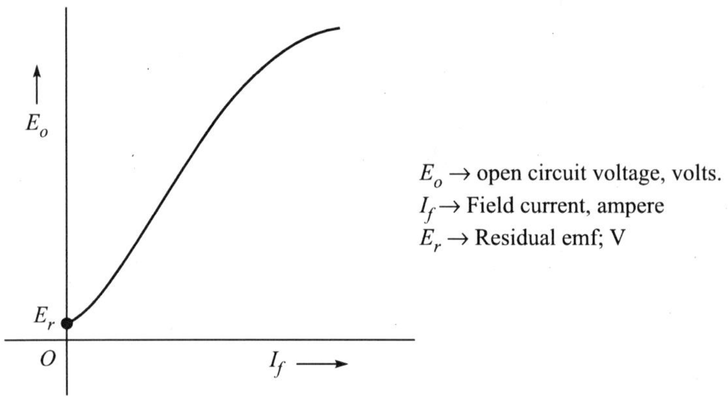

Fig. 6.26 OCC of separately excited d.c. generator.

The magnetization characteristic does not start from the origin but above it due to presence of residual flux under the field pole of the machine. The curve increases linearly for the machine till its pole is unsaturated but it bends as the machine gets saturated. In the linear portion it is considered that the magnetic flux intensity 'B' bears a direct relation with the magnetic field strength or magnetization intensity 'H'. After saturation of the machine it is not possible to further increase the open circuit voltage 'E_0'. The magnetization characteristic is obtained under no load or open circuit condition, so the magnetization characteristic is sometimes called the no load saturation characteristic or open circuit characteristic; in short it is represented as OCC.

The magnetization characteristic tells about the history of magnetic material used for the machine.

Magnetization Characteristic of Self Excited Generator

The magnetization characteristic of the self excited generators is shown in the Fig. (6.27) below.

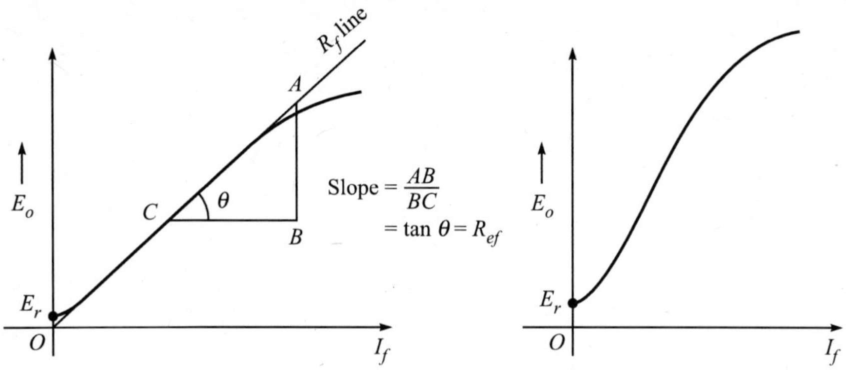

(a) OCC of shunt generator (b) OCC of compound generator

Fig. 6.27 Open circuit characteristic of a self excited generator.

The shape of the characteristic remains same for the d.c shunt and compound generator. The shape depends on the magnetic materials used. Since the machine is made up of ferromagnetic material. The magnetic property of the material varies and hence the characteristic also differs for different machines.

The series generator does not have any open circuit characteristic since under open circuit condition the current through field winding is zero and hence there will be no field flux and so it is not possible to obtain the OCC since the field flux control is not possible. From the OCC, the critical field resistance 'R_{cf}' can be determined which is the maximum value of field circuit resistance above which machine does not excite. The slope of the curve in the linear portion as shown in Fig. 6.27(a) corresponds to 'R_{cf}' (critical field resistance) of the d.c. machine.

Load Characteristic of Separately Excited D.C. Generator

The characteristic obtained under loaded condition is called load or external characteristic of the d.c generator. There are mainly two load characteristics, namely, internal characteristic and the external characteristic.

(a) Internal Characteristic: The internal characteristic of the generator is the curve drawn between the induced generated emf 'E' and the load current 'I_L'.

Since the induced generated emf 'E' cannot be realized practically, so another characteristic is used—called the external characteristic.

The internal characteristic of the d.c. separately excited generator is shown in the Fig. (6.28) below.

where, V_r = Rated voltage of the machine, volts
E = Induced emf, volts
I_L = Load current, amp

The external characteristic of the separately excited generator can be obtained by subtracting '$I_a r_a$' drop and armature reaction drop from the internal characteristic as shown in Fig. (6.28).

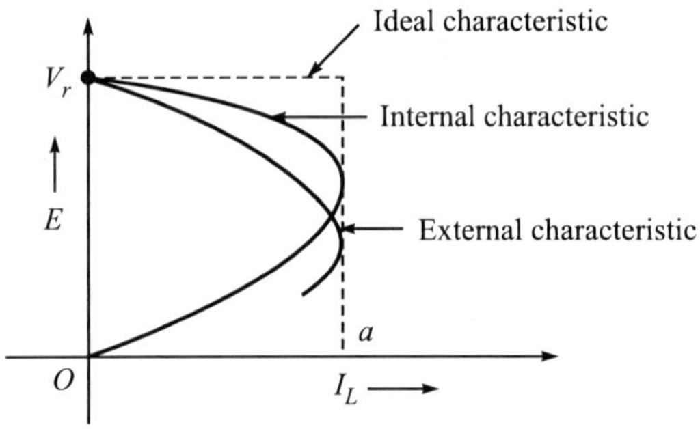

Fig. 6.28 Internal characteristic of separately excited generator.

Similarly, the load characteristic for a self excited d.c generator can be shown which is discussed in the coming section.

Load Characteristic of Self Excited D.C. Generator

The load characteristic is generally meant by the characteristic drawn by considering the external electrical quantities, namely, the load voltage 'V_L' and the load current 'I_L'. Since these two quantities can be measured easily using common measuring instruments. So this characteristic is sometimes referred to as the external characteristic or the performance characteristic of the d.c. generator. Since the performance parameters such as voltage regulation can be determined by the load or external characteristic.

Load Characteristic of D.C. Shunt Generator

The load characteristic of the d.c shunt generator is shown in the Fig. (6.29).

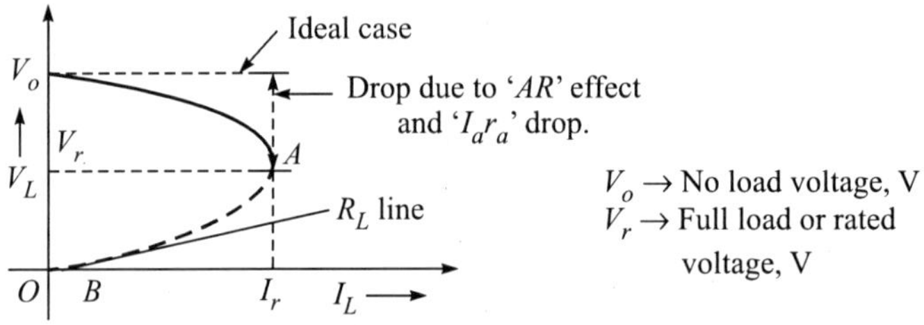

Fig. 6.29 Load characteristic of D.C shunt Generator.

There is a drop in terminal voltage from no load to full load condition. This is due to the armature reaction demagnetization effect and due to armature resistance voltage drop. After point 'A' the external characteristic bends in the backward direction due to rise of armature reaction flux weakening effect. It becomes so strong that the flux under the main poles decreases to a low value and the voltage across the terminal drops to zero value giving rise to dead short circuit.

The reversal of the current is practically not desired since it will lead to decrease in the residual magnetism to zero value. So the generator fails to build up voltage after the dead short circuit condition. The generator will fail to give voltage if restarted due to zero residual flux under the pole. That is, the poles of machine is completely demagnetized. So as to produce residual magnetism, the d.c generator is made to work as a motor and then it is stopped. The field poles are not demagnetized completely. Again if the machine is run as a generator then again the machine starts building voltage across its terminals. The load characteristic of the d.c. shunt generator can be realized using the voltage equation of d.c. generator.

The voltage generated in d.c. shunt generator is given by

$$E_g = V + I_a r_a + V_b, \text{ volt} \tag{6.27}$$

If the voltage across brushes is neglected, then the expression reduces to

$$E_g = V + I_a r_a \tag{6.28}$$

$\therefore$

$$I_a = I_L + I_f \tag{6.29}$$

I_f = constant, for d.c. shunt generator if the field circuit resistance is constant. Substituting the values of 'I_a' in the above expression, we have

$$E_g = V + (I_L + I_f)r_a \tag{6.30}$$

r_a = armature resistance, I_f, I_L are field and load current respectively. The terminal voltage is

$$V = E_g - (I_L + I_f)r_a \tag{6.31}$$

Since,

$$E_g \propto \phi N \tag{6.32}$$

$$\phi = \text{constant (for shunt generator)}$$

So,

$$E_g \propto N \tag{6.33}$$

If the value of 'N' is constant, then 'E_g' is also constant, so $E_g = E_o$(say) Substituting the values of 'E_g' in eq. (6.30), we have

$$V = E_0 - (I_L + I_f)r_a$$

or,

$$V = E_0 - I_f r_a - I_L \cdot r_a$$

$\because$

$$I_f r_a = \text{constant} = C$$

So,

$$V = \underbrace{E_0 - C} - I_L \times r_a$$

or,

$$\boxed{V = V_0 - I_L \cdot r_a} \tag{6.34}$$

At no load, $I_L = 0$, $V = V_0$ (no load terminal voltage)

The above expression is of a straight line having drooping nature and having slope equal to '$-r_a$'.

So the ideal characteristic is given by

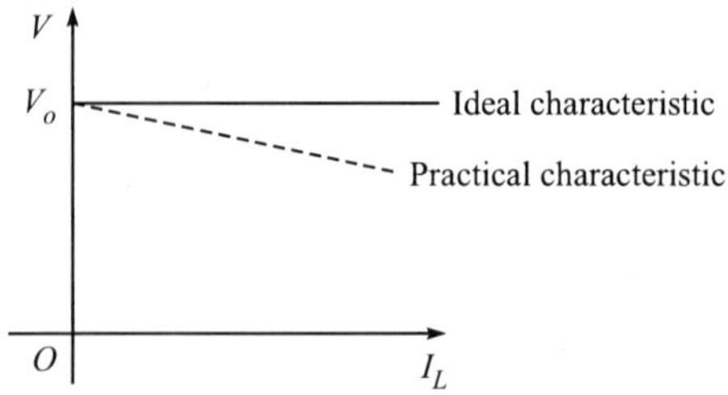

Fig. 6.30 Load characteristic of d.c. shunt generator.

Load Characteristic of D.C. Series Generator

The load characteristic of the d.c. series generator differs from that of d.c. shunt generator. In the d.c. shunt generator, the terminal voltage decreases with the increase of load current whereas in the series generator the terminal voltage increases with increase of load current. This could be understood by considering the voltage equation

$$E_g = V + I_L(R_a + R_{se}) + V_b \tag{6.35}$$

where, $V \rightarrow$ Terminal voltage, volts
$I_L \rightarrow$ Load current, amperes
$R_a, R_{se} \rightarrow$ Armature and field resistances, in ohms.
$V_b \rightarrow$ Brush drop, volt
$E_g \rightarrow$ Generated emf, volt

Rearranging eq. (6.35), we get

$$V = E_g - I_L(R_e + R_{se}) - V_b$$

The brush drop is constant, so it could be dropped from the above expression given by eq. (6.35), we get

$$V = E_g - I_L(R_a + R_{se}) \tag{6.36}$$

The generated emf is given by

$$E_g = \frac{NP\phi Z}{60A}$$

If the above expression, P, Z, A are constant, so it could be taken in the proportionality constant as written below

$$E_g \propto \phi N$$

If the speed of prime mover 'N' is constant, then the above equation reduces to

$$E_g \propto \phi$$

Since the flux 'ϕ' produced by the field winding depends on the load current 'I_L', so we can consider

$$\phi \propto I_L \tag{6.37}$$

or, $$E_g \propto I_L$$

or, $$E_g = K' I_L, \text{ volt} \tag{6.38}$$

Substituting the value of 'E_g' from eq. (6.38) in eq. (6.36), we get

$$V = K' I_L - I_L(R_a + R_{se})$$

Collecting the terms, we have

$$V = [K' - (R_a + R_{se})]I_L$$

Since, $$R_a + R_{se} = \text{constant}$$

so, $$V \propto I_L \tag{6.39}$$

The above equation is the linear relation between V and I_L and corresponds to straight line equation starting from origin.

If $$I_L = 0 \text{ then } V = 0 \qquad \text{(at no load)}$$

So the load characteristic starts from origin as shown below in Fig. (6.31).

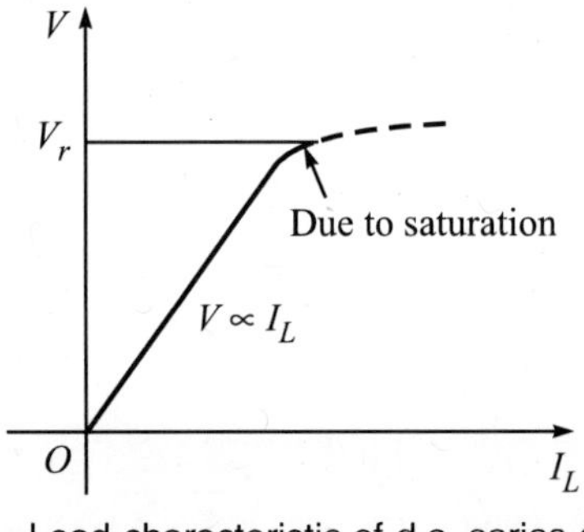

Fig. 6.31 Load characteristic of d.c. series generator.

The load characteristic bends for larger value of load current. As the magnetic material gets saturated for higher value of 'I_L'.

This type of generator is used for boosting the voltage in the case of long feeders where there is drop in the voltage from sending end to receiving end due to line drop.

Load Characteristic of Compound Generator

The load characteristic of compound generator can be divided into two categories, namely, one for differential compounding and the other for cummulative compounding. The cummulative compounded generator is further subdivided as over compound generator, under compound and level compound generator. So we will have three different load characteristics for the cummulative compounded generator. The characteristics are depicted in the Fig. (6.32).

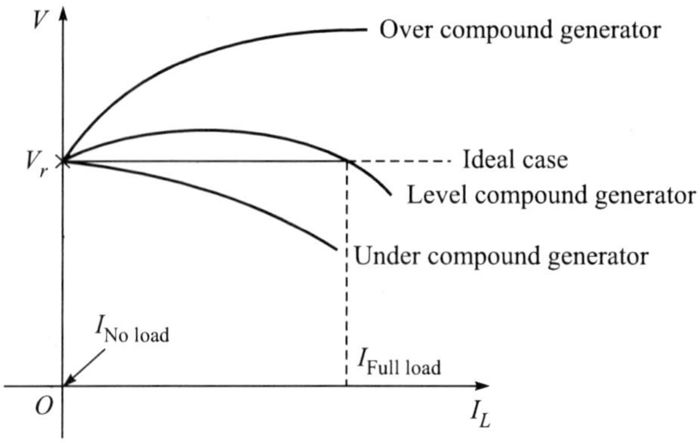

Fig. 6.32 Load characteristics of cummulative compounded generator.

The load characteristic for differential compound generator is shown in Fig. (6.33). There are only few practical uses of this type of generator.

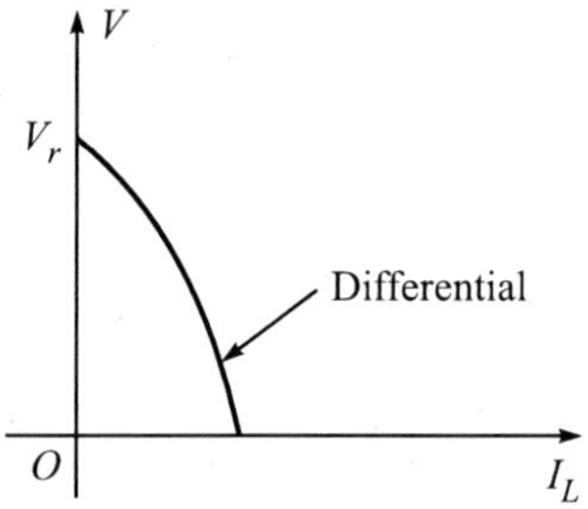

Fig. 6.33 Load characteristic of differential compound generator.

The differential compound generator is used for arc welding purposes and for research purpose, whereas the cummulative compound generator is used for varieties of applications such as boosters, for hotels, offices and for industries.

The performance characteristics of the d.c. generator gives the relation between the external parameters so sometimes the external characteristics is referred to as performance characteristic of the d.c. generator.

6.7 EMF EQUATION OF D.C. GENERATOR

The emf induced in the active length of armature conductor is given by the exp. (6.40) as per Faraday's law of electromagnetic induction as follows.

$$e = B\, l\, v \tag{6.40}$$

where, B = magnetic intensity flux, Wb/m^2

l = active length of armature conductor, m

v = linear velocity of conductor, m/s.

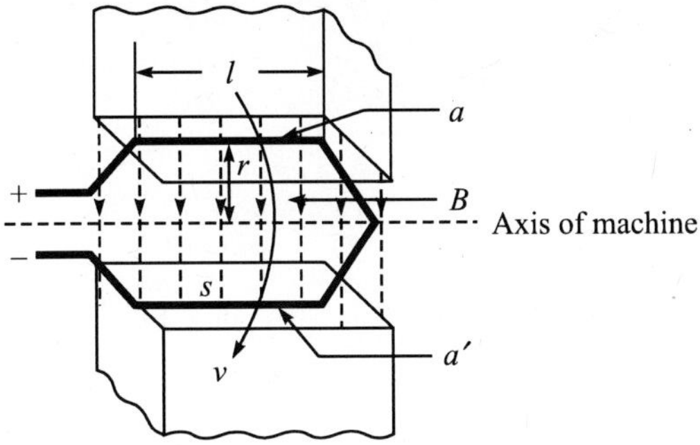

Fig. 6.34 EMF induced in armature coil.

The magnetic flux intensity 'B' is related to the flux per pole by the relation given by exp. (6.41).

$$\phi = B A_c, \text{ weber}$$

where, ϕ = Flux per pole, Wb

A_c = Cylindrical surface area, m^2

So, $A_c = 2\pi r l$, m^2

where, r = distance of armature conductor from axis of machine, m.

Thus,
$$B = \frac{\phi}{A_c} \quad (6.41)$$

The total emf induced in the armature coil is the sum of e.m.f induced in the armature conductors 'a' and 'a''. The emf's are in series so they get added. For two pole machine,

$$E = 2 B l v \quad (6.42)$$

The general expression of emf induced for 'P' pole machine can be obtained as follows

$$\text{Total flux inside the air gap} = \phi \times P \quad (6.43)$$

So, the average field in the air gap between stator and rotor is given by exp. (6.44).

$$B = \frac{\phi \times P}{2\pi r l} \quad (6.44)$$

So, the average emf induced in the armature winding is given by exp. (6.45).

$$E = \frac{\phi \times P}{2\pi r l} \times l \times r \times w \quad (6.45)$$

Since, $v = rw$, m/s

where, w = angular velocity of armature in rad/s

which is given by

$$w = 2\pi n, \text{ rad/s}$$

where, 'n' is in rev/s
So, on substituting 'w' in exp. 6.46, we get,

$$E = \frac{\phi \times P}{2\pi r l} \times l \times r \times 2\pi n$$

$$E = \phi \times P \times n, \text{ volts.}$$

Between the brushes, there are $\frac{'Z'}{A}$ armature conductors which are in series, so the net e.m.f induced in the armature between the brushes is given by exp. (6.46).

$$E_t = \frac{Z}{A} \cdot E, \text{ volts} \qquad (6.46)$$

$$E_t = \frac{Z}{A} \cdot \phi \cdot P \cdot n, \text{ volts.}$$

since
$$n = \frac{N}{60} \text{ rps}$$

where, 'N' is in rpm

So,
$$\boxed{E_t = \frac{Z\phi PN}{60\,A}} \rightarrow \text{Emf equation of d.c. machine}$$

$$E_t = E_g = \frac{NP\phi Z}{60\,A}$$

P, Z, A are constant for a d.c. generator and are regarded as machine constants

So,
$$E_g \propto N\,\phi \qquad (6.47)$$

The generated voltage is dependent on the speed of prime mover and the field flux. That is, with the increase of either quantity the generated voltage increases accordingly.

For increasing the power generation the prime mover speed is varied from stability point of view.

6.8 COMMUTATION

The commutation process regulates the direction of alternating current especially to make it direct current. This commutation is done by commutator which ensures that electric current flows as direct current. This commutation process could be visualized by considering the Fig. (6.35).

As the commutator segment moves in the forward direction the brush leaves the segment (2) and comes in contact with the segment (1) of the commutators. When the contact of brush with the segment (2) decreases the current collected by the brush from the copper segment (2) of the commutator decreases and the

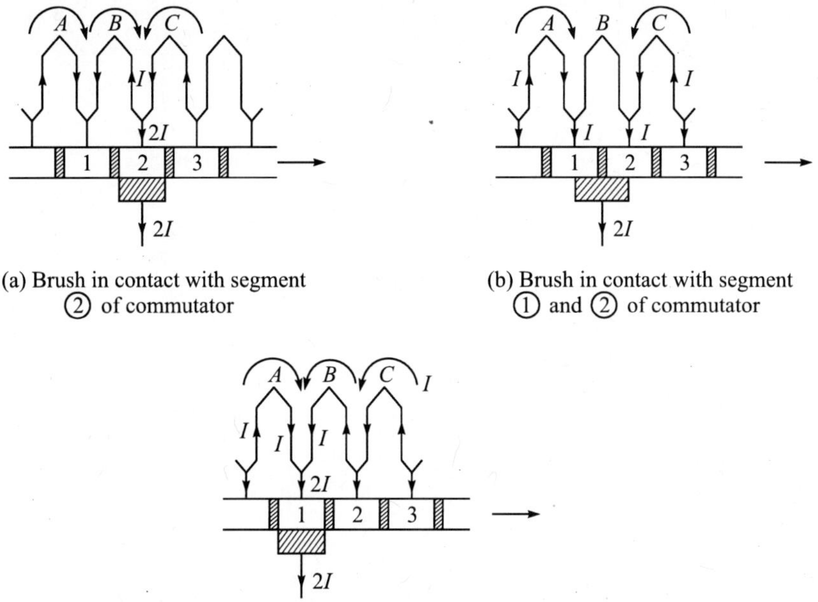

(a) Brush in contact with segment ② of commutator

(b) Brush in contact with segment ① and ② of commutator

(c) Brush in contact with segment ① of commutator

Fig. 6.35 Commutation process in d.c. machine.

current collected from the segment (1) of commutator increases. This is so because the armature windings are in contact with the commutator segments so as the armature coil leaves the north pole field and enter the south pole field then the current through the coil undergoing commutation should change from $+I$ to $-I$.

If this reversal does not occur, then an arc is produced between the copper segment of the commutator and the brush so that the amount of current collected by the brush could be equal to '$2I$'. The difference in the current from '$2I$' follows the arc path which is of low resistance and does not follow the high resistance path i.e., the brush which is having higher resistance due to presence of carbon.

If the commutation takes place in a linear manner as shown in Fig. (6.36) then that commutation is called linear commutation or normal commutation respectively.

In the Fig. (6.36), 't_c' is the time of commutation i.e., the time in which the current reversal takes place.

Except the case of linear commutation and sinusoidal commutation other commutation processes lead to sparking which eventually causes pitting of the commutator segments and hence there is a need of replacement of the copper commutator segments until some alternative methods are used to improve the commutation process.

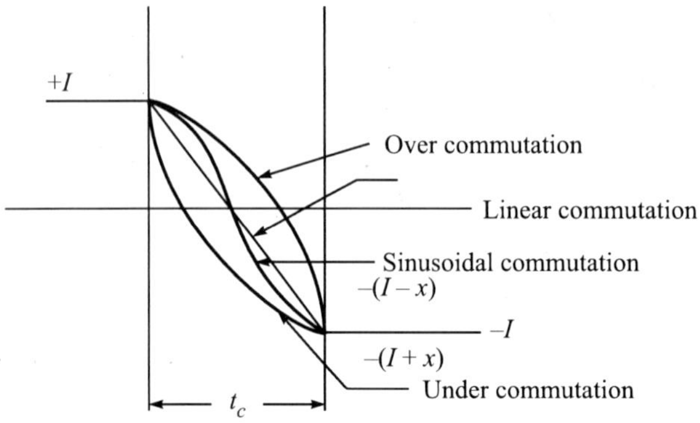

Fig. 6.36 Types of commutation.

Wherever there is a reversal of current in the coil undergoing commutation, a reactance voltage 'V_r' is produced due to the self-inductance of the coil say 'L_c'. Using the Faraday's law, this voltage can be expressed as

$$V_r = \frac{d\phi}{dt} \tag{6.48}$$

since,

$$\phi = L_c i_c \tag{6.49}$$

where, 'L_c' is the self inductance of armature coil undergoing commutation and 'i_c' the armature coil current.

Substituting the value of 'ϕ' in exp. (6.48), we have

$$V_r = \frac{d}{dt}(L_c i_c) = L_c \frac{li_c}{dt} \tag{6.50}$$

Since the armature coil current 'i_c' changes from a value of + 'I' to – 'I' as shown in the Fig. (6.36), so the change in current is nothing but equals to '+2 I' respectively. This change occurs in the period of 't_c' which is the commutation period. So the above exp. (6.50) can be written as

$$V_r = L_c \cdot \frac{2I}{t_c}$$

or,

$$V_r = \frac{2 L_c I}{t_c}, \text{ volts} \tag{6.51}$$

This exp. (6.51) of reactance voltage holds true only for linear or normal commutation.

The time of commutation 't_c' can be expressed mathematically, as follows.

$$t_c = \frac{W_b - W_m}{v_c}, \text{ sec} \tag{6.52}$$

where,

W_b = width of brush, m

W_m = width of mica layer, m

and, $$V_c = \pi D_c n \text{ m/sec}$$

where, 'n' is in rps

'D_c' is the diameter of commutator segment. Substituting

$$n = \frac{N}{60}$$

we get

$$V_c = \frac{\pi D_c N}{60}, \text{ m/s}$$

'V_c' is the linear velocity of commutator, m/s.

Methods of Improving Commutation

The commutation process could be improved by the following methods:

(a) EMF Commutation: In this method an additional emf is induced which opposes the reactance voltage so that the prevention of current reversal due to reactance voltage could be restricted. Since the reactance voltage 'V_r' is responsible for the poor commutation. Thus, to achieve reversal of current from +'I' to –'I' there is need of another field pole which decreases the interpolar space between the main poles and increases the field flux per pole. Other alternative method could be brush shift. But as the brush shift leads to uniform flux distribution below the main poles. So this method is not preferred and commutating poles are used which are called interpolar or compoles which are small poles which are placed in the interpolar region. The polarity of the interpole should be as that of the previous main pole in the direction of rotation.

The other commutation is also responsible for proper commutation called resistance commutation.

(b) Resistance Commutation: The high resistance carbon brushes are used for the resistance commutation. As the contact of the copper segment with the high resistance brush increases, then the part of contact having low resistance draws larger current from the armature coil as compared to the other part of commutator segment which could be understood from the Fig. (6.37).

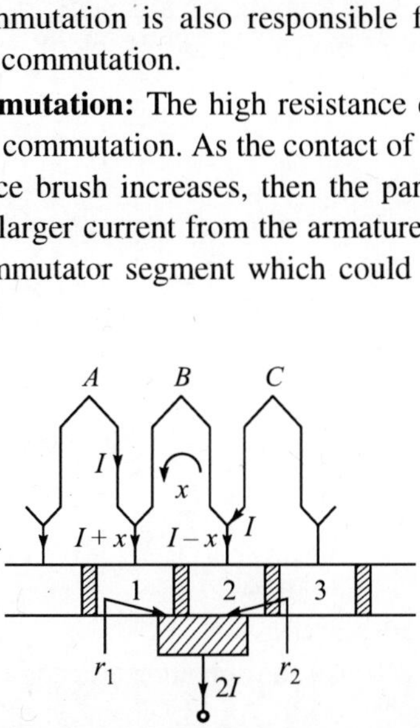

Fig. 6.37 Resistance commutation.

As shown in figure as $r_1 < r_2$ so, a small current 'x' preferred the path through coil 'B'. Higher current is drawn from the coil 'A'. But the total current available across the brush remains equal to '$2I$'. The carbon brushes are also used for lubrication of commutator segments. This method is used in small and medium rating machines. In large machines, both the methods are used.

6.9 ARMATURE REACTION

Whenever current passes through the field winding of the machine then a field flux 'ϕ_f' is produced in the air gap. The field flux under the main poles can be visualized as follows.

The magnetic neutral axis (MNA) passes through that axis in which no flux lines pass ideally, whereas the geometrical neutral axis (GNA) is that axis which divides the north pole region and the south pole region in the machine.

From the Fig. (6.38) it is shown that the flux under the pole shoe is not equal but decreases at one tip of pole shoe and increases at the other tip of the same pole. The Fig. (6.38) shows the armature reaction effect in the case of d.c generator in which the flux decreases under the leading pole tip and increases under the trailing pole tip of the machine main pole shoe. Also the magnetic neutral axis (MNA) of the machine also get shifted in the forward direction i.e. in the direction of rotation of the rotor. The reverse thing happens in the case of d.c. motor that is, the MNA gets shifted against the direction of the rotation of rotor. The resultant field distribution also gets shifted from the initial field distribution as shown in the figure in the forward direction. The brush remains in the previous MNA before the armature reaction effect and so the current reversal in the armature conductor which passes from the north pole field to south pole field can not able to change its direction. So as a result of this there is sparking near the brush it.

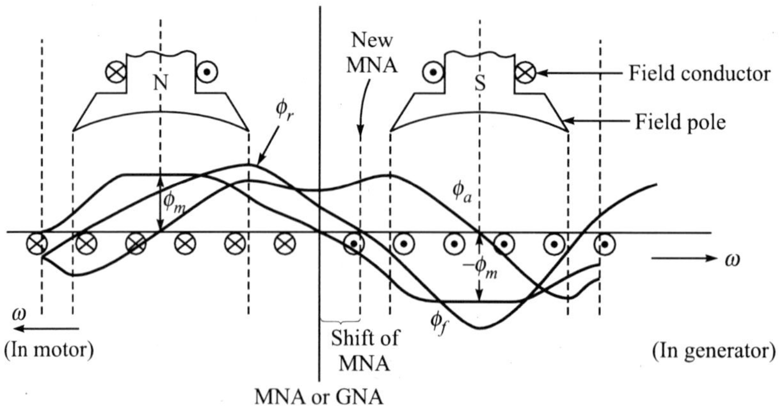

Fig. 6.38 Main pole field flux and armature flux pattern in the air gap of d.c. machine.

Thus, to prevent this sparking near the brushes there is a need of shifting the brushes along the direction of the rotation of the armature, in the case of d.c. generator and in the reverse direction in the case of d.c. motor.

In both d.c. generator and d.c. motor, the armature reaction effect takes place The effect is not desirable in the case of d.c. machines. So, there is a need of decreasing the effect of armature reaction.

Firstly, let's analyze the effect of armature reaction considering the brush shift along the direction of rotation of armature as shown in Fig. (6.39).

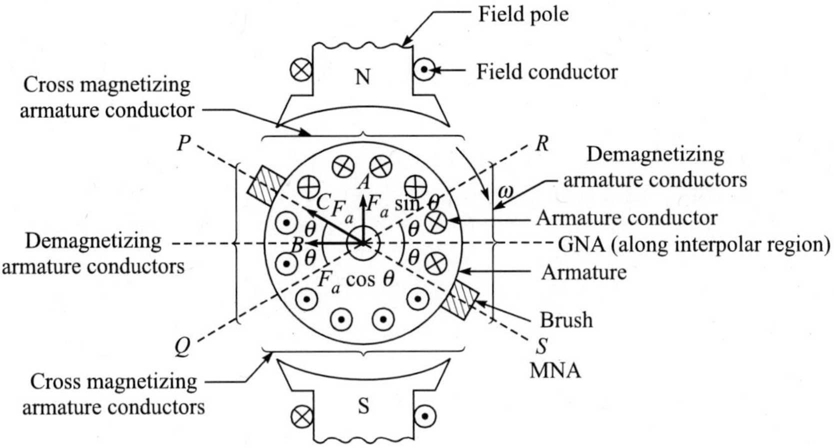

Fig. 6.39 Brush shifted by angle 'θ' along clockwise direction (In case of d.c. generator).

Initially, the mmf due to armature current 'I_a' is along the x-axis direction as shown in Fig. (6.40) shown below.

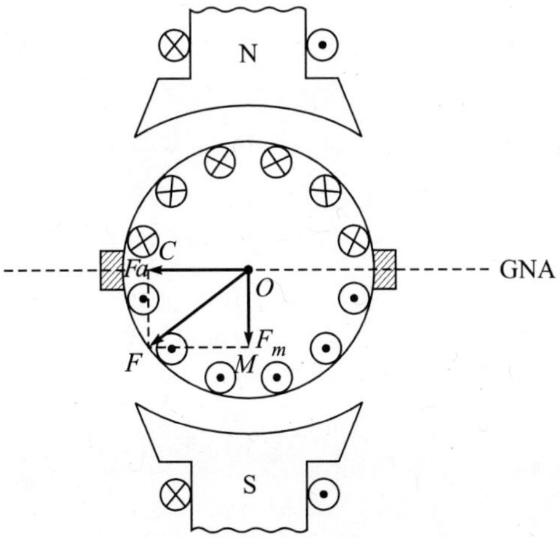

Fig. 6.40 Brush along GNA.

The mmf of the armature is given by 'F_a' which is equal to '$N_a I_a$' and points along the direction 'OC' as shown as per right hand palm rule, whereas the main field mmf 'F_m' points in the direction shown from 'N' pole to 'S' pole respectively in the downward direction. From the figure it is clear that the resultant mmf vector 'F' gets shifted from the GNA. This effect is called armature reaction. In other words, it could be defined as the effect of the armature mmf on the main field mmf. As a consequence of this armature effect the following things occur in the air gap flux distribution, namely, the demagnetizing effect and the cross-magnetising effect respectively. To get a proper understanding of the two, we consider these effects one by one.

Demagnetizing Effect of Armature Reaction (AR)

The demagnetizing effect of the 'AR' is shown in the figure by the vector 'OA' which has a magnitude equal to '$F_a \sin \theta$' which produces an effect which results in decreasing of the main pole flux magnitude under the pole shoe of the d.c. machine as it opposes main field flux. The demagnetizing effect of the 'AR' can be visualized by considering the Fig. (6.39). The demagnetizing effect of AR is produced by the armature conductors which lies between $\angle POQ$ and $\angle ROS$, so this number of armature conductors can be considered demagnetizing armature conductors respectively. Because these armature conductors produce a flux which is oriented in a direction which is opposite to the main field flux direction which is directed from 'N' pole to 'S' pole respectively. So the effect is reduction of main flux field magnitude so they are producing demagnetizing effect of armature reaction. Since the ampere turn or the mmf is responsible for the production of the flux, so there is a need of determining of the demagnetizing ampere turns or demagnetizing mmf. This could be determined as follows:

Number of demagnetizing armature conductor is equal to $= \dfrac{2\theta}{360} \times Z$

If the armature conductor carries a current 'I_c', then the demagnetizing ampere conductor is given by the exp. (6.53) i.e.,

$$\text{Demagnetizing ampere conductor} = \frac{2\theta}{360} \times Z \times I_c \tag{6.53}$$

where, $I_c = \dfrac{I_a}{A}$, I_a = armature current, A

and A = No. of parallel paths between brushes.

$A = P$ (For lap winding).

$A = 2$ (For wave winding)

Since one turn consists of two armature conductor so, the demagnetizing ampere turns is given by exp. (6.54).

$$AT_d = \frac{2\theta}{360} \times \frac{ZI_c}{2} \tag{6.54}$$

or, $$AT_d = \frac{\theta}{360°} \cdot ZI_c \tag{6.55}$$

The armature ampere turns or mmf is given by

$$AT_a = \frac{I_c \cdot Z}{2} \tag{6.56}$$

The armature ampere turn per pole is given by

$$AT_a/\text{pole} = \frac{I_c \cdot Z}{2P} \tag{6.57}$$

So the other effect of the armature reaction could be determined by considering the armature conductors which lie between $\angle POR$ and $\angle QOS$ respectively. The other effect of the armature reaction is cross magnetizing effect.

Cross Magnetizing Effect of Armature Reaction

The cross magnetizing ampere turns of armature is determined by taking the difference of the total armature ampere turns per pole and the demagnetizing armature turns per pole respectively. Mathematically, it is expressed as follows:

$$AT_c/\text{pole} = AT_a/\text{pole} - AT_d/\text{pole} \tag{6.58}$$

Substituting the values of AT_a/pole and AT_d/pole, we get

$$AT_c/\text{pole} = \frac{ZI_c}{2P} - \frac{\theta}{360°} \cdot ZI_c$$

$$AT_c/\text{pole} = ZI_c\left[\frac{1}{2P} - \frac{\theta}{360}\right] \tag{6.59}$$

where, Z = Total number of armature conductors
I_c = Armature conductor current, ampere
P = Number of poles of machine
θ = Electrical degree

We see from the above discussions that the cross magnetizing armature conductor produces a flux which crosses the main field flux thus produces distortion of the main field flux distribution. Thus, both the effects of armature reaction produces undesirable effects so there is a need of decreasing the magnitude of the demagnetizing and cross magnetizing effects. This could be minimized by producing mmf which opposes these two effects.

One is by shifting brushes from N_{c_1} (say) commutator segment it is previously present to the new MNA which corresponds to the next subsequent commutator segment N_{c2} (say).

So the number of the commutator segments by which the brush is shifted is '$N_{c2} - N_{c_1}$' respectively. If the total number of the commutator segments on the commutator is equal to 'N_c'.

Then, $$AT_d = \left(\frac{N_{c_2} - N_{c_1}}{N_c}\right) I_c Z \quad (6.60)$$

and, $$AT_c = \left[\frac{1}{2P} - \frac{(N_{c_2} - N_{c_1})}{N_c}\right] I_c Z \quad (6.61)$$

Other method to neutralize the demagnetizing effect of armature reaction is given by exp. (6.62) and (6.63).

$$AT_d = N_{sh}\, I_{sh} \text{ (In the case of D.C. shunt motor)} \quad (6.62)$$

and $$AT_d = N_{se}\, I_{se} \text{ (In the case of D.C. series motor)} \quad (6.63)$$

The demagnetizing effect of armature reaction is neutralized by using compensating winding on the field pole shoe slots such that the direction of the current through the compensating winding conductor under the main pole shoe produces an effect which opposes the demagnetizing effect of armature conductors present under the main pole shoe respectively. This could be easily understood by considering the Fig. (6.41).

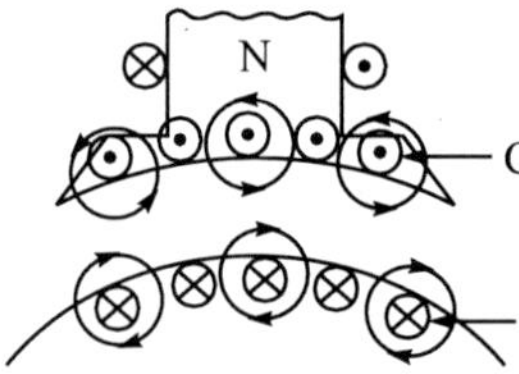

Fig. 6.41 Neutralizing of demagnetizing effect of armature conductor by conductor of compensating winding.

6.10 LOSSES IN D.C. MACHINES

The d.c. machine gets heated when it is rotated. This is also true for the a.c. machines. The heating is due to the temperature rise of the windings as well as the core which is made up of magnetic material. This is due to the losses which occur in the machines.

The losses may be divided as follows:

(a) Core losses
(b) Copper losses
(c) Stray load losses
(d) Dielectric losses
(e) Mechanical losses.

Core Losses

The core losses are mainly due to the hysteresis and eddy current losses, The hysteresis and eddy current losses are frequency dependent. Since the fluxes which pass through the magnetic material are not constant but changing in nature. So the magnetic material goes through many cycles of magnetization and demagnetization as a result power is consumed by the magnetic material which is given by the exps. (6.64) and (6.65) respectively.

$$\text{Hysteresis loss } P_h \propto f \tag{6.64}$$

$$\text{Eddy current loss } P_e \propto f^2 \tag{6.65}$$

The eddy current loss is more affected by the frequency of the pulsating flux wave. So there is need of minimizing it as compared to the hysteresis loss. The eddy current loss is due to the cyclic current which flows through the surface of the magnetic material. The eddy currents are minimized by laminating the core. That is, single core is split into large number of thin sheets which are insulated from each other. Thus, the eddy current path is disrupted and hence the eddy losses in the magnetic core are minimized. The hysteresis losses are minimized by using magnetic material having higher permeability. Thus, by using such material beside decreasing the hysteresis losses it will also decrease the size of core.

Copper Losses

The copper losses are due to the $I^2 R$ loss in the windings of the machine. Since the current will heat the windings due to Joules heating effect which is given by the exp. (6.66)

$$H = I^2 R t \tag{6.66}$$

The copper losses occur in the armature windings, interpole windings, main field windings, compensating windings and in the brushes.

The copper losses are expressed as follows:

$$P_{a_{cu}} = I_a^2 R_a;\ W \tag{6.67}$$

$$P_{f_{cu}} = I_f^2 R_f;\ W \tag{6.68}$$

where,

'R_f' can be 'R_{se}' or 'R_{sh}'

and 'I_f' can be 'I_{se}' or 'I_{sh}'

depending on the placement of windings with respect to the armature

$$P_{b_{cu}} = I_b^2 R_b;\ \text{watts}$$

$$P_{ip_{cu}} = I_{ip}^2 R_{ip};\ \text{watts}$$

$$P_{cw_{cu}} = I_{cw}^2 R_{cw};\ \text{watts}$$

where,

R_a = Armature resistance, Ω

R_f = Field resistance, Ω

R_b = Brush resistance. Ω

R_{ip} = Interpole winding resistance, Ω

R_{cw} = Compensating winding resistance, Ω
I_a = Armature current, amp.
I_f = Field current, amp.
I_b = Brush current, amp.
I_{ip} = Interpole winding current, amp.
I_{cw} = Compensating winding current, amp.

In each '$I^2 R$' loss, we see that there is a power loss in the winding or coil.

Stray Load Losses

The losses due to the stray field is termed the stray field losses. The stray field is always present in the machines which causes the distortion of the flux distribution in the air gap between the stator and rotor.

Dielectric Loss

The dielectric which is present on the windings and the core will lead to overheating of the rotating machines. This is due to high current flow due to the failure of dielectric. The dielectrics are electrical insulators which fail with the ageing effect or due to high voltages and high currents. The dielectrics get stressed if more voltage is applied across it and the dielectric gets brittle with the ageing of the machines. The high current flow will lead to the opening of the enamel or varnish insulation which gets vaporized after a certain temperature rise depending on the class of insulation used in the machine. The classes of insulations are *A*, *B*, *H*, etc. These classes are on the basis of the temperature range.

Mechanical Losses

These losses originate only when the machine rotates. It is sometimes called the rotating losses. The mechanical losses are due to friction or due to windage. They are termed frictional losses and windage losses respectively. The frictional losses are due to the friction of shaft and the bearings which will lead to heating and power loss. The windage loss is due to air friction. The air which is trapped in the case between the stator and rotor. The air exerts an opposing force to the direction of rotation of rotor and hence power is consumed by the machine to overcome it. They are sometimes known as air friction loss.

6.11 VOLTAGE REGULATION OF D.C. GENERATOR

The voltage regulation may be defined as the change in terminal voltage of generator from no load condition to full load condition. Voltage regulation is defined as follows

$$V.R = V_{NL} - V_{FL} \tag{6.69}$$

The percentage voltage regulation may be expressed as ratio of change in terminal voltage to the full load terminal voltage and expressed as follows.

$$\% V.R = \frac{V_{NL} - V_{FL}}{V_{FL}} \times 100\% \tag{6.70}$$

where, V_{NL} = No load terminal voltage, volts
V_{FL} = Full load terminal voltage, volts

The percentage voltage regulation is an important performance parameter of d.c. machine. The lower is its value, the better is the machine. The shunt generator has a low value of percentage voltage regulation that is about 5%. The voltage regulation can be positive or negative depending on the loading of a particular d.c. generator.

6.12 EFFICIENCY OF D.C. GENERATOR

The efficiency of d.c. generator is expressed as ratio of output power and input power. Mathematically, it is expressed as

$$\eta = \frac{\text{Output power}}{\text{Input power}} \tag{6.71}$$

The output power is electrical which constitutes voltage 'V' as well current 'I'. Whereas the input power is mechanical which comprises angular speed 'w' and torque 'T' respectively. Then such efficiency is called commercial efficiency and is denoted by 'η_c'. The expression of 'η_c' is given as follows.

$$\eta_c = \frac{VI}{\omega T} = \frac{P_e}{P_m} \tag{6.72}$$

If the parameters in the numerator and denominator are electrical quantities, then the efficiency is known as electrical efficiency and is denoted by 'η_e'; where subscript 'e' is used for electrical.

$$\eta_e = \frac{V \cdot I_L}{E_g \cdot I_a} \tag{6.73}$$

where, 'V' is the terminal voltage, in volts.
'I_L' is the line current, in amperes.
'E_g' is called the generated emf, in volts.
'I_a' is the armature current, in amperes.

The efficiency may be represented as ratio of the mechanical quantities as given below and is represented as 'η_m' called the mechanical efficiency and given by the exp. (6.74).

$$\eta_m = \frac{wT_{em}}{wT} = \frac{P_{em}}{P_{pm}}, \quad \text{where} \quad wT_{em} = E_g I_a \tag{6.74}$$

where, T_{em} = Electromagnetic torque, Nm

P_{em} = Mechanical power obtained, W

P_{pm} = Power supplied by prime mover, W

The commercial efficiency or the rated efficiency is the product of the two efficiencies, namely, 'η_e' and 'η_m' respectively as follows.

$$\eta_e = \eta_e \times \eta_m \tag{6.75}$$

The commercial efficiency is marked over the name plate of the machine and is considered rated quantity.

6.13 APPLICATIONS OF D.C. GENERATORS

D.C. Separately Excited Generator

1. Electroplating
2. Electrorefining of materials
3. Laboratory applications
4. Commercial testing
5. Special regulation sets.

The external source can be another d.c. generator, a controlled or uncontrolled rectifier or simply a battery. The external source is required for the field winding.

D.C. Shunt Generator

1. Lighting loads
2. Battery charging
3. Hotels
4. Offices.

The d.c. shunt generator are considered constant voltage generators. Since the drop in voltage is lesser from no load to full load condition.

D.C. Series Generator

1. For arc lighting (arc lamps)
2. As constant current generators
3. As boosters for d.c. feeders.

The d.c. series generator has typical load characteristic i.e., the voltage rises as the loading of the generator takes place.

D.C. Cummulative Compound Generator

1. For domestic lighting
2. For transmission over long distance
3. For boosting voltages for small feeders
4. For d.c. traction purpose.

D.C. Differential Compound Generators

1. For electric arc welding
2. For research purpose.

The differential compound generators are not used as the terminal voltage drops drastically with loading of generator terminals.

SOLVED QUESTIONS (EXAM)

Q.1 A 20 kW compound generator works on full load with a terminal voltage of 250 V. The armature, series and shunt windings have resistances of 0.05 Ω, 0.025 Ω and 100 Ω respectively. Calculate the total emf generated in the armature when the machine is connected as short shunt.

(2006 JMI)

A.1

$$P = 20 \text{ kW} \qquad E_g = ?$$
$$V = 250 \text{ volt}$$
$$R_a = 0.05\ \Omega$$
$$R_{se} = 0.025\ \Omega$$
$$R_{sh} = 100\ \Omega$$
$$\text{Brush drop} = \text{Neglected}$$

When connected as short shunt compound generator

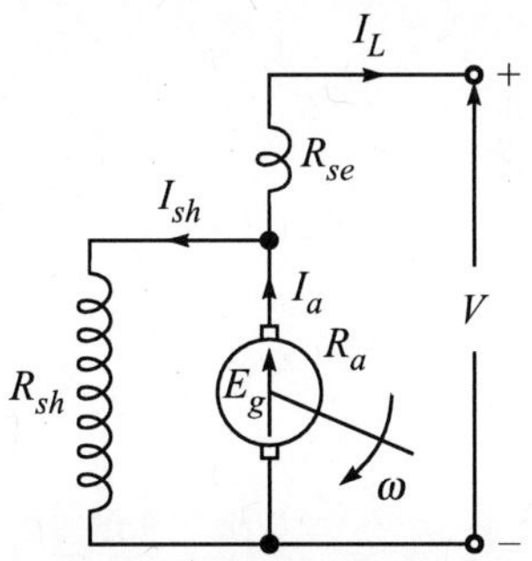

Fig. Short shunt CG.

$$E_g = I_a R_a + I_L R_{se} + V$$
$$I_a = I_{sh} + I_L$$
$$P = 20 \text{ kW} = V \cdot I_L$$
$$I_L = \frac{P}{V} = \frac{20 \text{ kW}}{250} = \frac{20000}{250} \text{ A} = 80 \text{ A}.$$

Power remains a source for long shunt and short shunt generator.

So, $$V = I_{sh} \cdot R_{sh} \text{ (considering long shunt)}$$
$$I_{sh} = \frac{V}{R_{sh}} = \frac{250}{100} = 2.5 \text{ A}$$

Thus, $$I_a = 2.5 + 80 = 82.5 \text{ A}$$

Total generated emf $$E_g = 82.5 \times 0.05 + 80 \times 0.025 + 250$$
$$= 4.125 + 2 + 250$$
$$= 256.125 \text{ Volt.}$$

Q.2 A shunt generator has a full load current of 196 A at 220 V. The stray losses are 720 W and the shunt field coil resistance is 55 Ω. It has a full load efficiency of 88 %. Find the armature resistance. Also find the load current corresponding to maximum efficiency.

(2006 JMI)

A.2

$$I_{FL} = 196 \text{ A} = I_L$$
$$V = 220 \text{ Volt}$$
$$R_{sh} = 55 \ \Omega$$
$$P_{stray} = 720 \text{ W}$$
$$\eta_{FL} = 88 \ \%$$
$$R_a = ?$$
$$I_L]_{\eta_{max}} = ?$$

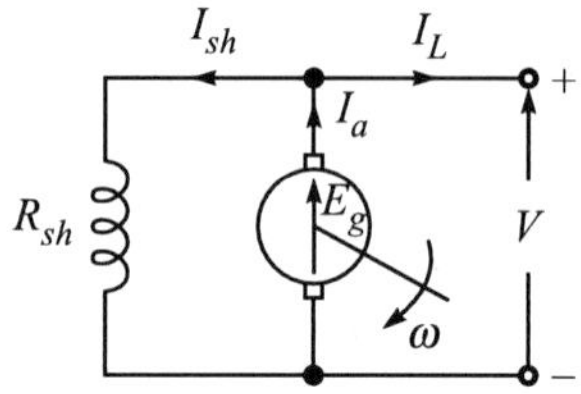

Fig. Shunt generator

$$I_a = I_{sh} + I_L$$
$$V = I_{sh} R_{sh}$$
$$I_{sh} = \frac{V}{R_{sh}} = \frac{220}{55} = 4 \text{ A}$$
$$I_a = 4 + 196 = 200 \text{ A}$$
$$E_g = I_a R_a + V$$
$$\eta_{FL} = 0.88 = \frac{V \cdot I_L}{E_g \cdot I_a}$$
$$E_g = \frac{V \cdot I_L}{0.88 \, I_a} = \frac{220 \times 196}{0.88 \times 200} = 245 \text{ volt}$$

So, $$R_a = \frac{E_g - V}{I_a} = \frac{245 - 220}{200} = \frac{25}{200} = 0.125\ \Omega$$

At $$\eta_{\max} : P_{cn} = P_i = P_s$$

$$I_a^2 R_a = P_s$$

$$I_a = \sqrt{\frac{P_s}{R_a}} = \sqrt{\frac{720}{0.125}} = 75.89\ \text{A}$$

Q.3 A 20 kW, 400 V, 6 pole d.c. generator has 720 lap wound conductors. It is given a brush lead of 2.5° (mechanical) from GNA. Calculate cross and demagnetizing ampere turns per pole.

A.3

$$P_0 = 20\ \text{kW} \qquad AT_c/\text{pole} = ?$$

$$V = 400\ \text{volt} \qquad AT_d/\text{pole} = ?$$

$$P = 6$$

$$Z = 720$$

$$A = P$$

$$\theta_m = 2.5^\circ$$

We know that $$\theta_e = \frac{P}{2}\theta_m = \frac{6}{2} \cdot 2.5 = 7.5^\circ \text{ (electrical)}$$

Armature conductor current is given by

$$I_c = \frac{I_a}{A} = \frac{I_a}{P} = \frac{I_L}{P}$$

Considering the generator a separately excited d.c. generator.

$$I_a = I_L$$

$$\text{Output power} = V \cdot I_L = 400 \cdot I_L = 20000\ \text{W}.$$

Power, $$I_L = \frac{20000}{400} = 50\ A.$$

So, $$I_c = \frac{50}{6} = 8.333\ A$$

Cross magnetizing ampere turns per pole is given as

$$AT_c/\text{pole} = ZI_c\left[\frac{1}{2P} - \frac{\theta_e}{180^\circ}\right],\ AT$$

$$= 720 \times 8.333\left[\frac{1}{2 \times 6} - \frac{7.5}{180}\right]$$

$$= 5999.76\left[\frac{1}{12} - 0.041667\right]$$

$$= 249.988\ AT/\text{pole}$$

$$AT_d/\text{pole} = Z \cdot I_c \cdot \left(\frac{\theta^\circ}{180^\circ}\right)$$

$$= 720 \times \frac{8.333}{1} \times \frac{7.5}{180}$$

$$= 249.99\ AT/\text{pole}$$

Q.4 A long compound generator delivers a load current of 50 A at 500 V and the resistances of armature series field and shunt fields are 0.05 Ω, 0.03 Ω and 250 Ω respectively. Calculate the emf generated and the armature current. Allow 1 volt per brush contact drop.

A.4

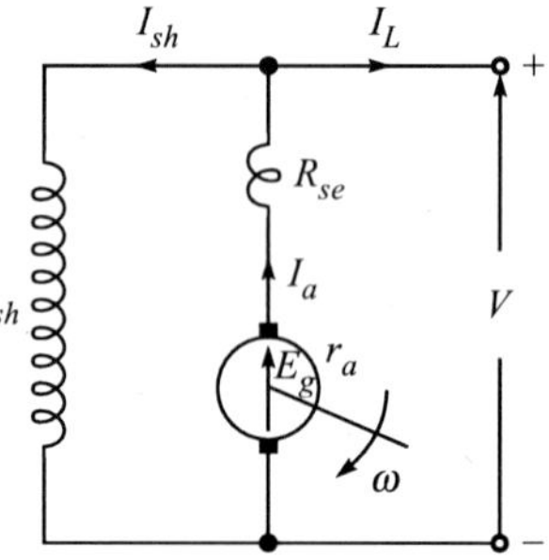

$$V = 500 \text{ volt}$$
$$I_L = 50 \text{ A}$$
$$R_a = 0.05\ \Omega$$
$$R_{se} = 0.03\ \Omega$$
$$R_{sh} = 250\ \Omega$$
$$E_g = ?$$
$$I_a = ?$$
$$V_b = \text{Brush drop} = 1 \text{ Volt/brush}$$
$$E_g = V - I_a R_a + V_b + I_a R_{se} \quad (1)$$
$$V = 500 = I_{sh} R_{sh}$$
$$I_{sh} = \frac{V}{R_{sh}} = \frac{500}{250} = 2 \text{ A}$$
$$I_a = I_{sh} + I_L = 2 + 50 = 52 \text{ A}$$
$$E_g = 500 + (52 \times 0.05) + 1 \times 2 + 52 \times 0.03$$
$$E_g = 500 + 2.6 + 2 + 1.56 = 506.16 \text{ volts.}$$

Q.5 The following data give the open circuit characteristics of a d.c. shunt generator at 300 rpm: (1999 AMU)

$I_f(A)$:	0	0.2	0.3	0.4	0.5	0.6	0.7
$V_{oc}(V)$:	7.5	93	135	165	186	202	215

The field resistance of the machine is given as 354.5 Ω.

(i) Determine graphically, the no load voltage.

(ii) Determine the critical field resistance.

(iii) Determine the critical speed for the given field resistance.

(iv) What additional resistance must be added in the field circuit to reduce the no load voltage to 175 V?

A.5

$$R_f = 354.5\ \Omega$$

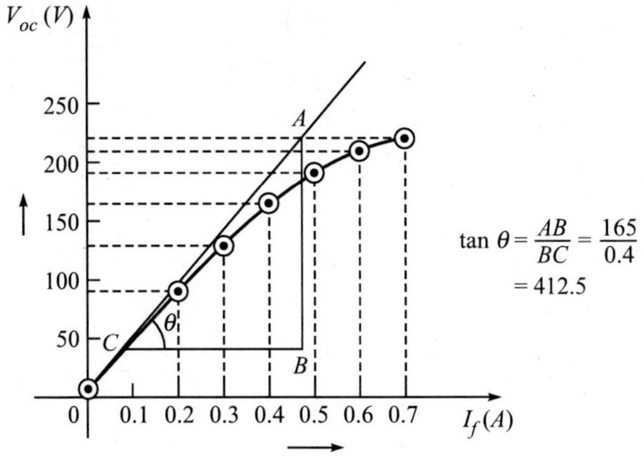

$$V = I_f R_f$$
$$E_g \propto N\phi$$
$$E_g = V + I_a R_a$$
$$\phi = \text{constant}$$

(i) V_{oc} = 215.Volts = E_{g1}

(ii) The critical field resistance = tan $\theta = R_{cf} = 412.5\ \Omega$

(iii) At no load, $I_L = 0$

So, $$I_{a_o} = I_f = \frac{V \approx E_{g1}}{R_f} = \frac{V \approx E_{g1}}{354.5\ \Omega} = \frac{215}{354.5} = 0.6\text{A}$$

$$E_{g1} \propto N_o\ ;\ E_g \propto N_r$$
$$\frac{E_{g1}}{E_r} = \frac{N_o}{N_r} \Rightarrow N_r = N_o \cdot \frac{E_r}{E_{g1}} = 300 \times \frac{7.5}{215} = 10.5 \text{ rpm}$$

∴ Critical speed of machine ≈ 11 rpm (Approx.)

(iv) $$Eg_1 = V + I_{a_o} R_a \quad (1)$$
$$E_{g2} = V + I_{a_o}(R_a + R) \quad (2)$$

∴ Subtracting (1) from (2), we get $E_{g2} - E_{g1} = I_{ao} \cdot R$

or, $$E_{g2} = (175 = V_o) + I_{ao}(R_a + R) \propto N_2$$
$$175 = I_f(R_f + R)$$
$$R = \frac{175}{0.28}$$
$$= 354.5$$
$$\boxed{R = 270.5\ \Omega}$$

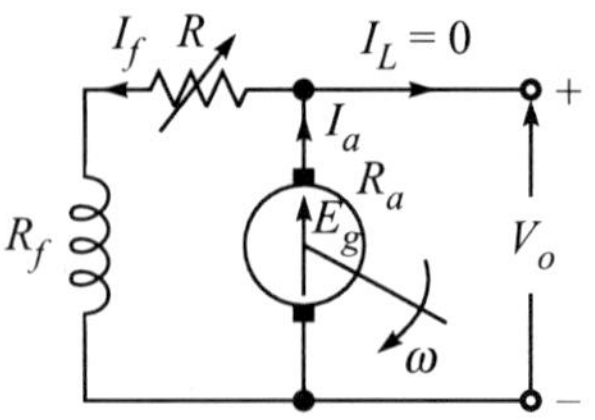

$$V_o = 175 \text{ V}, I_f = I_{ao} = 0.28 \text{ A (from graph)}$$

$$E_{g1} = (215 = V_o) + I_{ao} (R_a) \propto N_o R_f$$

when, $\quad E_{g1} = V_o = 215 \text{ V}, I_{ao} = I_f = 0.7\text{A}$

Q.6 The open circuit characteristic of a d.c. shunt generator at 800 rpm is given by following data.

$I_f(A)$	0	0.2	0.4	0.65	1.02	1.75	3.15	5.0
$V_{oc}(V)$	10	40	80	120	160	200	240	260

(a) Determine the critical field resistance at 800 rpm.

(b) Determine the critical speed if the field winding resistance is 55 Ω.

(c) With the field winding resistance of 55 Ω, find the range of external field resistance to vary the no load terminal voltage from 200 V to 250 V at 800 rpm.

A.6

(a) From graph, $R_{cf} = \tan\theta = \dfrac{AB}{BC} = \dfrac{150}{1.75 - 0.25}$

$$R_{cf} = 100\ \Omega$$

(b) $$\frac{E_{g1}}{E_r} = \frac{N_o}{N_{r_c}} \Rightarrow N_{r_c} = N_o \cdot \frac{E_r}{E_{g1}} = 800 \times \frac{10}{260} = 30.8 \text{ rpm}$$

Critical speed of machine $= N_{r_c} = 31$ rpm (approx.)

(c) $R_f = 55\ \Omega;\ V_{o1} = E_{o_2} = 250 \text{ V};\quad E_{o2} = V_{o2} = 200 \text{ V}, N_o = 800 \text{ rpm}.$

$$E_{g1} = V_{o1} + I_a R_a \propto N_o$$

$$E_{g2} = V_{o2} + I_a'(R_a + R_f') \propto N_1$$

$$I_a' = 1.75 \text{ A} = I_f' \text{ (corresponding to 200 V from graph)}$$

$$200 = 1.75\ R_f$$

or, $\quad R_f' = 114.3\ \Omega\ \text{‘}R_a\text{’} << R_f'$

So, external resistance required is

$$R = R_f' - R_f = 114.3 - 55 = 59.3\ \Omega$$

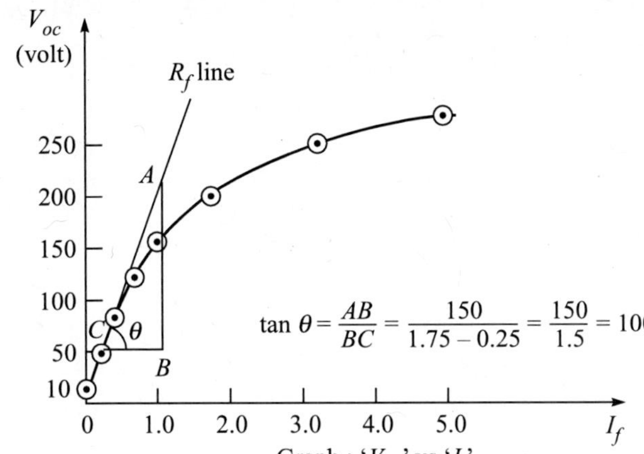

Graph : 'V_{oc}' vs 'I_f'

Q.7 The power output of 200 kW, 220 V; 600 rpm separately excited generator is delivering the rated load at rated voltage. The generator has $R_a = 0.01\ \Omega$. $R_f = 30\ \Omega$, $V_f = 100$ V. Find the voltage regulation of a generator if the induced emf at no load is 250 V.

A.7

$$V_{FL} = 220 \text{ V}$$

$$V_{NL} = E_{in}]_{NL} = 250 \text{ V}$$

Thus,

$$\text{V.R } \% = \frac{V_{NL} - V_{FL}}{V_{FL}} \times 100 \ \%$$

$$= \frac{250 \times 220}{220} \times 100 \ \%$$

$$= \frac{30}{220} \times 100 \ \%$$

$$= 0.1364 \times 100 \ \%$$

$$= 13.64 \ \%$$

TUTORIAL QUESTIONS (FOR EXAM)

Q.1 A 1500 kW, 600 V, 16 pole separately excited d.c. generator runs at 200 rpm. It has 2500 lap connected conductors and full load copper losses are 25 kW. Calculate the useful flux per pole.

[B. Tech II (UPTU) (2004 – 2005)]

[**Ans:** $\phi = 0.072$ Wb]

Q.2 A short shunt compound generator delivers a load current of 30 A at 220 V, and has armature, series and shunt field resistances of 0.05 Ω, 0.3 Ω and 200 Ω respectively. Calculate the induced emf and the armature current. Allow 1 V per brush for contact drop.

[UPTU (2003 – 04)]

[**Ans:** 232.6 V, 31.15 A]

Q.3 A 25 kW, 250 V, D.C shunt generator has armature and field resistances of 0.06 Ω and 100 Ω respectively. Determine the total armature power developed when working.

(i) As a generator delivering 25 kW output and

(ii) As a motor taking 25 kW input

[UPTU (2003 – 04)]

Ans: $\begin{bmatrix} P_g = 26.3 \text{ kW} \\ P_m = 23.8 \text{ kW} \end{bmatrix}$

Q.4 A 50 kW, 250 V, d.c. shunt generator has the following data. $R_a = 0.06\ \Omega$, $R_f = 125\ \Omega$. Calculate the induced emf in armature winding at rated load and terminal voltage. Take 2 V as the total brush contact drop.

[2005 – 06, UPTU]

[**Ans:** E = 264 V]

Q.5 A compound d.c generator is supplying a load of 120 A at 120 V. The shunt field, series field and armature resistances are 32 Ω, 0.07 Ω and 0.1 Ω respectively. Find the generated voltage in (a) long shunt connected generator (b) short shunt connected generator.

[UPTU 2005 – 06]

[**Ans:** E_{Lg} = 141.04 V, E_{sg} = 140.8 V]

Q.6 A 10 kW d.c. shunt generator has the following losses at full load.

Mechanical loss = 290 W, Iron loss = 410 W

Shunt copper loss = 120 W, Armature copper loss = 595 W. Calculate the efficiency of d.c generator at (i) No load (ii) 25% of full load.

(UPTU, 05 – 06)

[**Ans:** 92.42%, 74.6%]

Q.7 A 4 pole d.c. generator supplies full load current of 143 A. It has 492 armature conductors (i) wave connected, and (ii) lap connected. When delivering full load the brushes are given an actual lead of 10°, calculate the demagnetizing ampere-turns per pole. If the field winding is shunt connected and takes 10 A. Find the number of extra field turns necessary to neutralize this demagnetization.

[1045.5 *AT*; 105: 522.75 *AT*; 52]

Q.8 A 4 pole wave wound generator supplies a current of 50 A. The shunt field takes 2 A. There are 125 commutator segments. The brushes are given a lead of 3 commutator segments. Calculate (i) the demagnetizing, and (ii) cross magnetizing ampere turns per pole.

[156 *AT*/pole; 656.5 *AT*/pole]

REVIEW QUESTIONS (EXAM QUESTIONS)

Q.1 Derive the emf and torque equation of a d.c. machine.

[BSc, AMU (Engg) 98]

Q.2 Draw the external characteristics of d.c shunt, differentially compounded and cummulatively compounded d.c. generators. Comment on these characteristics.

Q.3 Derive emf equation of a d.c. machine.

[B.E., AMU, 99]

Q.4 Draw the external characteristic of a separately excited d.c. generator, shunt generator and compound generator and compare them.

[B.E., AMU, 99]

Q.5 What are the causes of delayed commutation in d.c. machine? How these can be minimised?

[B.Tech, AMU, 2001]

Q.6 Explain why does the terminal voltage fall more rapidly in a self excited d.c. shunt generator than in a separately excited generator.

[B.Tech, AMU, 2001].

Q.7 In a d.c. machine working under saturated condition, the effect of armature mmf is to demagnetize the main field. Explain.

[B.Tech, AMU, 1999]

Q.8 Explain the voltage buildup process in a d.c. shunt generator. Explain the conditions under which the generators fails to build up the voltage.

[B.Tech, AMU, 1999]

Q.9 Deduce an expression for the voltage induced in a d.c. generator. Why the air gap between the pole pieces and the armature is kept very small?

[B.Tech, JMI, 2006]

Q.10 What is the purpose and location of the following; (i) An interpole or commutating winding (ii) A compensating winding (iii) A series field.

[B.Tech, (JMI), 2006)]

Q.11 Sketch and explain the load characteristic of each of the following types of d.c. generators (i) shunt (ii) series, and (iii) compound. State the reason where each type is used.

[B.Tech, JMI, 2006]

Q.12 What is armature reaction ? Describe the effects of armature reaction on the operation of d.c. machines. How can the armature reaction be minimised?

[B.Tech, JMI, 2005]

Q.13 Explain the process of building up of voltage in a d.c. generator and give the conditions to be satisfied for voltage build-up.

[B.Tech, JMI, 2005]

Q.14 Mention the reasons for compounding in d.c. generators. Neatly sketch and explain the external characteristics of a d.c. compound generator.

[B.Tech, JMI, 2005]

Q.15 Discuss the characteristics of d.c. generators.

[B.Tech, UPTU, 04 – 05]

Q.16 Discuss the characteristics of a separately excited d.c. generator.

[B.Tech, UPTU, 2005]

Q.17 Discuss the load characteristics of a d.c. separately excited generator.

[B.Tech, UPTU, 2003, 2005]

Q.18 Discuss the voltage build up of a d.c. shunt generator.

[B.Tech, UPTU, 2005]

Q.19 Discuss the load characteristics of a d.c. series generator.

[B.Tech, UPTU, 2003]

Q.20 Discuss the load characteristics of a d.c. compound generator.

[B.Tech, UPTU, 2003]

Q.21 Discuss the constructional features and explain various parts of a d.c. generator.

[B.Tech, UPTU, 2003]

Q.22 Draw and explain the following characteristics of d.c. generators (i) No load and load magnetization characteristics (ii) External and internal characteristics.

[B.Tech, UPTU, 2004]

Q.23 With reference to no load characteristics of a d.c. generator, explain the followings:

(i) Why is emf not zero when the field current is zero?

(ii) Why does the slope of curve change, after a certain value of field current?

[B.Tech, UPTU, 2005]

Q.24 Distinguish between external and internal characteristics of a d.c. generator. How can internal characteristic be derived from external characteristics of separately excited generator?

[B.Tech, UPTU, 2005]

Q.25 Explain with labelled diagram the construction of d.c. machines.

[B.Tech, UPTU, 2005]

Q.26 What are the main causes of the voltage drop in generators? Explain in detail voltage build-up in self excited generators and their load characteristics.

[B.Tech, UPTU, 2005]

Chapter

7

D.C. Motors

7.1 Introduction
7.2 Operating Principle of Motor
7.3 Torque Equation of Motor
7.4 Speed Regulation
7.5 Types of D.C. Motor
7.6 Characteristics of D.C. Motor
7.7 Performance Characteristics of a Motor
7.8 Losses in Motor
7.9 Efficiency of D.C. Motor
7.10 Starting of D.C. Motor
7.11 Speed Control of D.C. Motor
7.12 Braking of D.C. Motor
7.13 Permanent Magnet D.C. Motor
7.14 Brushless Motor
7.15 Universal Motors
7.16 Applications of D.C. Motor
- Solved Exam Questions
- Unsolved Exam Questions
- Review Questions

7.1 INTRODUCTION

The rotating motors play a vital role in this modern world. The motors in FHP range are in use in CD drives in computers, in the machine tools operation, in the audio devices and in many home appliances. The large motors are used in the industries, traction, mining operation etc. Though the a.c. motors are fast replacing the d.c. motors in majority of industrial applications, still these d.c. motors are popular in the low power range.

The motors are available in a number of varieties from very small range to very high range in terms of cost and size. Thus, it is necessary to know about the details of the motor both the name plate ratings and the internal construction of the machine. Because in these motors if the polarity is wrongly connected then the desired function of the motor required for a particular work cannot be accomplished. Now small horsepower range brushless d.c. motor are also becoming popular. Moreover PMDC motor, PCB motors are also making their place in the market. In the home appliances such as audio cassette recorder motor if the speed of the motor is not adjusted then it affects the voice of the singer. So knowledge of the speed control is also necessary.

For the industrial and traction purpose there is a need of reducing the speed of motors so the knowledge of braking is necessary. Thus, various types of braking of motors are discussed in the coming section.

7.2 OPERATING PRINCIPLE OF A MOTOR

The motor works on a principle which differs from that of generator though the construction of motors and generators is same. The motor as opposed to the generator works on the basis of force which is exerted on the armature conductor due to which a torque is exerted on the rotor and thus the rotor starts revolving.

This force expression is given by

$$F = B I_c l \tag{7.1}$$

where, F = Lorentz force, N
B = Average value of field, tesla
l = Active length of armature conductor, m
I_c = Armature conductor current, A

Considering a single turn armature coil on the rotor as shown below

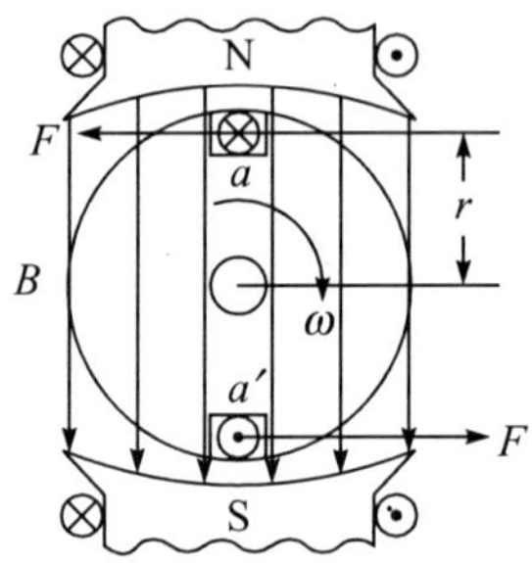

Fig. 7.1 Armature conductor 'a' and 'a'' (Sectional view).

From the centre of shaft upto the centre of armature conductor the distance is equal to 'r'. So a moment of force is exerted on the armature conductor 'a' and 'a''. Since the direction of current in armature conductor 'a' and 'a'' is in opposite direction so, the torque exerted on the armature conductor 'a' and 'a'' will be such that it will produce a turning moment as a result of which the rotor conductor starts moving.

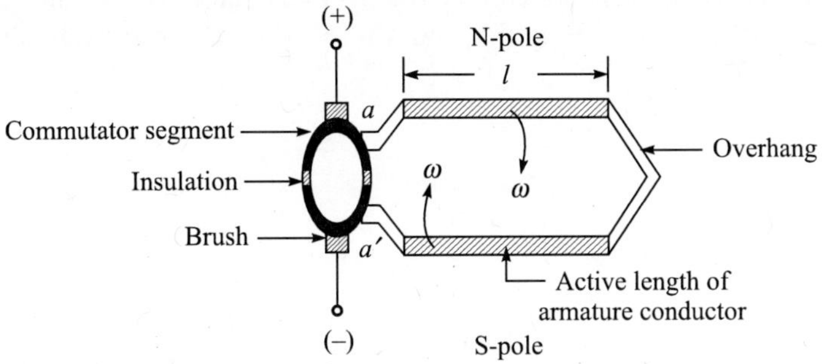

Fig. 7.2 Side view of armature conductor.

The direction of the force 'F' is given by Flemming's left hand rule which could be easily visualized if the left hand fingers are kept at right angle. Then the thumb will point the direction of motion of conductor i.e., force, the central finger points the direction of current and the forefinger points in the direction of field respectively. In the Fig. (7.2), the force and hence torque exerted on the armature coil will turn the coil in the clockwise direction.

7.3 TORQUE EQUATION OF MOTOR

The torque is an important mechanical parameter of the motor. The torque is responsible for the motoring operation. The torque can be expressed as follows:

$$T = F \cdot r, \text{ Nm} \tag{7.2}$$

Since, $F = B\,I_c\,l$, Newtons. Substituting the value of 'F' in the expression of 'T' we get,

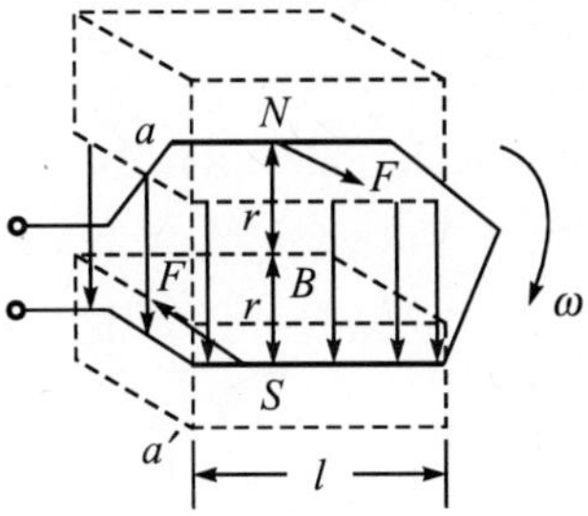

Fig. 7.3 Torque exerted on armature conductor.

$$T = B\,I_c\,l \cdot r, \text{ Nm} \tag{7.3}$$

Since, there are 'Z' number of armature conductors on the armature, we have,

$$T = Z \cdot B \cdot I_c \cdot l \cdot r, \text{ Nm} \tag{7.4}$$

The armature conductor current 'I_c' is given as $\dfrac{I_a}{A}$.

where, 'I_a' is called the armature current and 'A' is the number of parallel paths between the brushes. Substituting the values of 'I_c' in the above expression of torque given by (7.4).

we get,

$$T = Z \cdot B \cdot \frac{I_a}{A} \cdot l \cdot r \qquad (7.5)$$

'B' is the average value of field, in tesla which is given as

$$B = \frac{\phi_p \cdot P}{2\pi rl} = \frac{\phi}{A_c} = \text{tesla} \qquad (7.6)$$

where, ϕ is the total flux under the pole shoe between the air gap, Wb

'P' is the number of poles of machine

'A_c' is the cylindrical surface area, m^2

'ϕ_p' is the flux per pole, Wb

Thus, the final expression of the torque of motor can be obtained by substituting the values of 'B' in the expression of 'T',

we get

$$T = Z \cdot \frac{\phi_p \cdot P}{2\pi \not{r} l} \cdot \frac{I_a}{A} \cdot l \cdot \not{r}, \text{ Nm}$$

So,

$$\boxed{T = \frac{I_a \cdot P \cdot \phi_p \cdot Z}{2\pi A}}, \text{ Nm} \qquad (7.7)$$

The exp. (7.7) is called torque equation of the motor.

$$T \propto \phi_p \cdot P$$

7.4 SPEED REGULATION

The speed regulation is an important parameter which decides the selection of particular type of motor for a particular operation. The speed regulation is mathematically expressed as follows

$$\text{Speed regulation} = N_{NL} - N_{FL}, \quad \text{rpm} \qquad (7.8)$$

where,

N_{NL} = No load speed, rpm

N_{FL} = Full load speed, rpm.

The percentage speed regulation is generally used which is defined by a ratio in which the numerator is the change in speed from no load to full load condition, whereas the denominator is the full load speed value. Mathematically,

$$\%\ \text{Speed regulation} = \frac{N_{NL} - N_{FL}}{N_{FL}} \times 100\% \qquad (7.9)$$

The speed regulation of the d.c. shunt motor is about 5% whereas the d.c. series motor has a wide speed regulation. So the d.c. series motor is considered a variable speed motor whereas the shunt motor has a small value of speed regulation so it is considered a constant speed motor. The compound motor is also a variable speed motor since the value of speed regulation is higher.

7.5 TYPES OF D.C. MOTOR

The d.c. motors are divided into mainly two types, namely (a) PMDC motors (b) D.C. motors employing electro magnets. On the basis of brushes the d.c. motors are divided as brushless motor and motor employing brushes. On the basis of size or ratings of machine the motors are divided as small size, medium size and large size motor respectively. In this section, only the motors which have brushes and the field is produced by electromagnets are discussed. The d.c. motors are classified as shunt, series and compound motors on the basis of connection of field winding with respect to that of armature winding. If the field winding is connected in series with the armature winding then the motor is called series motor. If the motor field winding is connected across the armature winding i.e., in parallel then the d.c. motor is called shunt motor. If both series and shunt windings are present then the d.c. motor so formed is called compound motor. The different types of motor are shown in Fig. (7.4).

The different types of motor are required for different operations. For understanding it the knowledge of characteristics of the motors is necessary. The characteristics of the motor are again studied from no load to full load conditions which is discussed in the next section.

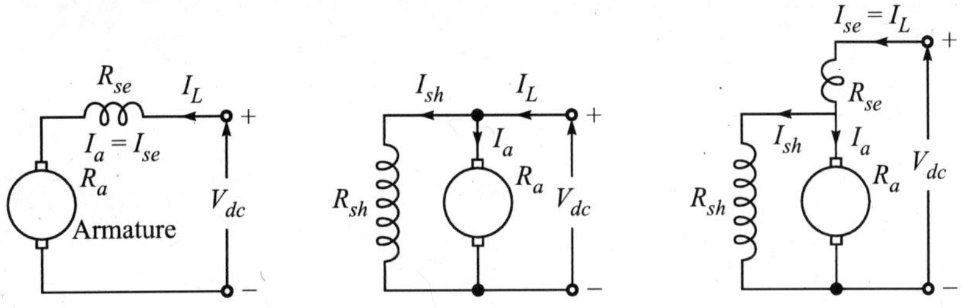

(a) (b) (c)

Fig. 7.4 Types of d.c. motor.

7.6 CHARACTERISTICS OF D.C. MOTORS

The characteristics of a d.c. motor are again classified as operating characteristics and performance characteristics. The operating characteristics are drawn between 'N' and 'I_a', 'T' and 'I_a' and 'N' and 'T' respectively. On the other hand, if the 'η', 'T', 'I_a' and 'N' are drawn with respect to output of motor then such characteristics are called performance characteristics. In this section only the operating characteristics are discussed for different motors.

Shunt Motor: In the shunt motor the terminal motor 'V' is expressed as follows:

$$V = E_b + I_a R_a \tag{7.10}$$

where, 'E_b' is the back emf due to self inductance 'L_a' of armature winding
'I_a' is the armature current, amp.
'R_a' is the armature resistance, ohms

The 'E_b' is expressed as follows:

$$E_b = \frac{NP\phi Z}{60A} \tag{7.11}$$

or,
$$E_b \propto N\phi$$

Since, the other parameters are constant for a particular design of d.c. machine.

If the field circuit current draws fixed field current, then ϕ = constant, so the exp. (7.11) can be expressed as

$$E_b \propto N \tag{7.12}$$

or,
$$E_b = kN \tag{7.13}$$

Substituting the value of 'E_b' in exp. (7.10), we get

$$V = kN + I_a R_a \tag{7.14}$$

or,
$$kN = V - I_a R_a \tag{7.15}$$

or,
$$\boxed{N \propto V - I_a R_a} \tag{7.16}$$

We see from the exp. (7.16) that, as 'I_a' is increased '$I_a R_a$' is increased hence the speed 'N' decreases. The exp. (7.16) is a linear relation so ideally the characteristic between 'N' and 'I_a' is a straight line having drooping nature.

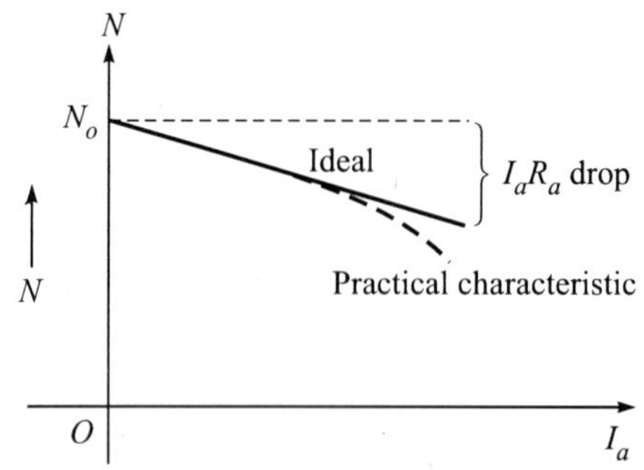

Fig. 7.5 N/I_a characteristic of d.c. shunt motor.

Where, 'N_0' is the no load speed. The no load speed is the highest speed since there is nothing attached to the shaft of motor and load torque is zero. So motoring torque is not opposed by load torque.

Series Motor: In the case of series motor, the terminal voltage is expressed as follows.

$$V = E_b + I_a (R_a + R_{se}) \quad (7.17)$$

$$E_b \propto N\phi \quad (7.18)$$

Since in the case of d.c. motor, $\phi \propto I_a$

So,

$$E_b \propto N I_a \quad (7.19)$$

$$V = kN I_a + I_a (R_a + R_{se}) \quad (7.20)$$

$$kN I_a = V - I_a (R_a + R_{se}) \quad (7.21)$$

$$\boxed{N \propto \frac{V}{I_a} - (R_a + R_{se})} \quad (7.22)$$

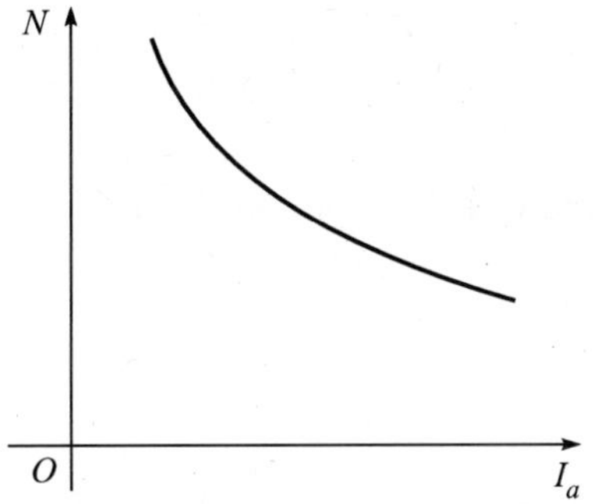

Fig. 7.6 N/I_a characteristic of d.c. series motor.

From the exp. (7.22) it is clear that as 'I_a' is increased the speed 'N' decreases accordingly. As the current approaches zero the speed is increased enormously which is undesirable. So the d.c. series motor is not preferred for no load operation and it is always operated under loaded condition.

Compound Motor: The characteristics of d.c. compound motor differ on the basis of the connection of series and shunt field winding connections. If the shunt and series winding connections are such that it draws current in opposite directions and produces fluxes which oppose each other then that is called differential compound motor and in which the fluxes aids each other then it is called cummulative compound motor. The differential compound motor is not used for practical operations as there is a increase in speed 'N' with increase of armature current 'I_a' as shown in the Fig. (7.7) will lead to unstable operation of motor.

The characteristic of d.c. compound motor is shown is shown in the Fig. (7.7).

The N/I_a characteristic of cummulative compound motor is in between the shunt and series motor characteristic which is shown by dotted lines.

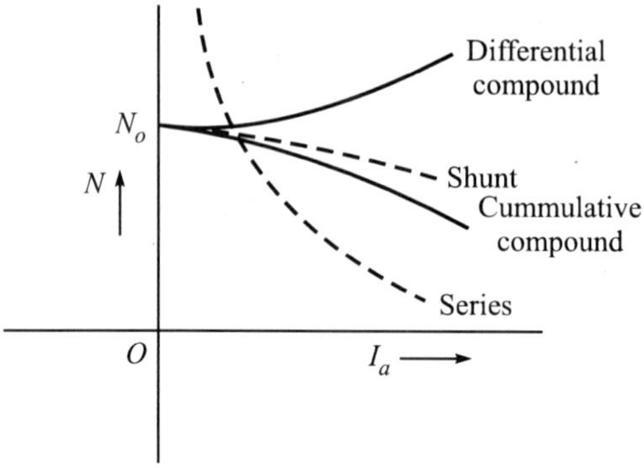

Fig. 7.7 *N*/I_a characteristic of d.c. compound motor.

The other operating characteristic of d.c. motor is between torque 'T' and current 'I_a'. From the torque eq. (7.7), we have

$$T \propto \phi I_a \tag{7.23}$$

Since the other parameters of torque expression are constant for a particular machine. So it is considered in the machine constant.

For d.c. shunt motor,

we have $\phi = \text{const}$

so, $T \propto I_a$ (7.24)

Hence, the T/I_a characteristic is a straight line starting from origin. Since the exp. (7.24) gives a linear relationship between 'T' and 'I_a', the characteristic is shown as

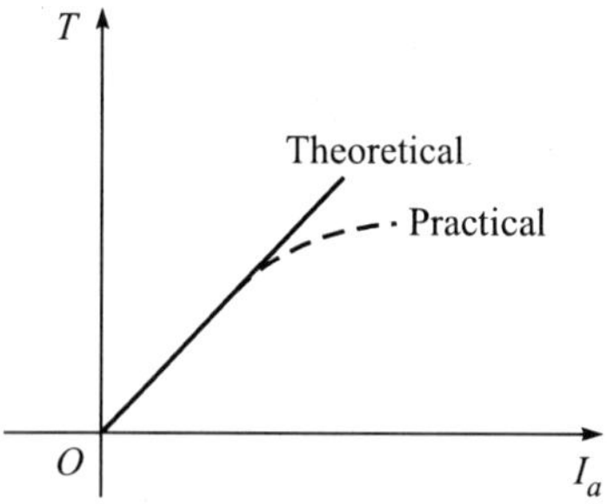

Fig. 7.8 *T*/I_a characteristic for d.c. shunt motor.

But actually the characteristic follows the dotted curve. This is due to saturation of poles of machines.

D.C. series motor: For the d.c. series motor, the field flux 'ϕ' is directly proportional to the armature current 'I_a' so the torque $T \propto \phi I_a$ reduces to

$$T \propto I_a^2 \tag{7.25}$$

The exp. (7.25) is a parabolic function. So the characteristic of d.c. series motor is non-linear and is depicted in Fig. (7.9) as follows.

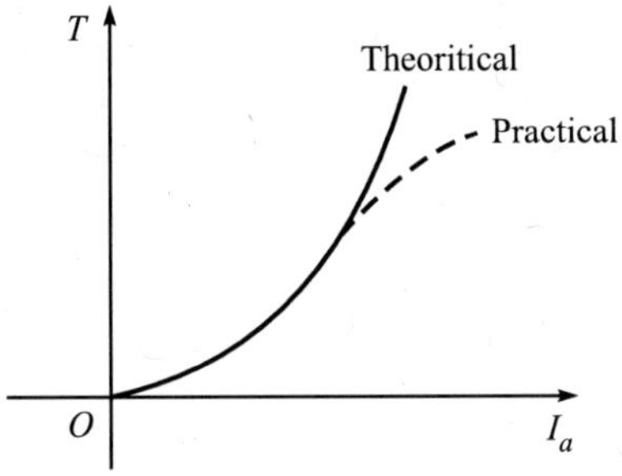

Fig. 7.9 T/I_a characteristic of d.c series motor.

The practical characteristic follows the dotted curve. The operating characteristic of d.c. compound motor T/I_a characteristic is given as follows:

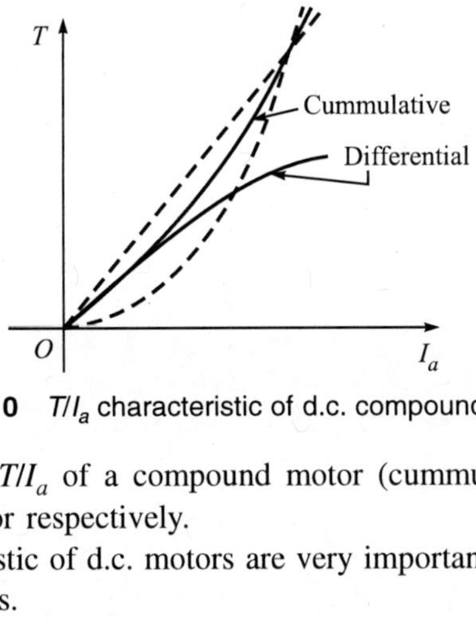

Fig. 7.10 T/I_a characteristic of d.c. compound motor.

The characteristic T/I_a of a compound motor (cummulative) is in between series and shunt motor respectively.

The N/T characteristic of d.c. motors are very important and are discussed in this section as follows.

D.C. shunt motor: The voltage equation of the motor is given as

$$V = E_b + I_a R_a$$

$$V = kN\phi + I_a R_a \qquad \text{(I)}$$

$\because T \propto I_a$ so,

$$V = kN\phi + k'TR_a$$

$$\phi = \text{constant}$$

so,

$$kN = V - K'TR_a$$

or,

$$\boxed{N \propto V - k''T}\ ;\ R_a = \text{constant}$$

so, at the torque 'T' is increased, the speed 'N' decreases accordingly linearly. At 'T' equal to zero, the speed 'N' is maximum equal to no load speed.

$$\boxed{N = N_0 - k''' T} \tag{7.26}$$

The *N/T* characteristic is depicted in Fig. (7.11).

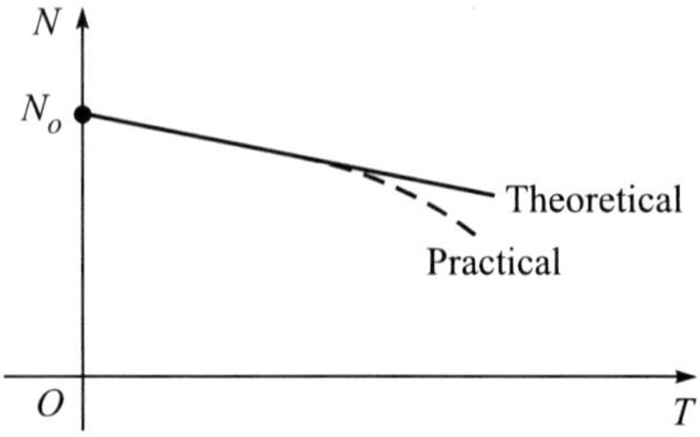

Fig. 7.11 *N/T* characteristic of d.c. shunt motor.

D.C. series motor: The voltage equation of d.c. series motor is given by

$$V = E_b + I_a (R_a + R_{se})$$
$$E_b \propto N\phi;\ \phi \propto I_a$$
$$T \propto I_a^2 \text{ so, } I_a \propto \sqrt{T}$$

Substituting the values of 'E_b' and 'I_a' in the exp. (I), we get,

$$V = kNI_a + I_a (R_a + R_{se}) \tag{7.27}$$
$$kN I_a = V - I_a (R_a + R_{se})$$
$$N \propto \frac{V}{I_a} - \frac{(R_a + R_{se})}{1}$$
$$\boxed{N \propto \frac{V}{\sqrt{T}} - \frac{(R_a + R_{se})}{1}} \tag{7.28}$$

As '*T*' increases the speed decreases rapidly. The exp. (7.28) follows hyperbolic function.

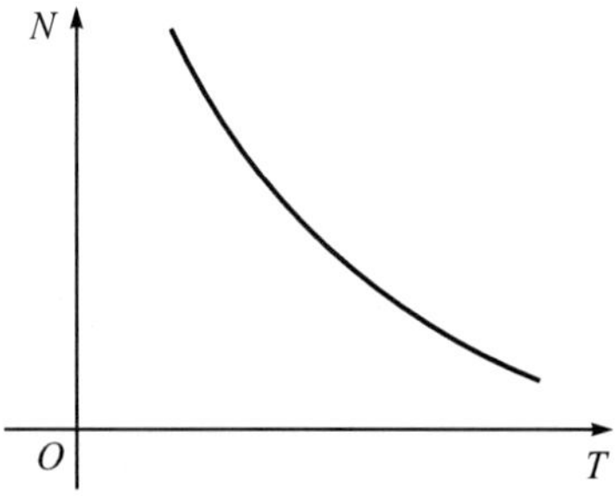

Fig. 7.12 *N/T* characteristic of a series motor.

For D.C. compound motor: The *N/T* characteristic is shown in Fig. (7.13) as follows.

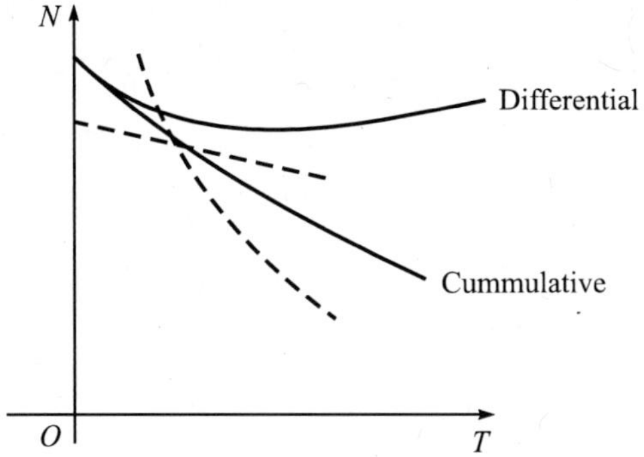

Fig. 7.13 *N/T* characteristic of a d.c. compound motor.

7.7 PERFORMANCE CHARACTERISTICS OF A MOTOR

The motor performance can be easily visualized by considering the motor parameter such as efficiency, armature current, speed and torque which is plotted against the output horsepower of motor available at the shaft of the motor. So the characteristics are called performance characteristics of motors. The performance characteristics of a d.c. shunt motor are shown in the Fig. (7.14).

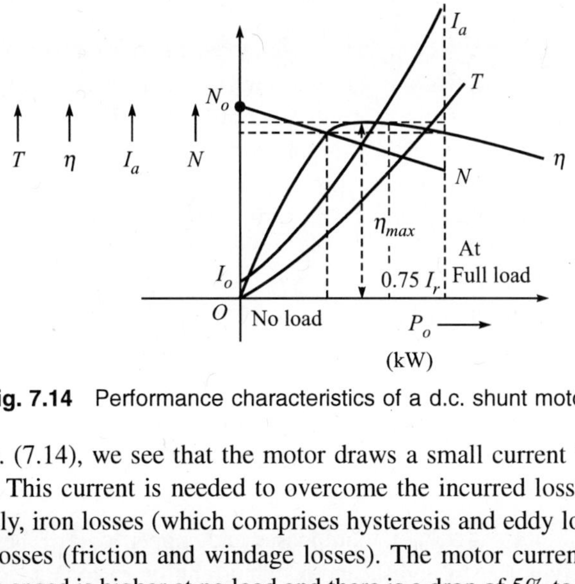

Fig. 7.14 Performance characteristics of a d.c. shunt motor.

In the Fig. (7.14), we see that the motor draws a small current 'I_0' called no load current. This current is needed to overcome the incurred losses in the machines, namely, iron losses (which comprises hysteresis and eddy losses) and the mechanical losses (friction and windage losses). The motor current is higher at full load. The speed is higher at no load and there is a drop of 5% to 10% from no load to full load condition depending on the machine rating. Further improvement in the speed regulation is possible by further improving the design. The efficiency of the machine is lesser at no load and increases at 25% no load speed or full load current and remains constant upto 75% of rated current then again decreases as the current drawn by motor exceeds the rated value due to increase of losses. The torque increases as the power drawn by shaft increases due to increase of motor current. From the characteristics of the motor, we see that applications for small torque applications and in such applications where there is need of constant speed this type of motor is most suitable, for example, in machine tool operations, lathes etc.

For d.c. series motor, the performance characteristics are given in Fig. (7.15) below.

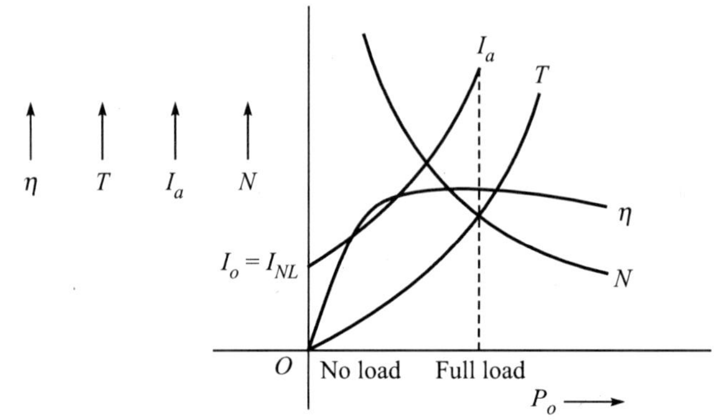

Fig. 7.15 Performance characteristic of a d.c. series motor.

The no load current 'I_0' is higher as compared to shunt motor but since the speed of the motors reaches dangerously high at no load so the motor is not operated under no load condition. The torque of the motor has parabolic function. For high accelerating torque applications at starting with less current requirement this type of motor is preferred. The efficiency is constant from 25% to 125% full load current or rated current value. The series motors are employed in cranes, trams, tractions etc.

7.8 LOSSES IN MOTOR

The motor losses are basically classified as:

(a) Electrical losses
(b) Mechanical losses
(c) Dielectric losses.

The electrical losses are due to core losses and copper losses. The core losses are further subdivided as hysteresis and eddy current losses and the copper losses are due to 'I^2R' loss.

The hysteresis loss is dependent on the volume of magnetic material used, its magnetic alignment and the cycles of magnetization which the magnetic material has undergone. The magnetic material is important in the sense that better the material lesser will be the heating of the core of machine. On the other hand, the eddy current losses depend on the thickness of laminations of the core of machine. The lesser is the thickness of core the lesser will be the eddy current losses. But the laminations could not be made thin below a certain limit. The thickness of lamination is from 0.35 mm to 2.5 mm depending on the size of the machine core diameter.

The mechanical losses are due to friction and windage. The frictional losses and windage losses are produced only when the motor rotor rotates. So these losses are sometimes called rotational losses. The frictional losses can be minimized by using frictionless bearings and proper lubrication of the bearing. The windage losses are always present but by cutting slit on the frame coverings of machines. This loss can be brought to minimum. The dielectric losses are due to deterioration of the insulating properties of the dielectrics which is used on the windings and on the laminations of the core. The dielectric losses lead to lessening the life of the machine. So after a given period of time, say 10 years, the insulations are placed over the machine by opening the machine and then it is dried and is again assembled and used.

7.9 EFFICIENCY OF D.C. MOTOR

The efficiency of a d.c. motor is also an important performance parameter of the d.c. motor. It decides the output which could be obtained from the machine. The higher the value of efficiency the better is the motor. The higher efficiency motor will work for a longer time. The efficiency which is marked over the machine name plate is the commercial efficiency denoted by 'η_c'. The 'η_c' is expressed as follows:

$$\eta_c = \frac{\text{Output power}}{\text{Input power}} \tag{7.29}$$

$$\eta_c = \frac{\omega T_{sh}}{V \cdot I_L} = \frac{P_{sh}}{P_i} \tag{7.30}$$

The percentage efficiency is given as follows:

$$\% \, \eta_c = \frac{\omega T_{sh}}{V \cdot I_L} \times 100\% \tag{7.31}$$

The commercial efficiency is defined as the product of the mechanical and electrical efficiency as given below.

$$\boxed{\eta_c = \eta_m \times \eta_e} \tag{7.32}$$

The mechanical efficiency 'η_m' is given as follows:

$$\eta_m = \frac{\omega T_{sh}}{\omega T_{em}} \tag{7.33}$$

where, ω = angular speed, vad/s

T_{sh} = shaft torque, Nm

T_{em} = Electromagnetic torque, Nm

The electrical efficiency 'η_e' is given as follows:

$$\eta_e = \frac{E_b \cdot I_a}{V I_L} \tag{7.34}$$

Since, $E_b I_a = \omega T_{em}$ (in armature) which is given by the power balance equation which do not violate the conservation of energy principle.

'E_b' is called the back emf produced in the armature winding, V

'I_a' is the armature current, A

The efficiency in terms of losses is expressed as follows:

$$\eta = \frac{P_0}{P_0 + \underbrace{P_{cu} + P_i + P_m}_{Plosses}} \times 100\%$$

$$\eta = \frac{P_i - \text{losses}}{P_i} = \left(1 - \frac{\text{Losses}}{P_i}\right) \times 100\% \tag{7.35}$$

where,

P_{cu} = copper losses

P_i = iron losses

P_m = mechanical losses

7.10 STARTING OF D.C. MOTOR

The d.c. motors are started by using starters. For understanding the need of starters the motor equation will be very useful which is given as follows.

Motor terminal voltage is given by

$$V = E_b + I_a R_a$$

given by $$I_a = \frac{V - E_b}{R_a}, \text{ amperes}$$

Armature current

Back emf $$E_b = \frac{NP\phi Z}{60A}, \text{ volts}$$

In the expression of back emf, $\left(\frac{PZ}{60\,A} = K_m\right)$ is a constant called machine constant which varies for different machine design.

So the above expression can be expressed as given below.

$$E_b \propto N\phi \tag{7.36}$$

Substituting the value of 'E_b' in the expression of 'I_a', we get

$$I_a = \frac{V - K_m N\phi}{R_a}$$

At starting,

The speed, $N = 0$ (since the rotor has large inertia)
So, the back emf, 'E_b' $= K_m N\phi = 0$

Hence,

$$I_a\big]_{\text{Starting}} = \frac{V - 0}{R_a}, \text{ amperes} \tag{7.37}$$

The voltage of the d.c. motor ~ 100's of volts, 'R_a' is of the order of ohms. So the current drawn by the motor is of the order of 100's of amperes which corresponds to few multiples of 10's of the rated current. This high value of starting current is sufficient to damage the armature windings and brushes.

So there is a need of decreasing this high value of starting current, though this starting current flows for a short interval of time which is of the order of few seconds. Therefore, if somehow the starting current is decreased to a value in the range of motor's rated current that is, less than rated value. This is accomplished by using starters.

In the coming section, the starters are discussed in detail. With the starting resistance 'R_{st}' in series with armature circuit, the starting current expression is expressed as follows:

$$\boxed{I_{st} = \frac{V - E_b}{R_a + R_{st}}} \tag{7.38}$$

At starting, $E_b = 0$

But with 'R_{st}', $I_{st} < I_{\text{rated}}$ (or full load current). Thus, the motor current is limited such that any damage to the machine is avoided.

The starters are different on the basis of its operation if it is operated manually then it is called manual starter and if it is operated automatically then the starter is called automatic starter.

The manual starter is classified as follows: (a) Two-point starter (b) Three-point starter and four-point starter respectively.

Two-Point Starter

The two point starter is used for starting series d.c. motor. The operation of the d.c. series motor is easily done using the Fig. (7.16) as given below.

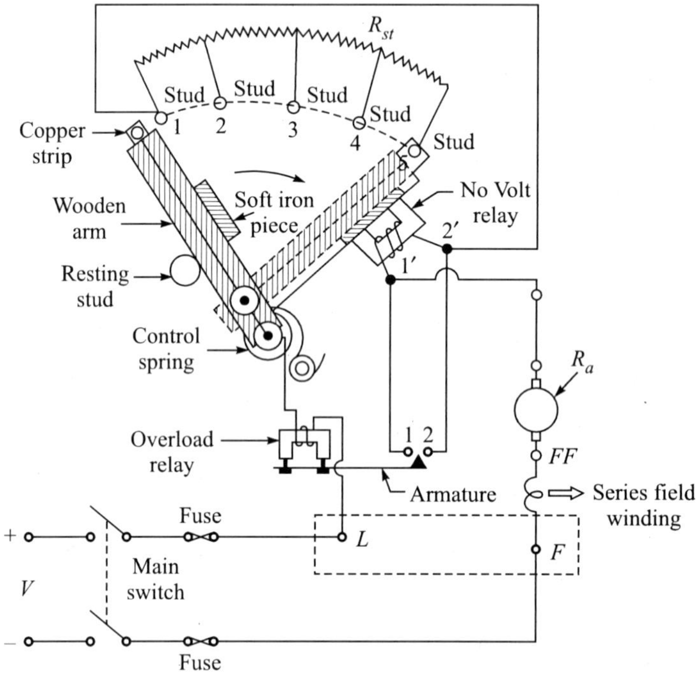

Fig. 7.16 Two point starter with d.c. series motor.

When the moving arm which is made of wood is moved clockwise after closing the main switch then the resistance 'R_{st}' comes in series with the armature circuit and the no volt relay coil. Then due to magnetic action of the no volt relay the wooden arm's soft iron piece gets attracted and the relay's magnetizing force is sufficient to overcome the restoring force produced by the control spring which is fixed at the end of the moving arm. Thus, with the addition of starting resistance the voltage equation of the d.c. series motor is modified as

$$V = E_b + I_a (R_a + R_{se} + R_{st})$$

so,

$$\boxed{I_a\Big]_{st} = \frac{V - E_b}{R_a + R_{se} + R_{st}}} \qquad (7.39)$$

At starting, $N = 0$ so, $E_b \propto N\phi$ Hence, $E_b = 0$

$$I_a\Big|_{\text{At starting}} = \frac{V - 0}{R_a + R_{se} + R_{st}} \sim I_{\text{rated}}$$

$$R_{se} \ll 1\Omega,\ R_a \sim 1\Omega,\ R_{st} \sim \text{variable}$$

The starting resistance 'R_{st}' varies as per the grading of the resistances between the brass studs.

As the back emf is built up i.e., with the increase of motor speed (speed pick up of motor) the difference between $(V - E_b)$ in the numerator of the exp. (7.40) is given as

$$\left. I_a \right]_{st} = \frac{V - kN\phi}{R_a + R_{se} + R_{st}} \tag{7.40}$$

As 'N' increases, $E_b = kN\phi$ increases.

So, '$V - kN\phi$' decreases. Hence, the armature current is decreased so to increase the speed of motor the starting resistance is decreased by moving the wooden arm towards the clockwise direction thus decreasing its value. Again the value of 'N' increases and hence 'E_b' increases 'R_{st}' is further decreased by clockwise motion of arm and current is increased. This process goes on till the moving arm is held by the no volt relay and all the starting resistance is eliminated completely and the motor acquires its full speed.

Three-Point Starter

The three-point starters have three points L, F and A which corresponds to line, field and armature respectively. The three point starter with studs is shown in Fig. (7.17) as shown below.

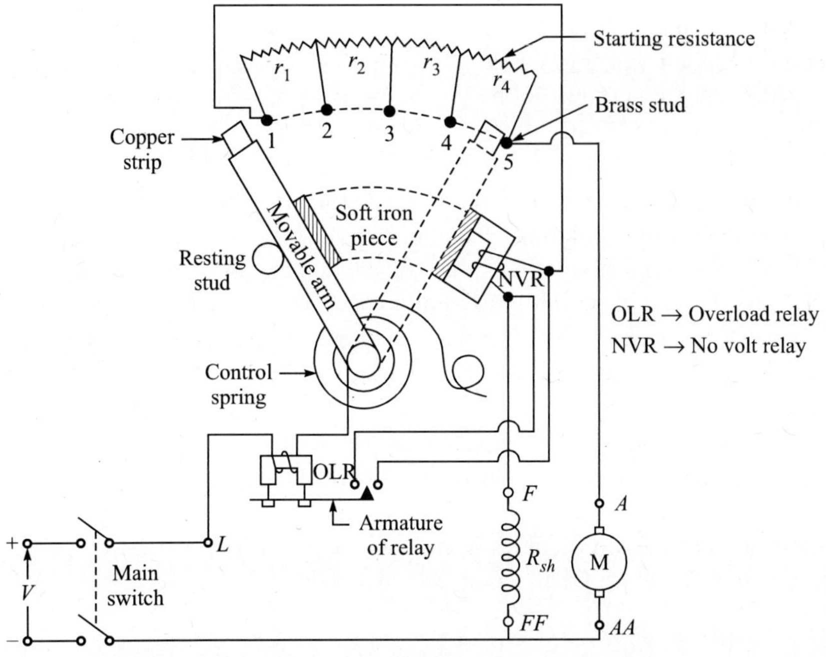

Fig. 7.17 Three-point starter and d.c. shunt motor.

In this starter, as the starting resistance is cut off from the armature circuit resistance. The cut out part gets added to the field circuit and so the flux 'ϕ' is decreased, so the e.m.f. 'E_b' is also decreased accordingly but this decrease does not affect the motor performance much so it is still used in the old installation. To decrease this effect brass arcs are used which are placed symmetrically with respect to the starting resistance such that the connections are such that this extra resistance and the motor speed reaches its maximum value equal to the designed tested value. In case the voltage fails then the no volt relay gets demagnetized thus releasing the soft iron piece which is fixed to the wooden arm and due to the spring effect (restoring force of spring) it returns to its normal position and rests on the resting stud. If the supply comes again then the motor will not start till the previously discussed method is not repeated again.

If there is overloading on the shaft of the motor then the motor draws high current which exceeds the rated (full load) current such that the overload relay operates if I > 1.3 I rated and the armature gets attracted by the magnetized part of relay. Hence, the armature starts moving upwards and thus the point '1' and '2' get short circuited thus the voltage across the two ends of no volt relay coil is equal to zero, hence the no voltage relay gets demagnetized and thus release the moving wooden arm and the motor is disconnected from the supply. Thus, the motor is saved from overloaded condition.

For the shunt and compound motor the three-point and four-point starters are used. The three-point starter is preferred for constant voltage applications whereas the four-point starter is used where there are fluctuations in the voltage applied to the motor where the three-point starter is unsuitable. The three-point starter with brass arc is shown in the Fig. (7.18) is preferred to that starter with studs as shown in Fig. (7.17).

By the arrangement as shown in Fig. (7.18), we see that the brass arc portion of resistance is nullifying effect of field circuit resistance with the decrease in the starting resistance during starting of d.c. motor.

But the three-point starter suffer from following demerits when compared to other starters. Firstly, when the voltage drops then the no volt relay magnetic property gets weakened as a result it releases the wooden arm which is holding the soft iron piece. So there is a need of starting again and again which is undesirable.

Secondly, if the motor field winding gets opened then the shunt motor behaves as a series motor which runs at very high speed which is not favourable and could cause an accident.

So to overcome these difficulties there is an another starter called four-point starter which is discussed in the coming section.

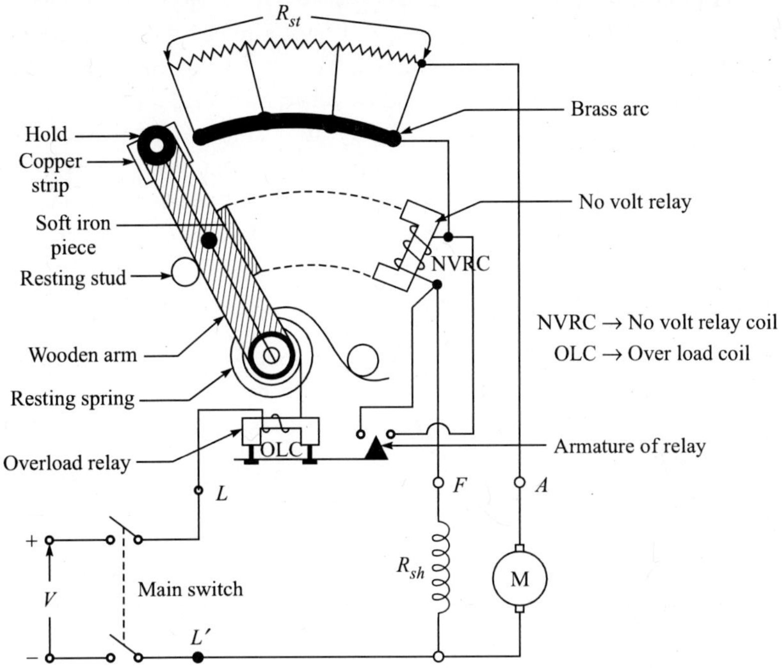

Fig. 7.18 Three-point starter with brass arc.

Four-Point Starter

The four points of starter are *L*, *L*, *F*, *A* which shows that two points are connected to the line supply and '*F*' denotes the field and '*A*' denotes the armature terminal respectively. The figure of the four-point starter is given in Fig. (7.19).

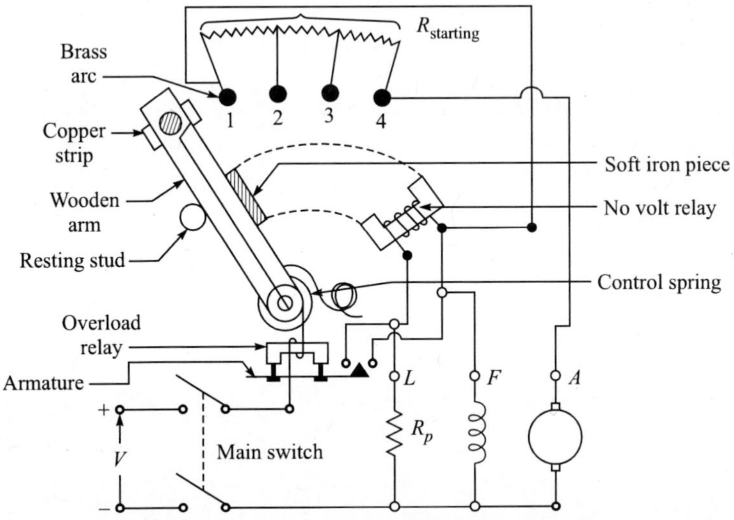

Fig. 7.19 Four-Point Starter for d.c. motor.

In the four-point starter it has a protective resistor 'R_p' which is connected to one end of the no volt relay coil. It serves as a protective measure in the case of opening of field circuit of the d.c. shunt motor. Thus restricting the operation of the d.c. shunt motor to operate as series motor. Since the series motor gets over speeded in the case of no load condition to very high value. Also the four point starter does not release the movable arm under small voltage fluctuations in the supply voltage. Thus, there is no need of starting the starter again and again. The brass arc improves the operation of starter as in the previous case.

Calculation of Starting Resistor Elements

The resistances of the starting resistor 'R_{st}' are graded so their values are not equal and are taken as r_1, r_2, r_3, ... During the shifting of starter arm towards right, the starting resistance value is decreased hence the current increases from 'I_1' to 'I_2' respectively as shown.

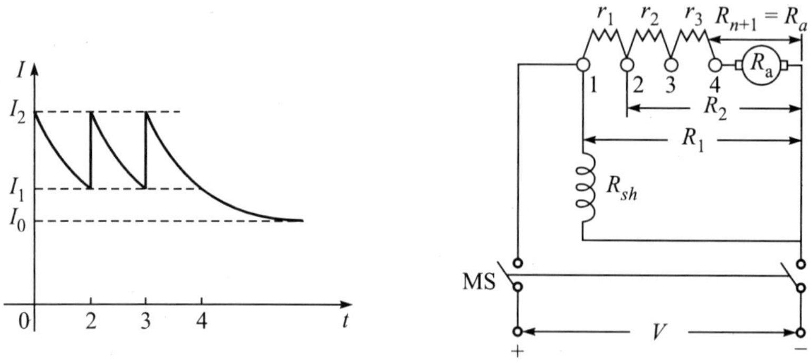

Fig. (a) Armature current variation (during starting) (b) Resistance elements of starter

If the starter contains 'n' elements, then number of studs is $(n + 1)$ respectively.

Initially, when the arm of starter towles studs (1), the current is brought to $I_2 < I_{\text{rated}}$. The current decreases from 'I_2', to 'I_1' due to increase of speed 'N', hence back emf 'E_b' of motor. The arm is moved to stud (2), the resistance decreases from 'R_1' to 'R_2' and again the current rises from 'I_1' to 'I_2'. This goes on till the resistor 'R_{st}' is eliminated and motor armature gets full voltage 'V' the motor runs at full speed drawing no load current 'I_0'.

Mathematically,
$$I_2 = \frac{V - |E_b|}{R_m} \quad N = N_1;\ R_m = m\text{th stud}$$

$$I_1 = \frac{V - |E_b|}{R_{m+1}} \quad N = N_1$$

The back emf 'E_b' can't change instantaneously because of rotor inertia

Thus, $$\frac{I_1}{I_2} = \frac{R_m}{R_{m+1}}$$

Substituting the values, $m = 1, 2, 3, \ldots$

$$\frac{I_1}{I_2} = K = \frac{R_1}{R_2} = \frac{R_3}{R_3} = \ldots = \frac{R_n}{R_{n+1}}$$

So, $$K^n = \frac{R_n}{R_{n+1}} = \frac{R_1}{R_a} \quad \text{or} \quad n \log_e K = \ln\left(\frac{R_1}{R_a}\right)$$

$$n = \ln \frac{R_1}{R_a} \Big/ \ln K$$

'n' no. of steps.

7.11 SPEED CONTROL OF D.C. MOTOR

The speed of d.c. motor is expressed as follows:

$$N \propto \frac{V - I_a R_a}{\phi} \qquad \text{(I)}$$

Where, N = Speed of motor, rpm

V = Terminal voltage of motor, volts

I_a = Armature current, amperes.

R_a = Armature resistance, ohms

ϕ = Field flux, weber.

From the exp. (I), we see that the motor speed is dependent on the voltage supplied to the motor, armature circuit resistance, armature current or line current and the field flux which the armature conductors cuts. The speed control of d.c. motor is classified in basically two types, namely,

(a) Flux control

(b) Armature control.

These two controls are implemented for both d.c shunt, d.c. series and d.c. compound motor. It could be implemented to d.c. separately excited motor also. This method can be applied to brushless d.c. motor as well.

Separately Excited d.c. Motor Speed Control

(a) *Field or Flux Control:* By placing a rheostat of value 'R' in the field circuit increases the field circuit resistance hence the field current 'I_f' is decreased. Since the field flux 'ϕ' depends directly in the magnitude of 'I_f'.

So with the decrease of the field current 'I_f' the flux 'ϕ' decreases correspondingly. Hence, the value of 'E_b' decreases. The speed 'N' of the motor increases. This decrease in the flux can be decreased to a value should not damage the dynamic balance of machine. This type of the speed control is called rheostatic control.

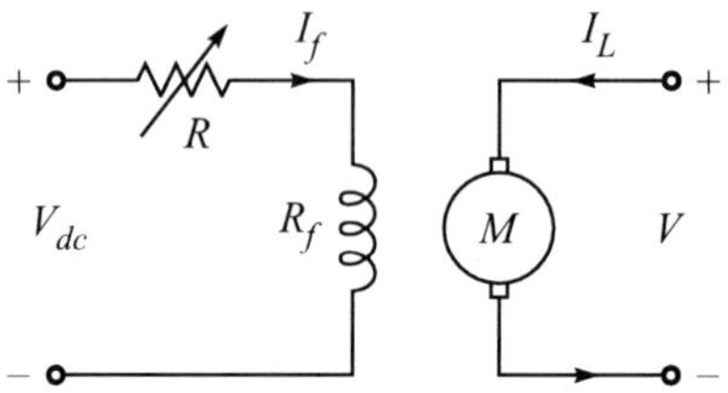

Fig. 7.20 Flux control.

(b) *Armature Control:* By changing the voltage applied to the motor the speed of the motor can be changed. But by this method the speed lesser than the rated speed can be obtained. This method is employed for small PMDC and brushless d.c. motors.

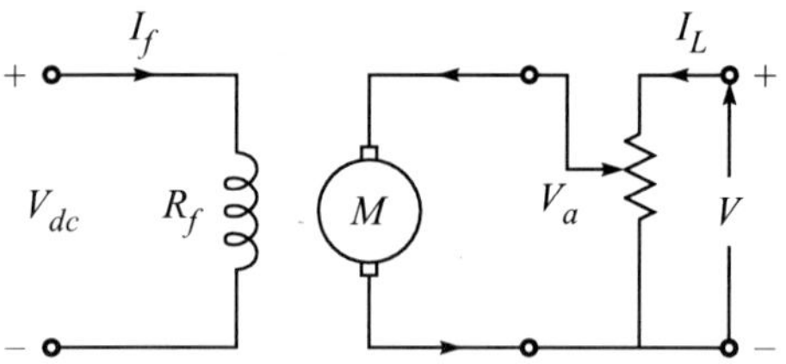

Fig. 7.21 Armature control.

The armature voltage 'V_a' can be decreased by a potential divider arrangement as shown in the Fig. (7.2).

In this control, the field flux is kept constant by keeping the field circuit resistance 'R_c' constant.

D.C. Shunt Motor

(a) *Flux Control (Field flux control):* As we know that the field flux is inversely proportional to the speed of motor. So as the resistance of field circuit 'R_f' increases. The field current 'I_f' decreases accordingly. Correspondingly, 'ϕ' decreases, so the speed above the rated speed of motor can be achieved by this method.

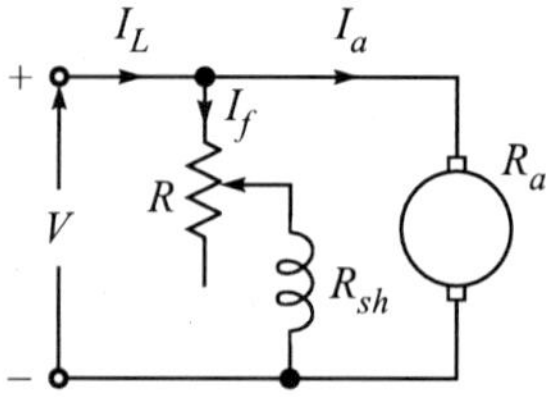

Fig. 7.22 Field control of d.c. shunt motor.

$$N \uparrow \propto \frac{V - I_a R_a}{\phi \uparrow} \qquad \text{(II)}$$

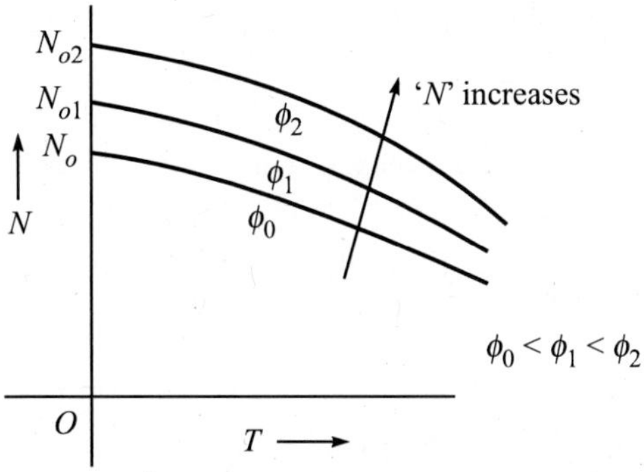

Fig. 7.23 *N/T* characteristic for field control of d.c. shunt motor.

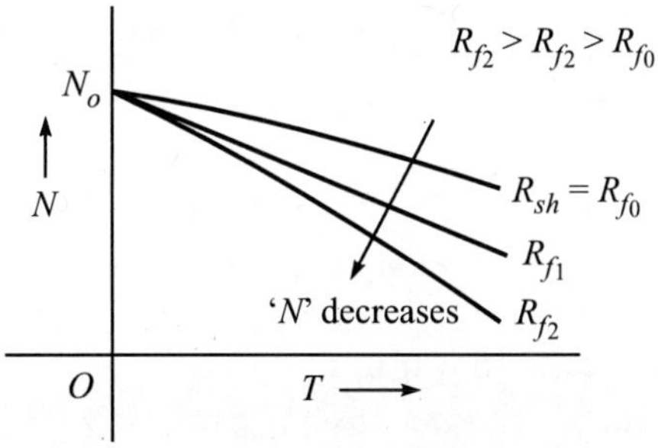

Fig. 7.24 Speed torque characteristic for armature control of d.c. shunt motor.

By field control, speed above the base speed or the rated speed can be obtained with small variation of field flux of the motor so it is the preferred method of speed control in industries. It is an efficient and economical method. The current in field circuit is small, so small value of rheostat is necessary. The control is also easy.

The flux control suffers from the following disadvantages, namely, unstable operation of motor after decreasing flux below a given value, it reduces the torque producing capability of motor.

(b) (i) *Armature Voltage Control or Rheostatic Control.* Speed control is achieved by putting a resistance 'R' in the armature circuit of the d.c. shunt motor. The field winding is connected across the supply voltage 'V' which is the rated motor voltage.

The additional resistance in the armature circuit of the motor decreases the armature current 'I_a'. Since the field flux 'ϕ' of the motor is

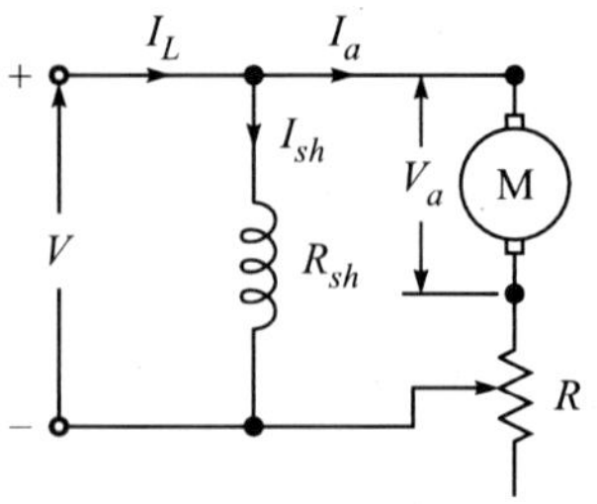

Fig. 7.25 Armature voltage control of d.c. motor.

constant so the torque 'T_m' of the motor is directly proportional to the armature current 'I_a'. That is,

$$\phi = \text{constant (since, } I_{sh} = \text{constant)}$$

$$T_m \propto I_a$$

At rated voltage, the motor has the maximum speed. But as the armature voltage decreases the motor speed decreases accordingly.

$$N \propto \frac{V - I_a(R_a + R)}{\phi} \qquad \text{(III)}$$

$$\phi = \text{constant}$$

So, $$N \propto V - I_a(R_a + R) \qquad \text{(IV)}$$

we see from the expression (IV), that the resistance of armature circuit ($R_a + R$) increases the speed of the motor 'N' decreases. The N/T characteristic is shown in Fig. (7.24).

(ii) *Armature Voltage control (using Potential Divider):* From the Fig. (7.26), we see that voltage across field winding is constant.

So $$\phi = \text{constant}$$

So, $$N \propto V_a - I_aR_a \qquad \text{(V)}$$

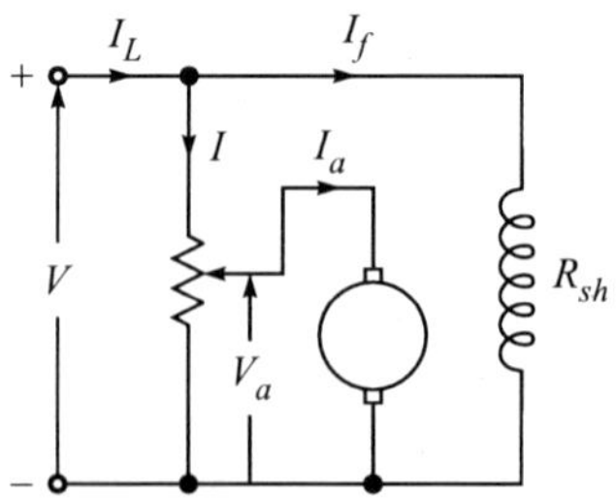

Fig. 7.26 Armature voltage control of d.c. shunt motor.

As the voltage across armature 'V_a' decreases so the speed decreases accordingly. This method is used for small size d.c. motor in audio instruments driving the magnetic tape or recorder.

D.C. Series Motor

(a) *Armature Resistance Control:* The speed relation is given by

$$N \propto \frac{V - I_a(R_a + R_{se} + R)}{\phi} \tag{VI}$$

The flux 'ϕ' is proportional to the current 'I_a' also $I_a = I_L$

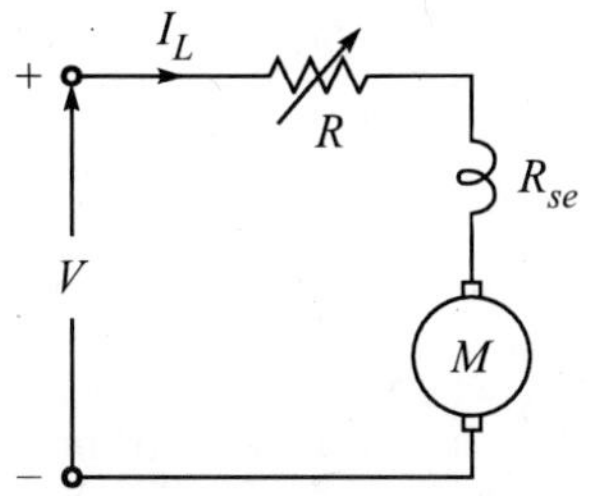

Fig. 7.27 Rheostatic control.

$$N \propto \frac{V - I_a(R_a + R_{se} + R)}{k' I_a}$$

$$\boxed{N \propto \frac{V}{I_a} - (R_a + R_{se} + R)} \tag{VII}$$

We see that 'R' is variable and it could only be increased. So with increase of this resistance the speed 'N' can be decreased. So speed below base or rated speed could be obtained by this method. With $R = 0$, the normal N/T can be obtained as follows

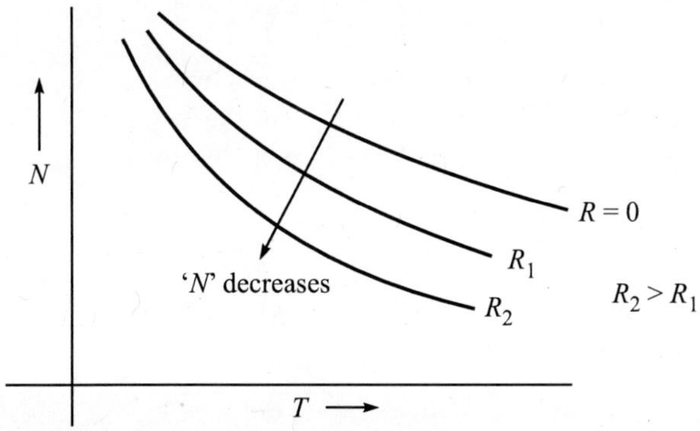

Fig. 7.28 *N/T* characteristics for armature rheostatic control.

(b) *Flux Control*

(i) *Diverter resistor control:* The diverter resistance maximum value is infinite (that is, it is not connected across the field circuit). So under that condition the *N*/*T* characteristic of the motor will be normal characteristic of the d.c. motor.

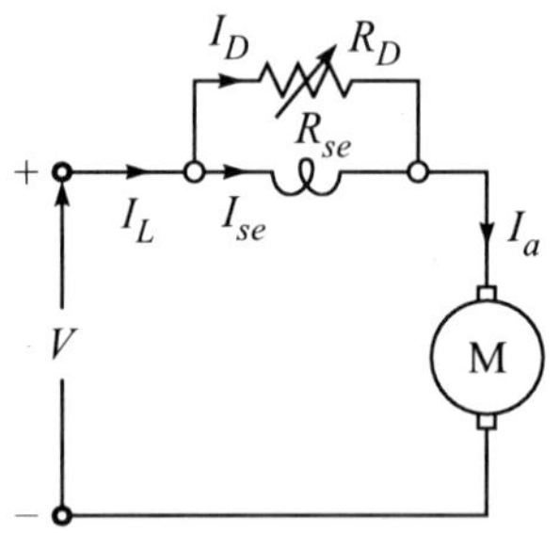

Fig. 7.29 D.C. series diverter resistor control.

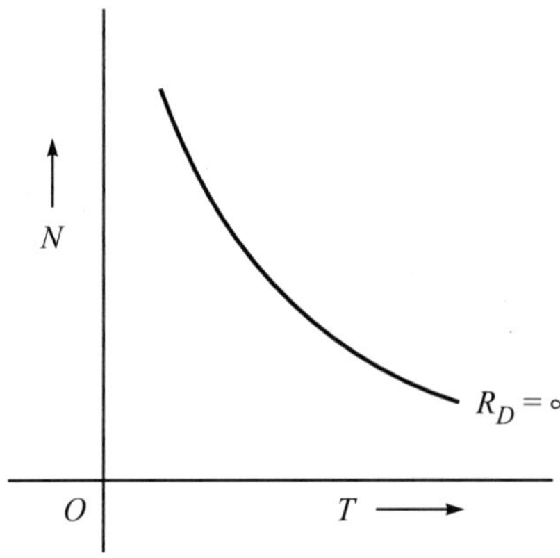

Fig. 7.30 N/T characteristic of flux control (with no. R_D).

So, with the decrease of diverter resistor value the current 'I_D' increases through the resistor 'R_D'. So the value of 'I_{se}' decreases, hence 'ϕ' decreases and the speed of the motor increases as per the relation

$$N \propto \frac{V - I_a\left[R_a + \dfrac{R_{se} R_D}{R_{se} + R_D}\right]}{\phi} \qquad \text{(VIII)}$$

$$\therefore \quad \phi \propto I_{se} \text{ and } I_{se} = \left(\frac{R_D}{R_D + R_{se}}\right) I_L = \frac{R_D}{R_D + R_{se}} \cdot I_a \quad (\because I_L = I_a)$$

or $$N \propto \frac{V - I_a\left[R_a + \dfrac{R_{se} R_D}{R_{se} + R_D}\right]}{\left(\dfrac{R_D}{R_D + R_{se}}\right)\cdot I_a}$$

or $$N \propto \frac{V}{I_a}\left(\frac{R_D + R_{se}}{R_D}\right) - \frac{R_a\,(R_{se} + R_D) + R_{se} R_D}{(R_{se} + R_D)} \times \frac{(R_D + R_{se})}{R_D}$$

or $$N \propto \left\{\frac{V}{I_a}\left[1 + \frac{R_{se}}{R_D}\right] - \left[\frac{R_a R_{se}}{R_D} + (R_a + R_{se})\right]\right\} \qquad \text{(IX)}$$

With decrease of 'R_D', speed 'N' rises and follow the characteristics as follows.

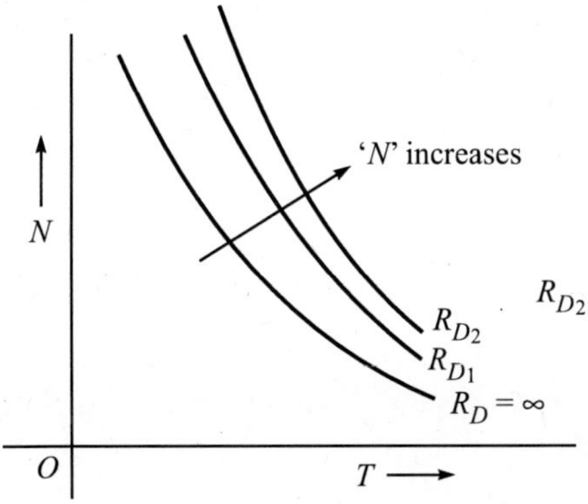

Fig. 7.31 *N/T* characteristic of d.c. series motor with diverter control.

(ii) *Tapped Field Control:* As the sliding control is moved in the clockwise direction over the various taps, the number of series turns decreases, hence the ampere turns of series winding decreases, so the series field flux 'ϕ' decreases and the motor speed 'N' increases. But the tapped field control can take few steps value as there are less number of series field winding turns of the motor.

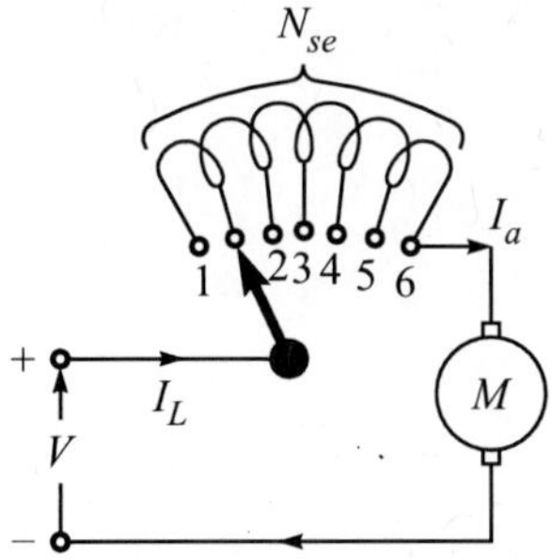

Fig. 7.32 Tapped field d.c. series motor control.

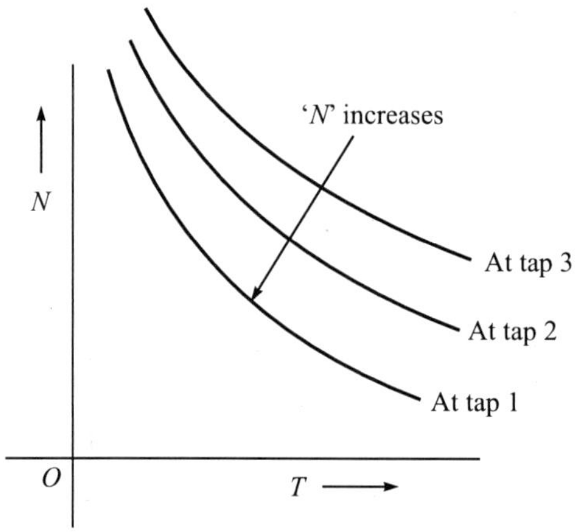

Fig. 7.33 *N/T* characteristic for tapped field control.

(iii) *Regrouping of field coils:* The field winding connection can be series or parallel as shown in Fig. (7.34). In the Fig. 7.34(a) the flux is decreased by halved as compared to the flux in the case of series connection as shown in Fig. 7.34(b). So the speed of the motor is two times in the parallel connection as compared to that of series connection. So by series parallel connections four speeds are possible with change in field connections. In this case the speed regulation of 4:1 is possible. This method is used in the case of traction purpose.

We see in the case of armature control only speed lesser than the base speed can be obtained, whereas the flux speed control is employed for speeds greater

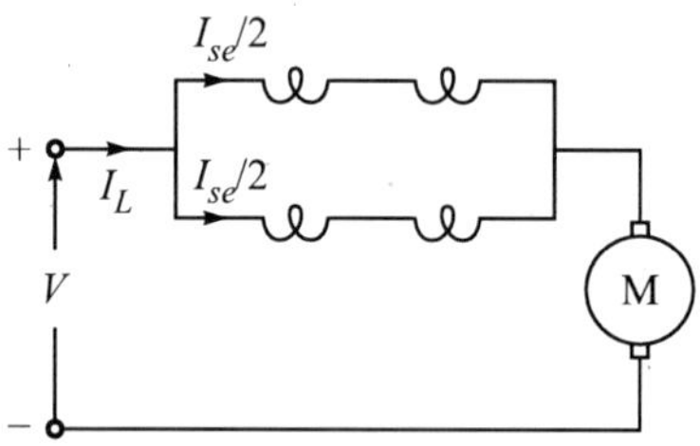

Fig. 7.34(a) Parallel connection of field winding.

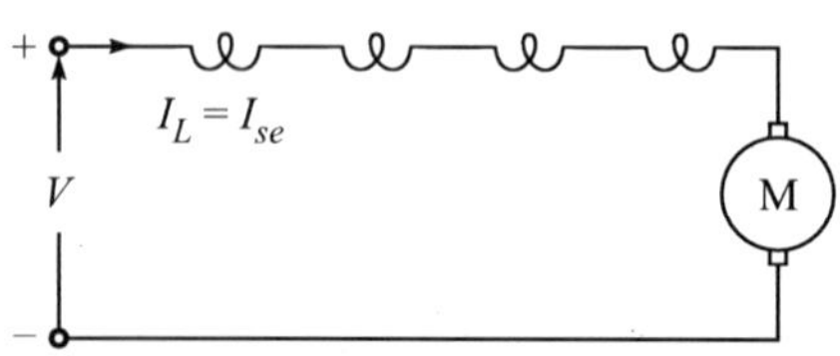

Fig. 7.34(b) Series connection of field winding.

than the base speed. But in some places it is required that by the same arrangement both speed above and below rated speed is necessary. So Ward Leonard speed control method is employed in such place. In which both step less speed control over wide range could be obtained.

Ward Leonard Method

It employs both flux control and armature control. This could be understood by using the Fig. (7.35).

By increasing the value of field current 'I_f' of generator 'G', the induced emf 'E_g' increases, so the armature voltage 'V_a' increases and hence speed of motor 'M' is increased above base speed. With the decrease of field current 'I_f', the voltage across the motor could be decreased accordingly and speed below the rated or base the speed could be obtained.

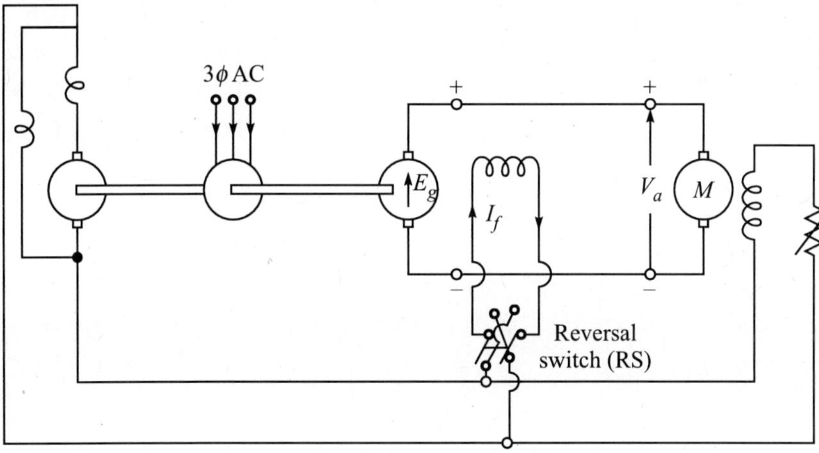

Fig. 7.35 Ward Leonard method of speed control.

For reversal several of the motor, the field current 'I_f' direction could be reversed using the reversal switch 'RS' and the voltage 'V_a' direction could be reversed and so the motor current direction is also reversed so the torque direction is changed accordingly. By the Ward Leonard control speed variation of few r.p.m. could be obtained.

One demerit of this method is the high cost due to large number of machines used and low efficiency. But still now it is used in the steel and aluminium industries for high power applications. Now by the advent of static semiconductor devices the static Ward Leonard scheme is becoming popular due to its low cost.

7.12 BRAKING OF D.C. MOTOR

The braking of d.c. motor occurs when there is a need of stopping it or decreasing the speed of the motor as in the industrial motor drives. The braking operations are classified under the three main types, namely,

(a) Plugging
(b) Dynamic Braking
(c) Regenerative Braking.

The braking operation may be mechanical or electrical. But due to the advantages of electrical braking, it is discussed in this section.

Plugging (Counter Current Braking)

In this operation the direction of armature current is reversed while the motor is in operation. Due to the large inertia of the motor, the motor could not reverse its direction instantly, so the braking torque results in a high braking current which should be taken care of. This is done by placing a plugging resistance 'R_p' during braking operation as shown in the Fig. (7.36).

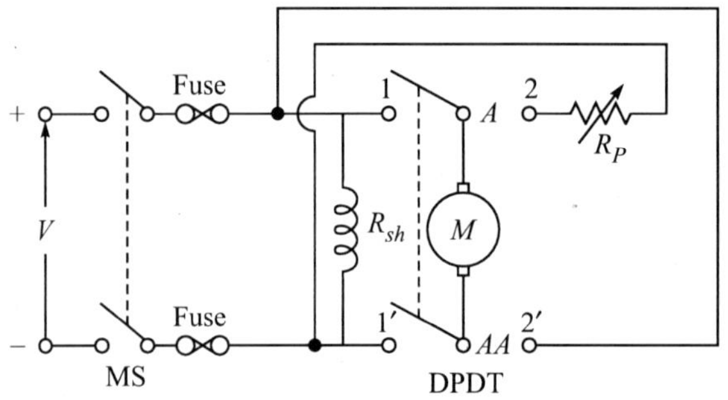

Fig. 7.36 Plugging operation.

When the motor armature terminals 'A' and 'AA' are in the positions '11′' normal motoring operation results. But during the braking operation i.e., when the DPDT is thrown to the position '22′', high braking current 'I_{b_p}' results which is restricted by the plugging resistor 'R_p' which is a high value of resistance. The braking torque is given by

$$T_{b_p} \propto \phi I_{b_p} \tag{7.41}$$

$$I_{b_p} = -\frac{(V + k\phi N)}{R_a + R_p} \tag{7.42}$$

$$T_{b_p} = -\text{ve (braking operation)}$$

$$T_m = +\text{ve (motoring operation)}$$

Plugging operation is a good method of bringing the motor to rest and to operate it in the reverse direction. The motor is cut out of supply once the motor speed becomes zero. The plugging operation can be implemented for the shunt and series case efficiently. The braking torque 'T_{bp}' during plugging operation opposes the motoring torque 'T_m' so it is expressed as negative.

Dynamic Braking (Rheostatic Braking)

The dynamic braking is employed in such operation where there is a need of bringing the motor to rest and then to reverse the direction of motor and then brought to rest and again rotated in its original direction.

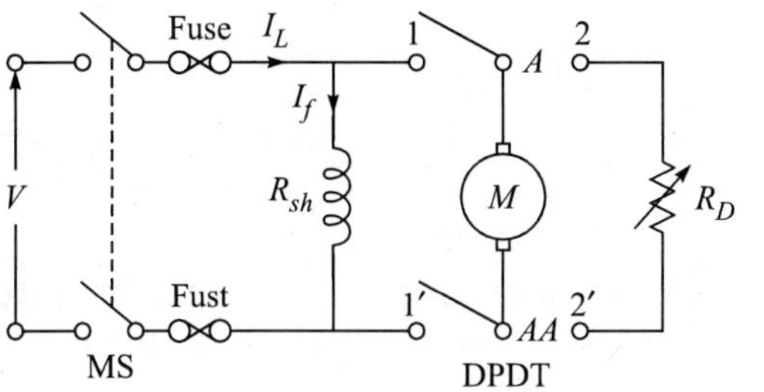

Fig. 7.37 Dynamic braking of d.c. shunt motor.

In this method, the armature circuit of the motor is opened and then connected across a variable dynamic resistance 'R_D'. Due to the inertia of motor rotor it will continue to rotate in the field, as the field is still connected to the supply and it is carrying field current 'I_f'. So the emf is induced in the rotor and a current 'I_{b_D}' is resulted which is directed out of the motor armature terminal and flows through the dynamic braking resistance 'R_{b_D}'. Since the direction of this current 'I_{b_D}' opposes that of the original motoring current 'I_a', so it will result in a torque 'T_{b_D}' which will oppose the motoring torque 'T_m' of the motor (responsible for the rotation of motor). This braking torque 'T_{b_D}' will result in decreasing the speed of motor. The speed of the motor is brought down by decreasing the value of resistance 'R_{b_D}' allowing large value of 'I_{b_D}' to pass through it and hence it results in higher braking torque. Once the motor is stopped the supply is put off from the motor circuit. This braking method brings the motor to rest quickly.

$$I_{b_D} = -\frac{E}{R_a + R_D} \tag{7.43}$$

$$T_{b_D} \propto \phi\, I_{b_D} \tag{7.44}$$

$$T_{b_D} = -\text{ve (braking torque)}$$

$$T_M \propto \phi\, I_a$$

$$T_M = +\text{ve (motoring torque)}$$

Regenerative Braking

The braking operation employed in the hilly areas traction purpose or the hoist while moving under gravity employs this type of braking operation.

This type of braking operation results when the speed of the motor is greater than the rated speed of motor. When the speed of motor exceeds that of base or rated speed then the back emf of the motor armature 'E_b' exceeds that of terminal voltage 'V' of the motor.

Since $$E_b \propto N\phi$$

$$N \propto \frac{V - I_a R_a}{\phi}$$

$$V = E_b + I_a R_a$$

$$\frac{V - E_b}{R_a} = I_a$$

As, $E_b > V$

So, $$I_a = -\text{ve}$$

The negative value of the armature current implies that the direction of the motor current has reversed. Since torque of the motor is directly proportional to the motor current 'I_a'. So,

$$T \propto I_a$$

So, $$T_{b_r} = -\text{ve (braking torque)}$$

So, the torque will also reverse implying that the braking operation starts. But in this type of braking the current is delivered by the motor back to the supply i.e., regeneration. So, this type of braking is called regenerative braking. In this braking about 20% – 30% of the electrical energy can be reused for some other work. In hilly areas, the train going downhill regenerates electricity for driving a train going uphill thus saving energy. Similar is the case of hoist operation. By the regenerative braking the motor can't be brought to rest only it could be slowed down.

7.13 PERMANENT MAGNET D.C. MOTOR (PMDC)

In the small motors the field windings are not used. The field is replaced by permanent magnet. In the field of permanent magnet present the armature conductors are placed. The commutator segments are only a few on which brushes are placed. All the audio equipment driver motor are permanent magnet type.

In most of the motors a printed circuit board consisting of an electronic circuitry is placed inside the cover of the d.c. motor. Only the terminal connection point are marked and placed outside the cover of machine.

The positive and negative are marked so that wrong polarity is not connected. There is a potentiometer (single coil type) for changing the terminal voltage of the motor is used for obtaining the speed regulation.

The main advantage of such motor is low cost, small size and no need of field excitation, high speed, easy control, almost no maintenance and reliable performance of such motors. The demerits are of low overloading capacity, weakening of field with aging effect; small life of motor etc.

7.14 BRUSHLESS MOTOR

Many of the limitations of the d.c. motor with brushes are discussed in the previous sections such as wear and tear of brushes, pitting of commutator segment, inability to keep in control at high speeds, sparking. So the rating of the machine is limited with reference to the speed of motor. The current density per unit area of the brushes limits the output of the motor. The imperfect contact of brush and commutator cause noise. There is a need of replacement of brushes due to wear and tear so box type holder is preferable for brush.

The d.c. motor having brush and commutator acts as a mechanical rectifier since the emf in armature is a.c. and current is also a.c. The static brush and the rotating commutator makes a.c. into d.c. In the brushless motor, commutator/brushgear assembly is replaced by an external switch (electronic) synchronised to the rotor's position. The efficiency of these motors are 85% – 90% whereas d.c. motor (with brushes) are 75% – 80% efficient.

Construction is similar as stepper motor often use a PM external rotor, three phases of driving coils, one or more Hall effect sensors to sense the position of the rotor and the associated drive electronics. The coils are activated, one phase after the other, by the drive electronics as queued by the signals from the Hall effect sensors. In effect, they act as a three-phase synchronous motors containing their own variable frequency drive electronics. A specialized class of brushless d.c. motor controllers utilize emf feedback through the main-phase connections instead of Hall effect sensors to determine the position and velocity. These motors are used extensively in electric radio-controlled vehicles for research purposes. DC motors are commonly used where precise speed control is necessary such as computer disk drives or in video cassette recorders, CD-ROM drives, fans laser printers and photocopiers.

Advantages of brushless motors

(1) Efficient, (2) Less noise, (3) Higher life, (4) Maintenance free, (5) Precise speed control applications, (6) No fire hazard, (7) Less heat, (8) Vibrations free.

The size is from fraction of watt to many kilowatt (as high as 100 kW in electric vehicles).

7.15 UNIVERSAL MOTORS

A variant of the wound field d.c. motor is the universal motor. It can operate on both a.c. and d.c. This motor is less efficient than an equivalent pure d.c. motor. Operating at normal 50 Hz the maximum output of universal motors is limited and motors exceeding 1 kW are rare. To keep efficiency high, the frequency is kept $16\frac{2}{3}$ Hz and 25Hz respectively.

Advantages

(1) High starting torque
(2) Compact
(3) Applicable on both a.c. and d.c.
(4) Speed compact using semiconductor devices possible.
(5) High speed (> 30,000 rpm)

Disadvantages

(1) Small life
(2) Maintenance required

Use: Food mixers, power tools, blenders, vacuum cleaners, hair dryers, weed trimmers, grinders etc.

The precaution of the motor is that the speed should not be greater than N_{rated} with no significant load. So load on shaft is must to keep motor speed within safe value.

7.16 APPLICATIONS OF D.C. MOTOR

D.C. Shunt Motor

1. Drilling machines
2. Milling machines
3. Leather machines
4. Paper machines
5. Centrifugal pumps
6. Reciprocating pumps
7. Blowers
8. Fans
9. Wood cutting machines.

The shunt motor has constant speed characteristic. That is, the drop in speed from no load to full load is low about 5%. So they are considered constant speed motors. But due to similar characteristic of squirrel cage IM, and low cost it is preferred over d.c. similar rated motor. Also the efficiency of IM is high. It has comparatively small size and is available in sealed type.

D.C. Series Motor

1. Trains
2. Diesel-electric locomotives
3. Cranes
4. Horists
5. Trolley cars.
6. Conveyers
7. Rapid transit systems
8. Mining motors
9. Lifts.

The series motor has variable speed range. It has high starting torque and preferred for regenerative braking operation. The speed characteristic of series motor shows that this type of motor are not worked under no load conditions.

Cummulative Compound D.C. Motors

1. Elevators
2. Shears
3. Punches
4. Rolling mills
5. Planers
6. Large CF pumps.

These motors have moderately high torque i.e., 250 % T_{fl} upto 350 % T_{fl} at the starting. Speed regulation is higher as compared to shunt motors.

Differentially Compound Motors

These motors are seldom used due to rise of motor speed with loading which results in unstable operation of motor.

SOLVED QUESTIONS

Q.1 A starter is to be designed for a 250 V d.c. shunt motor. The armature resistance is 0.15 Ω and full load armature current is 40 A. During starting, current should not exceed 200% of the rated. Current and should not fall below the rated current. Calculate:

(i) the total resistance of the starter.

(ii) the resistance to be cut out in each step in the starting operation.

A.1

$$V = 250 \text{ V}, R_a = 0.15 \ \Omega, I_{af} = 40 \text{ A}$$

$$I_{f_e} \le I_{st} \le 2 \ (I_r = I_{f_e})$$

$$(R_{st} + R_a) \ I_{a_{st}} + E_b = V.$$

$$I_{a_{st}} = \frac{V - E_b}{R_{st} + R_a}; \ E_b \propto N\phi$$

or,
$$2 \times 40 = \frac{250 - 0}{R_{st} + 0.15}$$

or,
$$(R_{st} + 0.15) = \frac{250}{80} = 3.125$$

or,
$$R_{st} = 2.975 \ \Omega = R_1$$

$$k = \left(\frac{R_1}{R_a}\right)^{\frac{1}{n}} = \left(\frac{2.975}{0.15}\right)^{\frac{1}{n}} = (19.833)^{1/n} = \frac{2 \times 40}{40} = 2$$

$$2^n = 19.833 \quad \text{or} \quad n \ln 2 = \ln 19.833$$

$$n = \frac{\ln 19.833}{\ln 2} = \frac{2.962}{0.693} = 4.273$$

$$n \approx 4$$

$$R_2 = \frac{R_1}{K} = \frac{2.975}{2} = 1.4875 \ \Omega$$

$$R_3 = \frac{R_2}{K} = \frac{1.4875}{2} = 0.74375 \ \Omega$$

$$R_4 = \frac{R_3}{K} = \frac{0.7438}{2} = 0.3719 \ \Omega$$

$$R_5 = \frac{R_4}{K} = \frac{0.3719}{2} = 0.18595 \ \Omega$$

∴
$$r_1 = R_1 - R_2 = 2.975 - 1.4875 = 1.4875 \ \Omega$$
$$r_2 = R_2 - R_3 = 1.4875 - 0.78375 = 0.70375 \ \Omega$$
$$r_3 = R_3 - R_4 = 0.78375 - 0.3719 = 0.41185 \ \Omega$$
$$r_4 = R_4 - R_5 = 0.3719 - 0.18595 = 0.18595 \ \Omega$$

The resistance to be cut out in each steps in the starting operations are r_1, r_2, r_3, r_4, resp.

Q.2 A starter is required for a 200 V shunt motor. The maximum allowable current is 55 A and the minimum current is 35 A. Find the number of sections of starter resistance required and the resistance of each section. Derive the formula used in the problem.

A.2
$$k = \frac{I_2}{I_1} = \frac{55}{35}, \quad R_1 = \frac{200\text{ V}}{55\text{ A}} = 3.636\ \Omega$$

$\therefore$
$$k = \left(\frac{R_1}{R_a}\right)^{\frac{1}{n}} \text{ suppose } R_a = 1\ \Omega$$

then
$$k = \left(\frac{3.636}{1}\right)^{\frac{1}{n}} = 1.571$$

$\Rightarrow$
$$(1.571)^n = 3.636 \quad \text{or} \quad n \ln 1.571 = \ln 3.636$$

No. of steps of starter = 4

$$n = \ln 3.636/\ln 1.571$$
$$n = 1.8925/0.452 = 4.10$$
$$n \approx 4$$

$$R_2 = \frac{R_1}{K} = \frac{3.636}{1.571} = 2.314\ \Omega$$

$$R_3 = \frac{R_2}{K} = 1.473\ \Omega$$

$$R_4 = \frac{R_3}{K} = 0.9376\ \Omega$$

$$R_5 = \frac{R_4}{K} = 0.5968\ \Omega$$

$$r_1 = R_1 - R_2 = 1.322\ \Omega$$
$$r_2 = R_2 - R_3 = 0.841\ \Omega$$
$$r_3 = R_3 - R_4 = 0.5354\ \Omega$$
$$r_4 = R_4 - R_5 = 0.341\ \Omega$$

Q.3 A 200 V d.c. shunt motor taken 22 A at rated voltage and runs at 1000 rpm. Its field resistance is 100 Ω and armature circuit resistance is 0.1 Ω. Compute the value of additional resistance required in the armature circuit to reduce the speed to 800 rpm. When the load torque is independent of speed.

[B.Tech, II AMU)

A.3
$$V_r = 200\text{ V}, \quad I_{Lr} = 22\text{ A}; \quad N_1 = 1000\text{ rpm}.$$
$$R_f = 100\ \Omega, R_a = 0.1\ \Omega, R = ? \quad \text{if } N_2 = 800\text{ rpm}$$

$$I_{a1} = I_L - I_f = I_L - \frac{V_r}{R_f}$$

$$= 22 - \frac{200}{100}$$

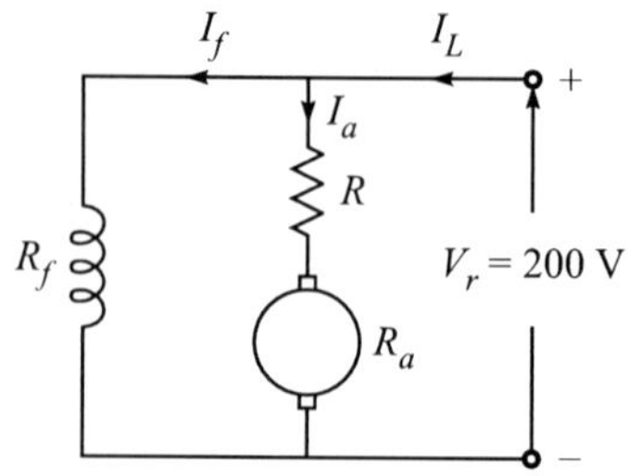

So, $$I_{a_1} = 22 - 2 = 20 \text{ A}$$

$$V = E_{b_1} + I_{a_1} R_a$$

or, $$E_{b_1} = 200 - 20 \times 0.1 \propto 198 \quad (1)$$

$$E_{b_1} \propto N_1$$

$$E_{b_2} = V - I_{a_2} (R_a + R) \propto N_2$$

$$N_2 \propto 200 - 20\,(0.1 + R) \quad (2)$$

Dividing (2) by (1), we get

$$\frac{N_1}{N_2} = \frac{198}{200 - 20(0.1 + R)} = \frac{1000}{800}$$

$$(198 - 20R) \times 10 = 198 \times 8$$

or, $$200\,R = 198 \times 2$$

or, $$R = 1.98\ \Omega$$

Value of additional resistance 'R' is 1.98 Ω which should be attached in series with 'R_a' to reduce the speed of motor to 800 rpm.

Q.4 A 230 V, d.c. shunt motor, takes an armature current of 3.33 Amp at rated voltage and at a no load speed of 1000 rpm. The resistance of the armature and field circuit are 0.3 Ω and 160 Ω respectively. The line current at full load and rated voltage is 40 Amps. Calculate the full load speed if the armature reaction weakens the no load flux by 4%.

[B.Tech, JMI, 2006]

A.4 $$V_r = 230 \text{ V},\ I_{a_0} = 3.33 \text{ A},\ N_0 = 1000 \text{ rpm}.$$

$$R_a = 0.3\ \Omega;\quad R_f = 160\ \Omega,\ \left.I_L\right]_{fe, V_r} = 40 \text{ A}$$

$$N_{fe} = ?;\ \phi_{fe} = 0.96\ \phi_0.$$

$$V = E_{b_0} + I_{a0} \cdot R_a = k \cdot N_0 \cdot \phi_0 + I_{a0} R_a$$

$$V = E_{b_f} + I_{a_f} \cdot R_a = k \cdot N_f \cdot \phi_{fe} + I_{a_f} R_a.$$

$$\frac{N_0\,\phi_0}{N_f \cdot \phi_{fe}} = \frac{V - I_{a_0} R_a}{V - I_{a_f} R_a}$$

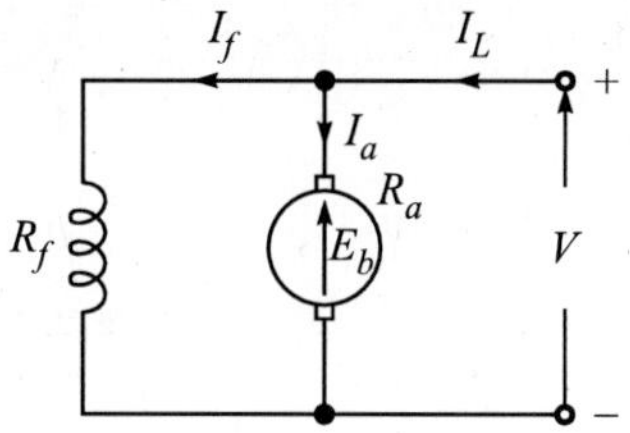

$$\frac{N_0 \cdot \phi_0 1}{N_f \cdot (0.96)\phi_0} = \frac{V - I_{a_0} R_a}{V - I_{a_f} R_a}$$

$$N_f = \frac{N_0}{0.96} \cdot \frac{(V - I_{a_f} R_a)}{(V - I_{a_0} R_a)}$$

$$I_{a_f} = I_{L_f} - I_f = I_{L_f} + \frac{V_r}{R_f} = 40 + \frac{230}{160} = 41.4375 \text{ A}$$

$$N_{fl} = \frac{1000}{0.96} \left[\frac{230 - 41.4375 \times 0.3}{230 - 3.33 \times 0.3}\right]$$

$$\boxed{N_{fl} = 989.7 \text{ rpm}}$$

The full load speed of motor is 989.7 rpm.

Q.5 A 250 V d.c. shunt motor having an armature resistance of 0.25 Ω carries an armature current of 50 A and runs at 750 rpm. If the flux is reduced by 10%, find the speed, if load torque remains the same.

[B.Tech, II JMI, 2006]

A.5 $V = 250$ volt, $R_a = 0.25\ \Omega$, $I_a = 50$ A, $N = 750$ rpm.

$\phi' = 0.9\ \phi$, $T_L =$ constant, $N' = ?$

$$N\phi \propto V - I_a R_a$$

$$N'\phi' \propto V - I_a' R_a$$

$$\frac{N}{N'} \cdot \frac{\phi}{\phi'} = \frac{V - I_a R_a}{V - I_a' R_a}$$

$$N' = \frac{V - I_a' R_a}{V - I_a R_a} \cdot \frac{\phi}{\phi'} \cdot N$$

or,

$$N' = \frac{V - I_a' R_a}{V - I_a R_a} \cdot \frac{\phi}{0.9\phi} \cdot N$$

or, $$N' = \frac{(250 - 50 \times 0.25)}{(250 - 50 \times 0.25)} \cdot \frac{750}{0.9} \quad \{I'_a = I_a. \because T_L = \text{same}\}$$

or, $$\boxed{N' = 833.3 \text{ rpm}}$$

Q.6 A 460 V series motor runs at 500 rpm taking a current of 40 A. Calculate the speed and percentage change in torque if the load is reduced so that the motor in torque if the load is reduced so that the motor is taking 30 A current. Total resistance of the armature and series field circuit is 0.8 Ω. Assume flux and field current to be proportional.

A.6
$$N = 500 \text{ rpm}; \; R_{\text{total}} = 0.8 \; \Omega$$
$$I_L = I_a = 40 \text{ A}$$
$$I'_a = 30 \text{ A} = I'_a$$
$$\phi \propto I_a$$

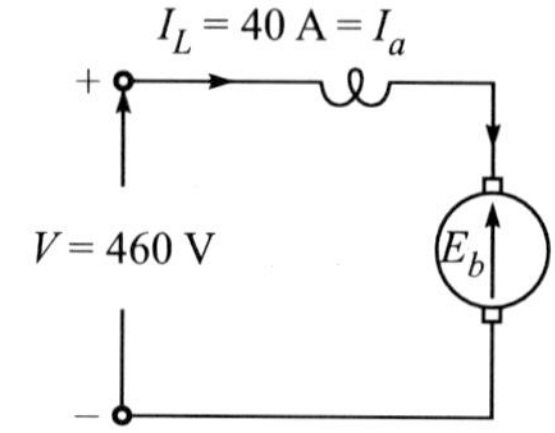

(a) $V = E_b + I_a\,(R_{\text{total}})$; $E_b \propto N\phi$ or $E_b \propto N \cdot I_a$.

$$V = k \cdot N\, I_a + I_a\, R_{\text{total}}$$
$$kN \cdot I_a = V - I_a\, R_{\text{total}}$$
$$N \propto \frac{V - I_a\, R_{\text{total}}}{I_a}$$
$$N' \propto \frac{V - I'_a\, R_{\text{total}}}{I'_a}$$
$$\frac{N'}{N} = \frac{V - I'_a\, R_{\text{total}}}{I'_a} \cdot \frac{I_a}{V - I_a\, R_{\text{total}}}$$
$$N' = N\left[\frac{I_a}{I'_a}\right]\left[\frac{V - I'_a\, R_{\text{total}}}{V - I_a\, R_{\text{total}}}\right]$$
$$= 500\left[\frac{40}{30}\right]\left[\frac{460 - 30 \times 0.8}{460 - 40 \times 0.8}\right]$$

$$\boxed{N' = 679.13 \text{ rpm}}$$

Speed of the series motor at $I'_L = 30$ A is 679 rpm.

(b) Percentage change in torque $\%T = \dfrac{\Delta T}{T_{f_1}} \times 100\%$

$$T \propto \phi I_a \qquad \because \phi \propto I_a.$$
$$T \propto I_a^2$$
$$= \frac{40^2 - 30^2}{40^2} = \frac{(4^2 - 3^2)}{4^2} \times 100\%$$
$$= 43.75\%$$

Q.7 The no load speed of a d.c. motor is 1500 rpm and the full load speed is 1300 rpm. Determine the % speed regulation of the motor.

A.7
$$N_{NL} = 1500 \text{ rpm}$$
$$N_{FL} = 1300 \text{ rpm.}$$

The % speed regulation is given by

$$\% \text{ speed regulation} = \frac{N_{NL} - N_{FL}}{N_{FL}} \times 100\%$$
$$= \frac{1500 - 1300}{1300} \times 100\%$$
$$= \frac{200}{1300} \times 100\ \%$$
$$= \frac{200}{13}\ \%$$
$$= 15.38\ \%$$

UNSOLVED QUESTIONS

Q.1 A d.c. shunt motor develops 12 bhp at 600 rpm when drawing line current of 20 A at 500 V. Find the efficiency at this load and the useful torque.

[B.Tech II, I.U., 2005]

Q.2 A 10 hp, 230 V shunt motor has an armature resistance of 0.5 Ω and field circuit resistance of 115 Ω. At no load and rated (full) voltage the speed is 1200 rpm and armature current is 2 A. If load is applied, the speed drops to 1100 rpm. Determine the armature current and the line current.

[B.Tech II, UPTU, 2002]

Q.3 A 230 V series motor is taking 50 A. Resistance of armature and series field winding is 0.2. Ω and 0.1 Ω respectively. Calculate:

(i) brush voltage.

(ii) back emf.

[B.Tech II, UPTU, 2003]

Q.4 A series motor takes 20 A at 400 V 10 drive a load whose torque is proportional to the square of the speed at 250 rpm. Determine the necessary applied voltage and current to drive the said load at 350 rpm. Assume straight line magnetising characteristic.

[B.Tech II, UPTU, 2004]

Q.5 A 120 V, d.c. shunt motor has an armature resistance of 0.2 Ω and a brush volt drop of 2 V. The rated full load armature current is 75 A. Calculate the current at the instant of starting and its value in terms of percentage of full load armature current. Assume field winding resistance of 120 Ω.

[B.Tech II, UPTU, 2004]

Q.6 A 10 kW, 250 V d.c. motor shunt type with an armature resistance of 0.8 Ω and a field resistance of 275 Ω takes 3.91 A when running light at rated voltage and speed. Calculate the efficiency when the machine runs as generator delivering an output of 10 kW at rated voltage and speed.

[B.Tech II, UPTU, 2004]

Q.7 A shunt motor runs at 500 rpm on a 220 volts d.c. supply. Its armature resistance is 0.5 Ω and the current drawn is 30 Amp in addition to field current. Calculate the extra value of the resistance that must be placed in the armature circuit in order that the speed is reduced to 300 rpm, keeping the armature current the same.

[B.Tech II, UPTU, 2000]

Q.8 A d.c. shunt machine develops an open circuit emf of 250 V at 1500 rpm. Find its developed torque for an armature current of 20 A.

[B.Tech II, UPTU, 2002]

Q.9 A 240 V shunt motor runs at 1450 rpm of full load with an armature current of 11 A. The load resistance of the armature and brushes is 0.6 Ω. If the speed is to be reduced to 1000 rpm with the same armature current, calculate the amount of resistance to be connected in series with the armature and power lost.

[B.Tech, UPTU, 2002]

Q.10 A 230 V d.c shunt motor is taking 5 A when running at no load. The armature resistance is 0.2 Ω and field resistance is 115 Ω. For an input current of 72 A, calculate the shaft output and efficiency.

[B.Tech II, UPTU, 2001]

Q.11 A 240 V, d.c series motor takes 40 A when giving its rated output at 1500 rpm. Its resistance is 0.3 Ω. Calculate the value of resistance that must be added to obtain the rated torque:

(i) at starting.

(ii) at 1000 rpm.

[B.Tech I, I.U., 2005]

REVIEW QUESTIONS

Q.1 Explain why a starter is needed for a d.c. motor. Explain with neat diagram the working of a three-point starter.

[B.Sc. Engg, AMU, 98]

Q.2 Obtain speed versus armature current and torque versus armature current characteristics of a d.c. series motor. Show the effect of saturation of magnetic circuit on these characteristics.

[B.Sc. Engg, AMU, 98]

Q.3 Obtain speed versus torque characteristic of a d.c shunt motor. How are these characteristics modified in commulatively compounded motor and differentially compounded motor? Comment on utilities of these motors.

[B.Tech, AMU, 2000]

Q.4 Explain how speed of a d.c. shunt motor can be controlled above the rated speed.

[B.Tech, AMU, 2000]

Q.5 Derive the step resistances of starter.

Q.6 Draw the speed torque characteristics of a d.c. shunt, series and compound motors in one figure and compare them.

[B.Tech, AMU, 1999]

Q.7 Explain with a neat sketch, the working of a three-point starter. What is its limitation and how is it overcome in a four point starter?

[B.Tech, AMU, 1999]

Q.8 How is torque produced in a d.c. motor? Deduce an expression for torque in a d.c. motor.

[B.Tech, JMI, 2006]

Q.9 What are the different methods of speed control? Discuss the methods with circuit diagram and state merits and demerits of all the methods.

[B.Tech, JMI, 2006]

Q.10 What are the general methods of speed control of d.c. motor? Explain their merits and demerits.

[B.Tech, JMI, 2005]

Q.11 Describe methods to control the speed of d.c. motors.

[B.Tech, UPTU, 2004]

Q.12 Based on field winding and armature winding connections and schematic diagram, explain different types of d.c. machines.

[B.Tech, UPTU, 2004]

Q.13 Define armature reaction and discuss its effects on the performance of d.c. machines.

[B.Tech, UPTU, 2004]

Q.14 Discuss the process of development of back emf in a d.c. shunt motor.

[B.Tech, UPTU, 2005]

Q.15 A d.c. shunt motor operates at almost constant speed irrespective of load. Why?

[B.Tech, UPTU, 2005]

Q.16 A d.c. series motor develops a high starting torque at the starting. Why?

[B.Tech, UPTU, 2005]

Q.17 Explain the Ward Leonard system for controlling the speed of d.c. motor.

[B.Tech, UPTU, 2005]

Chapter

8

Three-Phase Induction Motor

8.1 Introduction
8.2 Types of Polyphase IM
8.3 Construction of Polyphase IM
8.4 Principle of Operation
8.5 Rotating Magnetic Field
8.6 Equivalent Circuit
8.7 Power Flow Diagram
8.8 Power Expressions
8.9 Torque Expressions
8.10 Characteristic of Polyphase IM
8.11 Starting of 3-Phase IM
8.12 Speed Control of 3 ϕ IM
8.13 Cogging and Crawling of IM
8.14 Braking of IM
8.15 Induction Generator
8.16 Applications
- Solved Questions
- Unsolved Questions
- Review Questions

8.1 INTRODUCTION

The induction motor was first realized practically in 1885 by a scientist named Ferrari. He proposed that two alternating fields having phase displacement will produce a rotating magnetic field. Later in 1888, Tesla gave the idea of inductions which is physically realized as tesla motor and used commercially.

The three-phase induction motor gained popularity because of the acceptance in industries due to its robustness, low cost and high efficiency. Now in industries the induction motors are fast replacing the d.c. motor due to the availability of a.c. supply with respect to d.c. supply. Thus, it is very important to know about the details of this type of motor. The three-phase induction motor is the powerhorse of the modern industry. Now with the advent of power semiconductor devices such as SCR's, MOSFET, IGBT etc., the induction motors are fast replacing the d.c. series motors in the trains and in large industries. The induction motors are popular in the rural areas due to irrigation purpose for pumping water from deep below the soil.

The induction motor is available in various ratings depending on the power requirement. The induction motors are available in basically two types, namely squirrel cage induction motor and slip ring induction motor respectively. The polyphase induction motors are self-starting and are asynchronous type i.e., they cannot rotate at synchronous speed. 'n_s'. The squirrel cage induction motor is the simplest type of induction motor. It is different from that of d.c. type by having no brushes and commutators. Thus, the cost of polyphase induction motor is about one-third of that of corresponding rating d.c. motor. It can be used in a dusty and corrosive environment. The maintenance required is almost negligible.

8.2 TYPES OF POLYPHASE IM

The polyphase motors are available in following two types, namely, squirrel cage and wound rotor or slip ring type. The squirrel cage induction motor (SCIM) contains no windings on the rotor and has conducting bars placed in the slots of the motor rotor. The wound rotor has windings on both stator and rotor respectively so it is known as wound rotor IM. There are slip rings placed on the shaft of motor via which the external resistors are added, so it is also known as slip ring induction motor.

The two types have their own merits and demerits which will be discussed in the coming sections. They are shown in the Figs. (8.2) and (8.3) to clarify the concept of the machines.

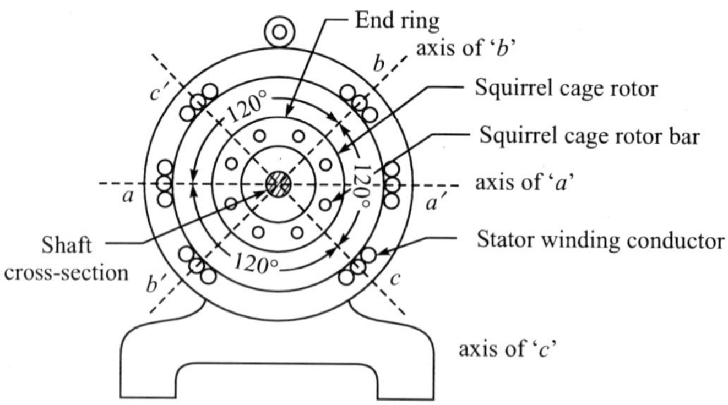

Fig. 8.1 Polyphase SCIM.

The magnetic axis of the three phase windings aa', bb' and cc' are phase displaced by 120° respectively as shown in Fig. (8.1). Since the currents flowing through the phases 'a', 'b' and 'c' are phase displaced by 120°. They are space displaced by the same angle, that is, 120° respectively. This is done so as to produce a rotating magnetic field in the air gap between the stator and rotor. The rotor consist of rotor bars made up of aluminium or bronze bars to provide conduction of induced currents through it.

The slip ring IM (SRIM) is shown in the Fig. (8.2) as shown below.

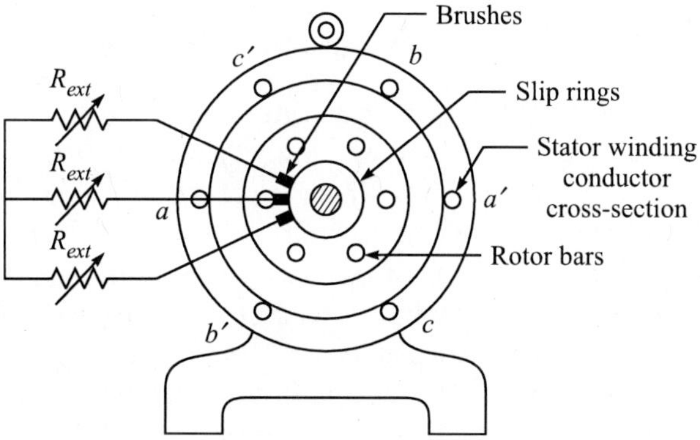

Fig. 8.2 Polyphase SRIM.

The polyphase SRIM consists of two windings in contrast to SCIM, its rotor resistance can be increased by variation of external resistance 'R_{ext}'. The squirrel cage induction motor comparatively has less cost with respect to that of slip ring induction motor so, where constant speed is required then the SCIM is preferred and where the speed control by rotor resistance variation is required then the SRIM is used.

8.3 CONSTRUCTION OF POLYPHASE IM

The constructional features of the polyphase IM can be explained by considering the Fig. (8.3), Fig. (8.4) and Fig. (8.5) respectively.

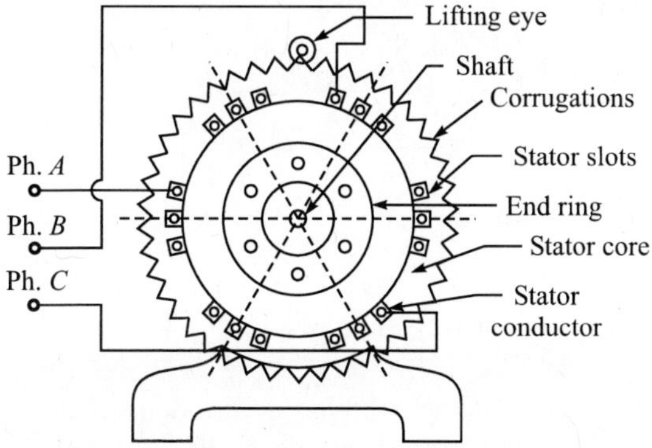

Fig. 8.3(a) Stator of polyphase SCIM.

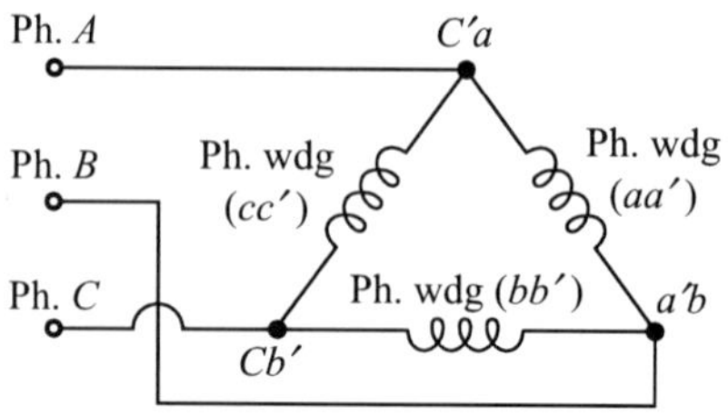

Fig. 8.3(b) Stator electrical equivalent of polyphase SCIM.

The polyphase IM consists of stator and rotor as in the d.c machines. The stator consists of excitation or stator winding which is made up of enamelled copper wire of fixed number of turns which is prewound and are placed in the stator slots after drying. The number of turns per phase of stator windings increases with the increase of rating of the machine and correspondingly the size of the machine increases. The stator core is made up of high grade steel preferably CRGO steel which is laminated and is used for reducing the eddy losses. Nowadays, a plenty of researches are going in the material science field to decrease the size of the machine. This is done by using material having higher permeability. In further large rating machines the cooling ducts are used. The fans are mounted on the shafts which cools the motor core and windings by taking the hot air (which is trapped inside the machine) and brings cool air inside the machine from outside as the fan is of exhaust type. The stator slots are of open type for large rating motors. The slots are semi-closed type for small rating motors. In the open type the windings are from wound whereas in later one winding is inserted in slots one by one.

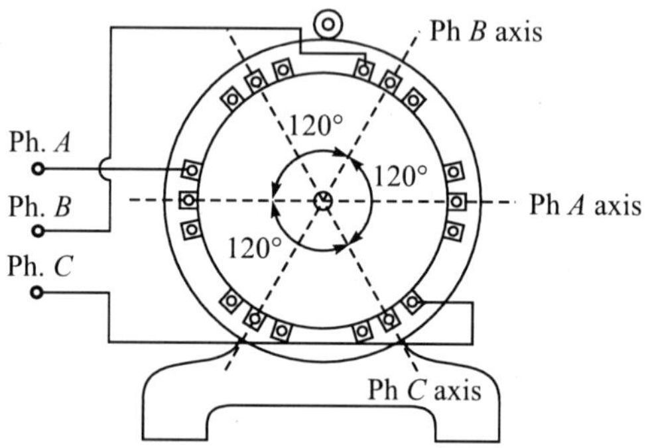

Fig. 8.4(a) Stator of polyphase SRIM.

The stator phase windings aa', bb', cc', are connected is delta (Δ) for SCIM and (wye) Y for SRIM respectively as shown in Fig. (8.3) and Fig. (8.4) respectively.

The external appearance of the machine differs depending on the operating conditions, namely, drip proof, splash proof, totally enclosed type, fan cooled, self-ventilated, separately ventilating type induction motors etc.

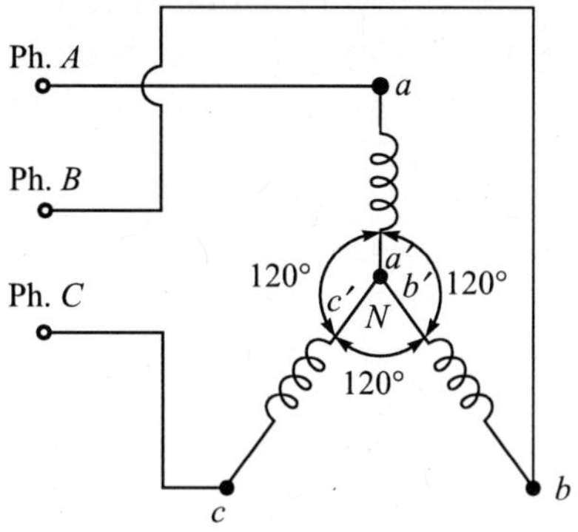

Fig. 8.4(b) Electrical equivalent of polyphase SRIM.

The rotor of SCIM has a cage type appearance having copper or aluminium bars placed in the slots as shown in Fig. 8.5(a).

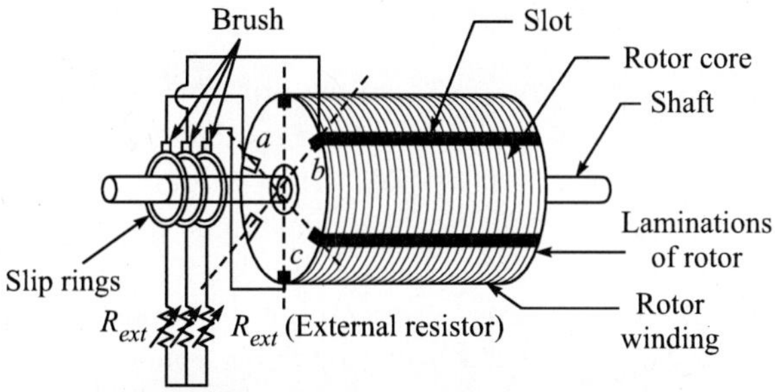

Fig. 8.5(a) Rotor of polyphase SRIM.

The number of stator and rotor teeth are prime to each other to prevent locking of stator and rotor at starting. The rotor laminations are skewed. This helps to prevent magnetic locking of the rotor and stator. Also it prevents teeth vibration and hence create low noise. The air gap between the rotor and stator is small (0.02 – 0.05 inches) depending on the size of machine. Similar air gap helps in better magnetic coupling of the two windings and are shown in Fig. 8.5(b).

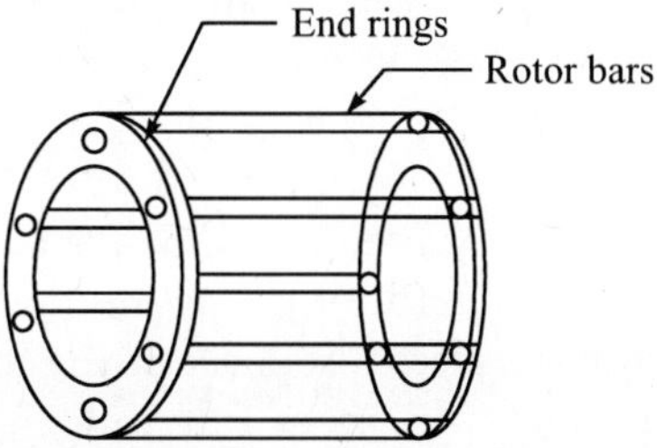

Fig. 8.5(b) Squirrel cage rotor of SCIM.

These bars are solidly brazed to end rings or welded. The bars are placed in slots which are closed on the periphery of rotor. In larger machines these bars are held by bolts to the end rings. But the bolts loosen due to temperature change due to loading.

In motor of small size the entire cage winding, bars and end rings together is made up of aluminium cast in place as an integral unit. This construction reduces the cost and eliminates the contact resistance. The induction motor rotor rotates at nearly synchronous speed. There is a small change of speed from no load to full load condition, so this type of motor is considered constant speed motor. For larger speeds, the SRIM is used as the rotor resistance control is easy in this type of motor.

The number of poles of stator and rotor are same. So the coils of SRIM must be properly taken. The terminals of the rotor windings are connected to the slip rings and are connected externally to the external resistors 'R_{ext}' respectively as shown in Fig. (8.5).

8.4 PRINCIPLE OF OPERATION

The three-phase induction motor works on the principle of e.m. induction that is, when the motor stator windings are excited by three currents having equal magnitude and a phase difference of 120°. Then a rotating magnetic field is produced in the air gap between the stator and rotor of machine. The rotor bars cut this flux at a constant rate. Since the flux linkage with stator coil differs at different instants of time, so according to Faraday's law of electromagnetic induction, emf is induced in the rotor bars and since the rotor bars are short circuited by the end rings, so this emf drives an induced current 'i_r' in the rotor bars. This current 'i_r' which produce rotor flux 'ϕ_r'. Since the stator rotating flux 'ϕ_s' and rotor flux 'ϕ_r' are phase displaced by 'δ' (say). This will lead to a torque which will be exerted on the rotor which is given by

$$\boxed{T \propto \phi_s \, \phi_r \sin \delta} \tag{8.1}$$

The rotor flux will follow the stator flux. Since the rotor flux is moving ahead of the rotor flux so in the cause of alignment action of two fluxes a torque will produce which will lead to rotation of the rotor. The direction of this torque is given by Flemming's left hand rule. For maximum torque, the angle 'δ' equal to 90°.

The $\delta = 0°$ corresponds to total alignment of the fluxes or magnetic locking. This corresponds to motor with zero torque or the motor stalls. The motor thus cannot run at synchronous speed 'n_s' since if the rotor runs at synchronous speed 'n_s', then the rotor will be at rest with respect to the rotating magnetic field of the stator and so there will be no cutting of stator flux by the rotor bars and so there will be no emf induced in the rotor bars as no flux cutting by rotor bars and hence no rotor currents and so the rotor field flux 'ϕ_r' will be equal to zero. So the torque exerted on the rotor will be equal to zero and so the motor fails to rotate. So the polyphase induction motor cannot run at synchronous speed but it rotates at a speed less than synchronous speed. Or we can visualize the motor as having a slip between the motor rotor and rotating stator flux respectively. The slip of the motor is given as follows:

$$\boxed{S = \frac{n_s - n}{n_s}} \tag{8.2}$$

where, s = slip

n_s = synchronous speed, rps

n = rotor speed, rps.

So the slip speed is given as '$s \cdot n_s$'. This is the speed at which the rotor bars are cut continuously by the stator flux. Since the frequency of stator flux is same as that of the stator current which is equal to 50 Hz. Since the stator flux is produced by the stator current flowing through stator phase windings. The rotor bars are cut at a frequency corresponding to slip speed, so the rotor frequency is termed slip frequency 'f_r' which is given as follows:

$$\boxed{f_r = s \cdot f} \tag{8.3}$$

where, f_r = rotor frequency, Hz

s = slip

f = stator frequency (50 Hz)

Since the slip of the induction motor varies between 0.01 and 0.05 so we can see that rotor frequency is a small percentage of the stator frequency. The frequency of the emf induced in rotor and rotor currents is same and it is about '2 Hz'.

The direction of the induction motor can be reversed by interchanging the phase winding connections. Since the rotating magnetic field is responsible for the motion of the rotor. So the direction of the rotor will also reverse accordingly. As $n < n_s$ always, so it is called asynchronous motor.

8.5 ROTATING MAGNETIC FIELD

The current in the phase windings aa', bb', and cc' are 120° respectively as shown in Fig. 8.6(b).

The current I_a, I_b and I_c flows in the phase windings of the stator of polyphase machine. The alternating current in the phase windings will lead to development of alternating fluxes ϕ_a, ϕ_b and ϕ_c respectively. The currents i_a, i_b and 'i_c' are expressed as follows:

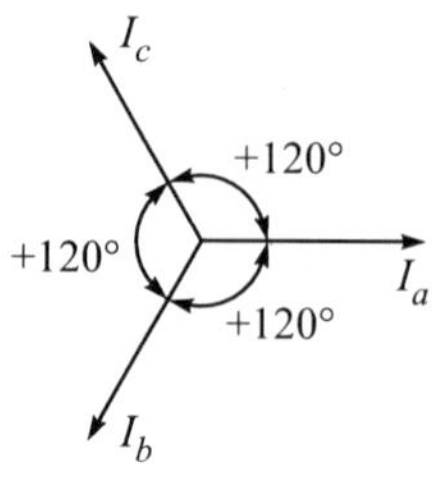

Fig. 8.6(a) Phasors of currents of polyphase IM.

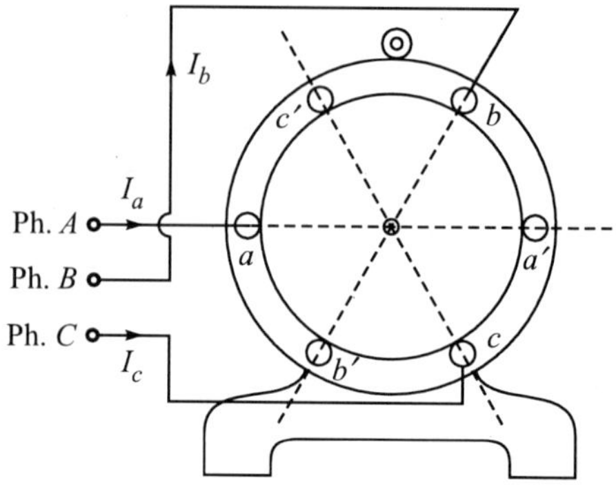

Fig. 8.6(b) Phase currents in phase windings of polyphase IM.

$$i_a = I_m \sin \omega t$$
$$i_b = I_m \sin(\omega t - 120°)$$
$$i_c = I_m \sin(\omega t + 120°) = I_m \sin(\omega t - 240°)$$

Since,
$$\phi_a \propto i_a \quad \text{or} \quad \phi_a = L_a\, i_a$$
$$\phi_b \propto i_b \qquad \phi_b = L_b\, i_b$$
$$\phi_c \propto i_c \qquad \phi_c = L_c\, i_c$$

Where, L_a, L_b and L_c are the self inductances of the phase windings aa', bb' and cc' respectively. Since L_a, L_b, L_c are scalar, so the phasor of ϕ_a, ϕ_b and ϕ_c will have phase differences as that of the currents i_a, i_b and i_c respectively.

The m.m.f's of the phase aa', bb' and cc' are $\mathcal{F}_a$, $\mathcal{F}_b$ and $\mathcal{F}_c$ respectively and expressed mathematically as follows:

$$\mathcal{F}_a = N_a\, I_a = \phi_a \mathcal{R}$$
$$\mathcal{F}_b = N_b\, I_b = \phi_b \mathcal{R}$$
$$\mathcal{F}_c = N_c\, I_c = \phi_c \mathcal{R}$$

Where, N_a, N_b and N_c are the number of turns of phase windings connected to phase A, B and 'C' respectively, ϕ_a, ϕ_b and ϕ_c are stator phase winding fluxes, '$\mathcal{R}$' is the reluctance of the magnetic circuit respectively. The fluxes ϕ_a, ϕ_b and ϕ_c are shown in Fig. (8.7).

At different instants the fluxes are expressed as follows:

At
$$t = t_1,\ (\theta = \pi/2)$$
$$\phi_a = \phi_m$$
$$\phi_b = -0.5\ \phi_m$$
$$\phi_c = -0.5\ \phi_m$$

where,
$$\phi_m = \text{Maximum flux, weber}$$

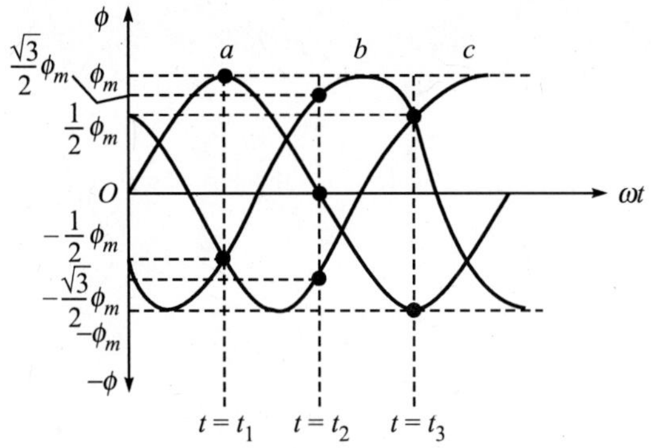

Fig. 8.7 Fluxes due to phase currents i_a, i_b and i_c.

So the resultant flux is given by

$$\phi_r = \sqrt{\left(-\frac{\phi_m}{2}\right)^2 + \left(-\frac{\phi_m}{2}\right)^2 + 2\left(\frac{-\phi_m}{2}\right)\left(\frac{-\phi_m}{2}\right)\cos 120^\circ + \phi_m}$$

$$\phi_r = 1.5\ \phi_m$$

This expression could be easily understood by the phasor diagram as shown in Fig. (8.8) below.

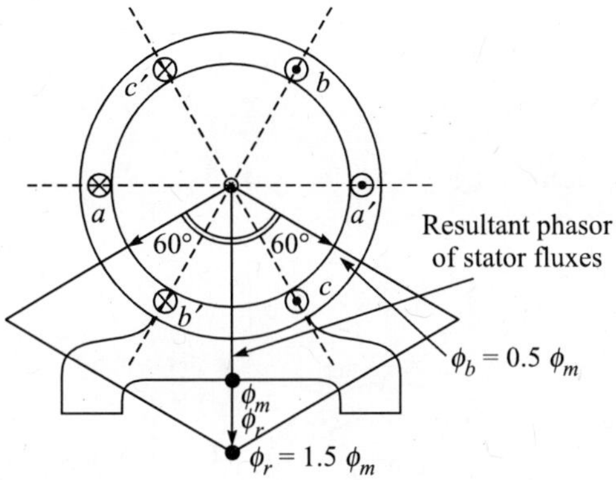

Fig. 8.8 Phasors of fluxes at $t = t_1$, at $\theta = \frac{\pi}{2}$ rad.

At $t = t_2$:

$$\phi_a(\theta = \pi) = 0$$
$$\phi_b(\theta = \pi) = 0.866\ \phi_m$$
$$\phi_c(\theta = \pi) = -0.866\ \phi_m$$

This could be visualized by using the Fig. (8.9).

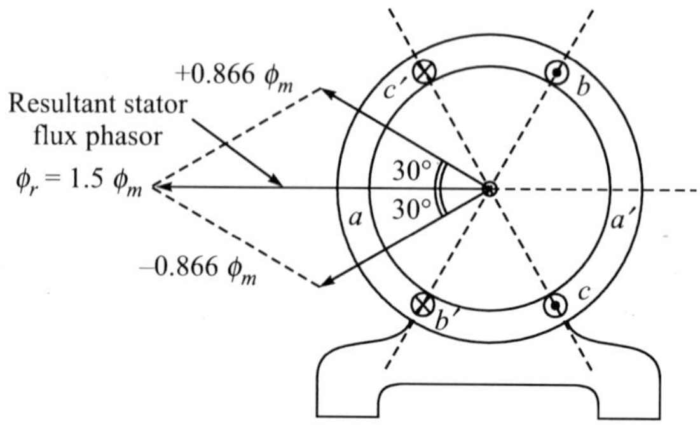

Fig. 8.9 Phasor of fluxes at $t = t_2$, at $\theta = \pi$ rad.

Similarly, at $t = t_3$

$$\phi_a = -\phi_m$$

$$\phi_b = +\frac{\phi_m}{2}$$

$$\phi_c = +\frac{\phi_m}{2}$$

So we see from the Figs. (8.8), (8.9) and (8.10) that the resultant stator flux phasor 'ϕ_r' rotates in the clockwise direction at synchronous speed 'N_s' which is given as

$$\boxed{N_s = \frac{120\,f}{P}}\text{, r.p.m.} \tag{8.4}$$

where, f = supply frequency (50 Hz)
P = number of poles of stator

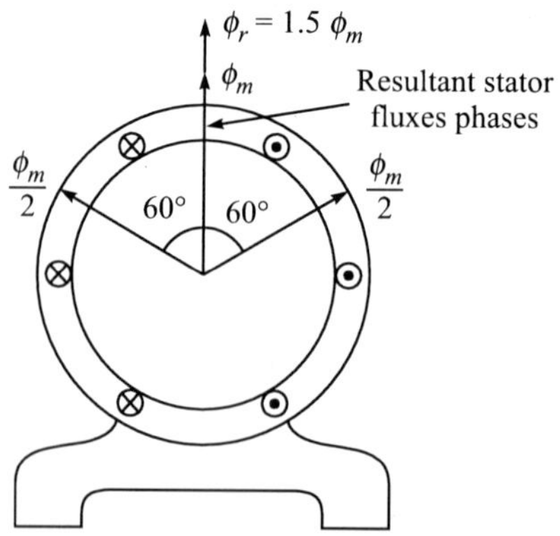

Fig. 8.10 Flux Phasors at $t = t_3 \left(\theta = \dfrac{3\pi}{2}\text{ radians}\right)$.

The resultant phasor direction could be reversed by changing any two phase connections and hence by changing the phase sequence of the stator currents respectively. It will start rotating in the anticlockwise respectively.

From the above, we conclude that the resultant flux phasor magnitude is constant equal to 1.5 ϕ_m and the speed of the phasor is also constant equal to synchronous speed in the clockwise direction.

8.6 EQUIVALENT CIRCUIT

The equivalent circuit is an electrical representation of a physical system i.e., induction motor in this case. The motor behaviour can be explained using equivalent circuit very easily. The visualization can be done using it. As in the case of transformer, the induction motor has two circuits one for the stator and other for the rotor. The only difference is that the rotor circuit of the induction motor is not stationary as in the case of transformer whose primary as well as secondary are stationary. So it can be represented as rotating secondary as in Fig. (3.12).

The induction motor equivalent circuit can be realized by considering one circuit at a time as follows:

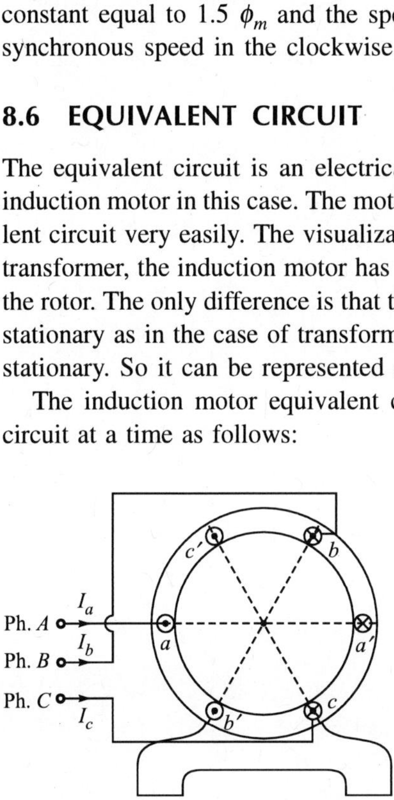

Fig. 8.11(a) Stator of polyphase IM.

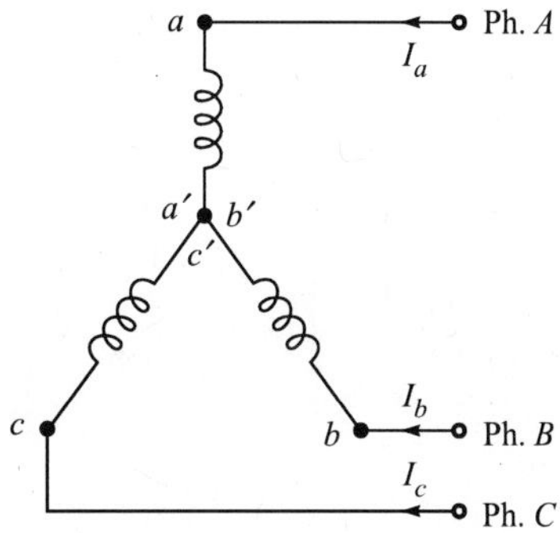

Fig. 8.11(b) Stator windings connection of polyphase IM.

The resistance and leakage reactances of the windings can be shown as follows for each phase of induction motor.

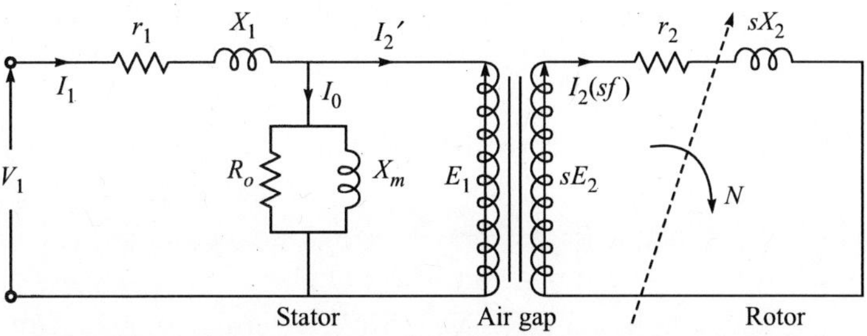

Fig. 8.12 Equivalent circuit of polyphase IM (per phase basis).

where the parameters of the stator and rotor side as discussed below.

V_1 = Stator applied voltage to each phase winding = $\frac{V_L}{\sqrt{3}}$ (*Y*), volts

I_1 = Stator phase winding current, $A = I_L$(Y), Amp.

r_1 = Stator phase winding resistance, Ω

x_1 = Stator phase winding leakage flux or equivalent leakage reactance, Ω

R_0 = Loss equivalent core resistance, Ω

X_m = Flux equivalent core reactance, Ω

I_0 = No load exciting current, Amp.

N = Speed of rotor, R.P.M

I_2' = Load equivalent of stator current, Amp.

r_2 = Rotor resistance, Ω

$s\,x_2$ = Rotor leakage reactance at rotor frequency, Ω

I_2 = Rotor current at rotor frequency, Amp.

S = Slip

E_1 = Stator emf, volts

$s\,E_2$ = Rotor emf, volts.

The rotor equivalent circuit electrical quantities are at frequency equal to '*sf*' where '*f*' is the stator frequency equal to 50 Hz. The rotor emf and current should be determined in terms of '*f*' i.e., supply frequency. Then only the rotor quantities could be transferred or referred to the stator side. This could be possible if we divide the rotor parameters by slip '*s*'. The equivalent circuit of the IM gets transformed to Fig. (8.13).

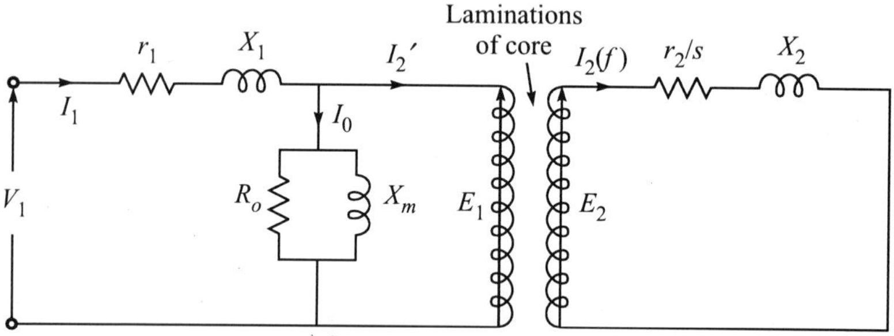

Fig. 8.13 Modified equivalent circuit of 3 phase IM.

The rotor equivalent circuit of the IM could be further rearranged as follows by splitting the resistance r_2/S as sum of two quantities as given by eq. (8.5).

$$\boxed{\frac{r_2}{s} = r_2 + \frac{r_2}{s}(1 - s)} \qquad (8.5)$$

The first part is the rotor resistance whereas, the second part is the electrical equivalent of the mechanical load attached to the shaft of IM. So the equivalent circuit could be further modified as follows.

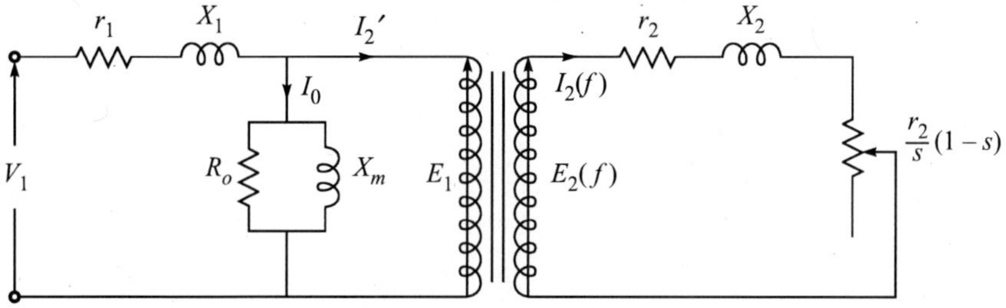

Fig. 8.14 Equivalent circuit of IM showing mechanical load.

The rotor sides quantities as shown by Fig. (8.14) can be transferred or referred to stator side and we get the complete equivalent circuit of three-phase IM whose one phase is represented in the Fig. (8.15)

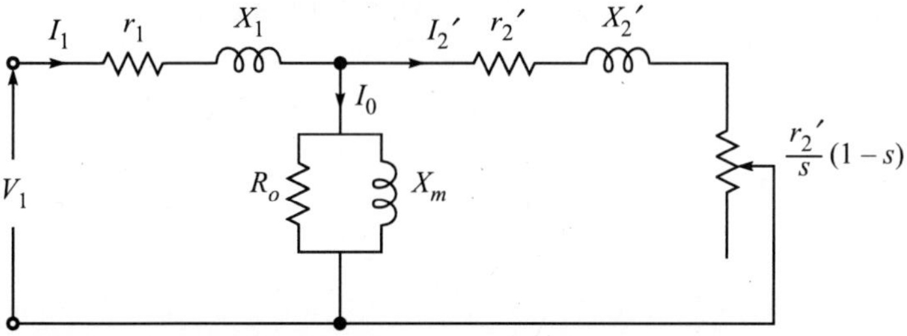

Fig. 8.15 Complete equivalent circuit of one phase of three-phase IM.

The resistance 'R_0' represents the core losses and thus the temperature rise of core due to heating. Whereas the 'X_m' corresponds to mutual reactance which represent the flux creation in the core. In the case of IM the rotor and stator turns are same for SRIM. The power received by the rotor was transferred to it via air gap sent by the stator, magnetic circuit.

In some cases, the resistance 'R_0' is also not considered and so the equivalent circuit is represented as follows.

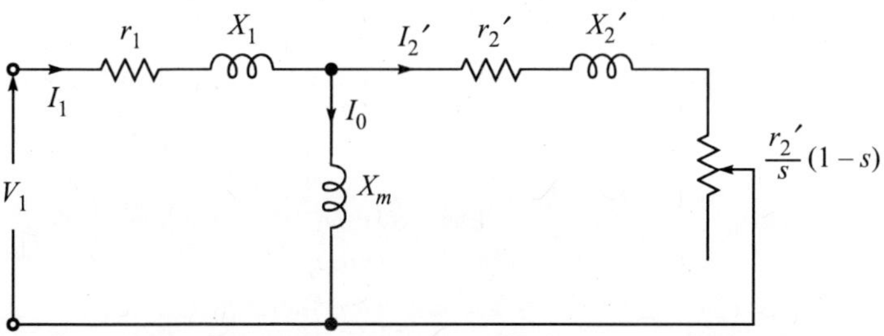

Fig. 8.16 Equivalent circuit of one phase of IM.

8.7 POWER FLOW DIAGRAM

The power flow from stator to rotor side can be shown using power flow diagram as follows:

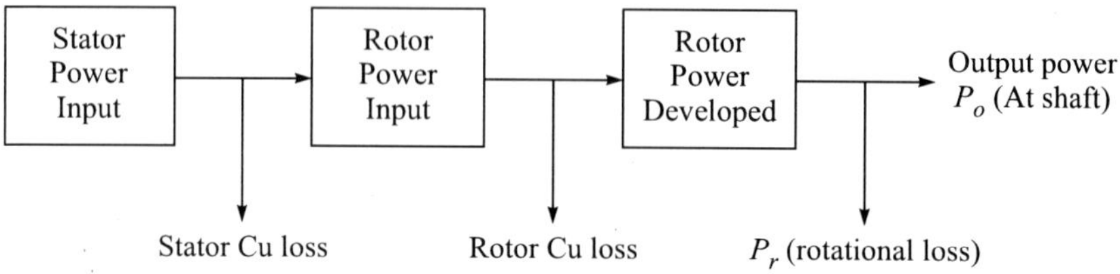

Fig. 8.17 Block diagram showing power flow in 3 ϕ IM.

The stator copper loss is greater than the rotor copper loss and rotational losses. The rotational losses are also termed mechanical losses which may be friction losses, windage losses, etc.

The stator power input is equal to $V_p I_p \cos\theta$. The stator copper loss is equal to $I_p^2 r_1$ the rotor power input is equal to the difference between the two and that is equal to $V_p I_p \cos\theta - I_p^2 r_1$, the rotor power input is also terminal as air gap power and represented by

$$P_g = \frac{I_2'^2 r_2'}{s} = Pr_i$$

The rotor copper loss is given by

$$P_{r_{cu}} = I_2'^2 r_2'$$
$$= s \cdot P_g$$

The rotor power developed at shaft is equal to $P_{r_i} - P_{r_{cu}}$ and is equal to $I_2' r_2' \left(\frac{1-s}{s}\right)$.

So it can be represented as $P_{sh} = P_{r_{cu}}\left(\frac{1-s}{s}\right)$. The rotor power at motor pulley (including rotational loss) $P_{sh} - P_r = \frac{P_0}{\phi}$; ϕ = Number of phases of IM. Thus the total power of IM is given by = 3 P_0.

8.8 POWER EXPRESSIONS

The power input is done at the stator side so sometimes it is known as stator input. The stator input power 'P_1' is given by expression as given below

$$P_1 = V_1 I_1 \cos\theta, \text{ watts}$$

where,

'V_1' = Phase voltage, V

'I_1' = Phase current, A

$\cos\theta$ = Input power factor

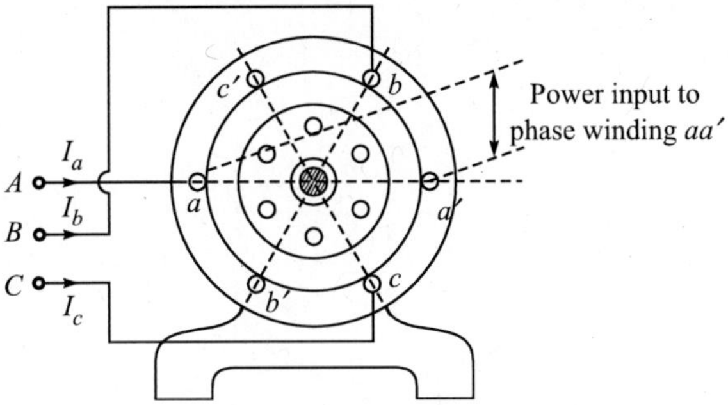

Fig. 8.18 3 ϕ SCIM.

The stator input power is not wholly available across the air gap between the stator and the rotor. A part of this power is lost in the form of stator copper loss in the stator winding resistance 'r_1'. Mathematically, the stator copper losses can be expressed as follows

$$P_{1_{cu}} = I_2^2 \, r_1, \text{ watts} \tag{8.6}$$

So the power available in the air gap is given by 'P_g' called air gap power and expressed by eq. (8.7).

$$\boxed{P_g = P_1 - P_{1_{cu}} = V_1 I_1 \cos\theta - I_1^2 r_1}, \text{ watts} \tag{8.7}$$

Alternatively, the air gap power can be expressed as

$$\boxed{P_g = \frac{I_2'^2 \, r_2'}{S}}, \text{ watts} \tag{8.8}$$

The rotor input power or air gap power is not wholly available at the shaft of motor which could be used for performing some useful work. A part of the rotor input power is lost in the rotor winding resistance in the form of heat which is not reversible and cannot be used. The rotor 'C_u' loss is given by

$$\boxed{P_{2_{cu}} = I_2'^2 \, r_2'} \tag{8.9}$$

So, in terms of air gap power 'P_g, the rotor copper loss is given as follows:

$$\boxed{P_{2_{cu}} = s\,P_g} \tag{8.10}$$

The mechanical power developed is given by 'P_m' which is given by equivalent circuit as follows:

$$\boxed{P_m = I_2'^2 \, r_2' \left(\frac{1-s}{s}\right)} \tag{8.11}$$

$$P_m = \frac{I_2'^2 r_2'}{s} - I_2'^2 r_2'$$

$$\boxed{P_m = P_g - P_{2_{cu}}} \quad (8.12)$$

$$= P_m = P_g - s P_g, \text{ watts}$$

$$\boxed{P_m = (1 - s) P_g} \quad (8.13)$$

where, P_m = mechanical power developed by motor

Thus, the output power 'P_0' is given by

Power at pulley $\boxed{P_0 = P_m - P_{f+w+s} = P_{sh}}$, watts (8.14)

P_{f+w+s} = friction, windage and stray losses

The rotor frequency 'f_2' is equal to 'sf' where 's' is between 0.01 and 0.04 and 'f' is equal to 50 Hz, so the core loss in rotor is almost negligible as compared to the stator core losses as f_2 or a few hertz. Since the mechanical losses originates when the motor rotates so these losses are sometimes called rotational losses.

8.9 TORQUE EXPRESSIONS

The torque of a polyphase induction motor is due to the interaction of the two fluxes, namely, rotating stator flux 'ϕ_s' and the rotor flux 'ϕ_r' respectively. The torque of induction motor can be expressed as follows:

$$\boxed{T \propto \phi_s \, \phi_r \sin \delta}$$

where, ϕ_s = resultant stator flux in air gap Wb

ϕ_r = rotor flux, Wb

δ = phase difference between 'ϕ_s' and 'ϕ_r'.

The total torque can be alternatively expressed as follows:

$$\boxed{T \propto 3 E_2' I_2' \cos \theta_2} \quad (8.15)$$

where, E_2' = Rotor emf referred to stator side, volts

I_2' = Rotor current referred to stator side, Amp.

$\cos \theta_2$ = Rotor circuit power factor

The rotor circuit at supply frequency '50 Hz' is given by

From Fig. (8.19), we have the rotor current 'I_2'' equal to

$$\boxed{\bar{I}_2' = \frac{\bar{E}_2'}{|\bar{Z}_2'| \angle \theta_2}} \quad (8.16)$$

$$\bar{Z}_2' = \frac{r_2'}{s} + j x_2'; \qquad \tan \theta_2 = \frac{x_2'}{(r_2'/s)}$$

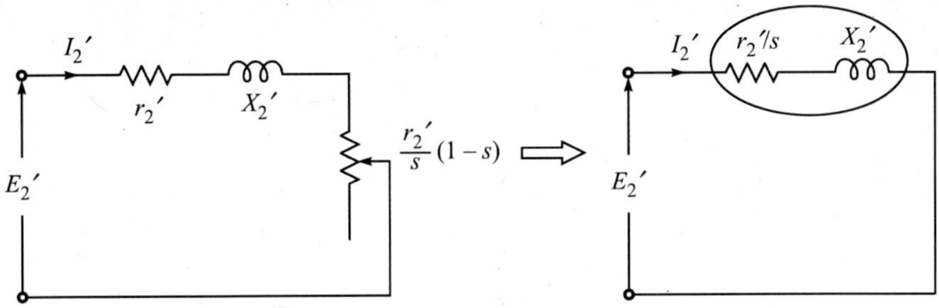

Fig. 8.19 Rotor circuit referred to stator side.

$$|\bar{Z}_2'| = \sqrt{\left(\frac{r_2'}{s}\right)^2 + x_2'^2} \qquad \theta_2 = \tan^{-1}\left(\frac{x_2'}{r_2'/s}\right)$$

so,

$$\boxed{\bar{I}_2' = \frac{\bar{E}_2'}{\left(\sqrt{\left(\frac{r_2'}{s}\right)^2 + x_2'^2}\right)\angle \tan^{-1}\left(\frac{s\,x_2'}{r_2'}\right)}} \tag{8.17}$$

Considering only the r.m.s value of rotor current, we get

$$\boxed{I_2' = \frac{E_2'}{\sqrt{\left(\frac{r_2'}{s}\right)^2 + (x_2')^2}}} \tag{8.18}$$

The power factor 'cos θ_2' of the rotor circuit is given by

$$\boxed{\cos\theta_2 = \frac{r_2'/s}{|Z_2'|} = \frac{r_2'/s}{\sqrt{\left(\frac{r_2'}{s}\right)^2 + (x_2')^2}}} \tag{8.19}$$

Substituting the values of 'I_2'' and 'cos θ_2' in the exp. (8.15), we get

$$T \propto 3\,E_2' \cdot \left(\frac{E_2'}{\sqrt{\left(\frac{r_2'}{s}\right)^2 + (x_2')^2}}\right)\left(\frac{r_2'/s}{\sqrt{\left(\frac{r_2'}{s}\right)^2 + (x_2')^2}}\right)$$

or,

$$\boxed{T \propto \frac{3\,E_2'^2\, r_2'/s}{\left(\frac{r_2'}{s}\right)^2 + (x_2')^2}} \tag{8.20}$$

Multiplying 's^2' in numerator and denominator of exp. (8.20), we get

$$T \propto \frac{3\,s\,E_2'^2\,r_2'}{r_2'^2 + (s\,X_2')^2}$$

$$\boxed{T_{3\phi} = k \cdot \frac{3\,s\,E_2'^2\,r_2'}{r_2'^2 + (s\,X_2')^2}}, \text{ Nm} \qquad (8.21)$$

$T_{3\phi} \rightarrow$ Torque expression for a three-phase induction motor.

Thus, we see that the torque which is exerted on the rotor of IM depends on the slip 's' which is given by $(N_s - N)/N$, the e.m.f. induced in rotor circuit 'E_2'', rotor resistance 'r_2'' and the leakage reactance 'X_2'' respectively.

So as to obtain high torque the rotor resistance and leakage should be optimally selected.

The maximum value of torque that could be applied at the shaft is limited by maximum value called breakdown torque. If a torque of more than this value is applied at the shaft, then the motor stops after going in the unstable region of N/T characteristic. So we should know about this value of torque which is very important from point of view of industrial applications.

The maximum value of torque is determined by taking the derivative of exp. (8.21) with respect to 's' and equate it to zero.

$$\frac{\partial T_{3\phi}}{\partial s} = 0 \qquad (8.22)$$

$$\frac{[r_2'^2 + (s\,X_2')^2]\,3k \cdot E_2'^2\,r_2' - (3k \cdot s\,E_2'^2\,r_2')\,2 \cdot s\,X_2' \cdot X_2'}{[r_2'^2 + (s\,X_2')^2]^2} = 0$$

$$[r_2'^2 + (s\,X_2')^2] - s \cdot 2\,s\,X_2'^2 = 0$$

$$r_2'^2 + (s\,X_2')^2 - 2\,s^2\,X_2'^2 = 0$$

$$r_2'^2 - s^2\,X_2'^2 = 0$$

or,

$$s]_{T_{max}} = \frac{r_2'}{X_2'} = S_m$$

Thus, the slip at maximum torque is given by

$$\boxed{S_m = \frac{r_2'}{X_2'}} \qquad (8.23)$$

So, the maximum torque value is given by

$$T_{max} \propto \frac{3\,S_m\,E_2'^2\,r_2'}{r_2'^2 + (S_m\,X_2')^2} \qquad (8.24)$$

Substituting the value of 'S_m' in exp. (8.24), we get

$$T_{max} \propto \frac{3 \cdot \frac{r_2'}{x_2'} \cdot E_2'^2 \cdot r_2'}{r_2'^2 + \left(\frac{r_2'}{x_2'} \cdot x_2'\right)^2}$$

$$\propto \frac{3 r_2'^2 \cdot E_2'^2}{x_2' \cdot 2 r_2'^2}$$

$$\propto \frac{3\, E_2'^2}{2\, X_2'}$$

So,

$$\boxed{T_{max} = \frac{3k\, E_2'^2}{2\, X_2'}} \quad (8.25)$$

The starting torque 'T_{st}' is also important since this value decides when will the motor pick up its speed. We can compare different motors on this value as per which motor will attain its maximum speed quickly.

At the starting, the speed of motor is zero. So the slip of the motor is unity.

At $$N = 0, \quad S = \frac{N_s - N}{N_s} = 1$$

So the torque of the motor at starting is given by

$$T_{st} \propto \frac{3 \cdot 1 \cdot E_2'^2\, r_2'}{r_2'^2 + (1 \cdot x_2')^2}$$

or, $$T_{st} \propto \frac{3 \cdot E_2'^2\, r_2'}{r_2'^2 + x_2'^2}$$

or, $$\boxed{T_{st} = \frac{3k\, E_2'^2\, r_2'}{r_2'^2 + x_2'^2}} \quad (8.26)$$

The torque expression given from the power expression given by (8.26) as follows

Torque developed $$T_d = \frac{3 P_g (1 - s)}{\omega_r}, \text{ Nm} \quad \text{at slip '}s\text{'} \quad (8.27)$$

Torque (e.m) $$T_{em} = \frac{3 P_g}{\omega_s}, \text{ Nm} \quad (8.28)$$

Alternatively, $$T_{em} = 3\, P_g, \text{ Synchronous Watts} \quad (8.29)$$

8.10 CHARACTERISTIC OF POLYPHASE IM

The characteristic of polyphase IM could be easily realized using the torque equation of the motor given by exp. (8.21) as given below.

$$T_{3\phi} = T = \frac{k \cdot 3\, s\, E_2'^2\, r_2'}{(r_2'^2) + (s\, X_2')^2}$$

There are two main parts of the characteristic mainly defined by the following slip values:

(i) At low slip (S $\to$ 0)
(ii) At high slip (S $\to$ 1)

Normally all the induction motors operate at low value of slip (0.01) to 0.4). Analysing the behaviour of the motor at these extreme values. We have for very small value of slip $(s\, X_2')^2$ becomes negligible as compared to that of $(r_2')^2$ and could be neglected from the above expression. So the torque expression reduces to

$$T \propto \frac{s\, E_2'^2\, r_2'}{(r_2')^2}$$

or,

$$T = k' \frac{s\, E_2'^2}{r_2'}$$

For, fixed supply voltage 'V_1' and no leakage in primary or stator side, 'E_2'' is constant. If we consider that rotor resistance 'r_2'' is also constant, then,

$$\boxed{T \propto s} \qquad (8.30)$$

The above exp. (8.30) is a linear relation which gives a straight line starting from origin.

Secondly, if we consider the high slip values then $(s\, X_2')^2$ is no more negligible. Since 'r_2'' is of small value so the term $(r_2')^2$ is further smaller. So the term $(r_2')^2$ could be considered to be negligible and can be neglected from the above expression of torque. We have,

$$T \propto \frac{s\, E_2'^2\, r_2'}{(s\, X_2'^2)}$$

$$\boxed{T \propto \frac{E_2'^2\, r_2'}{s\, X_2'^2}}$$

If we consider E_2', r_2' and x_2' as constant, then the torque expression reduces to

$$\boxed{T \propto \frac{1}{s}} \qquad (8.31)$$

The above relation is that of hyperbola. Thus, by combining the two cases which is analyzed i.e., low slip and high slip we get the complete torque slip characteristic of the three phase induction motor as follows.

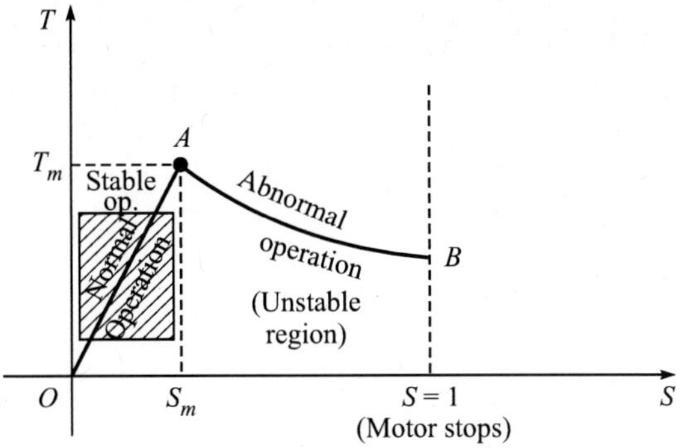

Fig. 8.20 Torque-slip characteristic of the 3 ϕ induction motor.

The motor normally operates in the shaded portion of the curve. In the portion 'OA' the motor has a slip between 1% to 4% depending on the size of motor. For larger motor, the slip varies between 1% to 2% and the lower rating motors have slip greater than 2% and less than 5%. So the speed of the motor is higher for lower slip and speed decreases as the slip value increases. Since the speed of the motor is governed by the equation as given below.

$$\boxed{N = N_s(1 - s)} \tag{8.32}$$

where, N = Rotor speed, rpm

N_s = Synchronous speed, rpm

$$N_s = \frac{120 f}{P}$$

$$s = \text{slip}$$

The N/T characteristic is also an important characteristic which is used by the machine designers and technologists. The speed torque characteristic is given by the curve shown in Fig. (8.21).

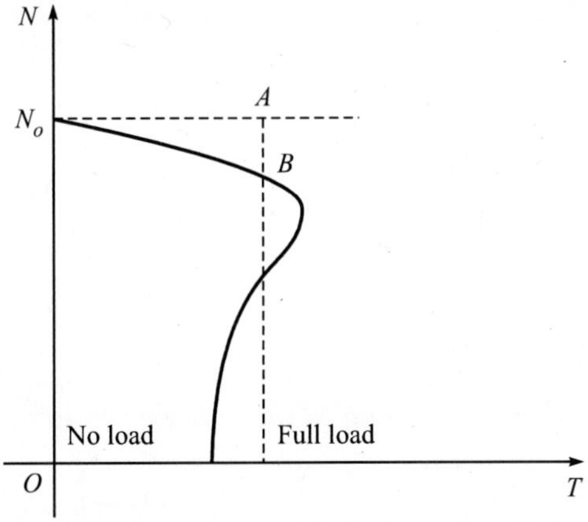

Fig. 8.21 N/T characteristic of 3 ϕ IM.

Since the speed at no load 'N_0' is higher than the full load condition. We see from the curve as shown in Fig. (8.21) that there is a drop in the speed of motor from no load to full load. This is called the speed regulation of motor. We see that the drop is very small in these types of motors. So these are considered constant speed motors and widely used in the industries. The characteristic of three-phase IM is similar to that of d.c shunt motor, so in the place of d.c. shunt motor the three-phase induction motor can be invariably be used.

8.11 STARTING OF 3-PHASE IM

The three phase induction motor can be started on line but as the induction motor draws lagging current from supply and that too of the order of 5 to 8 times that of rated current of the motor. So deteriorates the voltage regulation of distribution system as a result, the voltage dips occur during starting. Moreover, due to high value of starting current though for a transient period deteriorates the insulation of the windings. As a result the three-phase induction motors are not started directly. The problem becomes disastrous as we go for the high rating motors. So to prevent this problem the induction motors are started using starters. Depending on the ratings we have various type of starters for three-phase IM. They are:

(a) DOL starters (< 3 hp)
(b) Star delta starters (< 20 hp)
(c) Auto transformer starters (> 20 hp)

Depending on the operation, the starters are classified as follows:

(a) Manual
(b) Automatic
(c) Semi-automatic.

The starters are provided with protective devices which protect the motor from high voltage, high current, high speed, etc. The starters are employed at the input of motor. The starters thus supply the safe current to the motor terminals as shown below.

Direct on Line Starters (DOL)

The DOL is used for low rating IM. The starter is provided with overload, under voltage and short circuit protection respectively. If the motor is overloaded at the shaft, then the input current drawn by the motor increases. If the current increases above rated value, then the OL relay operates (which is designed for a fixed current value). The OL relay contacts which is made up of bimetallic strip opens its contacts as a result the motor is disconnected from the supply. Thus, the motor windings are protected of the value of current.

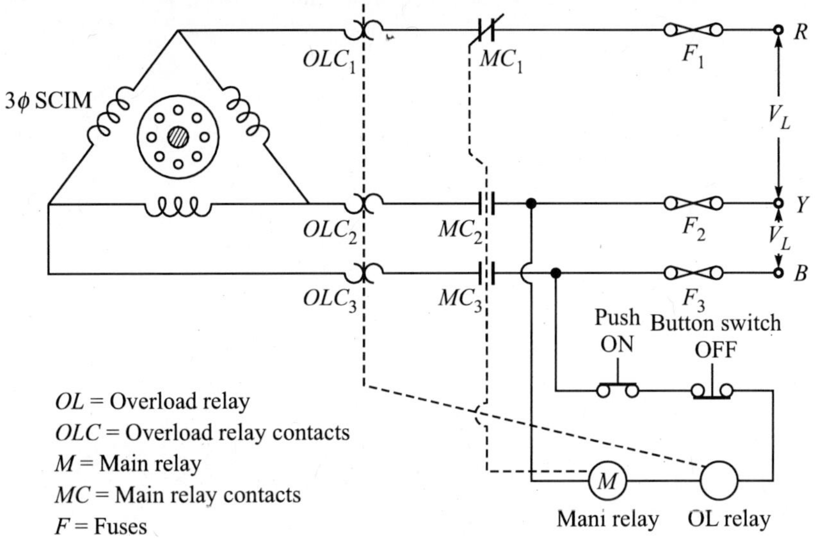

Fig. 8.22 DOL starter with 3 ϕ IM.

Fuses are employed for short circuit protection i.e., if any short circuit in any phase winding of motor occurs then the fuse attached to that particular phase gets blown up as a result the motor starts operating as unbalanced two-phase motor. This is called single phasing of three-phase motor. The motor is stopped as soon as single phasing occurs the sound of motor changes as a result of it.

When the motor supply voltage drops below a given value the main relay 'M' get deenergized as a result the main relay contacts 'MC's' gets opened and the motor gets disconnected from the supply. Because if a motor is operated at reduced voltage for long time then the current drawn by motor increases and so heating of the motor occurs which will ultimately lead to damage of motor.

The starting of IM using DOL starter is easily understood using Fig. (8.23) given below.

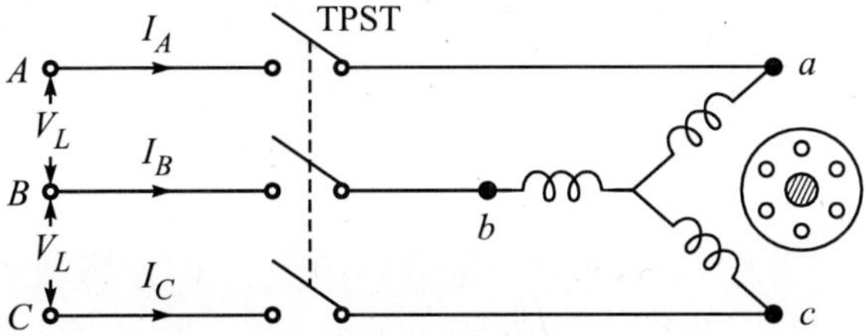

Fig. 8.23 SCIM. Started with full rated voltage.

The ratio of starting torque 'T_{st}' to full load torque 'T_{fe}' is given by

$$\frac{T_{st}}{T_{fl}} = \frac{I_{2_{st}}'^2 \cdot \frac{r_2'}{1}}{I_{2_{fl}}'^2 \cdot \frac{r_2'}{s_{fl}}} \qquad \because T = \frac{s}{\omega_s} \cdot I_2'^2 \cdot \frac{r_2'}{s}$$

$$T = \frac{3}{\omega_s} \cdot P_g$$

or,

$$3\,P_g = \omega_s T = P_G$$

$$\boxed{\frac{T_{st}}{T_{fl}} = \left(\frac{I_{2st}'}{I_{2fl}'}\right)^2 \cdot s_{fl}} \tag{8.33}$$

If no load current is neglected, then

$$\frac{I_{2st}'^2}{I_{2fl}'} \approx \frac{I_{st}}{I_{fl}} \tag{8.34}$$

Hence, torque ratio is given by

$$\boxed{\frac{T_{st}}{T_{fl}} = \left(\frac{I_{st}}{I_{fl}}\right)^2 \cdot S_{fl}} \tag{8.35}$$

On DOL starting, the starting current is equal to short circuit current 'I_{sc}' i.e., at rated voltage of motor (Full Voltage Starting Current).

So,

$$\boxed{\frac{T_{st}}{T_{fl}} = \left(\frac{I_{sc}}{I_{fl}}\right)^2 \cdot S_{fl}} \tag{8.36}$$

The direct on line starting can be employed for 1000's of horse power motor which is used for gate opening of large dams but the voltage rating of such motor are large of the order of kilovolts.

Reduced Voltage Starting

The IM can be started at reduced voltage i.e., less than its rated voltage 'V'(say). The voltage is reduced by 'x', the current at starting reduced by same proportion. Ultimately the torque which depends on the current decreases accordingly. There are various methods of reduced voltage starting. They are given as follows:

(a) Reactor starting
(b) Auto transformer starting
(c) Star-delta starting.

The reactor starting reduces the voltage of stator at starting as some voltage of total rated voltage appears across the reactor and rest across the phase winding.

So a part of rated voltage appears across the phase winding of motor. The reactor starting is shown in Fig. (8.24)

The voltage 'V_1' represents stator voltage per phase and is the rated value.
'x' → fraction of voltage appearing across the phase windings.

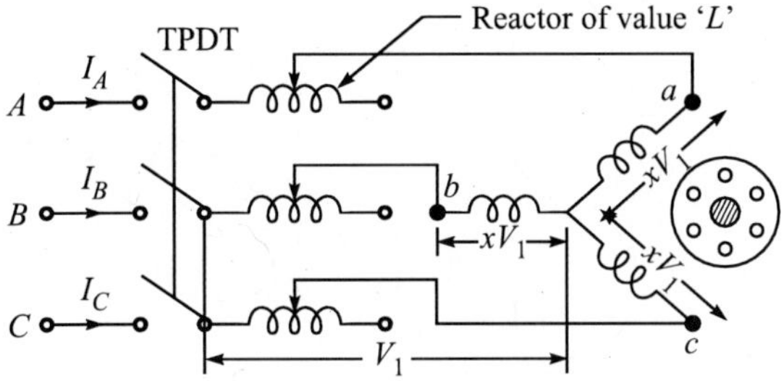

Fig. 8.24 Reactor starting.

The current at starting 'I_{st}' of any phase is given by

$$\boxed{I_{st} = \frac{x V_1}{Z_{sc}}}, \text{ Amp.} \tag{8.37}$$

where, Z_{sc} = short circuit impedance per phase of motor Ω.

The short circuit current $\boxed{I_{sc} = \frac{V_1}{Z_{sc}}}$, Amp is the current which will flow due to direct on line starting of the motor. Substituting the value of 'I_{sc}' in the exp. (8.37), we get

$$I_{st} = x \cdot I_{sc} \tag{8.38}$$

So from the exp. (8.35)

$$\frac{T_{st}}{T_{fl}} = \left(\frac{I_{st}}{I_{fl}}\right)^2 \cdot S_{fl}$$

$$\boxed{\frac{T_{st}}{T_{fl}} = x^2 \left(\frac{I_{sc}}{I_{fl}}\right)^2 \cdot S_{fl}} \tag{8.39}$$

So, the torque ratio differs from eq. (8.35) by x^2

So, we can say that

$$\boxed{\frac{[T_{st}]_{\text{reactor}}}{[T_{st}]_{\text{DOL}}} = x^2} \tag{8.40}$$

That is, the starting torque with reactor starting gets reduced by a factor x^2, where $x < 1$.

Auto Transformers Starting

At starting, the reduce voltage about 40%–60% of the rated voltage is applied to the motor terminals. After the starting period which rests for a few seconds the voltage is raised to its full voltage 'V_1'.Thus, the current at starting is reduced as per the equation given below

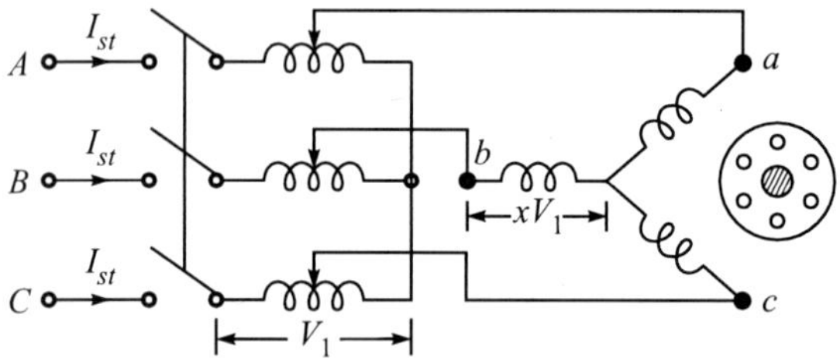

Fig. 8.25 Auto transformer starting.

$$I_{st} = \frac{x V_1}{Z_{sc}} = x I_{sc}$$

If $('I_0')_{\text{Auto tr.}}$ is neglected, then...

$$(\text{VA})_1 = (\text{VA})_{\text{auto}} \tag{8.41}$$

Hence, $$I_1 V_1 = x V_1 \cdot I_{st}$$

$\therefore$ $$I_{st} = \frac{I_1}{x} \tag{8.42}$$

$$I_1 = x^2 I_{sc} \tag{8.43}$$

Thus, $$\frac{T_{st}}{T_{fl}} = \left(\frac{I_{st}}{I_{fl}}\right)^2 \cdot S_{fl}$$

$$\left.\frac{T_{st}}{T_{fl}}\right]_{\text{auto}} = x^2\left(\frac{I_{sc}}{I_{fl}}\right)^2 \cdot S_{fl}$$

$\therefore$ $$\boxed{\frac{[T_{st}]_{\text{auto}}}{[T_{st}]_{\text{DOL}}} = x^2} \tag{8.44}$$

Thus, we see that the starting torque of the motor gets reduced by a factor 'x^2' using auto transformer starting. A reduction of voltage of 50% of 'V rated' results in 25% reduction in torque value at starting.

Star Delta Starting

In this method reduce voltage starting is accomplished by placing the 'R.S.' (reversal switch) position on (1) side i.e., Y side. Thus, the voltage is reduced by a factor $\frac{1}{\sqrt{3}}$ with respect to the line voltage or rated voltage of motor.

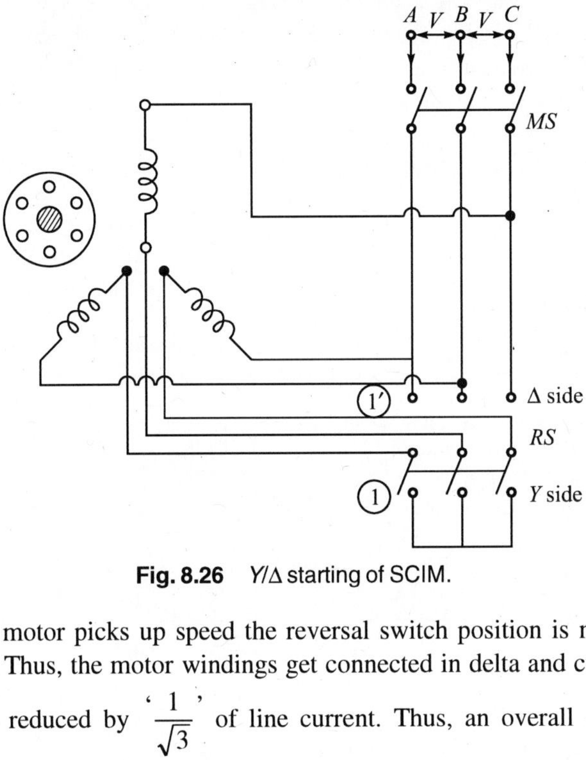

Fig. 8.26 *Y/Δ* starting of SCIM.

After the motor picks up speed the reversal switch position is moved to the position (1). Thus, the motor windings get connected in delta and current drawn by motor is reduced by '$\frac{1}{\sqrt{3}}$' of line current. Thus, an overall $\frac{1}{3}$ or 33.3% reduction in the torque value can be obtained. The phase current with *Y* and Δ connections are

$$I_y = \frac{V_y}{Z_{sc}} = \frac{V}{\sqrt{3}\,Z_{sc}} \tag{8.45}$$

$$I_d = \frac{V}{Z_{sc}} \tag{8.46}$$

$$\frac{T_y}{T_{fl}} = \left(\frac{I_y}{I_{fl}}\right)^2 \cdot S_{fl}$$

$$\frac{T_d}{T_{fl}} = \left(\frac{I_d}{I_{fl}}\right)^2 \cdot S_{fl}$$

$$\therefore \quad \frac{T_y}{T_d} = \left(\frac{I_y}{I_d}\right)^2 = \left(\frac{1}{V_3}\right)^2 = \frac{1}{3}$$

Thus,
$$\boxed{[T_{st}]_y = \frac{1}{3}[T_{st}]_\Delta} \tag{8.47}$$

Thus, the *Y/Δ* starter behaves as an auto transformer starter with x = 57.7 % tapping.

The advantage is that it is a cheap method. It is used for starting motors, generator sets, pumps, etc. The switching from Y to Δ should be made quickly to prevent high current transients due to temporary loss of power hence, spring loaded snap action TPDT switches are used and not knife iron clad switches.

Starting of SRIM (Slip Ring Induction Motor)

The rotor resistance of slip ring induction motor can be varied by adding resistances to each phase externally via slip rings. The rotor windings are Y connected. The stator windings are also star connected as shown in the Fig. (8.27) below

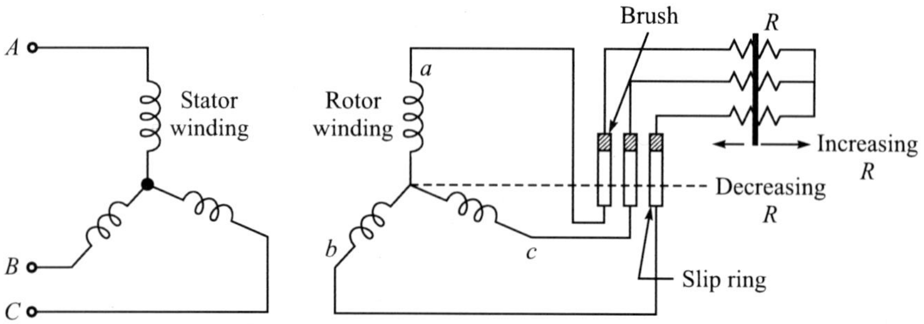

Fig. 8.27 Rotor resistance control of SRIM.

The torque of the motor is given by

$$T \propto \frac{3V_1^2 r_2' \cdot S}{r_2'^2 + (s x_2')^2} \tag{8.48}$$

Since the rotor induced emf 'E_2'' is directly proportional to the applied stator voltage 'V_1'. At starting the value of slip is unity so the torque expression is given by

$$T_{st} \propto \frac{3 V_1^2 r_2'}{r_2'^2 + x_2'^2} \tag{8.49}$$

The above expressions are derived using the equivalent circuit of the induction motor. Assuming that the supply voltage 'V_1' to the stator is constant, then the above expression is given by

$$T_{st} \propto \frac{3 r_2'}{r_2'^2 + x_2'^2} \tag{8.50}$$

Thus, we see that the starting torque 'T_{st}' which is responsible for the rotation of the motor depends on the value of rotor circuit resistance 'r_2' and rotor leakage reactance 'x_2'. The power factor of the motor rotor circuit is given by

$$\cos\theta_2 = \frac{r_2' + R}{\left[(r_2' + R)^2 + x_2'\right]^{1/2}} \tag{8.51}$$

where, 'R' is the external resistance added to rotor circuit.

We see that, as 'R' is added, the denominator rises so the value of '$\cos\theta_2$' decreases if the value of $R >> r_2'$ but if 'R' is of value comparable to that of 'r_2'' then the power factor of the motor rotor circuit increases at the starting. Now let us see what is the effect of increase of rotor resistance on the current of rotor circuit.

The current in the rotor circuit is given by

$$I_2' = \frac{s\,E_2'}{\sqrt{(r_2' + R)^2 + x_2'^2}} \tag{8.52}$$

Thus, as the value of 'R' increases, the value of rotor current decreases accordingly. The value of the rotor emf at starting is also high due to high value of 's'. That is,

$$s\,E_2' = \text{high for large slip value.}$$

So, as the torque $\quad T \propto E_2' \cdot I_2' \cdot \cos\theta_2$

Hence, the torque value will also be high at starting. The torque slip and torque current characteristic is shown in the Fig. (8.28) below.

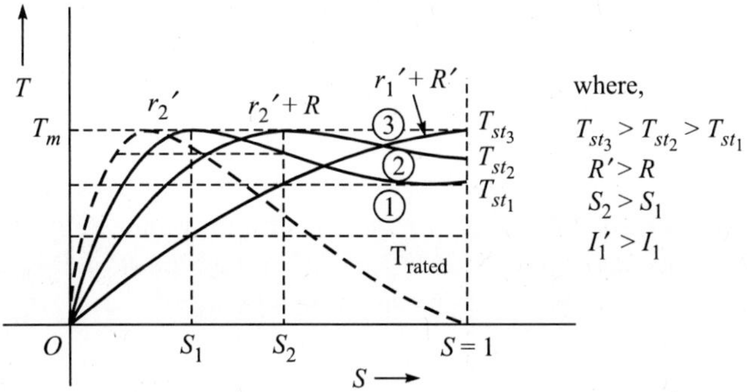

Fig. 8.28 *T/S* characteristic of SCIM showing starting torques.

We see that the slip at which the maximum torque or breakdown torque occurs also gets shifted with the increase of rotor resistance. Since the slip at maximum torque is given by

$$S_m = \frac{r_2'}{x_2'}$$

So, when external resistance 'R' is added to the rotor resistance 'r_2'' then the expression becomes

$$\boxed{S_m = \frac{r_2' + R}{x_2'}} \tag{8.53}$$

We see from the Fig. (8.28), the starting torque 'T_{st_3}' rises to about 50% the starting torque value 'T_{st_1}' having rotor resistance 'r_2'' only. Whereas the 'T_{st_3}' is for rotor having higher rotor resistance equal to $(r_2' + R')$ since the rotor power factor rises at standstill. If we consider $R \approx r_2'$, then the power factor '$\cos \theta_2$' is equal to $\frac{1}{\sqrt{2}}$ or 0.707.

With further increase of 'r_2'' the power factor of rotor circuit increases but the starting torque as well as the maximum developed torque also decreases. The starting current decreases as well. Thus, by increasing resistance of rotor the maximum value of starting torque is equal to the breakdown torque of the motor. Above which starting torque can't be increased.

For most applications, the starting torque of 1.5 to 2 times breakdown or maximum or pull out torque is sufficient. It is not necessary to start the motor with starting torque equal to maximum torque of motor.

At starting it is desirable that the motor should not draw lagging current from supply much otherwise it will affect the other loads attached to the lines, such as lamps, motors, etc. But added resistance to rotor circuit decreases the speed regulation of motor. That is, we can't vary the speed of the motor much by variation of the rotor resistance.

8.12 SPEED CONTROL OF 3 ϕ IM

The three phase induction motor is a constant speed motor and has a small drooping nature from no load to full load. It can be compared with the *N/T* characteristic of d.c. shunt motor so where d.c. shunt motor is used we can replace it with that of three-phase induction motor.

But normally the motor speed control is necessary in industries for numerous applications such as rolling mills, paper mills, textile mills, steel industries, machine tool operations, etc. So there is a need of speed control of IM. Different methods of speed control are used in which some control speed while others deliberately affect the power factor of the motor.

In the speed control methods discussed below there is no special provision for compensation of low power factor. In squirrel cage induction motor the rotor resistance control is possible at the design stage such as double cage rotor IM and deep slot rotor IM. But it is not possible to increase the rotor resistance during the running operation. So the rotor resistance speed control is only for the wound rotor IM also called slip ring IM or SRIM in short.

Rotor Resistance Control

At normal speeds, the slip is of the order of 5% so 's^2' is negligible compared to x_2^2. So the torque of the induction motor reduces to

$$T \propto \frac{s\,x_2}{r_2} \propto \frac{s}{r_2} \text{ (for constant } x_2)$$

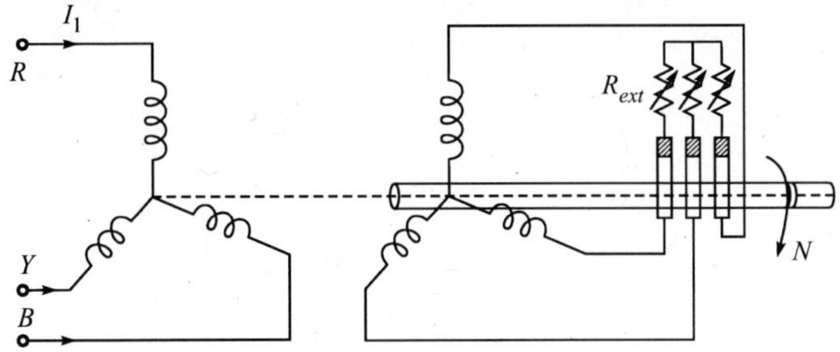

Fig. 8.29 3 ϕ SRIM (with rheostatic control).

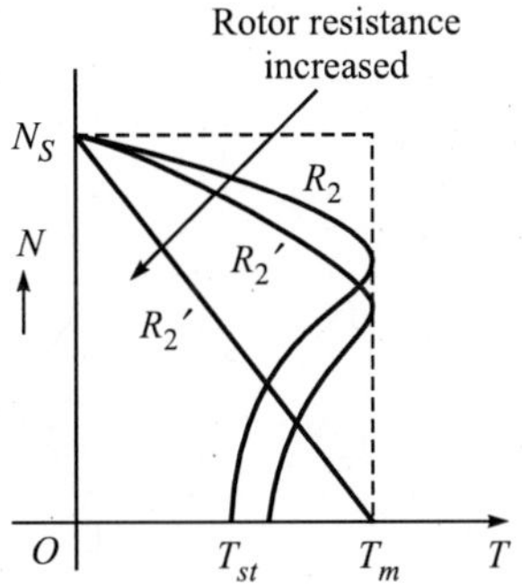

Fig. 8.30 *N/T* characteristic of 3 ϕ IM (Rotor resistance control).

For a given torque, we have $s \propto r_2$ so as the rotor resistance is increased by addition of resistance via slip rings then the speed of the motor decreases accordingly. Since input to the motor is proportional to the torque it follows that the input for a given torque is same regardless of the speed although there is a decrease of speed implies corresponding decrease of the output or shaft power of the motor as shown.

If the slip speed 'sN_s' increases or the difference $(N_s - N)$ increases, then as the slip increases, implies that the mechanical power of the motor decreases and the copper losses of rotor increases hence the efficiency of the motor decreases. Thus, the losses correspond to much heating of motor and so comparable less efficiency occurs and thus with the advent of the power semiconductor devices the conventional methods are not used in the industries.

For small motors the stator impedance control is employed by using rheostat attached to each stator phase windings of the motor. The lesser speed can be obtained by this method. The regulation and efficiency are not good by this method. So we go for other methods of speed control such as pole changing, frequency control, slip control, etc.

Pole Changing

By providing the IM with two different sets of windings designed for different number of poles such that only one winding is connected at a time. By this method multiple synchronous speeds revolving flux can be obtained in the air gap of the machine. Since the rotor speed depends on the synchronously revolving field. So, different speeds of motor are possible. The switching of windings from one to another could be electronic control or by using e.m. relays. The separate stator winding for each speed desired, rated to unit the particular applications giving performance on each speed substantially independent of the others. The use of such control is in lifts operation where the changeover from one winding to another has to be made without losing control of the motor.

Speed in the ratio 2:1 can be obtained by such methods.

In the case of SCIM the rotor adjusts automatically to any number of poles and is equal to that of stator number of poles. The second method based on pole changing is called consequent pole method. In this the stator windings connection are changed such that number of poles can be increased or developed and decreased respectively. A machine whose stator windings are designed for 4 poles as shown below.

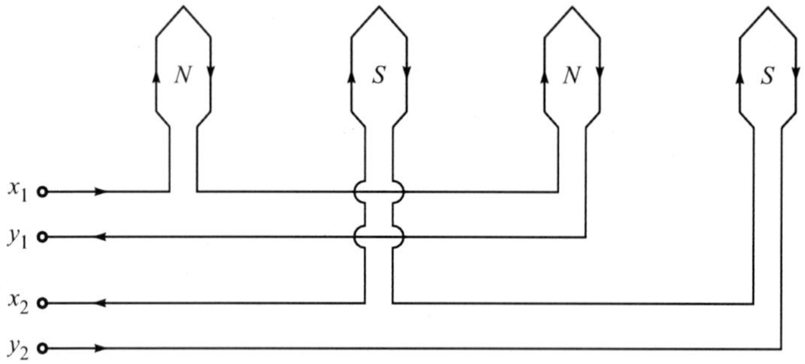

Fig. 8.31 Four-pole IM.

x_1, y_1, x_2, y_2, are four groups of coils which are magnetized in the same direction.

If the connections of 'x_2' and 'y_2' are reversed such that the current direction is reversed, then the consequent (extra) south poles are generated simultaneously as shown in the Fig. (8.32).

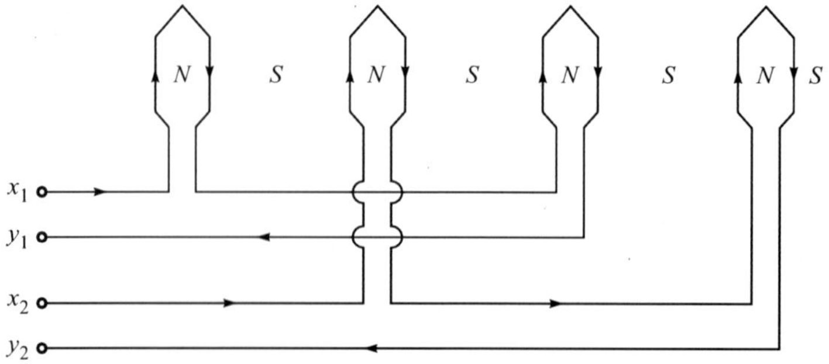

Fig. 8.32 Eight-pole IM (with 4 consequent poles).

Thus, we see that the number of poles of the induction motor is increased to eight, so the speed of the IM is decreased to half. Since $N = \frac{120f}{P}(1 - s)$

The possible combinations of the two circuits of each phase are

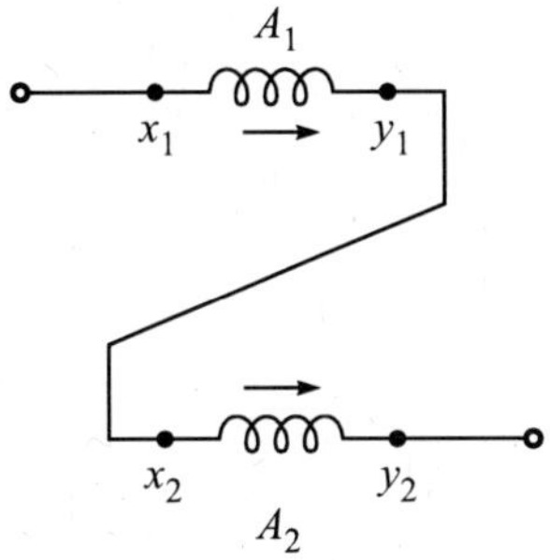

(a) Low speed

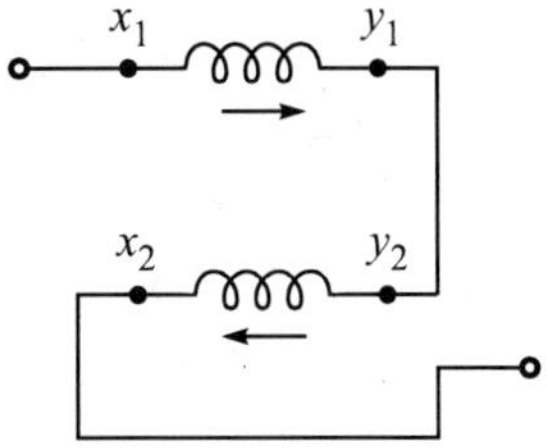

(b) High speed

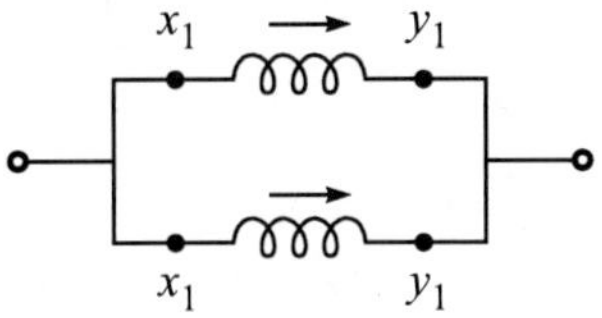

(c) Low speed

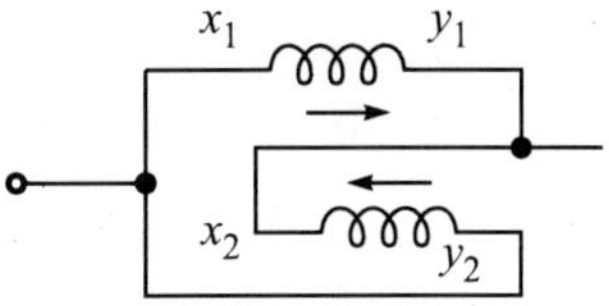

(d) High speed

Fig. 8.33 Connections of phase windings of IM.

For constant torque connections are shown in Fig. (8.34).

The efficiency of the machine is affected by the change of connections. The power factor is similarly affected, i.e., by increasing the speed, the power factor increases for constant power output motors. It is used in high power drives.

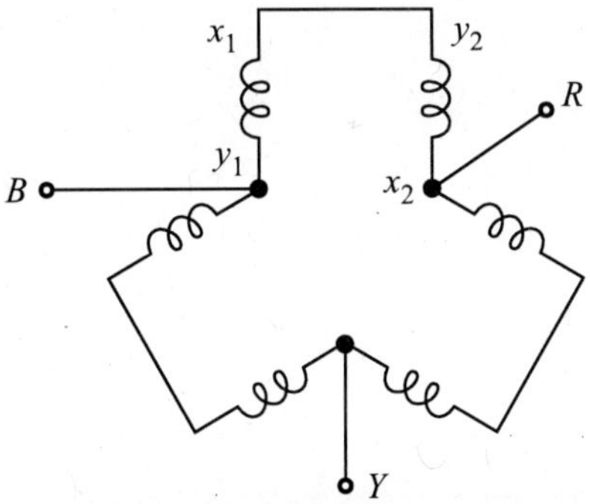

Fig. 8.34(a) Low speed, series connected in Δ.

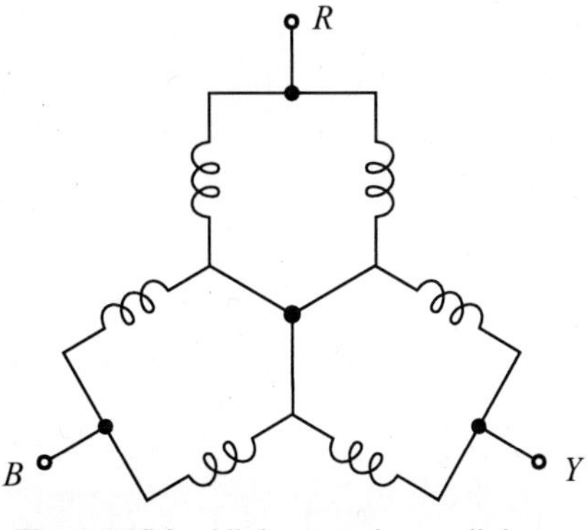

Fig. 8.34(b) High speed, parallel connected *Y*.

Constant Horse-power Connections

The starting torque for low speed connection is two times that of starting torque for high-speed connection. It is used for driving machine tools.

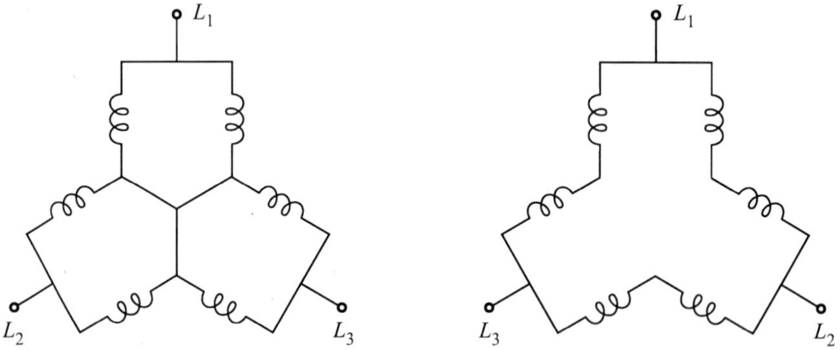

(a) Low speed (b) High speed

Fig. 8.35

Variable Torque Connections

The starting torque with low speed connections is 70% that of starting torque of high speed connection. It is used for blowers and fans.

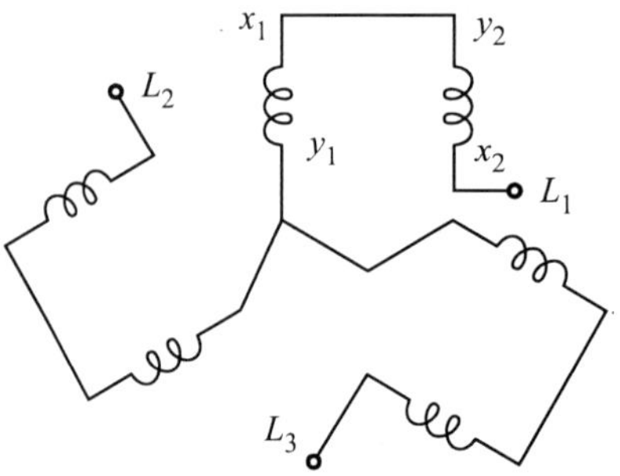

(a)

Fig. 8.36(a) Low speed (series).

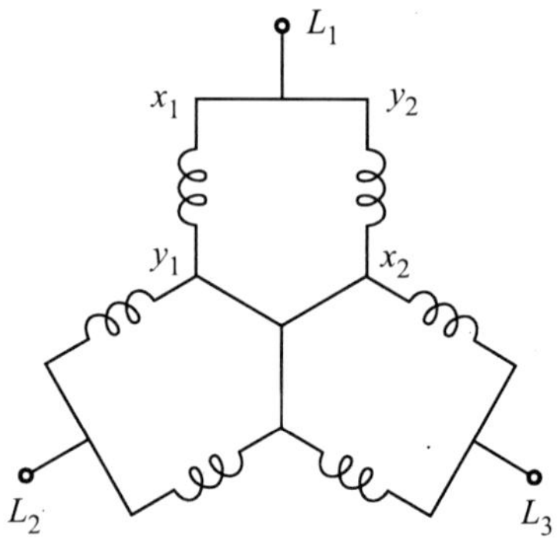

(b)

Fig. 8.36(b) High speed (parallel).

Speed Control (by Changing Frequency)

The motor speed is dependent on the supply freq. 'f' as per the relation given by

$$\boxed{N = N_s(1-s) = \frac{120f}{P}(1-s)} \quad (8.54)$$

This control is normally employed for the ship propulsion motors for driving the ships. It needs variable speed alternator for each motor to be independently control. The driving motor are directly connected to the propeller shaft is supplied by its own turbine driven generator.

If the excitation of the generator is held constant, a change in the speed results in the change of voltage and frequency in same proportion provided the impedance drop in the armature neglected.

$$T \propto \frac{V_1^2\, r_2' \cdot s}{(s\, r_1 + r_2'^2)^2 + s^2 (x_1 + x_2')^2}$$

The main flux of the driven motor is nearly constant. The torque of the motor with simultaneous voltage and frequency control when the slip is low is given by

$$\boxed{T \propto \frac{V_1^2}{f}} \qquad (8.55)$$

If the motor is started at low voltage and frequency

$T_{st} \propto 1/f$

Speed Control of IM by Concatenation

When different speeds are required then two machines of equal ratings are connected such that the input of second machine is given by the first machine's rotor circuit as shown in Fig. (8.37).

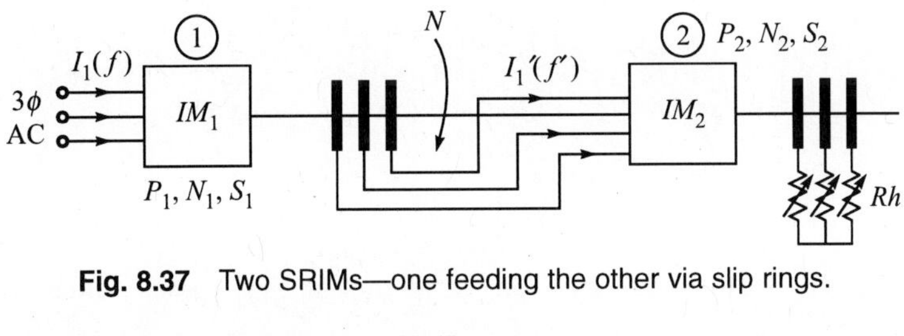

Fig. 8.37 Two SRIMs—one feeding the other via slip rings.

where, f = Supply frequency = 50 Hz
f' = Slip frequency = sf, Hz
P_1, P_2 = Machine poles
N_1, N_2 = Speed of machines, RPM
S_1, S_2 = Slip of machines

The speed of machines (1) and (2) are given by

$$N_1 = N_{s_1} \times (1 - s_1);\ N_{s_1} = \frac{120 f}{P_1} \quad \text{(synchronous speed of machine (1))}$$

$$N_1 = \frac{120 f}{P_1}(1 - s_1)$$

$$N_2 = N_{s_2}(1 - s_2); \quad N_{s_2} = \frac{120 f'}{P_2}$$

$$= \frac{120 f'}{P_2}(1 - s_2); \quad f' = s_1 f$$

$$= \frac{120 s_1 f}{P_2}(1 - s_2)$$

since the two machines are coupled.

So,
$$N_1 = N_2 = N \text{ (say)}$$

$$\frac{120 f}{P_1}(1 - s_1) = \frac{120 s_1 f}{P_2}(1 - s_2)$$

So,
$$s_1 = \frac{P_2}{P_1(1 - s_2) + P_2}$$

when the rheostat '*Rh*' of second machine is short circuited then, $S_2 \to 0$

So, the expression of S_1 reduces to

$$S_1 \approx \frac{P_2}{P_1 + P_2}$$

$\because$
$$s_1 = \frac{N_{s_1} - N}{N_{s_1}}$$

So,
$$N_1 = N_{s1}\left(\frac{P_1}{P_1 + P_2}\right); \; N_{s_1} = \frac{120 f}{P_1}$$

Substituting the value of 'N_{S_1}' in the above exp., we have

$$\boxed{N_1 = \frac{120 f}{P_1 + P_2}} \tag{8.56}$$

So by this connection, we have four different values of poles which corresponds to four different speeds, namely,

$$N_1 = \frac{120 f_1}{P_1 + P_2}, \; N_1 = \frac{120 f_1}{P_1 - P_2}, \; N_1 = \frac{120 f_1}{P_1}, \; N_2 = \frac{120 f_1}{P_2}$$

The poles $(P_1 - P_2)$ are obtained by changing two-phase connections of any one machine. The frequency of the driven machine is given by

$$\boxed{f' = f\left[1 \pm \frac{N_1}{N_{s1}}\right]} \tag{8.57}$$

where,
$$N_{s_1} = \frac{120 f}{P_1} \text{ synchronous speed}$$

$$N_1 = \text{speed of machine} \quad (1)$$
$$f = \text{synchronous frequency} = 50 \text{ Hz}$$

Thus, the first machine acts as a variable frequency source for second machine.

8.13 COGGING AND CRAWLING OF INDUCTION MOTOR

It happens that sometimes the induction motor fails to start rotating even when the stator field is excited with a.c. supply. This happens due to the magnetic locking of the poles of stator and rotor respectively. Since the number of poles of the rotor is kept as integral multiples of the poles of stator to prevent this cogging or jamming of the induction motor.

It happens that at sub-synchronous speeds there is a strange behaviour of IM i.e., vibrations due to harmonics (5th and 7th) respectively. This is due to harmonic induction torques due to superimposition of 5th and 7th harmonic with the fundamental. The *N*/*T* characteristic is given by Fig. (8.38) below.

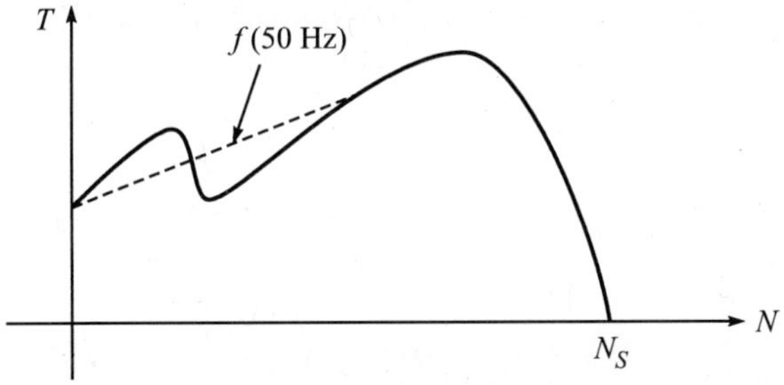

Fig. 8.38 Crawling in 3 ϕ IM.

To avoid the vibrations, the rotor slots should not exceed 50% of that of stator slots. The cogging is due to harmonic synchronous torques due to which at one point the torque value becomes equal to zero as shown in Fig. (8.39) below.

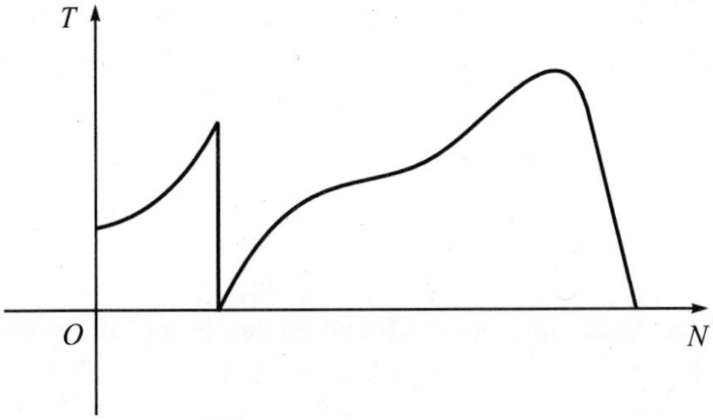

Fig. 8.39 Cogging of 3ϕ IM.

To prevent cogging the rotor bars are slightly skewed to prevent the magnetic locking of the two fields i.e., stator and rotor respectively.

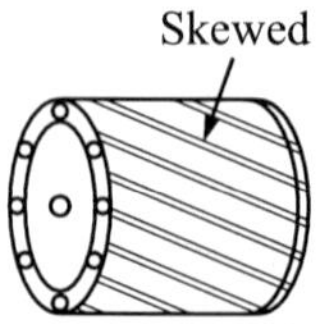

Skewing of rotor bars of IM.

8.14 BRAKING OF INDUCTION MOTOR

The stopping of motor while in motion is called braking. This can be employed by mechanical brake shoe, by electrical method or by using electronic switches employing power devices.

In this section only the electrical method of braking is discussed. They are discussed as follows.

(a) *Plugging:* The plugging operation can be initiated by reversing the connections of phases while the motor is revolving. The rotating magnetic field reverses simultaneously but the motor due to inertia could not reverse with the reversal of rotating field in air gap. So the torque called braking torque originates as soon as the phase reversal takes place hence the motor speed decreases and it stops. There is a contactor which opens the motor circuit with the supply.

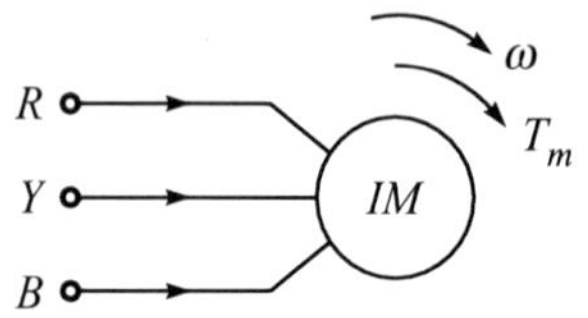

Normal operation of IM.

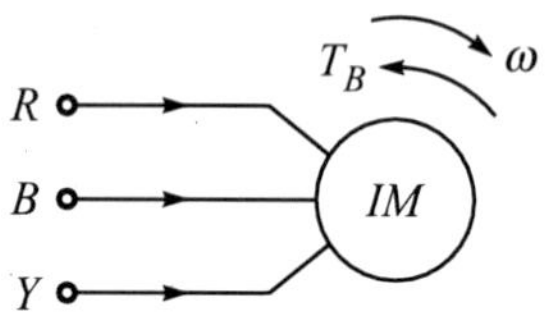

Plugging of IM.

During plugging operation, the power factor of the motor decreases. Motors of rating below 25 hp are plugged using Y/Δ starter. For stopping the motors plugging is not used, so faster method of braking is rheostatic method.

(b) *Dynamic (or Rheostatic braking):* In this braking the stator windings connections is changed from a.c. to d.c. The field in the air gap becomes steady and not rotating with d.c. So a braking torque originates in the rotor which opposes the motoring torque and so the motor comes to halt quickly and smoothly. In this method the loss is also less.

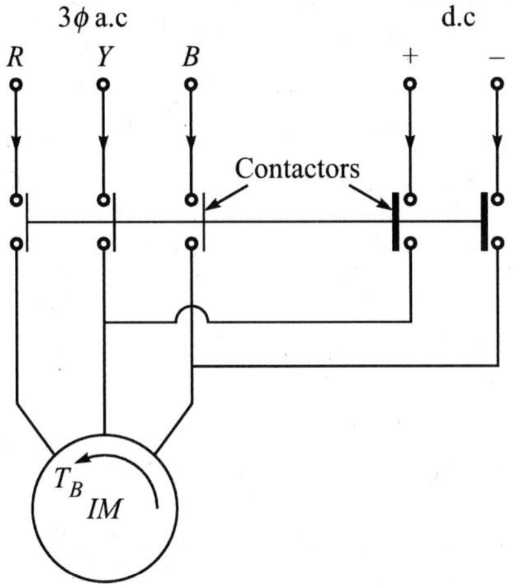

Dynamic braking of IM.

The applications of this method is in mine winders and hoists.

(c) *Regenerative braking:* In the regenerative braking the motor speed increases and exceeds the rated speed of motor so the motor operates as generator and starts giving current from stator windings to the supply. Since the current direction reverses so there originates a torque in the opposite direction with respect to the motoring torque hence this acts as a braking torque as a result, the motor speed is decreased. But complete stopping of the motor is not possible with this method. This type of braking normally occurs in the case of active loads such as hoist or lifts. In which the speed of the motor increases due to gravity.

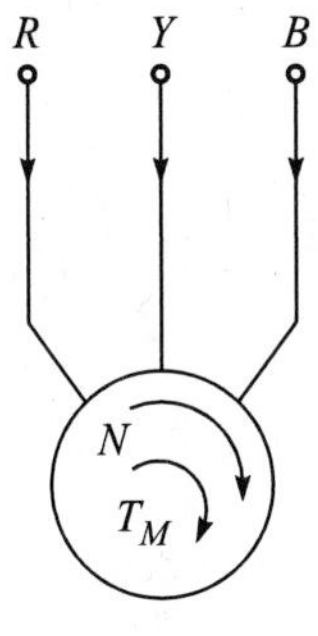

Normal operation of IM.

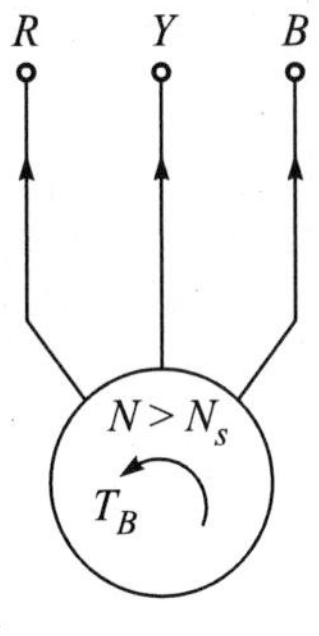

Regenerative operation of IM.

8.15 INDUCTION GENERATOR

The induction generator has similar construction as that of induction motor. Only the difference is that the rotor is moved by a prime mover at a speed greater than that of the speed rated on the motor. Thus, the motor starts generating and gives electrical energy. Though the induction generator has limited use. It is used in the windmills for generating alternating current directly.

The prime mover can be steam powered or diesel by other non-electrical means.

The induction generator has negative slip, that is, the speed of the prime mover and hence the rotor is greater than the synchronous speed so using relation (8.58)

$$s = \frac{n_s - n}{n_s} \tag{8.58}$$

since, $n_s < n$, so $s = -\text{ve}$

So the torque which is proportional to slip is also negative and the induction motor goes in the generating mode as shown by the characteristic given by (8.40) shown below.

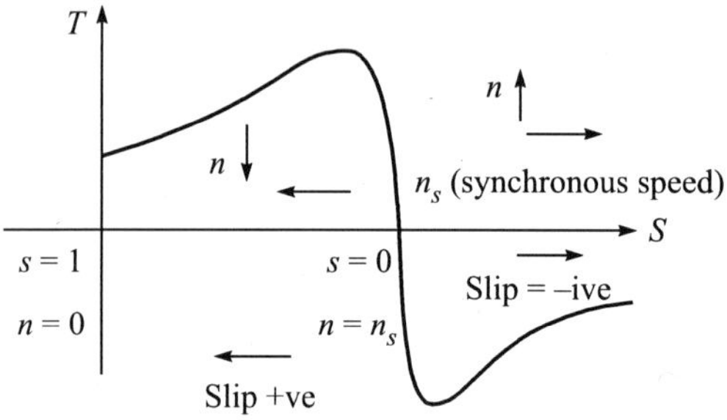

Fig. 8.40 Induction machine *T/N* characteristic (In diff. modes).

Any induction motor can be driven at super-synchronous speed and can be operated as induction generator irrespective of the design. Although the low resistance types of rotors produce higher currents for the same induced voltage 'E_2' in the rotor. The induction generator is not a self-excited generator it can be realized by connecting the stator terminals to a source of fixed voltage and frequency and then driven at a speed greater than its synchronous speed for which it is designed.

The induction generator always generates current which leads the generated voltage. As the negative slip value increases, that is, it is driven at higher speeds then the current value also increases accordingly. So it can be used for increasing the power factor. But it is not preferred to that of synchronous condensor which do not require any prime mover.

8.16 APPLICATIONS

The three-phase induction motors are used for a variety of applications from agriculture, transportation, domestic and industrial purposes. The three-phase motors find wide applications in the irrigation pumps of ratings between 5 hp and 20 hp. Mostly 5 hp and 10 hp motors are used. The three-phase induction motor is widely used as centrifugal pumps, blowers, winders, etc. in the industries.

They found wide applications in traction purposes. It is used in ships. It is used in the mine hoists. In domestic purpose to draw water from ground where depth of water underground is very high.

It is used for a number of applications depending on the torque requirements. The SCIM is used for fans and blowers, grinders, drilling machines, blowers, pumps for puming water, lathe machines and printing machines.

The SRIM is used for high starting requirement such as hoist, elevators, lifts cranes and compressors. The SCIM are available in different classes such as *A*, *B*, *C*, *D*, *F*. The class *A* type motor is normal or standard motor for constant speed use such as type loads. Size generally less than 5 hp. Class *B* type SCIM is used for general purpose. Use for centrifugal pumps, machine tools and blower. It can be rated greater than 5 hp which uses Y/Δ starter.

The class *C* SCIM has poor speed regulation. It is used for pumps and piston type compressors where the starting is not easy.

Class *D*, SCIM are used for shear and punch presses class *F*, SCIM are employed for size greater than 25 hp with full voltage starting.

Comparison of Various SCIM

Table

NEMA (Class)	T_{st} ($X\ T_{rated}$)	I_{st} (times rated)	% speed regulation	Class
A	1.5 – 1.75	5 – 7	2 – 4	Normal
B	1.4 – 1.6	4.5 – 5	3 – 5	General purpose
C	2 – 2.5	3.5 – 5	4 – 5	Double cage, high torque
D	2.5 – 3.0	3 – 8	5 – 8	High torque, high resistance
F	1.25	2 – 4	>5	Double cage, low torque, low, starting current

SOLVED QUESTIONS

Q.1 A 4 pole, 50 Hz, 3 phase IM running at full load develops a useful torque of 200 Nm when the rotor emf makes 120 rpm. If the mechanical torque lost in friction and rotor core loss is 15 Nm. Calculate

(i) Slip (ii) Shaft power output (iii) Rotor Cu loss (iv) Input to stator (v) Efficiency.

A.1 $P = 4, \quad f = 50 \text{ Hz}, \quad T_{sh} = 200 \text{ Nm}, \quad f_2 = \frac{120}{60} \text{rps} = 2 \text{ Hz}$

$T_{\text{loss}} = T_{f+c} = 15 \text{ Nm}.$

(i) $f_2 = sf$ so, $s = \frac{f_2}{f} = \frac{2}{50} = 0.04 = 4\%$ slip $= 4\%$

(ii) $N_s = \frac{120f}{P} = \frac{120 \times 50}{4} = 1500$ rpm

$$N = N_s(1 - 5) = 1500(1 - 0.04) = 1440 \text{ rpm}$$

$$\omega = \frac{2\pi N}{60} = \frac{2 \times \pi \times 1440}{60} = 150.8 \text{ rad/sec}$$

Shaft power $P_{sh} = \omega T_{sh}$
$= 150.8 \times 200$
$= 30.2$ kW

(iii) $T_m = T_{sh} + T_{\text{loss}}$
$= 200 + 15 = 215$ Nm

Rotor I/P $P_m = \omega T_m = 150.8 \times 215 = 32.4$ kW

$$\frac{\text{Rotor Cu loss}}{\text{Rotor input}} = \frac{s}{1-s} = \frac{P_{r_{cu}}}{P_m}$$

$$P_{r_{cu}} = \frac{0.04}{1 - 0.04} P_m = 1350.9 \text{ W}$$

$= 1.35$ kW

(iv) Stator I/P = stator loss + Rotor Cu loss + Rotor O/p

Assume stator loss = 1000 W
$= 1000 + 1350.9 + 32.419$
$= 34.7$ kW

(vi) Efficiency $\eta = \frac{P_0}{P_i} \times 100$

$$= \frac{30.2}{34.7} \times 100$$

$= 87.03\%$

Q.2 A 5 hp, 230 V, 50 Hz IM has a rated full load speed of 950 rpm. The induced voltage per phase of rotor at standstill is 100 V. Calculate

(i) Number of poles and percentage of full load slip.

(ii) Rotor induced voltage and its frequency at full load.

[B.Tech II, UPTU, 2004]

A.2 $P_{sh} = 5$ hp, $V_r = 230$ V, $f = 50$ Hz, $N_r = 950$ rpm,

$E_2/ph|_{s=1} = 100$ V

(i) $N_s = \dfrac{120f}{P}$ = 1000 rpm (nearer to 950 rpm than 800 rpm)

so, $$P = \frac{120f}{N_s} = 6$$

$$s = \frac{N_s - N}{N_s} \times 100\% = \frac{1000 - 950}{1000} \times 100\% = 5\%$$

(ii) $E_2|_s = s\ E_2|_{s=1}$

$$= 0.05 \times 100 = 5 \text{ V}$$

Slip frequency $f_2 = sf = 0.05 \times 50 = 2.5$ Hz

Q.3 A six-pole, 50 Hz, three-phase IM has rotor resistance of 0.4 Ω/phase, maximum torque is 200 Nm at 850 rpm. Find

(i) Torque at 4% slip.
(ii) Additional rotor resistance to get 2/3rd of maximum torque at starting.

[B.Tech II, UPTU, 2005]

A.3 $$P = 6, f = 50 \text{ Hz}, r_2 = 0.4\ \Omega/\text{ph}, T_{max} = 200 \text{ Nm}$$

$$N_m = 850 \text{ rpm}$$

(i) $S|T_{max}\ \dfrac{r_2}{x_2} = S_m \quad \therefore \quad x_2 = \dfrac{r_2}{s_m} = \dfrac{0.4}{0.15} = 2.67\ \Omega/\text{ph}$

$$S_m = \frac{N_s - N_m}{N_s} = \frac{1000 - 850}{1000} = 0.15$$

$$N_s = \frac{120f}{P} = \frac{120 \times 50}{6} = 1000 \text{ rpm}$$

(ii) $\dfrac{T_L}{T_{max}} = \dfrac{25 \cdot s_m}{s^2 + s_m^2}$

$$T_L = T_{max} \frac{2 \times 0.15 \times 0.4}{(0.15)^2 + (0.04)^2}$$

$$= 200 \times \frac{2 \times 0.15 \times 0.4}{(0.15)^2 + (0.04)^2}$$

$$T_L = 99.6 \text{ Nm}$$

$$T_{st} = \frac{2}{3} T_{max}$$

$$\frac{T_{st}}{T_{max}} = \frac{2}{3} = \frac{2 r_2' \cdot x_2}{r_2'^2 + x_2^2}$$

$$r_2'^2 + 7 = 8\, r_2'$$

$$r_2'^2 - 8\,r_2' + 7 = 0$$

$$r_2' = 7\ \Omega \text{ or } 1\ \Omega$$

Neglecting larger value, $r_{ext} = r_2' - r_2 = 1 - 0.4 = 0.6\ \Omega/\text{ph}$

Q.4 A three-phase IM is wound for four-pole and is supplied from 50 Hz system. Calculate

(i) Synchronous speed.

(ii) Rotor speed when slip is 4%.

(iii) Rotor frequency when rotor runs at 600 rpm.

[B.Tech II, 2003 UPTU]

A.4 $P = 4, f = 50$ Hz

(i) $N_s = \dfrac{120f}{P} = \dfrac{120 \times 50}{4} = 1500$ rpm

(ii) Rotor speed at 4% slip

$$\therefore \quad N = N_s(1 - s) = 1500\,(1 - 0.04)$$
$$= 1500 \times 0.96$$
$$= 1440 \text{ rpm}$$

(iii) Rotor frequency when rotor runs at 600 rpm

$$S = \frac{N_s - N}{N_s} = \frac{1500 - 600}{1500} = 0.6$$

$$\therefore \text{Rotor frequency} = f_2 = s\,f$$
$$= 0.6 \times 50$$
$$= 30 \text{ Hz}$$

Q.5 A 400 V, three-phase, four-pole, 50 Hz IM has a rotor resistance and reactance per phase of 0.01 Ω and 0.1 Ω respectively. Determine

(i) Maximum torque in Nm and the corresponding slip.

(ii) The full load slip and power output in watts if maximum torque is twice the full load torque and the ratio of stator to rotor turns is 4.

[B.Tech II, UPTU, 2003]

A.5

$$E_2 = \frac{N_2}{N_1} \cdot E_2 = \frac{1}{4} \times 400 = 100 \text{ V}$$

$$E_{2\text{ph}} = \frac{100}{\sqrt{3}} = 57.7 \text{ V}$$

$$\therefore \quad T_m = \frac{k\,E_2^2}{2\,X_2} = \frac{3}{\omega_s} \cdot \frac{E_2^2}{2\,X_2}$$

$$\omega_s = \frac{2\pi N_s}{60} = \frac{2\pi}{60}\left(\frac{120f}{P}\right) = 2\pi \times 25 = 50\pi \text{ rad/s}$$

$$\therefore \quad T_m = \frac{3}{50\pi} \cdot \frac{(57.7)^2}{2 \times 0.1} = 318.3 \text{ Nm}$$

$$\text{Slip}|_{T_m} = \frac{r_2}{x_2} = \frac{0.01}{0.1} = 0.1 = 10\%$$

$$\frac{T_f}{T_m} = \frac{2 s_f s_m}{s_f^2 + s_m^2} = \frac{2 \times s \times 0.1}{s^2 + (0.1)^2} = \frac{1}{2}$$

$$0.5 = \frac{0.2 s}{0.01 + s^2}$$

$$\therefore \quad 0.05 + 0.5 s^2 - 0.2 s = 0$$

$$0.5 s^2 - 0.2 s + 0.005 = 0$$

$$\therefore \quad s = \frac{0.2 \pm \sqrt{0.03}}{1}$$

$$s = 0.03 \quad \text{or} \quad 0.04$$

Considering lower value of 's'

$$T_m = 2 T_{fl}$$

$$\therefore \quad T_{fl} = \frac{T_m}{2}$$

$$= \frac{318.3}{2} = 159 \text{ Nm}$$

$$\therefore \quad P_{sh} = \omega T = \frac{2\pi N}{60} \cdot T$$

$$= \frac{2\pi}{60} N_s(1 - s) \cdot T$$

$$= \frac{2\pi}{60} \times \frac{120 f}{P}(1 - s) \cdot T$$

$$= \frac{2\pi}{60} \times 1460 \times 159. \text{ N m}$$

$$\boxed{P_{sh} = 48.7 \text{ kW}}$$

Q.6 The result of no load and full load tests on a three-phase delta connected motor were as follows. The intake as measured by the two wattmeter method gave 3 kW and –5 kW on no load and 39 kW and 23.5 kW on full load. The motor had 4 poles, the supply frequency was 50 Hz, the full load speed 1460 rpm and the stator resistance per phase (hot) 0.13 Ω. No load current, 25 A, full load current 92 A per line.

A.6 Stray losses + No load stator I^2R loss = 3000 – 550 = 2450 W

$$NL \text{ loss (stater Cu – loss)} = 3 \times \left(\frac{I_{N_L}}{V_3}\right)^2 \times R_1$$

$$= 3 \times \left(\frac{25}{\sqrt{3}}\right)^2 \times 0.13 = 82 \text{ W}$$

$\therefore$ Stray losses = 2450 – 82 = 2370 W (say)

P_1 = Full load intake = 39000 + 23500 = 62500 W

$$\text{Stator } I^2 R \text{ loss (at full load)} = 3 \times \left(\frac{I_{fl}}{\sqrt{3}}\right)^2 \times R_1$$

$$= 3 \times \left(\frac{92}{\sqrt{3}}\right)^2 \times 0.13 = 1100 \text{ W}$$

$\therefore$ Rotor intake = $P_g = P_2 = P_1$ – (FL stator loss + stray losses)

= 62,500 – (1100 + 2370)

= 59,030 W

Synchronous speed, N_s = 1500 rpm

Actual speed, N = 1460 rpm

$$\therefore \quad s = \frac{N_s - N}{N_s} = 2.66\%$$

$$\therefore \quad \text{Rotor } I_2^2 R_2 \text{ loss} = P_g \times s = \frac{2.66}{100} \times 59030 = 1570 \text{ W}$$

$\therefore$ Output = Rotor intake – Rotor Cu loss

$= P_g - s\,P_g$ = 59030 – 1570 = 57,460 W

Intake = 62500 W

$$\therefore \quad \eta = \frac{\text{Output}}{\text{Input}} = 91.9\%$$

Q.7 The power input to a 3 phase, 50 Hz IM is 50 kW. The total stator loss is 800 W. Find the total mechanical power developed and the rotor copper loss per phase; if it is observed that the rotor emf makes 90 rpm.

A.7

$$f_2 = \frac{90}{60} = 1.5 \text{ Hz}$$

$$s = \frac{f_2}{f} = \frac{1.5}{50} = 0.03 = 3\%$$

$$P_g = P_i - P_{\text{stater}} = 49200 \text{ W}$$

$$P_m = (1 - s)\,P_g = 47.724 \text{ kW}$$

Total rotor copper loss = $s\,P_g$ = 1476 W

$$\therefore \text{ Rotor Cu power loss/phase} = \frac{1476}{3}$$

$$= 492 \text{ W}$$

Q.8 The power input to the rotor of a 430 V, 2 pole, 50 Hz, 3 phase IM is 60 kW. The rotor emf makes 105 cycles per minute. Calculate (a) slip (b) mechanical power developed (c) rotor speed (d) rotor copper loss/phase.

A.8

$$V_2 = 430 \text{ V}, \quad P = 2, \quad f = 50 \text{ Hz}, \quad P_g = 60 \text{ kW}, f_2 = 105 \text{ rpm}$$

$$= \frac{105}{60} = 1.75 \text{ Hz}$$

(a) $\text{Slip} = \frac{f_2}{f} = \frac{1.75}{50} = 0.035 = 35\%$

(b) $P_m = (1 - s)\, P_g = (1 - 0.035) \times 60 \times 10^3 = 57900 \text{ W}$

$$= 77.614 \text{ HP} \quad (\because 1 \text{ HP} = 746 \text{ W})$$

(c) $s = \frac{N_s - N}{N_s} = \frac{3000 - N}{3000} = 0.035 \left(\because N_s = \frac{120 f}{p}\right)$

$$\therefore \qquad N = 2895 \text{ rpm}$$

(d) $P_{2cu}/\text{ph} = I_2'^2\, r_2' = \frac{s\, P_g}{3} = \frac{0.035 \times 60 \times 10^3}{3}$

$$= \frac{2100}{3} = 700 \text{ W}$$

Q.9 A four-pole, three-phase, IM delivers 37 hp at the shaft at a speed of 1425 rpm on 500 V, 50 Hz supply. The mechanical losses total 3 hp and the power factor is 0.9. Calculate for this load (i) the slip (ii) the rotor Cu loss (iii) the total power I/P if the stator losses are 2500 W (iv) the efficiency (v) the line current (vi) the number of complete cycles per minute of the rotor emf.

A.9

$$P = 4, P_0 = 37 \text{ hp}, N = 1425 \text{ rpm}, V = 500 \text{ volts}, f = 50 \text{ Hz},$$
$$P_{f+w} = 3 \text{ h.p.}, \cos\theta_1 = 0.9$$

(a) $N_s = \frac{120 f}{P} = 1500 \text{ rpm}$

$$N = N_s (1 - s) \;\therefore\; s = 1 - \frac{N}{N_s} = 1 - \frac{1425}{1500} = 0.05 = 5\%$$

(b) $P_{2cu} = I^2 R_2 = s\, P_g = 0.05 \times 31.41 \times 10^3 = 1570.5 \text{ W}$

$$P_g = P_m + P_{2cu} \quad \Rightarrow \quad P_g (1 - s) = P_m$$

$$\therefore \qquad P_g = \frac{P_m}{1 - s} = \frac{29840}{1 - 0.05} = 31.41 \text{ kW}$$

$$P_m = P_{sh} + P_{w+f} = 40 \text{ hp} = 40 \times 746 = 29840 \text{ W}$$
$$= 29.84 \text{ kW}$$

(c) Total power $I/P = P_g + P_{\text{stator loss}}$

$$= 33.91 \text{ kW}$$

(d) $\eta = \dfrac{P_0}{P_i} = \dfrac{37 \times 746}{33.91 \times 10^3} = 81.4\%$

(e) $P_i = \sqrt{3}\, V_1 I_1 \cos\theta_1 \quad \Rightarrow \quad I_1 = \dfrac{33910.5}{\sqrt{3} \times 500 \times 0.9}$

$$= 43.5 \text{ A}$$

(f) $f_2 = sf = 0.05 \times 50 = 2.5$ Hz

Q.10 The power input to a 440 V, 50 Hz, 4 pole, 3 phase IM running at 1455 rpm is 50 kW. The stator losses are 1.2 kW and the mechanical losses total 2.5 kW. Calculate: (a) slip (b) the rotor copper loss (c) the shaft torque (d) efficiency

A.10 $V_1 = 440$ Volt, $f = 50$ Hz, $P = 4$, $N = 1455$ rpm,

$P_1 = 50$ kW, $P_{\text{stator}} = 1200$ kW, $P_{\text{mech}} = P_{w+f} = 2500$ W

$$N_s = \frac{120 f}{P} = \frac{120 \times 50}{4} = 1500 \text{ rpm}$$

$$N = N_s (1 - s)$$

(a) $s = \dfrac{N_s - N}{N_s} = \dfrac{1500 - 1455}{1500} = 0.03 = 3\%$

(b) P_{rotor} (Cu) $= I_2^2 R_2 = s P_g = 0.03 \times 48800$ W

$$P_g = P_1 - P_{1_{\text{cu}}} = 50 \times 10^3 - 1200 = 48.8 \times 10^3 \text{ W}$$
$$= 48800 \text{ W}$$

(c) $P_{\text{sh}} = P_m - P_{\text{mech}}$

$$= (1 - s) P_g - P_m$$
$$= 47336 - 2500 = 44836 \text{ W}$$
$$= 44.836 \text{ kW}$$

$$\omega T_{sh} = P_{sh} \quad \Rightarrow \quad T_{sh} = \frac{P_{sh}}{\omega} = \frac{P_{sh}}{\frac{2\pi N}{60}}$$

$$= \frac{44836}{152.29} \text{ W}$$
$$= 294.41 \text{ Nm}$$

(d) $\eta = \dfrac{P_{sh}}{P_i} = \dfrac{44836}{50000} = 89.67\%$

UNSOLVED QUESTIONS

Q.1 An eight-pole, 50 Hz, three-phase, induction motor has an equivalent rotor resistance of 0.07 Ω per phase. If its stalling speed is 630 rpm, how much resistance must be included per phase to obtain maximum torque at starting? Ignore magnetising current.

[B.Tech III, JMI, 2005]

Q.2 The Steinmetz model of a 3 phase, star connected 220 V, 7.5 kW, 60 Hz, 6 pole induction motor referred to stator at standstill condition is shown below. Calculate the torque developed at 2% slip.

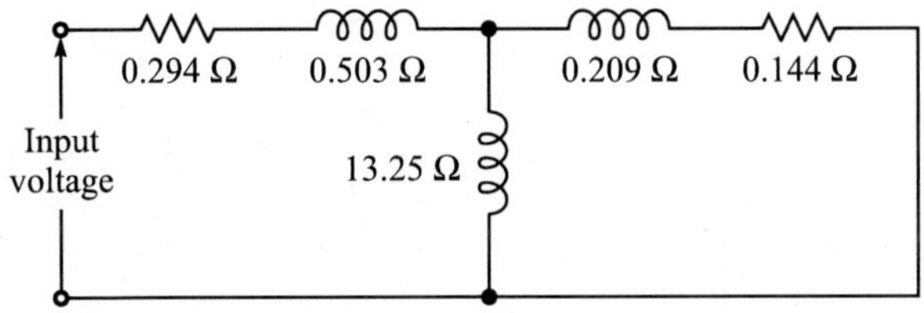

[B.Tech III, JMI, 2005]

Q.3 A three-phase, 50 kW, 1475 rpm, 420 V, 50 Hz, four-pole, star connected induction motor has the following data: $R_s = 0.4\ \Omega$, $R_r = 0.21\ \Omega$, $X_s = 0.95\ \Omega$, $X_r = 0.85\ \Omega$ and $X_m = 32\ \Omega$, all quantities being referred to the stator side. If the frequency is increased to 58 Hz by frequency control. Determine

(a) Slip at maximum torque.
(b) Speed at maximum torque.
(c) Breakdown torque.

[B.Tech IV, JMI, 2006]

Q.4 The power input to the rotor of a 400 V, 50 Hz, six-pole, three-phase induction motor is 80 kW. The rotor emf is observed to make 100 complete alternations per minute. Calculate slip, mechanical power developed and rotor copper loss.

[B.Tech II, UPTU, 2002–03]

Q.5 A three-phase squirrel cage type induction motor, when started by means of Y/Δ starter takes 200% of full load current (line) and develops 44% of full load torque at starting. Calculate the starting torque and current if an auto transformer with 75% tapping was employed.

[B.Tech II, UPTU, 2002–03]

Q.6 A 3 ϕ induction motor is wound for 4 poles and is supplied from 50 Hz system. Calculate

(i) Synchronous speed.

(ii) Rotor speed when slip is 4%, and

(iii) Rotor frequency when rotor runs at 600 rpm.

[B.Tech II, UPTU, 2003–04]

Q.7 A 400 V, four-pole, three-phase, 50 Hz induction motor has a rotor resistance and reactance per phase of 0.01 Ω and 0.1 Ω respectively. Determine:

(i) Maximum torque in N – m and the corresponding slip.

(ii) The full load slip and power output in watts, if maximum torque is twice the full load torque and the ratio of stator to rotor turns is 4.

[B.Tech II, UPTU, 2003–04]

Q.8 The rotor of a four-pole, 50 Hz, slip ring induction motor has a resistance of 0.25 Ω per phase and runs at 1440 rpm at full load. Calculate the external resistance per phase, which must be added to lower the speed to 1200 rpm, the torque being same as before.

[B.Tech II, UPTU, 2003–04]

Q.9 A 5 hp, 230 V, 50 Hz induction motor has a rated full load slip of 950 rpm. The induced voltage per phase of rotor at standstill is 100 V. Calculate

(i) Number of poles and percentage full load slip.

(ii) Rotor induced voltage and its frequency at full load.

[B.Tech II, UPTU, 2004–05]

Q.10 The power input to a 3 phase induction motor is 60 kW. The stator losses total 1 kW. Find the total mechanical power developed and rotor copper losses per phase, if the motor is running with a slip of 3%.

[B.Tech II, UPTU, 2003]

Q.11 A eight-pole, 50 Hz, three-phase induction motor is running at 4% slip when delivering full load torque. It has standstill rotor resistance and reactance of 0.1 Ω and 0.6 Ω per phase respectively. Calculate the speed of the motor if an additional resistance of 0.5 Ω per phase is inserted in the rotor circuit. The full load torque remains constant.

[B.Tech II, UPTU, 2001]

Q.12 A 4 kW, 500 V, three-phase induction motor develop its full current at 955 rpm. The starting torque at normal voltage is 1.5 times the full load torque. What will be the starting torque if the supply voltage is 440 V only?

[B.Tech II, UPTU, 2002]

Q.13 A ten-pole, 50 Hz, SRIM runs at 576 rpm on full load. The rotor resistance per phase to be inserted in the rotor circuit if the speed is to be reduced to 480 rpm for full load torque.

[B.Tech II, I.U., 05]

Q.14 The output of a three-phase, 50 Hz, four-pole IM is 12 h.p. at 1440 rpm. The maximum torque is developed at 1250 rpm. Calculate the starting torque assuming mechanical losses of 1 hp.

[B.Tech II, I.U., 07]

Q.15 The stator impedance and equivalent rotor resistance of a 400 V, three-phase, Y connected IM are (0.05 + j 0.2) Ω and (0.05 + j 0.24) Ω respectively. Neglecting 'I_0'; determine the maximum gross power and the slip at which it occurs.

[B.Tech II, I.U., 06]

Q.16 A six-pole, 50 Hz, three-phase IM running on full load develops a useful torque of 162 Nm and it is observed that the rotor emf makes 90 rpm. Calculate the mechanical BHP of motor.

[B.Tech II, I.U.,06]

REVIEW QUESTIONS

Q.1 With single line diagram, describe power flow in a three-phase induction motor. Also explain why it is advantageous to operate induction motor with a slip as small as possible.

[B.Tech II, UPTU, 2004]

Q.2 Derive equation for the torque developed by an induction motor. Draw a typical torque-slip curve and deduce the condition for maximum torque.

[B.Tech II, UPTU, 2000]

Q.3 What is the effect of introducing resistance in the rotor circuit of an induction motor? In which type of motor it is possible? Under which conditions it is done and why?

[B.Tech II, UPTU, 2000]

Q.4 Discuss the principle of operation of three-phase induction motor.

[B.Tech II, UPTU, 2002]

Q.5 State the effects of increasing rotor resistance on starting current, starting torque, maximum torque and full load slip of an induction motor.

[B.Tech II, UPTU, 2002]

Q.6 Derive an expression for the torque developed in an induction motor and obtain the condition for maximum torque.

[B.Tech II, UPTU, 2001]

Q.7 Draw the equivalent circuit of a three-phase induction motor.

[B.Tech II, UPTU, 2001]

Q.8 Show that a rotating magnetic field is produced in a three-phase winding fed by three-phase supply.

[B.Tech II, UPTU, 2001]

Q.9 Explain in brief, the different methods of speed control of a three-phase induction motor.

[B.Tech II, UPTU, 2001]

Q.10 State the effects of increasing the supply frequency on starting torque, synchronous speed, full load speed and full load current of 3 phase induction motor.

[B.Tech II, UPTU, 2003]

Q.11 Explain the torque/slip characteristic of an induction motor.

[B.Tech II, UPTU, 2003]

Q.12 Describe methods of speed control of three phase induction motors. Also mention advantages, limitations and application areas of each.

[B.Tech II, UPTU, 2004]

Q.13 With single line diagram, describe power flow in a three phase induction motor. Also explain why it is advantageous to operate induction motor with a slip as small as possible.

[B.Tech, UPTU, 2004]

Q.14 Explain the working principle of 3 phase induction motor. The rotor of induction motor cannot run at synchronous speed. Explain, why?

[B.Tech II, UPTU, 2003]

Q.15 Derive torque, mechanical power and rotor output equations of a three phase induction motor connected from A.C. mains.

[B.Tech II, UPTU, 2003]

Q.16 Why are starters needed to start the induction motors? Explain various starters used for cage and wound rotor type induction motors.

[B.Tech II, UPTU, 2003]

Q.17 Describe the methods to control the speed of induction motors.

[B.Tech II, UPTU, 2003]

Q.18 Briefly explain how the rotation of the indiction motor is produced.

[B.Tech II, UPTU, 2001]

Q.19 What is the effect of introducing resistance into the rotor circuit of a slip ring induction motor while it is running on a constant load torque.

[B.Tech II, UPTU, 2001]

Q.20 Develop an expression to show that the torque developed in a 3 ϕ induction motor depends upon the supply voltage, rotor reactance and slip. Also obtain the conditions of maximum torque.

[B.Tech II, UPTU, 2001]

Q.21 Show that the magnetic field produced in the air gap of a three phase induction motor is of rotating nature. Give the expression of this speed in terms of poles and frequency of supply.

[B.Tech II, UPTU, 2002]

Q.22 By means of a power flow diagram, show the flow the power in a 3 phase induction motor from the electrical source to the mechanical load at the motor shaft. Based on above, show that $P_g : P_{cu} : P_m = 1 : s\ (1 - s)$ where,

P_g, P_{cu}, P_m are air gap power, rotor copper loss and mechanical power developed respectively and 's' is the per unit slip.

[B.Tech II , UPTU, 2002]

Q.23 Draw and explain the torque-slip characteristic of a typical three-phase induction motor. Mark the starting torque and maximum torque and explain how the starting torque and maximum torque vary with rotor circuit resistance.

[B.Tech II, UPTU, 2002]

Q.24 Derive from the Thevenin's model of a three-phase induction motor, the expression for electromagnetic torque. Explain with reasons why the air gap of induction motors is kept very low.

[B.Tech III, JMI, 06]

Q.25 Discuss the stator voltage control with constant slip frequency as applied to an induction motor. What are its merits and demerits?

[B.Tech IV, JMI, 06]

Q.26 Show that the ratio of motor torque to maximum torque is $\frac{T}{T_{max}} = \frac{2}{\frac{s_m}{s} + \frac{s}{s_m}}$, where $S_m \rightarrow$ slip at maximum torque.

[B.Tech II, IU, 2006, 2005]

Q.27 Write short notes on.

(a) Rotating magnetic field
(b) Cogging and crawling of induction motor.
(c) V/f constant speed control.

[B.Tech II, I.U., 2006]

Q.28 Explain torque-slip characteristics of IM.

[B.Tech II, I.U., 2007]

Q.29 Derive the expression for maximum torque of induction motor.

[B.Tech II, I.U., 2007]

Q.30 Why is the starter required in three-phase IM? Explain star-delta starter.

[B. Tech II, I.U., 2007]

Q.31 Compare the different starting method of three-phase induction motor.

[B.Tech II, I.U., 2005]

Q.32 With the help of neat diagram, explain the construction of a three-phase SCIM. What are the advantages and disadvantages of this motor over a wound rotor motor?

[B.Tech II, AMU, 2006]

Q.33 Draw torque speed characteristic of an induction machine. Explain how the machine can be operated in three different modes.

[B.Tech II, AMU, 2006]

Q.34 Show that when a three-phase a.c. supply is given to a three-phase balanced stator winding, a synchronously rotating magnetic field is produced.

[B.Tech II, AMU, 1996]

Q.35 Obtain expression for torque, slip for maximum torque, maximum torque, starting torque and mechanical power output of a three-phase induction motor. Also draw and explain torque speed characteristic of the motor and effect of change of rotor winding resistance on the characteristics.

[B.Tech II, AMU, 1996]

Chapter

9

Synchronous Generators

9.1 Introduction
9.2 Constructional Features
9.3 Winding Factor 'K_w'
9.4 EMF Equation
9.5 Armature Reaction
9.6 Phasor Diagram on Load
9.7 Characteristics of Alternators
9.8 EMF or Synchronous Impedance Method
9.9 MMF Method
9.10 Potier's Method
9.11 Capability Curve
9.12 Blondel's Two Reaction Theory
9.13 Parallel Operation of Synchronous Generators
9.14 Hunting
9.15 Transients in Synchronous Machines
9.16 Synchronous Machine Dynamics
• Solved Questions
• Unsolved Questions
• Review Questions

9.1 INTRODUCTION

The electrical energy needs of the countries are fulfilled by the conventional sources mainly such as coal, hydro, nuclear, oil etc. The non-electrical quantities are changed into electrical energy via alternators or synchronous generators. The synchronous generators of large sizes are mainly three-phase a.c. machines. They are normally rated at 11 kV and 33 kV in India. The older installation uses 11 kV only. The output is sinusoidal quantities so they are known at alternators. Due to the merits of three-phase systems over single phase systems it is popular. The alternators are mainly three phase machines. The alternators size depends on its output power and on the basis of poles of machines.

The generators are available in giant sizes of order of thousands of megawatts. But in India the largest size of a.c. generator operated till now is 500 MW. That too is located in the super thermal power plants. The alternators are mainly located in the generating stations irrespective of the type of prime movers used. The mechanical power may be provided by the non-electrical quantities such as water, gas, steam, diesel or oil. The oil is the cleanest source of generating electrical energy. But as the oil prices are high so they are not preferred in India. Most of the alternators prime mover use coal as source of energy. Then the hydropower plant is the next large source of electrical energy. The generators which are used in the steam power plants are called steam turbine operated generators or turbo generators. Whereas, the generators used in the hydropower plants is called hydro generators.

9.2 CONSTRUCTIONAL FEATURES

The constructional features of the synchronous generators differ depending on the type of field. That is, if the field is stationary then it is called stationary field synchronous generator, whereas if the field is rotating then it is called rotating field synchronous generator. The stationary field alternators are for small sizes less than 5 kVA and the rotating field alternators are preferred for higher ratings greater than 5 kVA upto thousands of kVA ratings. The complexity increases as the size of generator increasers. But in general, the a.c. generators (synchronous type) looks as depicted in Fig. (9.1) as shown below.

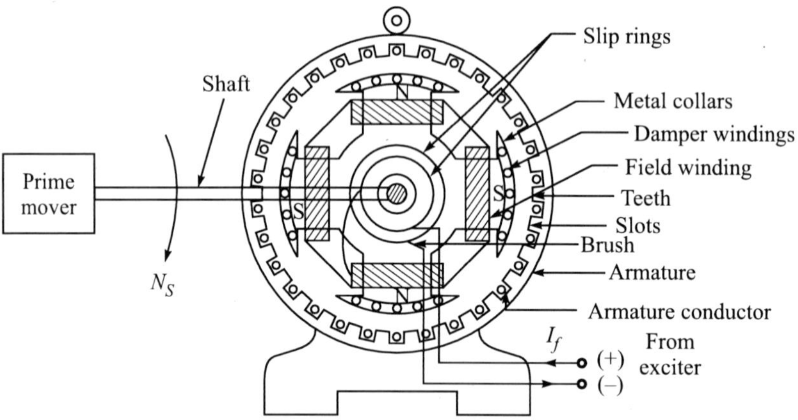

Fig. 9.1 Four-pole synchronous generator (Rotating field type).

The rotating field is popular as it is cheaper, safer and more reliable for higher power ratings. But then too the stationary field alternators are used as rotating converters from which a.c. to d.c. and vice versa can be obtained. The Fig. (9.2) depicts the stationary field alternator.

By this type rotatory converters both a.c. and d.c. can be obtained simultaneously. But nowadays this type of rotatory converters are not used. Now due to

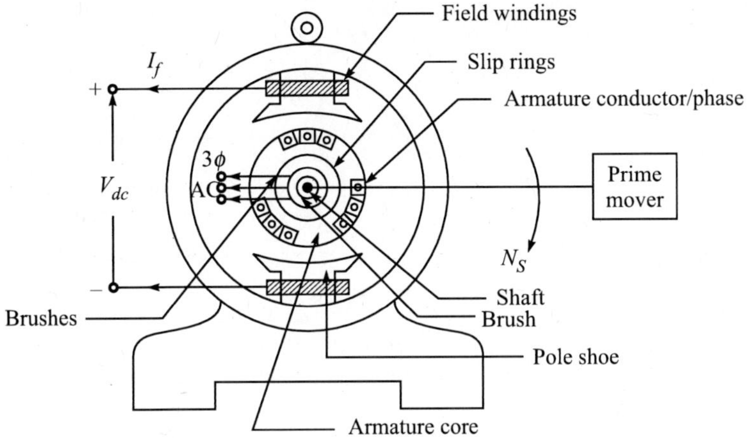

Fig. 9.2 Synchronous generator (Stationary field type).

the low cost and high efficiency of the power devices operated converters and inverters are preferred.

A stationary field synchronous generator has the appearance d.c. generator. But the difference is that the three-phase winding whose terminals are connected to three slip rings mounted on the shaft of machine. A set of three brushes sliding on the slip rings enables the armature to be connected to an external three-phase load. The armature is driven by petrol engine. The frequency of voltage depends on the speed and poles of the field core. In the rotating field type alternators the armature windings are placed in the armature which is stationary so the current collection is easier as compared to rotating armature type alternators. The three-phase stationary windings are directly connected to the load (three-phase type). So the slip rings used in the rotating field type machines is less and is equal to two. The purpose of the slip rings is to collect the d.c. current from external d.c. voltage source. The field winding is usually excited by a small d.c. generator mounted on the same shaft and is called exciter. The purpose is to create poles on the rotor. The brushes are used for current collection from external d.c. source. The Fig. (9.1) is having poles which is protruded out of the rotor core. So the poles are visible or salient. So such type of a.c. generator is called salient pole synchronous generator. The other type of generator is having poles which are not salient, they are called non-salient or cylindrical pole machines as shown in the Fig. (9.3).

The number of poles on a synchronous generator depends upon the speed of prime mover and hence the speed of rotor of machines and the frequency required. The frequency 'f' of voltage generated is given by

$$n = \frac{2f}{P} \tag{9.1}$$

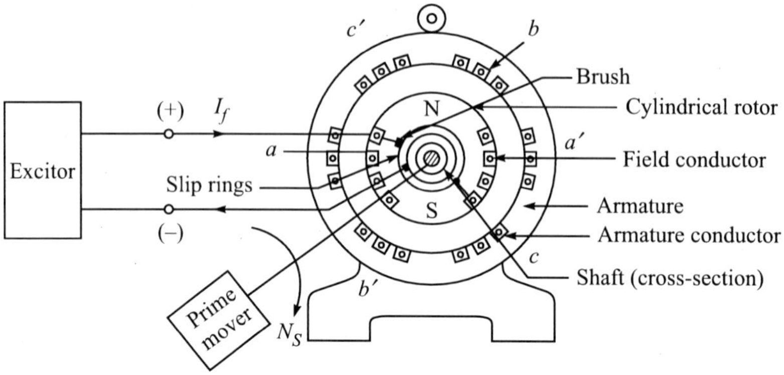

Fig. 9.3 Cylindrical rotor synchronous machines (two pole).

or, $$f = \frac{Pn}{2}$$

where, 'n' is in rps, 'P' is number of poles of machine.

In r.p.m.,

$$f = \frac{p \cdot N}{120} \tag{9.2}$$

where, 'N' is in rpm.

So if a machine shaft rotates at 1500 rpm by prime mover and the induced voltage is having a frequency of 50 Hz, the machines have '4' poles or '2' pole pairs that is two north and two south poles respectively as shown in Fig. (9.1) respectively. The schematic diagram and cross-sectional view of a modern high rating generator is shown below.

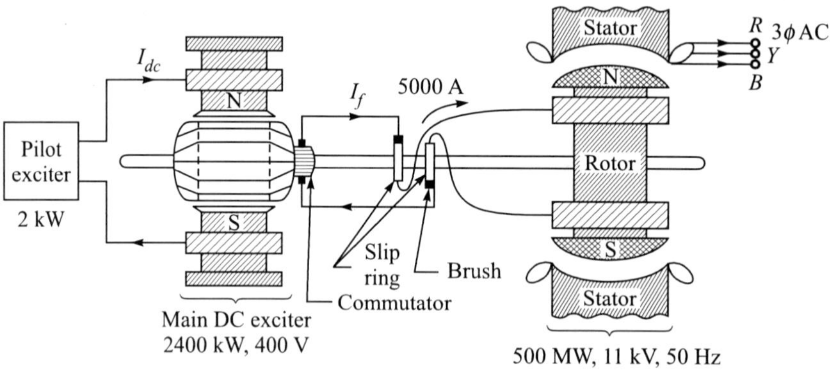

Fig. 9.4 500 MW synchronous generator.

The pilot excitor permits variable field control of the main d.c. excitor which in turns controls the field current.

The stator is having three-phase armature windings as that of three-phase induction motor if consists of cylindrical laminated core containing slots and teeth alternately carrying three-phase lap or wave windings, generally lap winding is used for high current machine. The armature windings are always connected in Y and the neutral is connected to ground. This is preferred to Δ connection as the voltage per phase is $\frac{1}{\sqrt{3}}$ or 58% of line voltage. So the insulation requirement will be less so the conductor diameter could be increased which corresponds to larger current and hence the power output of the machine. The line voltages remain sinusoidal under all load conditions as the harmonics (third) gets cancelled with each other. In the case of Δ the harmonics remains present so the $I^2 R$ losses increase. The voltage rating of alternators are limited due to higher insulation requirement. So the current carrying armature conductor thickness is limited as a result of it.

The rotor is smooth for cylindrical pole machine and projected type for salient pole machine. The salient pole machine employs low speed hydraulic turbines whereas the cylindrical rotor machine employs high speed steam turbines. The low speed rotors have larger diameter to accommodate large number of poles. The salient poles are mounted on a circular steel frame which is fixed to a revolving vertical shaft. For large rating machines the field coils are replaced by bars made up of copper insulated from each other by mica strips. The north pole winding is connected to south pole windings in series.

In addition to the d.c. field winding, a squirrel cage winding called damper windings are placed in the rotor pole faces to damp the oscillations due to hunting. The damper winding maintains balanced three-phase voltage between the lines even when the load currents are unbalanced due to unequal loadings of lines.

The cylindrical rotors are used with the well known high speed steam turbines which are efficient than the low speed turbines. Generally, smaller number of poles are used, namely 2 or 4 which corresponds to speeds of 3000 rpm and 1500 rpm respectively. The rotor length is long made up of steel. Concentric field coils are placed in the slots of steel cylinder firmly wedged in the slots and retained by high strength end rings serves to create north and south poles respectively. The speed limits the diameter of rotor. The diameter of rotor is maximum upto one metre.

The exciters are d.c. generators which provide d.c. is small and medium alternators to the field winding of the synchronous generator. There is a pilot excitor and main excitor are for large alternators used. The voltage rating of exciter is between (10 V – 400 V) respectively for small alternators. The power rating of exciter increases with increase of rating of generator.

The exciter maintains the a.c. voltage across a.c. generator terminals constant by increasing the excitation to the field winding. The response time to varying load condition is less than one second.

Brushless excitation is employed for modern a.c. generators which has additional advantages over the conventional generators as it requires no brush, slip rings, commutators, etc.

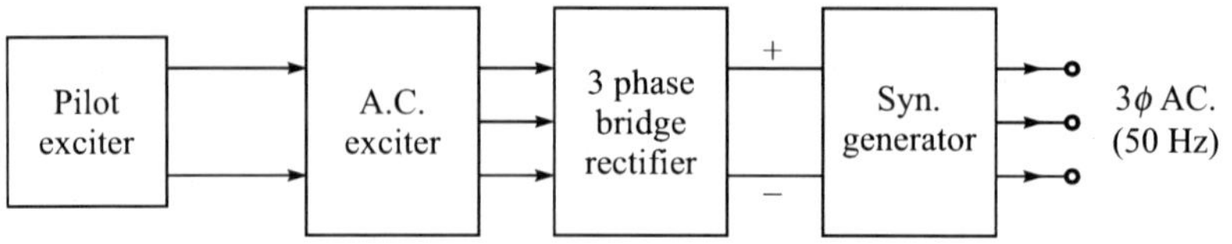

Fig. 9.5 Block diagram of brushless excitation system.

The rectifier and a.c. exciter (3 ϕ generator) is mounted on the main shaft of machine. The frequency of a.c. exciter is 2-3 times the generated frequency (50 Hz). The frequency of supply is increased by increasing the poles of a.c. exciter. The efficiency of generator increases with the size of generator.

9.3 WINDING FACTOR 'K_W'

The rotor of the synchronous generator rotates at a speed n_s (rps) given by $n_s = \frac{2f}{P}$, where 'f' is the frequency of rotation and 'P' is the number of poles of machine if $P = 2$ then $n_s = f$ (rps) or Hertz. Consider the Fig. (9.6).

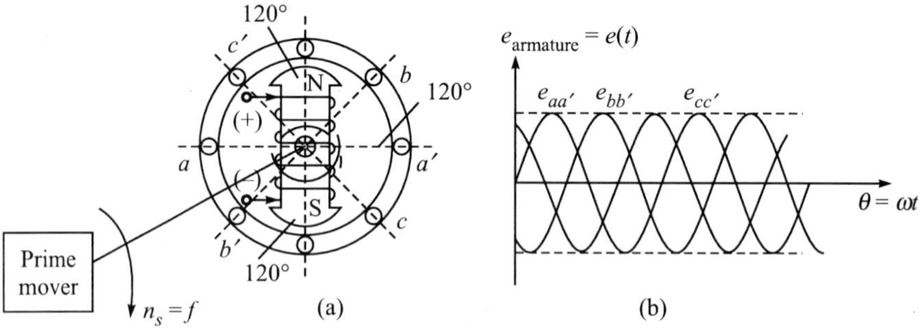

Fig. 9.6 Emf generated in synchronous generators.

If a d.c. is fed to field winding of the synchronous generator and is rotated at a speed of 'n_s' (synchronous speed), then it establishes a sinusoidal distributed flux in the air gap. The rotor prime mover may be a turbine, diesel engine or d.c. or induction motor. The rotating flux so produced will change the flux linkages of armature windings aa', bb' and cc' so an induced voltage $e_{aa'}$, $e_{bb'}$ and $e_{cc'}$ is generated. These induced emf's have equal magnitudes but are shifted by 120° (elec.). The rotor speed is fixed i.e., n_s and the frequency 'f' of the generated voltage is equal to 'n_s' for two-pole machines. So the frequency of the generated voltage is also fixed and is equal to 50 Hz in India.

The r.m.s. value of the induced emf is given by

$$E = 4.44\, f \phi_f \cdot N_t \cdot K_w \qquad (9.3)$$

where, ϕ_f = Flux per pole due to 'I_f'

N_t = Number of turns in each phase
K_w = Winding factor

$\because$ $$f = \frac{NP}{120}$$

So, $$E \propto N \cdot \phi_f \tag{9.4}$$

E = r.m.s. value of emf generated, volts

Since, $$\phi_f \propto I_f$$

So, the induced or generated emf 'E' depends on the speed 'N' of prime mover and 'ϕ_f' or 'I_f' in field winding.

Induced voltage 'E_r' at $I_f = 0$ is due to 'ϕ_r' called residual flux which is always present in the machine. As we increase 'I_f' the flux 'ϕ_f' increases but after some value of 'I_f' the flux 'ϕ_f' gets constant due to saturation of the magnetic circuit and therefore 'E' is fixed or gets constant.

For controlling the active power generated by the alternators, governors are used to maintain the speed of the rotor constant equal to synchronous speed. The higher the speed the greater will be the active power 'P' of the generator which is mentioned on the name plate rating of machine. The increase of field flux 'ϕ_f' increases the reactive power of the alternator and so the power factor of the machine decreases and also the system efficiency decreases accordingly.

The generated voltage depends on the winding factor 'K_w' which is defined mathematically as product of 'K_p' (pitch factor) and K_d (distribution factor) as follows:

$$\boxed{K_w = K_p \cdot K_d} \tag{9.5}$$

The value of 'K_w' is less than unity in the practical alternators.

For full pitched coil, $K_p = 1$

For concentrated winding, $K_d = 1$

So we can find the values of 'K_p' by considering the chording or short pitched coil in which the two coil sides of the coil (armature) are less than 'π' radians shown dotted in Fig. (9.7). Since the coil pitch is equal to 'π' radians for full pitched coil. If the induced emf in the coil sides of armature coil is equal to 'e', then its r.m.s. value is equal to 'E'. Consider a single turn armature coil as shown in Fig. (9.7).

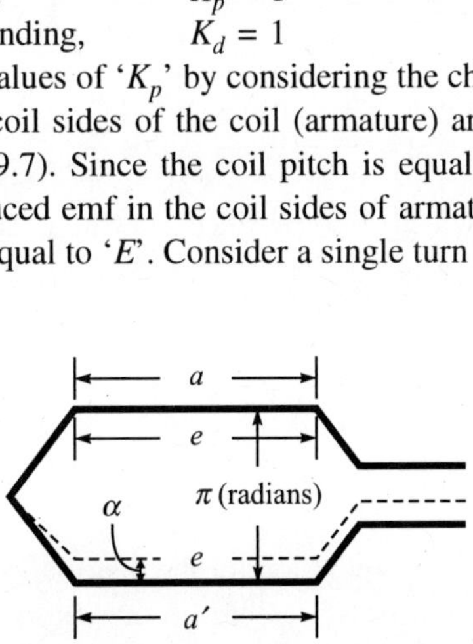

Fig. 9.7 Single turn armature coil.

The phase diagram of the induced voltages for full pitched and chorded coil is given as follows:

Fig. 9.8(a) For full pitched coil (EMF phasors).

If the chorded or short pitched coil by an angle 'α' radians, then the phasor is represented by Fig. (9.8)(b) shown below.

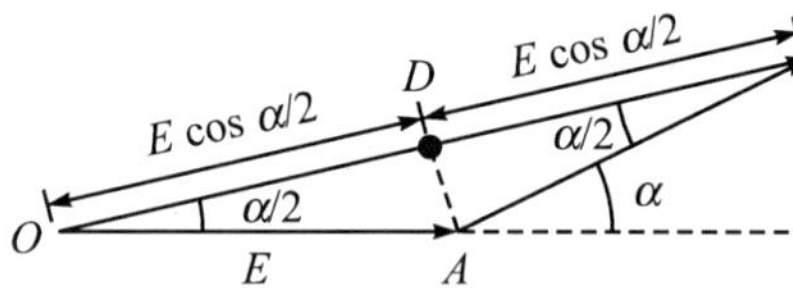

Fig. 9.8(b) For chorded or short pitched coil (Phasors or emf).

The resultant phasor magnitude is given by

$$\boxed{OB = 2\,E \cos \frac{\alpha}{2}} \qquad (9.6)$$

where, α = Chording angle, radians

E = r.m.s. value of induced voltage 'e' in the coil sides of armature coil, volts

$|\overline{OB}|$ = Phasor sum of voltages induced.

The chording or pitch factor (K_p) is given by the formula as written below.

$$K_p = \frac{\text{Phasor sum of induced voltages}}{\text{Arithmetic sum of induced voltages}}$$

So,
$$K_p = \frac{2\,E \cos \dfrac{\alpha}{2}}{2\,E} = \cos \frac{\alpha}{2}$$

Thus, the pitch factor which is also called chording factor is given by the cosine of the chording angle $\frac{\alpha}{2}$ as given below.

$$\boxed{K_p = \cos \frac{\alpha}{2}} \qquad (9.7)$$

For full pitched coil $\alpha = 0$ so, $K_p = 1$

For chorded or short pitched coil $\alpha \neq 0$ so, $K_p < 1$

Similarly, we can derive the expression for the distribution factor of the armature conductors in the slots of armature core of alternators. Consider the distribution of the armature conductors as shown in Fig. (9.9).

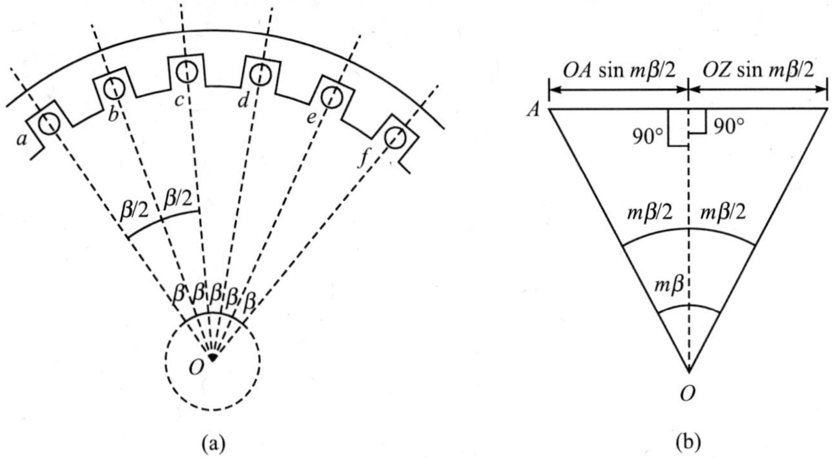

Fig. 9.9 Armature conductor placed at a slot angle 'β'.

where, m = Total number of slots in the armature

β = Slot angle, radians

$OZ = OA$ (since the distance of conductor from centre of shafts is equal)

For determining the arithmetic sum of the voltage induced in the armature conductor. Consider Fig. (9.10).

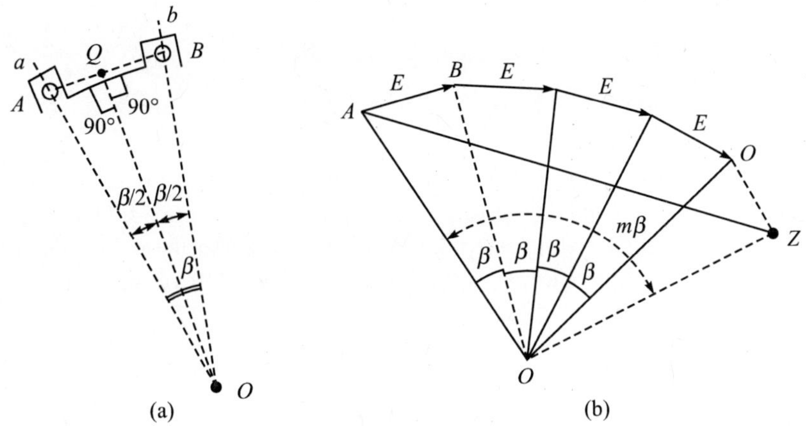

Fig. 9.10 Conductors in two slots of armature and phasors of induced voltages in it.

$$AQ = OA \sin \frac{\beta}{2} = QB$$

So,
$$AB = 2\ OA \sin \frac{\beta}{2}$$

Thus, the emf induced (r.m.s value) in a conductor is given by

$$E = 2\ OA \sin \frac{\beta}{2}$$

So, as there are 'm' conductors corresponding to the armature slots. So the total e.m.f. generated is given by

$$E_t = 2\ m \cdot OA \cdot \sin \frac{\beta}{2} \tag{9.8}$$

where, 'E_t' corresponds to the arithmetic sum of the induced voltages in the armature conductors.

The phasor sum of voltage induced is given by 'AZ' which is equal to '$2\ OA \sin\left(\frac{m\beta}{2}\right)$ from Fig. (9.10).

The distribution factor of the armature winding is given by the ratio of the phase sum to the arithmetic sum of the voltages as given by the expression (9.8).

$$K_d = \frac{\text{Phasor sum of voltages induced}}{\text{Arithmetic sum of voltages induced}}$$

$$K_d = \frac{\cancel{2}\,\cancel{OA} \sin\left(\frac{m\beta}{2}\right)}{\cancel{2}\, m \cdot \cancel{OA} \sin \frac{\beta}{2}}$$

$$\boxed{K_d = \frac{\sin m \frac{\beta}{2}}{m \sin \frac{\beta}{2}}} \tag{9.9}$$

For concentrated winding, $K = 1$

For distributed winding, $K_d < 1$

As harmonics are also present in the induced voltages due to the non-linearity of the magnetic material and due to various other parameters. So we should take into account the harmonics while determining the distribution factor and the pitch factor of the coil. With the harmonics, the exps. (9.7) and (9.9) get modified as follows.

For n^{th} harmonics,

The chording factor is given by

$$K_{c_n} = \cos \frac{\alpha n}{2} \tag{9.10}$$

The distribution factor is given by

$$K_{d_n} = \frac{\sin\left(m \frac{n\beta}{2}\right)}{m \sin\left(\frac{n\beta}{2}\right)} \tag{9.11}$$

In order to eliminate a particular harmonic component, the expressions are equated to zero such as to eliminate third harmonic component, we have $n = 3$

So,
$$K_{c3} = \cos\frac{\alpha n}{2} = 0$$

$$\frac{n\alpha}{2} = \frac{\pi}{2}$$

or,
$$\alpha = \frac{\pi}{n}$$

∵
$$n = 3,\ \alpha = \frac{\pi}{3}$$

So, by chording the coil by 60° the 3rd harmonic components can be fully eliminated. Similarly, the other harmonic components could be removed from the generated voltages. The generated voltage could be made sinusoidal thereby.

Similarly, the harmonic effects could be minimized by selecting proper slot angle 'β' for the machines. For eliminating the 3rd harmonics

$$n = 3$$

$$K_{d_3} = \frac{\sin\frac{3m\beta}{2}}{m\sin\frac{3\beta}{2}}$$

To eliminate 3rd harmonics, $K_{d_3} = 0$

So,
$$\frac{\sin\frac{3m\beta}{2}}{m\sin\frac{3\beta}{2}} = 0$$

or,
$$\sin\frac{3m\beta}{2} = \sin\pi\cdot n$$

or,
$$\frac{3m\beta}{2} = \pi\cdot n,\quad n = 1, 2, 3\ldots$$

or,
$$\beta = \frac{2\pi n}{3m}$$

If there are 30 slots consider $n = 1$

$$\beta = \frac{2\times\pi\times 1}{3\times 30} = \frac{2\times 180^2}{90} = 4°$$

Thus, if the slot angle is kept '4°' then the 3rd harmonics can be totally eliminated from the generated voltage of an alternator.

9.4 EMF EQUATION

Consider an armature conductor 'a' per phase of a three phase alternator which is cut by a flux 'ϕ' at a rate of 'N_s' r.p.m. or 'n_s' r.p.s.

Since, $$n_s = \frac{2f}{P}$$

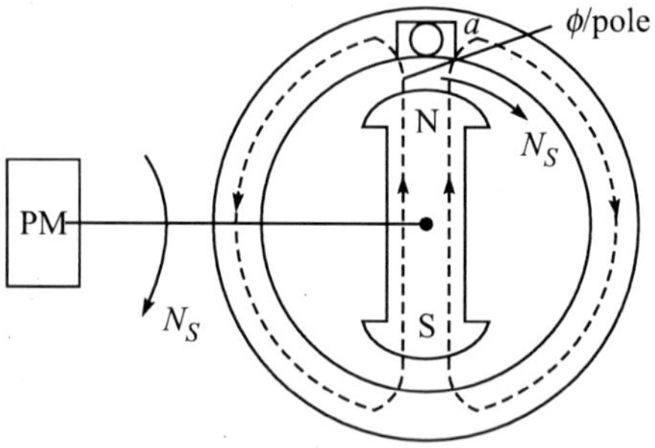

Fig. 9.11 Alternator (flux path).

For two pole machine, as shown

$$P = 2,$$

So, $$n_s = f, \text{ Hz or rps.}$$

Since the rotor of the synchronous generator rotates at synchronous speed only and so, the frequency of the generated voltage is also fixed and is equal to 'f'. In India, this frequency is fixed and is equal to 50 Hz which is called the synchronous frequency.

The flux linked with the armature conductor changes as the rotor moves. So according to Faraday's law of electromagnetic induction the emf is induced in the conductors or coil sides. The emf's in series get added to give a large value of generated emf across the armature terminals.

$$e = -N\frac{d\phi}{dt} \tag{9.12}$$

$$N = \text{no. of armature turns}$$

Let $$\phi = \phi_m \sin \omega t \text{ (say)},$$

as both the north pole and south pole fluxes are being cut by the armature conductor 'a' per revolution. So the flux wave is given by

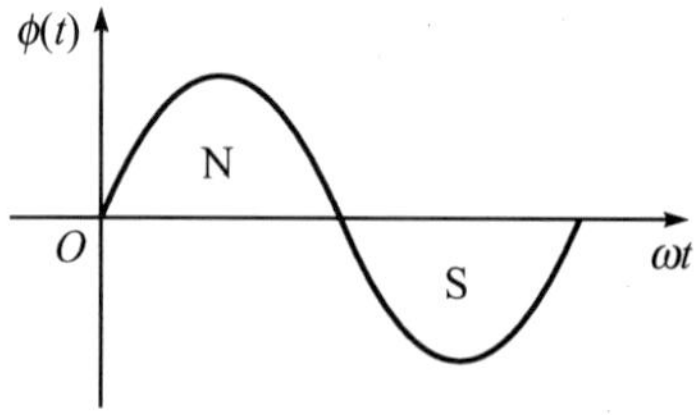

Fig. 9.12 Flux wave under field poles of alternator in a cycle of rotation.

where, $$\omega = 2\pi\, n_s = 2\pi \cdot \frac{2f}{P} = \frac{4\pi f}{P} \tag{9.13}$$

So, substitute the value of 'ω' in the exp. (9.12)

We get, $$e = -N\frac{d}{dt}\left[\phi_m \sin\left(\frac{4\pi f}{P} t\right)\right]$$

$$e = -N\,\phi_m \cdot \frac{4\pi f}{P} \sin\left(\frac{4\pi f}{P}\right)t$$

The maximum value of emf is given by

$$E_m = N\,\phi_m \cdot \frac{4\pi f}{P}$$

So, the r.m.s. value is given by

$$E_{\text{rms}} = E = \frac{E_m}{\sqrt{2}} = \frac{4\pi f}{\sqrt{2}} \cdot \frac{N\phi_m}{P}$$

$$E = 2\sqrt{2} \times \pi \frac{f \cdot N\phi_m}{P} \tag{9.14}$$

For two pole alternator, $P = 2$

So, $$\boxed{E = \sqrt{2}\,\pi f N \phi_m = 4.44 f \cdot N \cdot \phi_m} \tag{9.15}$$

where, f = 50 Hz
N = Number of stator or armature turns
ϕ_m = Maximum value of flux, weber.

The emf equation of the synchronous generator if we consider K_p and K_d equal to unity is given by exp. (9.16). For practical alternators, we have to consider another factor called winding factor 'K_w' in the emf equation of the alternator as given below

$$\boxed{E = 4.44\, K_w \cdot f \cdot N \cdot \phi_m} \tag{9.16}$$

where, $K_w < 1;\ K_w = K_p \cdot K_d$

For good designed alternators 'K_w' is greater than 0.95.

9.5 ARMATURE REACTION

The armature reaction (AR) may be defined as the effect of the armature flux on the main field flux of the machine. So the flux distribution is the main parameter responsible for the electromechanical energy conversion. So as the flux distribution in the air gap between the stator and rotor gets changed the e.m.f. also changed accordingly.

The armature current in the case of alternators flows only when a load is attached to the armature phase windings terminals. So as the load type changes the phase angle between the induced voltage generated 'E' and the load current

'I' also changes accordingly. So the flux distribution changes accordingly. Similarly, when the load current 'I' increases the armature reaction increases.

Consider the armature reaction for different type of loads one by one so as to predict the space behaviour of flux wave which is the sum of the field flux 'ϕ_f' and armature flux 'ϕ_a' respectively, given by $\phi = \phi_f + \phi_a$ where, 'ϕ' is the resultant flux in the air gap, Wb. At same speed 'N_s', the phasor diagram of the mmf's corresponding to field and armature are given by

$$\mathcal{F}_f = N_f \cdot I_f = \phi_f \mathcal{R} \tag{9.17}$$

$$\mathcal{F}_a = N_a \cdot I_a = \phi_a \mathcal{R} \tag{9.18}$$

where, $\mathcal{F}_f$ = mmf of field, AT
N_f = Number of turns of field windings
I_f = Field current, amperes
ϕ_f = Field flux, Wb
$\mathcal{R}$ = Reluctance of magnetic circuit of alternator flux 'ϕ_f' and 'ϕ_a'
N_a = Armature field winding turns
I_a = Armature current, amperes
ϕ_a = Armature flux, weber

For inductive load ($R - L$ type), the phasor diagram is given by

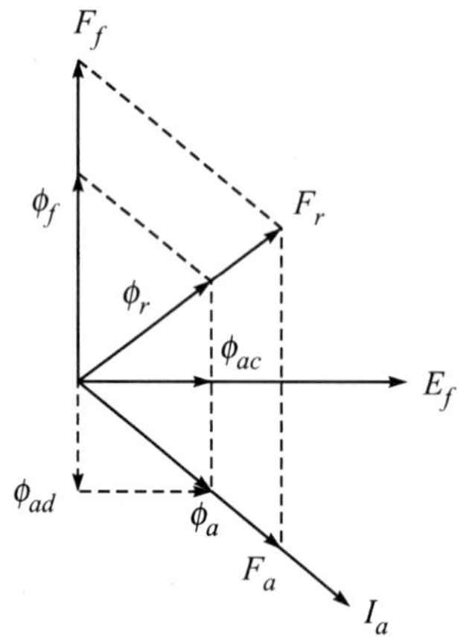

Fig. 9.13 Phasor diagram showing the effects of armature reaction.

where, E_f = emf produced due to field, volts
ϕ_r = Resultant flux, Wb
ϕ_{ad} = Demagnetizing effect of armature reaction, Wb
ϕ_{ac} = Cross magnetising effect of armature reaction, Wb

The armature flux 'ϕ_a' can be resolved as 'ϕ_{ad}' which is opposing the main field flux 'ϕ_f' so it is called demagnetising flux of armature reaction. Other component is given by 'ϕ_{ac}' is called the cross magnetizing flux of armature reaction. The armature flux 'ϕ_a' is expressed as

$$\phi_a = \phi_{ad} + \phi_{ac} \tag{9.19}$$

The flux 'ϕ_{ad}' results in decreasing the magnitude of field flux inside the air gap produced by the field poles and so it is a demagnetizing effect of armature reaction.

The flux 'ϕ_{ac}' causes distortion of the air gap field flux so it is called the distorting or cross magnetizing effect of armature reaction. In the time domain, it could be represented as

The flux 'ϕ'_f' represents the resultant flux due to demagnetizing effect of AR, (without considering 'ϕ_{ac}' effect) we see from phasor diagram that the resultant flux 'ϕ_r' is decreased due to 'ϕ_{ad}' as shown by 'ϕ_r' in Fig. (9.14).

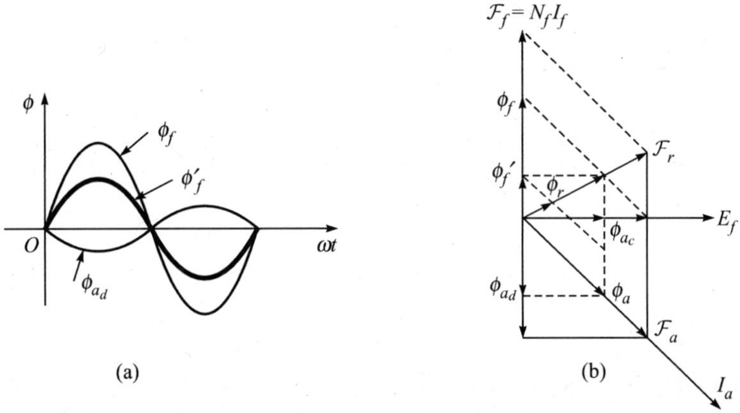

Fig. 9.14 Demagnetizing effect of AR.

For capacitive load, the current 'I_a' leads the voltage as shown in Fig. (9.15).

If the armature current 'I_a' leads the voltage by some angle 'θ'. Since $\phi_a \propto I_a$ so, the phase of 'ϕ_a' and 'I_a' remains same as shown. If we resolve the compo-

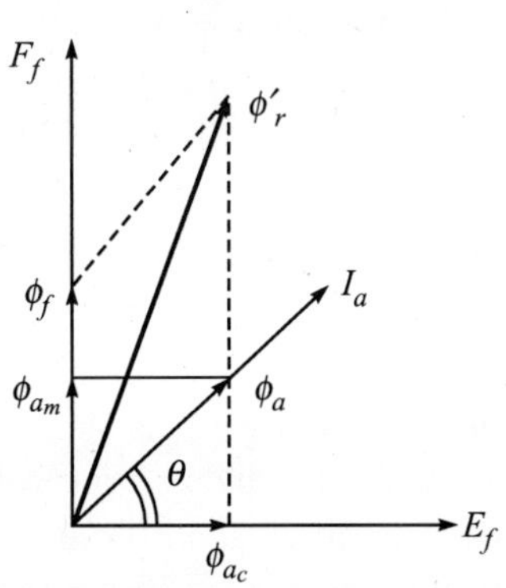

Fig. 9.15 Phasor diagram for capacitive load.

nent of the armature flux 'ϕ_a' into two components 'ϕ_{am}' and 'ϕ_{ac}'. We see that the resultant flux inside the air gap is increased due to the magnetizing effect of armature reaction. The resultant flux 'ϕ_r' is shown in Fig. (9.15) respectively.

Since pure resistive load is not used for practical applications. So the phasor diagram of it is not considered. Practically the load is $R - L$ type or $R - C$ type of $R - L - C$ type respectively. Mostly the nature of load is inductive.

9.6 PHASOR DIAGRAM ON LOAD

The behaviour of the alternator depends on the load type, the alternator need to supply.

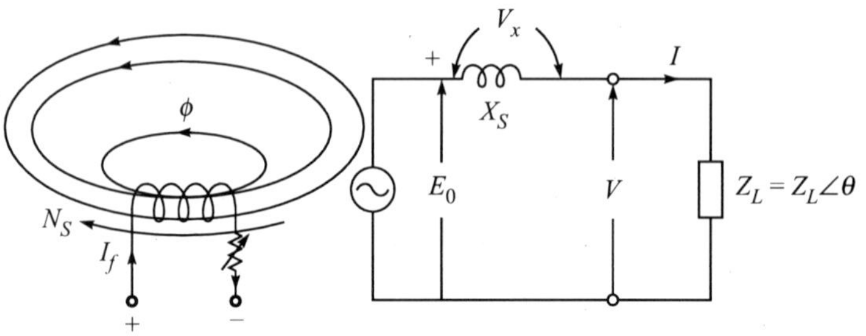

Fig. 9.16 Equivalent circuit of alternator under load.

The above Fig. (9.16) represents one phase of alternator which is shown supplying a load having impedance 'Z_L' having phase angle of θ. In the Fig. (9.16) shown above,

where, E_0 = Open circuit voltage, volts
X_s = Synchronous impedance, Ω
V_x = Voltage drop across 'X_s', volts
V = Load voltage or terminal voltage, volts
I = Load or phase current, amperes

The phase windings of the alternators are wye (Y) connected as it is mostly used. For inductive load 'Z_L', the current 'I' lags voltage 'V' by an angle 'θ'. The power factor of load is given by 'cos θ' respectively. The voltage drop across 'X_s' is given by

$$V_x = j\,I\,X_s \qquad [\because\ j = \angle +90°]$$

$$V_x = V_x\,\angle{+90°}$$

So 'V_x' leads 'I' by + 90° (anticlockwise direction)

The generated voltage 'E_0' is equal to '$V + V_x$' and is generated by rotating flux 'ϕ' produced by field winding. Since, 'E_0' and 'V_x' are voltages that exist inside the synchronous generator windings and cannot be measured directly. So the phasor diagram of the alternator under load is given by

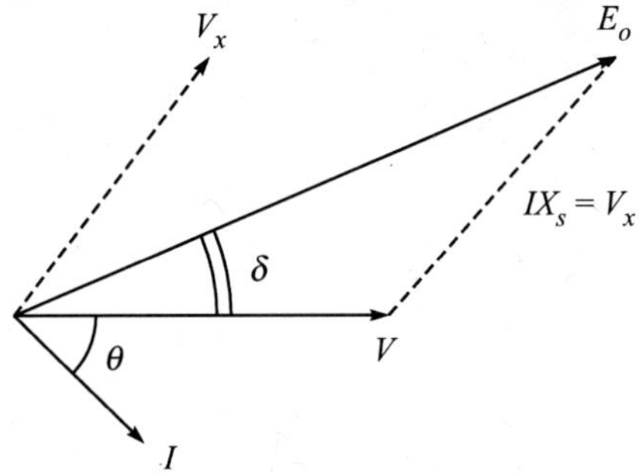

Fig. 9.17 Phasor diagram for a lagging power factor load. (neglecting 'R_a').

The voltage 'E_0' leads 'V' by an angle 'δ'. For lagging load, the open circuit voltage of generator 'E_0' is greater than 'V'. So with the loading of generator with lagging power factor load, the voltage at the terminals of generator drops. So the voltage regulation is always positive.

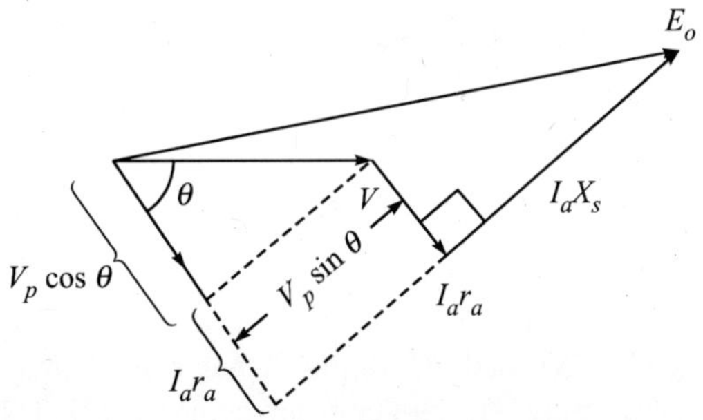

Fig. 9.18 Synchronous generator phasor diagram (with 'R_a').

For Capacitive Loads

The current 'I_a' leads the voltage 'V' by some angle 'θ' (say) then the phasor diagram of alternator is given by

From Fig. (9.19), we show that the terminal voltage 'V' is greater than the generated voltage 'E_0'. The inductive reactance 'X_s' resonates with capacitive reactance of load thus this does not means greater power output.

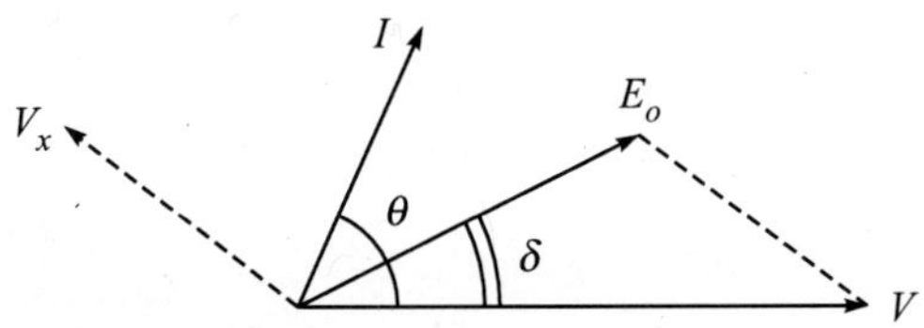

Fig. 9.19 Phasor diagram for a leading power factor load (capacitive) (neglecting 'R_a').

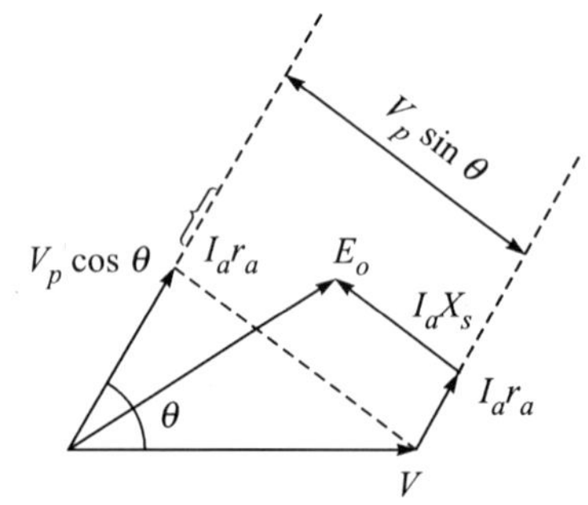

Fig. 9.20 Phasor diagram (for capacitive loads) of an alternator with armature resistance 'r_a'.

9.7 CHARACTERISTICS OF ALTERNATORS

The characteristics of alternators are important for determining the performance parameter of the machine such as voltage regulation as in the case of d.c. machines the important characteristics of alternators are given by:

(i) Open circuit characteristics (O.C.C.)
(ii) Short circuit characteristic (S.C.C.)
(iii) Zero power factor characteristic (ZPFC)
(iv) Regulation characteristic or load characteristic.

Open Circuit Characteristic

The O.C.C. is also called no load saturation characteristic. While plotting the O.C.C. the speed of prime mover is kept constant equal to 'N_s'. The terminals are kept open circuit as shown in Fig. (9.21).

For different values of field current 'I_f', we will have different values of open circuit voltage across the phase winding 'E_{op}'as shown in the Fig. (9.21).

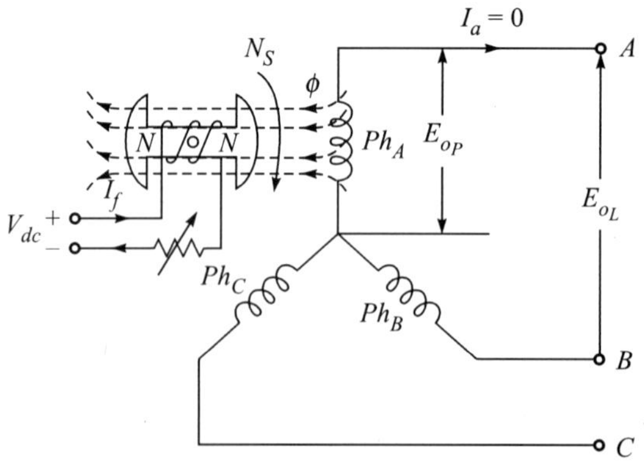

Fig. 9.21 O.C voltage of alternator.

The plot between 'E_{op}' and 'I_f' is called open circuit characteristic and is given by Fig. (9.22).

After saturation is reached, with large increase of current the voltage rise is small.

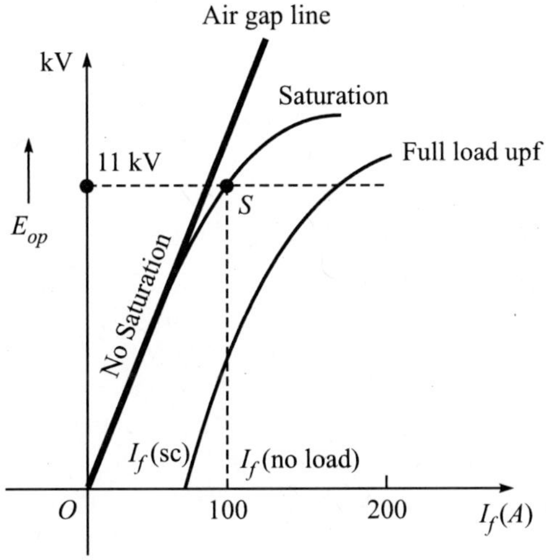

Fig. 9.22 OCC of a three-phase alternator.

The OCC characteristic represents the *B*/*H* curve of the magnetic material of the alternator.

As $\quad E_{op} \propto \phi \propto B \quad$ [Since, $E_{op} \propto \phi N_s$

and $\quad I_f \propto H \quad\quad N_f I_f = \mathcal{F}_f = H \cdot l$]

So, the OCC characteristic has similar shape as that of *B*/*H* curve of the magnetic material of the core of alternator. Higher saturation value is preferred.

Along the air gap line the voltage increases if no saturation of magnetic material takes place.

I_f (sc) is measured under unsaturated condition at extremely low power factor and is the s.c current in the armature winding. Whereas I_f (no load) corresponds to the value of 'I_f' corresponds to the rated voltage of the alternator at no load or open circuit. It is seen that OCC starts not from the origin but from the above due to residual flux in the alternator.

Short Circuit Characteristic (SCC)

By shorting the three armature windings of the synchronous generators. By keeping the speed of prime mover constant equal to N_s (synchronous speed). By varying field current 'I_f' and measuring the armature winding current 'I_a'.

The plot of 'I_a' versus 'I_f' is called short circuit characteristic (SCC) of the alternator.

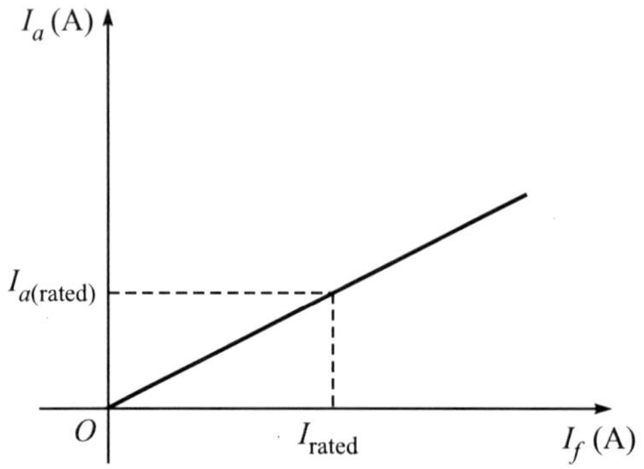

Fig. 9.23 SCC of an alternator.

The SCC is a straight line as shown above. This is due to the fact that under short circuit conditions the magnetic circuit does not saturate because the air gap flux remains at a low level.

The short circuit test is performed by connecting the armature windings as shown in Fig. (9.24).

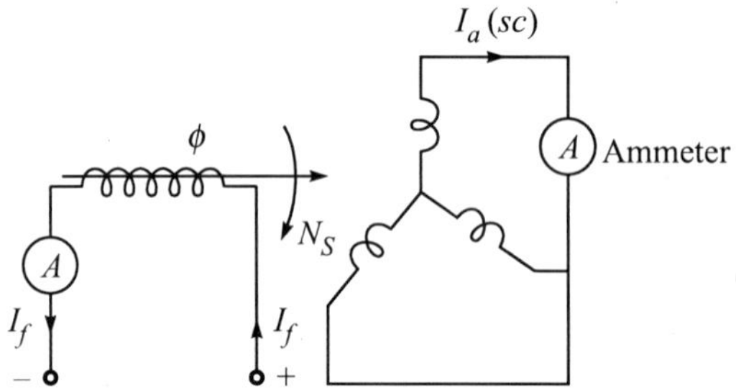

Fig. 9.24 S.C. Test on alternator.

Zero Power Factor Characteristic

The characteristic of the synchronous generator drawn by operating the alternator at zero power factor lagging such that by varying the field current the voltage varies accordingly and when the field windings draw short circuit current 'I_{fsc}' such that the voltage across the phase windings are equal to zero. Since for lagging power factor 'E_0' is equal to zero due to total demagnetising effect of armature reaction. But as the field flux is increased the voltage rises and gets constant after the field flux rises above a certain value which leads to saturation of the pole shoes.

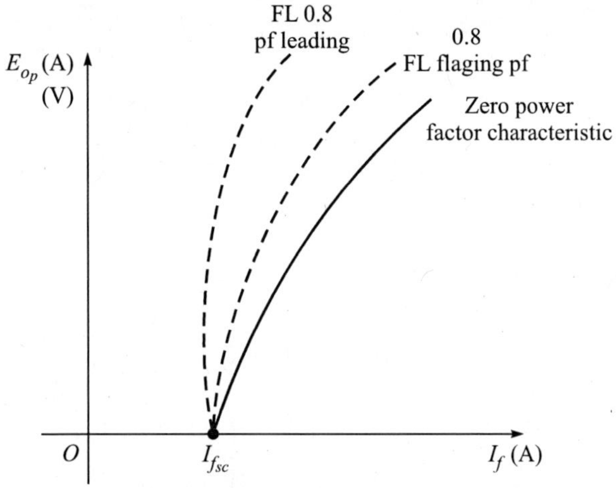

Fig. 9.25 ZPFC of alternator.

Regulation Characteristic

The regulation characteristics are very important to see the variation of the terminal voltage of the alternator with the increase of load. For different power factors the load characteristics of the alternator are shown as given below.

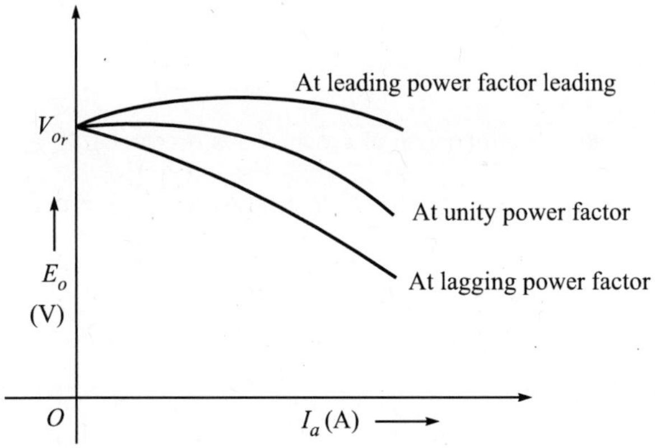

Fig. 9.26 Regulation characteristic of alternator (at constant I_f) and at no load.

The starting point of the characteristic for all the curves will be rated line to neutral voltage at rated line current as shown above.

9.8 EMF OR SYNCHRONOUS IMPEDANCE METHOD

The value of the synchronous reactance 'X_s' of the alternator is determined by using the OCC and SCC of the alternator as follows.

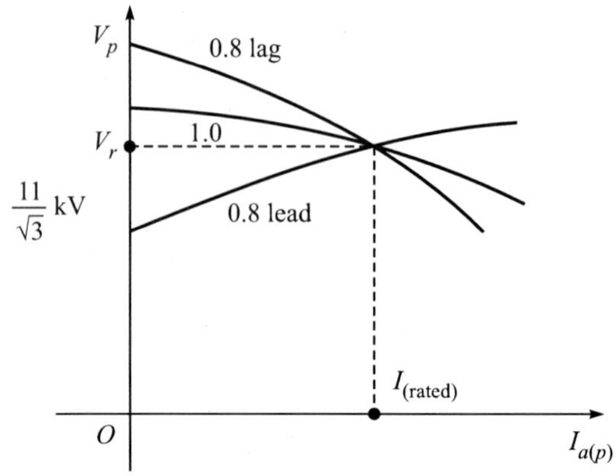

Fig. 9.27 Regulation characteristics at different power factors of load.

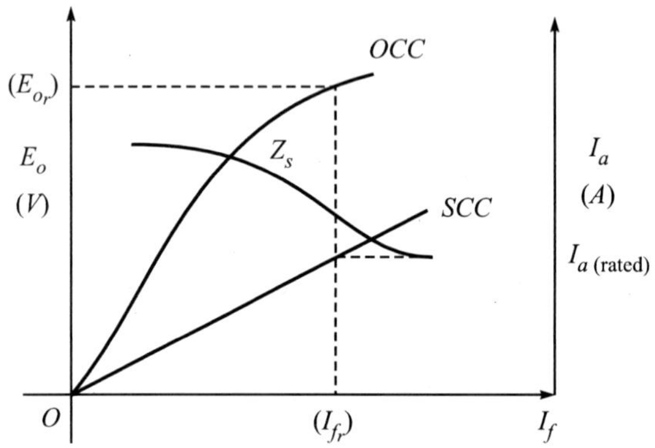

Fig. 9.28 Determination of synchronous reactance 'X_s'.

So, the synchronous reactance 'X_s' is determined using the reaction

$$\boxed{X_s = \frac{(E_0)_r}{(I_a)_r}} \tag{9.20}$$

where, X_s = Synchronous reactance per phase (Ω)
$(E_0)_r$ = Rated O.C. phase voltage, (volts)
$(I_a)_r$ = Short circuit current per phase using the same field current $(I_f)_r$, that is required to produce $(E_0)_r$, in amperes.

The synchronous reactance is not constant, but varies with saturation. So the value of voltage regulation as determined by the synchronous impedance method is approximate. The synchronous reactance takes into account the armature reaction effect and armature leakage reactance. The synchronous impedance 'Z_s' is given by

$$Z_s = \sqrt{R_{a_{eff}}^2 + X_s^2} \tag{9.21}$$

$$R_{a_{eff}} = \text{armature resistance, } \Omega \text{ armature}$$
$$R_{a_{eff}} = 1.5 \times R_{dc}$$

So, the armature current '*I*' is given by

$$I = \frac{(E_0)_r}{Z_s}$$

The total impedance of circuit including load is (say) '*Z*' then

$$Z = Z_s + Z_L = R + jX$$

So, the voltage across the load is given by

$$V = IZ$$

So, the voltage regulation can be determined using the relation

$$\text{V.R.} = E_0 - V, \text{ volts}$$

The % voltage regulation is given by

$$\boxed{\%\text{V.R.} = \frac{E_0 - V}{V} \times 100\%} \tag{9.22}$$

The %V.R. value is positive for lagging power factor loads and negative for leading power factor loads.

As the %V.R. determined by using EMF method differs from the actual value of voltage regulation. So we go for other methods such as MMF or AT method.

9.9 MMF METHOD

The mmf method gives the value of voltage regulation better than the EMF method though it also does not give the exact value of the voltage regulation of the alternator. In this method the ampere turns (AT's) of the field and armature winding is considered. As the ampere turns are the basic source of the flux produced in the air gap. The total ampere turn is given by

$$AT = AT_a + AT_f$$
$$\mathcal{F} = \mathcal{F}_a + \mathcal{F}_f \tag{9.23}$$

The mmf's are represented in the phasor diagrams for different types of load i.e., resistance, inductive and capacitive loads respectively.

We see from the Fig. (9.29) that the resultant ampere turns phasor '$\mathcal{F}$' increases for lagging load and decreases for leading load. Since the ampere turn is the product of current and turns. So as the turns are fixed so the increase in the AT's implies increase of the current.

The $AT_{resultant}$ '$\mathcal{F}$' causes 'E_0' which is produced by the total flux inside the air gap. The $AT_f\,(F_t)$ corresponds to 'I_f' required to produce 'E_0'.

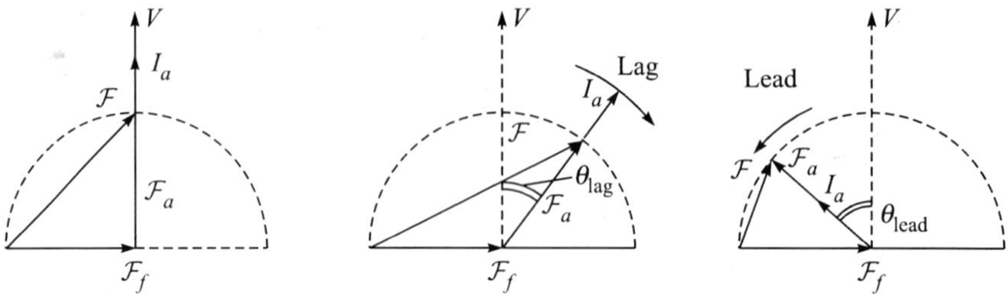

Fig. 9.29 Ampere turns of alternator windings.

So, from the OCC corresponding to '$\mathcal{F}_f$' the field currents which are determined as follows

$$I_f = \frac{\mathcal{F}_f}{N_f}$$

Then corresponding to 'I_f' from O.C.C. the value of voltage on y-axis corresponds to 'E_0'.

Corresponding to 'F_a', the armature current 'I_a' is determined as follows

$$I_a = \frac{\mathcal{F}_a}{N_a}$$

where, N_a, N_f are number of turns of armature and field winding. From the SCC, corresponding to 'I_a' the value of the field current 'I_f' is determined. From OCC corresponding to 'I_f' the value of the voltage 'V' is determined corresponding to 'I_{fr}' the value of 'E_0' is determined. So the voltage regulation can be determined as per the relation (9.24) given below

$$\boxed{\%\text{V.R.} = \frac{E_0 - V}{V} \times 100\%} \quad (9.24)$$

In this AT method also the synchronous reactance takes into account of both leakage reactance and armature reaction collectively and not separately. So the value of the V.R. determined by this method remains approximate. So to determine the exact value of the voltage regulation some other methods need to be used.

9.10 POTIER'S METHOD

In the Potier's method the synchronous reactance 'X_s' is not used but 'X_a' and 'X_l' are used. So the result obtained is accurate. In this method, a curve called zero power factor characteristic (ZPFC) is used along with the OCC and SCC of the alternator as shown in the Fig. (9.30)

$$A'D' = AD = I X_l \text{ (leakage reactance drop)}$$
$$D'B' = DB = I X_{ar} \text{ (armature reaction drop)}$$

The triangle $O'A'B'$ is called potier's triangle which is slipped in the downward direction keeping side '$O'A'$' remains in touch with OCC.

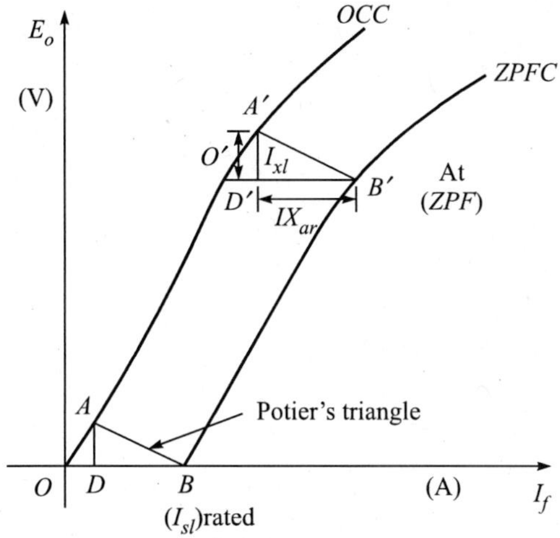

Fig. 9.30 Potier's triangle.

Potier's Regulation Diagram

'V_{rated}' is measured and also 'I_{rated}' at a given power factor angle 'θ' of the load (inductive in this case).

$I X_l$, $I X_{ar}$ are the leakage and '*AR*' drop which is measured from the Potier's triangle.

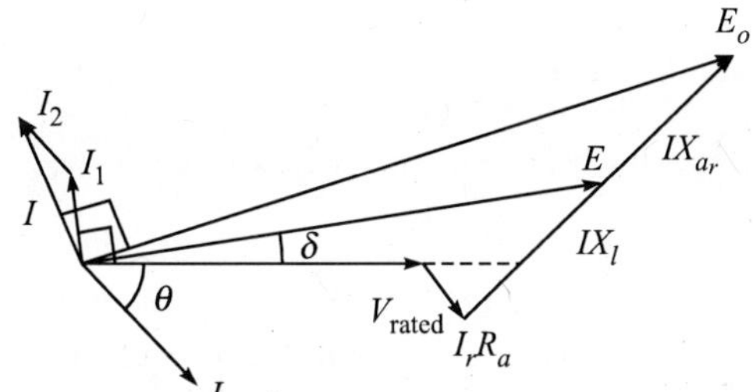

Fig. 9.31 Potier's regulation phasor diagram.

Corresponding to the equation of V.R., we can determine the %V.R. as follows

$$\% V.R. = \frac{E_0 - V_{\text{rated}}}{V_{\text{rated}}} \times 100\%$$

9.11 CAPABILITY CURVE

A synchronous machine can't be operated at all points inside the region bounded by the circle shown without exceeding the machine rating. The region of operation is restricted by considering

(i) Steady state limit of stability
(ii) Heating of armature winding ($I_a^2 R_a$ loss)
(iii) Heating of field winding ($I_f^2 R_f$ loss)

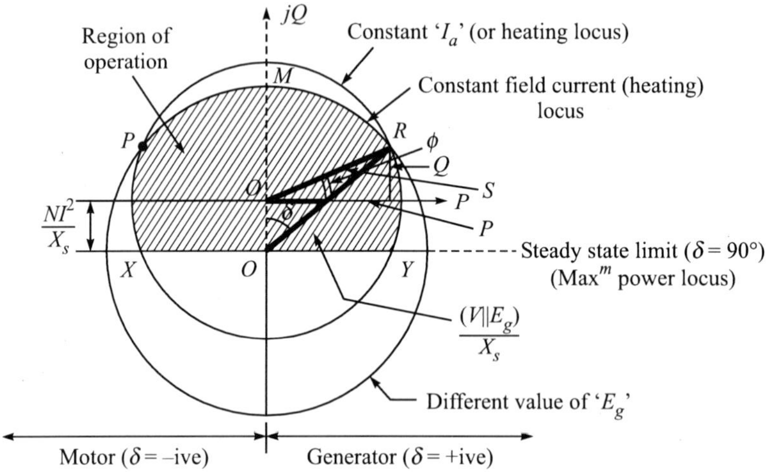

Fig. 9.32 Capability curve of a synchronous machine (3 ϕ).

The capability curve gives the limiting region of operation of synchronous generator in a complex plane. For constant armature voltage 'V', the region of operation is limited to pt 'M' and 'R' and '$O'Y$' respectively.

The circle has centre at 'O' and radius is $(V \cdot I_a)$ defines the region of operation for which armature heating will not exceed a specified limit as shown by shaded portion. The circle with centre at O' $(O, -|V|^2/X_s)$ and radius $\frac{|V||E_g|}{|X_s|}$ defines the region of operation for which field winding heating will not exceed the steady state limit $X'OY'$. The shaded portion bounded by these three capabilities curve defines the restricted region of operation for the synchronous machine. The intersection points 'R' denotes the optimum operating points because operation at these points makes the maximum utilization of the armature and field circuits of alternator.

9.12 BLONDEL'S TWO REACTION THEORY

In case of salient pole machines, the air gap is non-uniform around the inner periphery of the armature. That is the reluctance is minimum along the polar axis of d-axis and the reluctance is maximum along the interpolar axis or q-axis. This could be easily understood by considering the Fig. (9.33).

The armature reaction mmf '$\mathcal{F}_a$' can be split into two as follows:

$$\bar{\mathcal{F}}_a = \bar{\mathcal{F}}_{ad} + \bar{\mathcal{F}}_{aq} \qquad (9.25)$$

where, '$\mathcal{F}_{ad}$' is the armature mmf due to in-phase current, AT
and '$\mathcal{F}_{aq}$' is the armature mmf due to quadrature current, AT

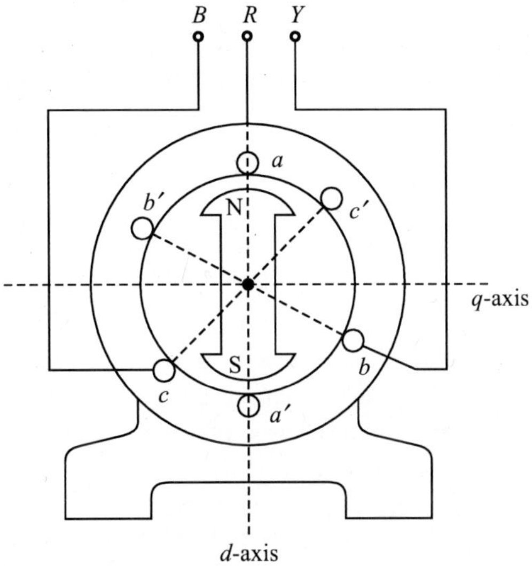

Fig. 9.33 Salient pole synchronous machine.

It is not correct to combine the actual field 'AT' of the distributed armature winding. They can be combined vectorially only when they act upon the same magnetic circuit.

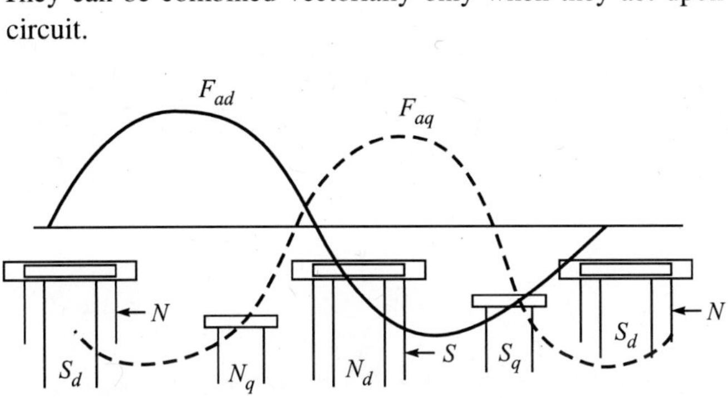

Fig. 9.34 Fictitious field poles equivalent to armature mmf.

Consequently, the methods for finding out the regulation of cylindrical rotor machines, when applied to salient pole machine give results which are not correct.

In case of salient pole machines regulation is found by Blondel's two-reaction theory. According to this theory, two axes are recognized in the machine, namely, as axis along polar axis or d-axis (direct axis) and an axis along the interpolar of quadrature or q-axis which is displaced from the d-axis by 90° in space.

The armature mmf '$\mathcal{F}_a$' consists of two components, namely, quadrature axis component '$\mathcal{F}_{aq}$' which is responsible for the cross magnetizing effect of armature reaction which causes distortion of the field pole flux distribution in the air gap. The other component is the direct axis component '$\mathcal{F}_{ad}$' which is responsible for the demagnetizing effect or reduction of flux under the pole in the air gap between rotor and stator of the synchronous machine. This occurs when the armature current 'I_a' lags the generated voltage 'E_0' and the mmf '$\mathcal{F}_{ad}$' exerts magnetizing effect if the current 'I_a' leads the voltage 'E_0'. Thus, the armature current 'I_a' can be considered a sum of two components d-axis and q-axis components respectively as follows:

$$I = I_a = I_d + j\,I_q$$

where,

$$I_d = I \cos \psi \text{ (in phase component)}$$
$$I_q = I \sin \psi \text{ (quadrature component)}$$

The synchronous reactance of the synchronous machines can be resolved as 'X_d' (d-axis synchronous reactance) and X_q(q-axis synchronous reactance).

Both the components of armature reaction rotates at synchronous speed 'N_s' with respect to the armature winding. These may be looked upon as due to a fictitious set of poles, the set producing the direct effect (demagnetizing or magnetizing) is marked as N_d, S_d and is in alignment with the main poles respectively. The other set produces the magnetizing effect and is marked as N_q, S_q and lies mid-way between the main poles.

The phasor diagram of the various current and voltages of synchronous machines are depicted as follows.

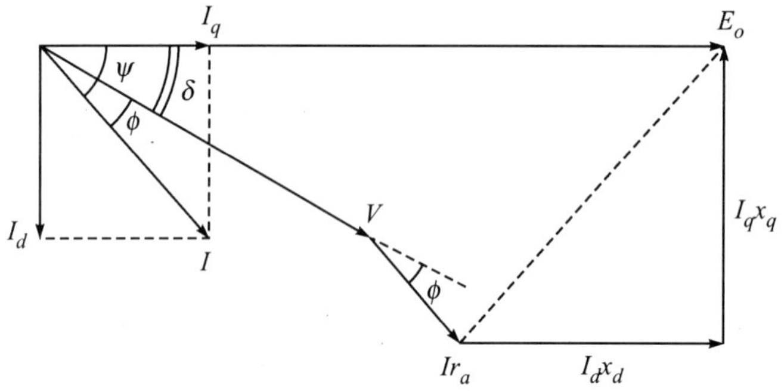

Fig. 9.35 Phasor diagram of 3 ϕ synchronous machine.

where, 'δ' is the power angle

'ϕ' is the phase angle of synchronous machines's load

'V' is the terminal voltage, volts

I = Load current, amperes

I_d = Direct axis current component, amperes

I_q = Quadrature axis current component, amperes

$I \cdot r_a$ = Armature resistance drop, volts

$I_q \cdot x_q$ = Quadrature axis synchronous reactance drop, volts

E_0 = Induced voltage in armature, V

x_d, x_q = Reactances due to fluxes which have d-axis and q-axis paths.

The direct axis component of '$\mathcal{F}_a$' i.e., '$\mathcal{F}_{ad}$' acts on the main winding circuit of the machine whereas the quadrature axis component has a magnetic circuit which has air gap in its magnetic path.

Hence, the q-axis reactance 'x_q' is smaller than 'x_d' (d-axis reactance) and so it is less affected by the saturation.

In non-salient pole machine i.e., cylindrical rotor machine. The reluctance of magnetic circuit path is almost constant. So the d-axis component 'X_d' is equal to that of 'X_q'.

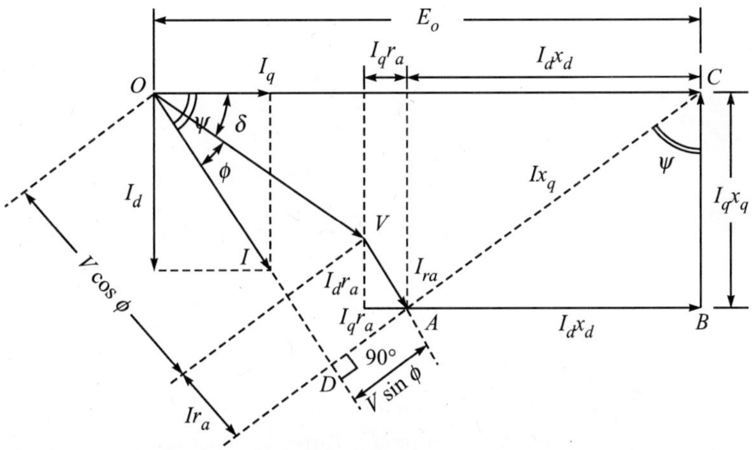

Fig. 9.36 Phasor diagram for lagging p.f. load.

From the Fig. (9.36)

$$\psi = \phi + \delta$$

AC is $\perp$ to current phasor $\bar{I}$

CB is $\perp$ to voltage phasor E_0

$$I_d = I \sin \psi \quad (9.26)$$

$$I_q = I \cos \psi \quad (9.27)$$

$$\therefore \quad I = I_q / \cos \psi$$

In ΔABC

$$\cos \psi = \frac{BC}{AC} = \frac{I_q x_q}{AC}$$

$$AC = \frac{I_q x_q}{\cos \psi} = I x_q$$

In ΔODC

$$\tan \psi = \frac{DC}{OD} = \frac{V \sin \phi + I\, x_q}{V \cos \phi + I\, r_a} \tag{9.28}$$

If 'ψ' is known, then 'E_0' can be determined using the relation

$$\boxed{E_0 = OC = V \cos \delta + I_q\, r_a + I_d\, x_d} \tag{9.29}$$

The reluctance of the magnetic path for flux on which the emf acts is different along the d-axis and q-axis. The reluctance of the d-axis magnetic circuit is due to yoke, teeth of the stator, air gap pole and core of the rotor. The reluctance in q-axis is mainly due to large air gap in the inter pole axis. The two reactance concepts of salient pole machine replaces the effect of armature reaction by two fictitious voltages '$I_d\, x_{ad}$' and '$I_q\, X_{aq}$', where 'I_d' and 'I_q' are the d-axis and q-axis component of 'I_a' respectively.

Synchronous reactance along d-axis; $X_d = X_{ad} + X_l$
Synchronous reactance along q-axis; $X_q = X_{aq} + X_l$

where, 'X_l' is the armature leakage reactance, Ω caused by armature leakage flux 'ϕ_l'.

The values of 'x_d' and 'x_q' are determined by applying a balanced reduced external voltage say 'V' volts to an unexcited machine at a speed little less than the synchronous speed 'N_s' (slip being less than 1%).

Applied voltage to the armature, armature current and voltage induced in the field winding are measured by oscillographs

$$x_d = \frac{\text{Maximum voltage}}{\text{Minimum current}} \tag{9.30}$$

$$x_q = \frac{\text{Minimum voltage}}{\text{Maximum current}} \tag{9.31}$$

where, 'ϕ' is the phase angle
'δ' is the power or torque angle
'ψ' is the internal power factor angle.

$$\delta = \psi - \phi \text{ (for generator)} \tag{9.32}$$

$$\delta = \phi - \psi \text{ (for motor)} \tag{9.33}$$

If armature resistance 'r_a' is neglected, then

$$\boxed{E_0 = V \cos \delta + I_d\, x_d} \tag{9.34}$$

and

$$\boxed{\tan \delta = \frac{I\, x_q \cos \phi}{V + I\, x_q \sin \phi}} \tag{9.35}$$

'E_0' is the phase voltage.
The line voltage is given by the exp. (9.36).

$$\boxed{(E_0)_{\text{line}} = \sqrt{3}\, E_0} \tag{9.36}$$

The power per phase is given by $P_{ph} = VI \cos \phi$

$$I \cos \phi = I \cos(\psi - \delta)$$
$$= I (\sin \psi \sin \delta + \cos \psi \cos \delta)$$
$$= (I \sin \psi) \sin \delta + (I \cos \psi) \cos \delta$$
$$= I_d \sin \delta + I_q \cos \delta \tag{9.37}$$

Substituting the value of '$I \cos \phi$' in the expression of 'P_{ph}',

$$P_{ph} = V I_d \sin \alpha + V Iq \cos \alpha \tag{9.38}$$

Substituting the value of I_d, I_q in eqn. (9.38)

$$I_d x_d = E_0 - V \cos \delta \tag{9.39}$$
$$I_q x_q = V \sin \delta \tag{9.40}$$

$$P_{ph} = V \cdot \frac{E_0 - V \cos \delta}{X_d} \cdot \sin \delta + \frac{V^2 \sin \delta \cos \delta}{x_q}$$

$$= \frac{E_0 V}{x_d} \sin \delta - \frac{V^2 \cos \delta \sin \delta}{x_d} + \frac{V^2 \sin \delta \cos \delta}{x_q}$$

$$= \frac{E_0 \cdot V}{x_d} \sin \delta + V^2 \left[\frac{1}{x_q} - \frac{1}{x_d}\right] \sin \delta \cos \delta$$

$$= \frac{E_0 V}{x_d} \sin \delta + \frac{V^2}{2} \left[\frac{1}{x_q} - \frac{1}{x_d}\right] \sin 2 \delta \tag{9.41}$$

$$P_{3ph} = 3 P_{ph} = \underbrace{\frac{3 E_0 V}{x_d} \sin \delta}_{\text{(I)}} + \underbrace{\frac{3 V^2}{2} \left[\frac{1}{x_q} - \frac{1}{x_d}\right] \sin 2 \delta}_{\text{(II)}} \tag{9.42}$$

The (I) term is due to field excitation and (II) term is due to saliency or reluctance power. For cylindrical rotor machines the (II) term is zero as

$$x_d \approx x_q \tag{9.43}$$

So,

$$P_{3ph} = \frac{3 E_0 V}{X_d} \sin \delta \tag{9.44}$$

9.13 PARALLEL OPERATION OF SYNCHRONOUS GENERATORS

As the load increases at the utility end of power system network, the limit of the alternator ratings limits further increase of the load. So some alternative methods are required for take care of the increase of load. One method is to increase the power rating of the alternators. But this increases the complexity and cost of the alternators. The other method can be to operate two alternators of same ratings in parallel and thus load is shared between the two. So we should analyze the conditions needed for connecting two alternators in parallel as shown below in Fig. (9.37).

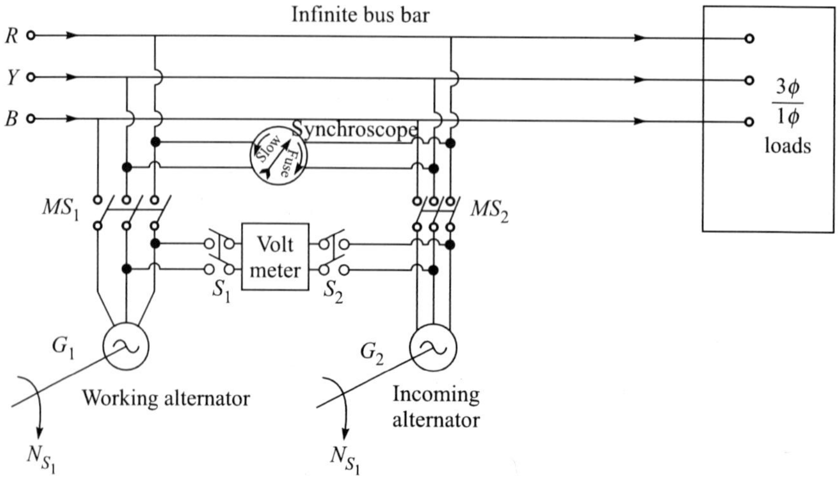

Fig. 9.37 Parallel operation of alternators.

The generators G_1 and G_2 are two synchronous generators having identical ratings. When the generator 'G_2' is to be connected across the load generating power at the same frequency is tested for the following parameters as given below.

(i) The speed of the prime movers should be same i.e., $N_{s1} = N_{s2} = \frac{120 f_1}{P}$ such that $f_1 = f_2$.

(ii) The generator 'G_2' voltage should be equal to voltage of generator G_1'.

(iii) The phase sequence of the two generators should be same.

(iv) The generator voltage should be in phase with the system voltage.

If the conditions as written above are fulfilled by generator 'G_2', then it is said to be synchronized and then it can be connected in parallel across the load or bus bar.

For synchronization, firstly the speed of the prime mover is regulated by speed governors such that the generator frequency is equal to that of system frequency. It is measured by vibrating reed frequency meter. Secondly, the current through the field winding 'I_f' is adjusted such that the generator voltage 'E_0' is equal to the system voltage 'V'. Thirdly, the phase angle between 'E_0' and 'V' are measured by a synchroscope (weston type). This instrument has a pointer that moves continuously indicating the phase angle between the two voltages from 0° to 360°. The pointer starts moving slowly as the phase angle between 'E_0' and 'V' decreases. When it is equal to zero, that is, when they are in phase, the pointer stops.

If the generator frequency is higher than the system frequency the pointer starts moving in the clockwise direction and when the generator frequency is lower than the system frequency then the pointer moves in the counter-clockwise direction. The speed of the generator is adjusted such that the pointer speed decreases and it reaches the vertical position the main switch of alternator (2), MS_2 is closed. Thus, the frequency of the two generators are fixed as that of system frequency. In modern generating station, the synchronization is done automatically. When the alternator is connected to an infinite bus (V, f are fixed) then it becomes a part of a large network comprising hundreds of alternators that delivers power to large loads. So the load type can be predicted on any single alternator. The alternator power can be varied by varying I_f (field excitation) and mechanical torque (T) exerted by the prime mover. When the generator 'G2' is connected to an infinite bus then the induced voltage 'E_0' is equal to 'V'. Since $E_0 = V$, so $I = 0$, so the generator drains no power and is said to be floating on bus bar.

If now 'I_f' is increased, 'E_0' increases and so, voltage drop across the synchronous reactance occurs i.e., $I X_s = E_0 - V$

So, $$I = \frac{E_0 - V}{X_s} \tag{9.45}$$

$\therefore$ 'X_s' is inductive. Thus, 'I' lags 'E_x' by 90°.

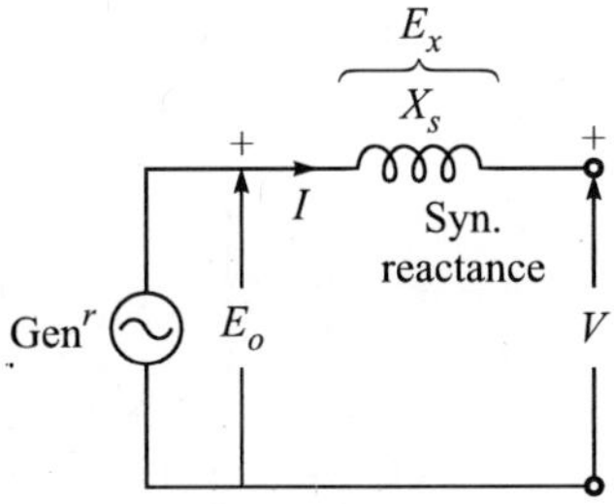

Fig. 9.38 Equivalent circuit of SG.

The generator looks as if it is inductive. When the alternator is over excited, then it supplies reactive power to the infinite bus.

If we decrease the field current ($E_0 < V$)

So, $$I = \frac{E_0 - V}{X_s} = \text{negative}$$

so, the current 'I' leads 'V' by 90° which means the alternator sees as if a capacitor is attached. The alternator draws reactive power from the bus.

If the torque of prime mover is increased, then rotor accelerates and 'E_0' rises. 'E_0' leads 'V' by 'δ'. Although 'E_0' is equal to 'V'. The 'δ' produces $E_0 - V$ across 'X_s'. A current 'I' flows through X_s and almost along 'V' such that the generator feeds active power in the system. This will accelerate as 'δ' increases till the electrical power delivered by the generator is equal to the mechanical power supplied by prime mover and thus rotor will cease to accelerate. The synchronous generator runs at 'N_s' and 'δ' becomes constant.

All Dark Lamp Method for Synchronization

In all dark lamp method, the bulb glows together and all the bulbs get dark at the same time. When synchronizing of generator 'G_2' is to be done to bus bars then the mid of dark period is taken as the instant of synchronization. So expert personnel is needed to synchronize the generator. An easy method of synchronization is available in which two bulbs glow while one is dark. The method is called two bright lamp and one dark lamp method.

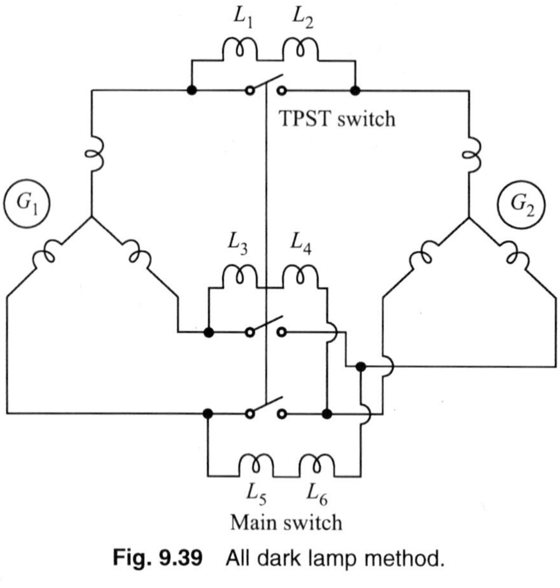

Fig. 9.39 All dark lamp method.

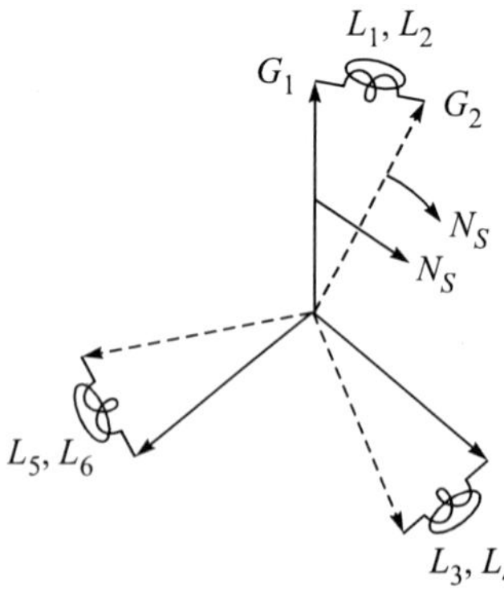

Fig. 9.40 Phasor diagram of dark lamp method.

The stator windings of two generators 'G_1' and 'G_2' are shown connected via a TPST switch. Three bulbs are attached, two are cross connected (L_2, L_3) while one bulb (L_1) is connected in series with phase. The phasor of voltage induced is given by Fig. (9.42).

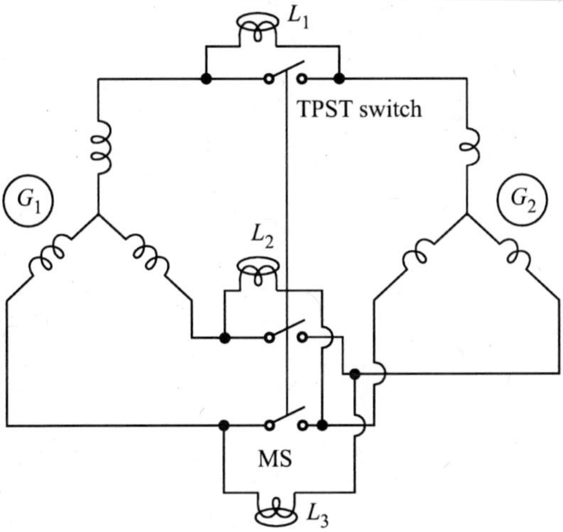

Fig. 9.41 Two bright lamp and one dark lamp method.

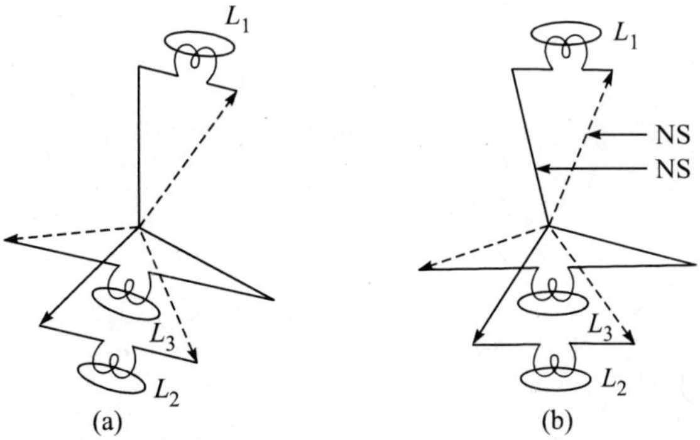

Fig. 9.42 Phasor diagram of two bright lamp one dark lamp method.

The synchronizing instant is the instant when lamp 'L_1' is dark and L_2, L_3 are equally bright. The switch TPST is closed at that instant then G_1 and G_2 are synchronized. When the bold line and dotted line overlaps lamp 'L_1' is dark and at that instant L_2 and L_3 are equally bright.

Power Angle Characteristic

The power is given by

$$P_{em} = \frac{E}{Z_s}[E\cos\theta - V\cos(\theta + \delta)]$$

where,

$$\theta = \tan^{-1} \frac{X_s}{r}$$

'r' is the resistance of armature phase winding, ohms

$$P_0 = \frac{V}{Z_s} [E \cos (\theta - \delta) - V \cos \theta], \text{ in motoring mode}$$

If 'r' is very small, $r << x_s$, neglecting 'r' then

$$P_{em} = P_0 = \frac{V \cdot E}{Z_s} \sin \delta \simeq \frac{V \cdot E}{X_s} \sin \delta \tag{9.46}$$

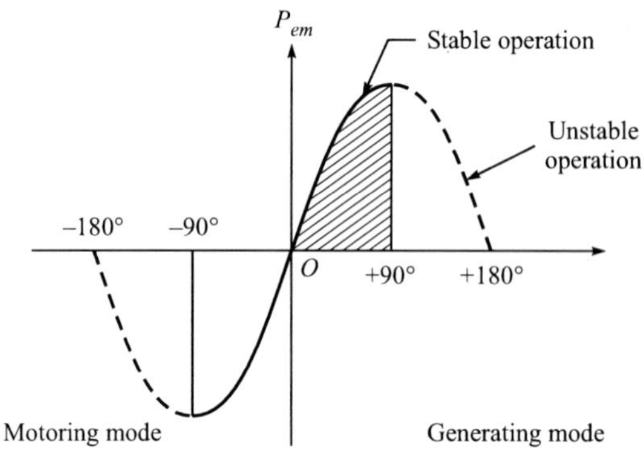

Fig. 9.43 *P*/δ curve of synchronous machine.

The expression (9.46) is called the power angle expression of generator if 'V' terminal voltage and 'E' is the generated voltage in the armature winding.

For given value of 'I_f', $\quad \phi_f$ = constant

So, $\quad E$ = constant

For given value of V, the power = $P = VI \cos \phi$ = constant

till the value of power angle 'δ' remains same.

If 'δ' changes the power 'P_0' changes. The maximum value of power = P_{max} $= \frac{V \cdot E}{X_s}$, when $\delta = 90°$.

Thus, for a given value of 'I_f', the limit of stability of alternator is $\delta \leq 90°$. If machine is loaded, 'δ' increases, thus there is greater need of power for stable running and generation of synchronizing power.

Terminal Voltage of two Alternators in Parallel

If the two generators G_1 and G_2 have identical characteristics, then

$$V = E_1 - I_1 Z_1 = E_2 - I_2 Z_2$$

$$E_1 - E_2 = I_1 Z_1 - I_2 Z_2$$

$$I = I_1 + I_2$$

$$V = IZ$$

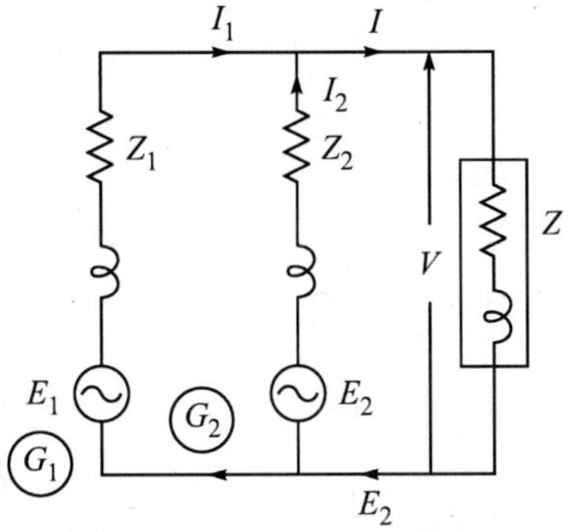

Fig. 9.43(a) Two generators connected in parallel across load 'Z'.

$$E_1 = I_1 Z_1 + IZ = I_1(Z + Z_1) + I_1 Z$$

$$E_2 = I_2 Z_2 + IZ = I_2(Z + Z_2) + I_2 Z$$

or,
$$I_1 = \frac{(E_1 - E_2)Z + E_1 Z_2}{Z(Z_1 + Z_2) + Z_1 Z_2} \tag{9.47}$$

and
$$I_2 = \frac{(E_2 - E_1) + E_2 Z_1}{Z(Z_1 + Z_2) + Z_1 Z_2} \tag{9.48}$$

$\therefore$
$$I = \frac{E_1 Z_2 + E_2 Z_1}{Z(Z_1 + Z_2) + Z_1 Z_2} \tag{9.49}$$

$$V = \frac{E_1 Z_2 + E_2 Z_1}{Z_1 + Z_2 + \dfrac{Z_1 Z_2}{Z}}$$

and
$$I_1 = \frac{E_1 - V}{Z_1} \quad \text{and} \quad I_2 = \frac{E_2 - V}{Z_2}$$

Z_1, Z_2 are synchronous impedances of generators G_1 and G_2 respectively. The circulating current 'I_c' under no load condition is

$$I_c = \frac{E_1 - E_2}{Z_1 + Z_2} \tag{9.50}$$

The synchronous power $P_s \simeq \dfrac{E \cdot V}{Z_s} \sin(\theta + \delta) \cdot \sin \alpha$ (9.51)

If the power angle 'δ' is changed to '$\delta + \alpha$' an additional current 'I_s' is needed.

$$I_s = \frac{E_s}{Z_s} = \frac{2E \sin \alpha/2}{Z_s}$$

'P_{em}' changes to

$$P'_{em} = \frac{E}{Z_s}[E \cos\theta - V \cos(\theta + \delta \pm \alpha)] \tag{9.52}$$

Increased power developed if 'α' leads and accelerating power when 'α' lags.

$P_s = P_{em}' - P_{em}$ which approximates to exp. (9.53) In large alternators, Z_s is reactive i.e., $\theta \simeq 90°$.

So, $$\cos(\theta + \delta) \simeq \cos\delta$$

So, $$P_s \simeq \frac{EV}{Z_s} \cdot \cos\delta \cdot \sin\alpha \qquad (9.53)$$

'P_s' depends on load.

$\because$ 'α' is very small so, $$P_s \simeq \frac{E \cdot V}{z_s} \cdot \sin\theta \cdot \alpha$$

$$\sin\theta \to 1 \quad \because \quad \frac{x_s}{Z_s} \simeq 1 \quad \text{and} \quad \frac{E}{Z_s} = I_s$$

So, $$\boxed{P_s \approx V I_s \cdot a} \text{ per phase} \qquad (9.54)$$

where, I_s = SC current with same 'I_f' as reqd. by 'E' and obtained from SCC

$$P_s = \frac{d P_{em}}{d\delta} = \text{slope of } P_{em} \ \& \ \delta \text{ curve} \qquad (9.55)$$

Synchronizing power tells whether the machine will run fast or slow during sudden changes of load.

$$\text{Synchronizing torque } T_s = P_s \text{ syn. watts}$$
$$= \frac{P_s}{2\pi n} \text{ N.m.} \qquad (9.56)$$

9.14 HUNTING

As in the previous section, we see that as the field excitation is increased or the prime mover speed is increased then a synchronizing current is produced which will produce synchronizing power that causes the alternator generating the power to drop back into synchronism and the alternators receiving the power to advance ahead of synchronize. The system will be stable and in equilibrium if the speed of prime mover is constant over a cycle. The prime mover of the alternators have not constant speed over a cycle. So the synchronous speed of the alternator varies as $n_s = \dfrac{2f}{P}$.

But considering the poles of all machines operating in synchronism is equal to two then $n_s = f$ (Hertz). So the change in the frequency will affect the operation of all the alternators attached to the bus bars. As a result the generators will lose synchronism and as a result the alternators search or hunt its equilibrium position as we had discussed for synchronous motors. As a result there is oscillations of rotor which is called hunting or surging of the alternators respectively. The huntings if not checked then it would be disastrous and it will result in complete shut down of the grid containing 100's of alternators. So some remedial measures need to be taken which should check it from growing or building to large value.

For controlling or limiting hunting of alternators damper bars are placed in the slot which is present in the rotor poles, which provide the necessary torque and maintains the synchronous speed and hence the frequency f = 50 Hz.

Sometimes in alternators, a flywheel is mounted over the shaft to store the excess energy and to supply the same while the alternator needs it.

By detecting speed changes as small as 0.01% an automatic control system sensitive to small speed changes immediately modifies the valve opening of the turbine so as to maintain the constant speed and constant power output. As steam value opens, steam jet speed increases, the rotor accelerates about 50% greater than normal for 4 to 5 seconds. Steam values are immediately closed off during such emergencies. The pressure rises in boilers must be relieved and the fuel burners must be shut off.

9.15 TRANSIENTS IN SYNCHRONOUS MACHINES

A synchronous generator connected to a system is subjected to unpredictable load changes that occurs very quickly. For sudden load changes the synchronous reactance 'X_s' is replaced by another reactance X' whose value varies when the generator terminals are short circuited suddenly as shown in Fig. (9.44).

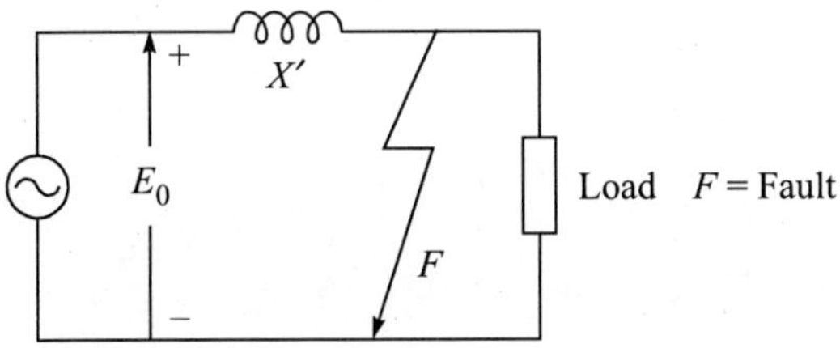

Fig. 9.44 Short circuited alternator by a fault.

The synchronous reactance 'X_s' is for steady state when operating normally. At the short circuit instant, the reactance 'X' falls to 'X_d'' (transient reactance).

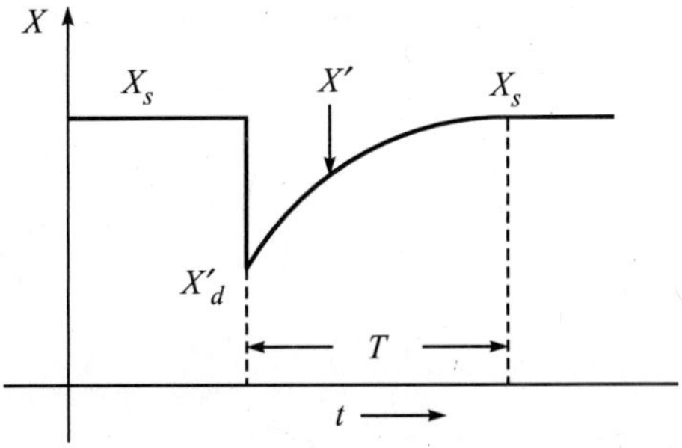

Fig. 9.45 Transient behaviour of alternator.

After a short period 'T' it again attains the value 'X_s'. The time 'T' depends on the size of the alternator. For alternators below 100 kVA it is a fraction of a second and for 1000 MVA alternators it is upto 10 seconds.

$$X_d' = 0.85\ X_s$$

As the value of 'X_d''' is less, so the current drawn by the alternator load is very high during transient operation under fault. The C.B attached at the generator output takes care of it ($X' < X_s$). As 'I_f' increases, 'E_0' increases and thus stabilizes the voltage 'V' of the alternator. The current variation during sudden short circuit are shown in Fig. (9.46).

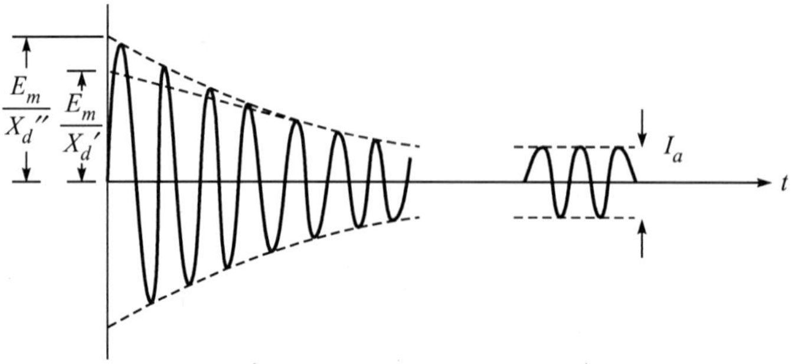

Fig. 9.46 Short circuit current.

$X_d' \rightarrow$ Transient reactance
$X_d'' \rightarrow$ Subtransient reactance
The 'd' subscript stands for d-axis component of 'I_a'.

9.16 SYNCHRONOUS MACHINE DYNAMICS

Whenever there is sudden change in the speed of alternator, to overcome this change it takes some time called transient time before the steady state operation of generator takes place. The electrical transients are like short circuiting of the alternator terminals as discussed in the previous section. The symmetry of armature currents (i_a, i_b and i_c) above the x-axis is lost.

The transient periods are short specially subtransient but transient period is little bit higher and rate of decrease of current is lesser.

Flux linkage of field windings is $= N_f \phi_f = L_f I_f$

where, $L_f \rightarrow$ Self-inductance of field winding
$I_f \rightarrow$ Field winding current

Since 'L_f' is the effective inductance of field winding, so

$$L_f = L_{lf} + L_{af}$$

$L_{lf} \rightarrow$ Leakage inductance of field winding, H
$L_{af} \rightarrow$ Mutual inductance due to 'I_a', H

Since, we consider the alternator is operating under no load before the short circuit, the flux linkage due to armature winding is

$$\lambda_a = 0 \quad \text{as} \quad I_a = 0$$

Consider the phase winding 'a' and field windings are at 90° space displaced as shown, then for a rotor movement of 'α' with respect to phase 'a' magnetic axis.

The flux linkages for phase 'a' winding is given by

$$\lambda_a = i_a(L_{la} + L_{af}) + (I_f + i_f)\, L_{af} \sin\,(90° - \alpha) = 0$$

where, $L_{la} \rightarrow$ leakage inductance of phase 'a'

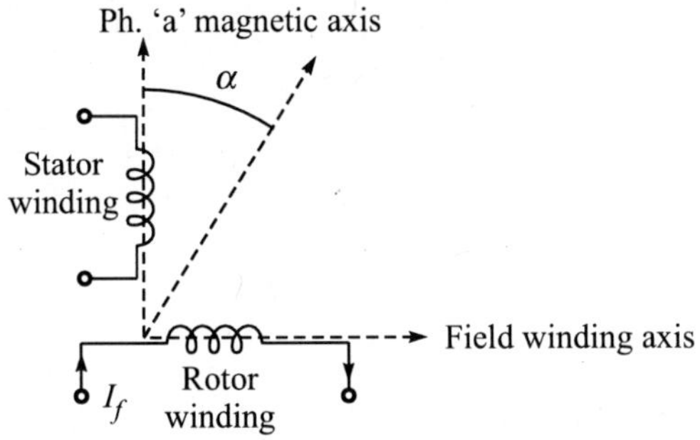

Fig. 9.47 Position of stator and rotor at $t = 0$.

The flux linkages associated with field winding are

$$\lambda_f = (I_f + i_f)\,(L_{lf} + L_{af}) + L_{af}\, i_a \sin\,(90 - \alpha) \tag{9.57}$$

Solving, we get

$$i_a = \frac{L_{af} I_f\,(L_{lf} + L_{af}) \sin(90 - \alpha)}{L_{af}^2 \sin^2(90 - \alpha) - (L_{lf} + L_{af})(L_{la} + L_{af})} \tag{9.58}$$

and

$$i_f = \frac{L_{af}^2 \sin^2(90 - \alpha) \cdot I_f}{L_{af}^2 \sin^2(90 - \alpha) - (L_{lf} + L_{af})(L_{la} + L_{af})} \tag{9.59}$$

The current is maximum if $\alpha = 0$, which corresponds to severe fault condition. The voltage prior to short circuit is

$$E = I_f X_{af} \text{ (under no load)}$$

The transient reactance X_d' and subtransient reactance X_d'' are given by

$$X_d' = X_{la} + \frac{X_{af}\, X_{lf}}{X_{af} + X_{lf}} \tag{9.60}$$

and $$X_d'' = X_{la} + \frac{X_{af} X_{lf} X_{ld}}{X_{lf} X_{ld} + X_{af} X_{ld} + X_{af} X_{lf}} \quad (9.61)$$

which could be understood by using the equivalent circuits.

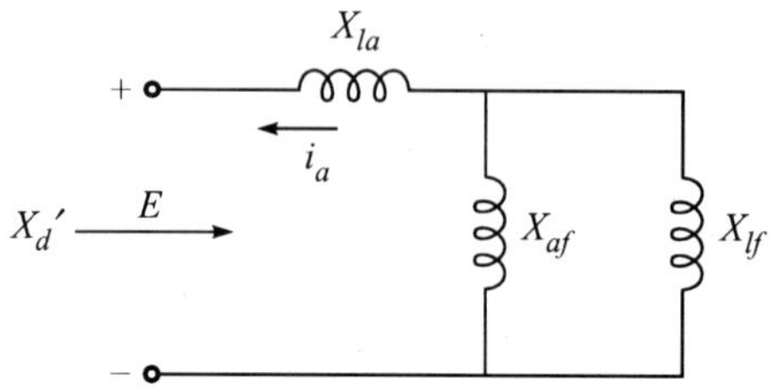

Fig. 9.47(a) Equivalent circuit for X_d'.

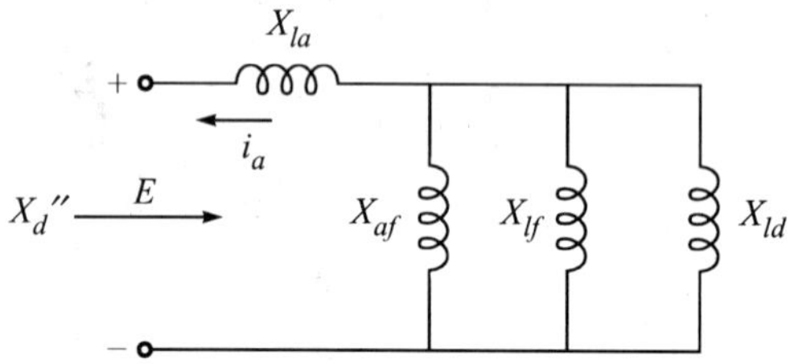

Fig. 9.48 Equivalent circuit for X_d''.

since $X_d'' << X_d'$ the $I_d'' >> I_d'$

i.e, $$I_d'' = \frac{E}{\sqrt{2}\, X_d''} \quad \text{and} \quad I_d' = \frac{E}{\sqrt{2}\, X_d'} \quad (9.62)$$

at steady state condition $$I_d = \frac{E}{\sqrt{2}\, X_s} \text{ respectively} \quad (9.63)$$

SOLVED QUESTIONS

Q.1 A part of an alternator winding consists of 6 coils in series, each coil having an emf of 10 V rms induced in it. The coils are placed in successive slots and between each slot and the next, there is an electrical phase displacement of 30°. Calculate the emf of six coils in series.

A.1 $$\beta = 30°,\ m = 6$$

$\therefore$ $$K_d = \frac{\sin m \frac{\beta}{2}}{m \sin \frac{\beta}{2}} = \frac{\sin 6 \times 15°}{6 \sin 15°} = \frac{1}{6 \times 0.2588}$$

Arithmetic sum of voltage induced in 6 coils

$$= 6 \times 10 \text{ V} = 60 \text{ V}$$

Vector sum of voltage induced in 6 coils

$$= K_d \times \text{Arithmetic sum}$$

$$= \frac{60}{6 \times 0.2588} = 38.64 \text{ V}$$

Q.2 A three-phase, Y-connected alternator supplied a load 1000 kW at 0.8 p.f. lagging and at terminal voltage of 11 kV. Its resistance is 0.1 Ω/phase and synchronous reactance is 0.66 Ω/phase. Calculate the line emf generated.

[BSc Engg, AMU, 1989]

A.2

$$\text{Rated current} = I = \frac{1000 \times 10^3}{\sqrt{3} \times 11 \times 10^3 \times 0.85} = 618 \text{ A}$$

$$\text{Resistive drop, } IR_a = 618 \times 0.1 = 61.8 \text{ V};$$

$$\text{Reactive drop, } IX_L = 618 \times 0.66 = 408 \text{ V}$$

$$\text{Phase voltage, } V_p = \frac{V_L}{\sqrt{3}} = 6351 \text{ V} \quad \because V_L = 11 \times 10^3 \text{V}$$

$$E_0 = [(V \cos \phi + I R_a]^{2/3} + (V \sin \phi + I X_s)^2]^{1/2}$$

$$= \{[6351 \times 0.85 + 61.8]^2 + [6355 \times 0.527 + 408]^2\}^{1/2}$$

$$= 6628 \text{ V}$$

$\therefore$ Line emf $= E_L = \sqrt{3} \times E_0 = 11.48$ kV

Q.3 Find the synchronous reactance and impedance of an alternator in which a given field current produces an ammeter current of 250 A on short circuit and a generated emf of 1500 V on open circuit. The armature resistance is 2 Ω. Hence, calculate the terminal voltage, when a load of 250 A at 6600 V at 0.8 p.f. (lagging) is switched off.

A.3 Synchronous impedance $Z_s = \frac{E_0}{I_a} = \frac{1500}{250} = 6 \ \Omega$

$$Z_s = R_a + j X_L = 2 + j X_L = 6$$

$$\therefore \quad X_L = \sqrt{6^2 - 2^2} = 5.657 \ \Omega$$

$$\text{Resistive drop} = I R_a = 250 \times 2 = 500 \text{ V}$$

$$\text{Reactive drop} = I X_s = 250 \times 5.657 = 1414 \text{ V}$$

$$\therefore \quad \text{Phase voltage } V_p = \frac{V_L}{\sqrt{3}} = \frac{6600}{\sqrt{3}} = 3811 \text{ V}$$

$$\cos \phi = 0.8 \text{ so, } \sin \phi = 0.6$$

$\therefore$ Terminal voltage at no load is equal to

$$E_0 = \sqrt{(V_P \cos \phi + I R_a)^2 + (V_P \sin \phi + I X_s)^2}$$

$$E_0 = \sqrt{(3811 \times 0.8 + 500)^2 + (3811 \times 0.6 + 1.414)^2}$$

$$E_0 = 4632 \text{ V}$$

Q.4 A 3 ϕ, 16 pole alternator has a *Y*-connected winding with 144 slots and 10 conductors per slot. The flux per pole is 0.03 Wb and the speed is 375 rpm. Find the frequency and the phase and the line electromotive free.

A.4

$$\phi = 30 \times 10^{-3} \text{ wb}, \qquad f = \frac{NP}{120} = 50 \text{ Hz}$$

$$S/P = \frac{144}{3} = 48, \quad Z/ph = 48 \times 10 = 480$$

$$\text{.Turns per phase} = \frac{Z/ph}{2} = \frac{480}{2} = 240$$

$\therefore$ Pitch factor, $K_c = 1$ (Assuming full pitch coil)

$$\text{No. of slots/ pole} = \frac{144}{16} = 9$$

$$\text{Angular displacement,} \quad \beta = \frac{180^\circ}{9} \ 20^\circ$$

$$S/P/ph = m = \frac{9}{3} = 3$$

$\therefore$ Distribution factor, $K_d = \dfrac{\sin(m\beta/2)}{m \sin \beta/2} = 0.96$

$$\therefore \quad E_{ph} = 4.44 \times K_c \times K_d f \phi T$$

$$= 4.44 \times 1 \times 0.96 \times 50 \times 0.03 \times 240$$

$$= 1534 \text{ V}$$

$$\therefore \quad E_2 = \sqrt{3}\ E_{ph} = 2657 \text{ V}$$

Q.5 A 3 ϕ, 800 kVA, 11 kV, *Y*-connected alternator has resistance of 1.5 Ω/phase and synchronous reactance of 25 Ω/phase. Find the percentage regulation for a load of 600 kW at 0.8 leading power factor.

A.5

$$V_{ph} = \frac{V_L}{\sqrt{3}} = \frac{11 \times 10^3}{\sqrt{3}} = 6351 \text{ V}$$

$$R_a = 1.5\ \Omega,\ X_a = 25\ \Omega/ph,\ I = \frac{600 \times 10^3}{\sqrt{3} \times 11 \times 10^3 \times 0.8}$$

$$I = 39.4 \text{ A}$$

$$I R_a = 39.4 \times 1.5 = 59.1 \text{ V}$$

$$I r_s = 39.4 \times 25 = 985 \text{ V}$$

$\therefore$ For leading power factor

$$E_0 = \sqrt{(V\cos\phi + IR_a)^2 + (V\sin\phi - IX_s)^2}$$

$$E_0 = \sqrt{(6351\times 0.8 + 59.1)^2 + (6351\times 0.6 - 985)^2}$$

$$E_0 = 5865 \text{ V}$$

$$\therefore \quad \text{Regulation} = \frac{E_0 - V_{ph}}{V_{ph}} = -7.65\%$$

For capacitive load, V.R. is always negative.

Q.6 A 11 kV, three-phase, Y-connected turboalternator delivers 200 A at unity power factor when running on constant voltage and constant frequency bus bars. If the induced electromotive force is raised by 25%, find the current and power factor at which the machine now works. Assume a constant steam supply and an unchanged efficiency. Armature resistance is negligible, synchronous reactance Ω.

A.6 $V = 11 \text{ kV}, \quad X_s = 10\ \Omega, \quad I_a|_{\text{upf}} = 200 \text{ A}, \quad E'_f = 1.25\ E_f,$

$I' = ?;\ \text{p.f} = \cos\phi' = ?;$

$$E_f = \sqrt{V^2 + (Ir_s)^2}$$

$$E_f = \sqrt{\left(\frac{11\times 10^3}{V^3}\right) + (200\times 10)^2}$$

$$E_f = 6.658 \text{ kV}$$

$$\sin\delta = 0.179;\ \cos\delta = 0.984$$

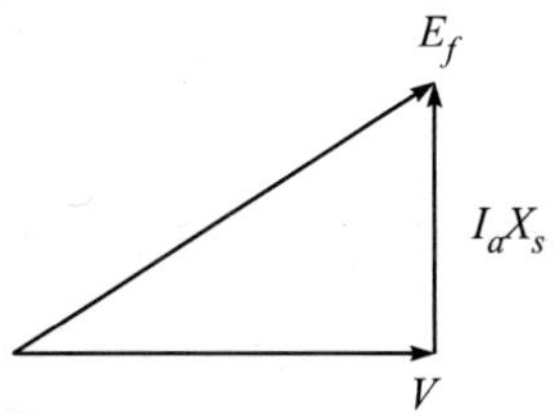

with increased excitation

$$E'_f = 1.25 \times 6.658 = 8322.9$$

For an unchanged input

$$E'_f \sin\delta' = E_f \sin\delta$$

$$\sin\delta' = \frac{6658\times 0.179}{8322.9} = 0.143$$

$$\cos\delta' = 0.99$$

$$I'\sin\phi' = \frac{E'_f\cos\delta' - E_f\cos\delta}{X_s} = \frac{8323\times 0.99 - 6658\times 0.984}{10}$$

$$= \frac{8735 - 6552}{10} = 218.4$$

$$I'\sin\phi = 218.4$$

$$I'\cos\phi' = I\cos\phi = 200 \text{ A} \times 1 = 200 \text{ A}$$

$$I' = 296.14 \text{ A}$$

$$\cos\phi' = 0.675 \text{ (lag)}$$

Q.7. An 11 kV, 3 ϕ, star connected turboalternator delivers 200 A at upf when running on infinite busbars. If the induced e.m.f. is raised by 25%, find the current and power factor at which the machine now works. Assuming constant efficiency, the armature resistance is 0.5 Ω and synchronous reactance of 8 Ω.

A.7 3 ϕ. 11000 V, Y-connected, $R_a = 0.5\ \Omega$, $X_s = 8\ \Omega$

$$I|_{\text{upf}} = 200\text{ A},\ E_f' = 1.25\ E_f,\ I' = ?,\ \cos\phi' = ?$$

$$\theta = \tan^{-1}\frac{8}{6.5} = 86.424^\circ = \tan^{-1}\frac{x_s}{R_a}$$

At 200 A, $I_{\text{p.f}}$

$$E_f = \sqrt{\left(\frac{11000}{\sqrt{3}} + 200 \times 0.5\right)^2} = 6646.32\text{ V} = 6.65\text{ kV}$$

$$\sin\delta = 0.241$$
$$\cos\delta = 0.971$$
$$\delta = 13.93^\circ$$

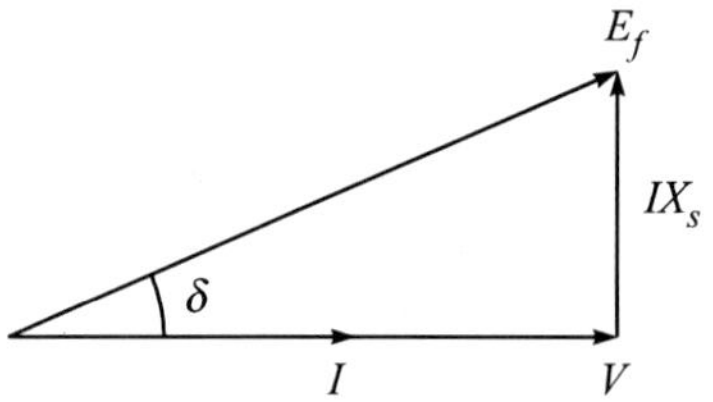

with increased excitation and constant

$$E_f' = 1.25 \times 6.65 \times 10^3 = 8.31\text{ kV}$$
$$E_f' \cos(\psi - \delta') = E_f \cos(\psi - \delta)$$
$$8310 \cos(86.424 - \delta') = 6650 \cos(86.424 - 13.93)$$
$$\delta' = 10.35^\circ$$
$$I' \sin\phi' = 215.2$$

For constant active power

$$I \cos\phi' = I\cos\phi = 200\text{ A}$$
$$I' = 293.8\text{ A}$$
$$\cos\phi' = 0.681\text{ lag}$$

UNSOLVED QUESTIONS

Q.1 A 3.5 MVA, slow speed, three-phase, Y-connected synchronous generator rated at 6.6 kV has 32 poles. Its direct and quadrature axis synchronous reactances are 9.6 Ω and 6 Ω respectively. Neglecting armature reactance, determine the voltage regulation and excitation emf needed to maintain 6.6 kV at the terminals when supplying a load of 2.5 MW at 0.8 p.f lagging. What maximum power can the generator supply at the rated terminal voltage if the field becomes open circuited?

[Bsc. Engg III, AMU, 1998]

Q.2 A 6.6 kV, 3 phase, 50 Hz, star connected alternator gave the following data for open circuit, short circuit and full load zero power factor tests:

I_f (in A)	3.2	5.0	7.5	10.0	14.0
V_{oc} (in kV) line	3.1	4.9	6.6	7.5	8.24
Zero power factor Voltage kV (line)	–	1.85	4.24	5.78	7.0
I_{sc}(A)	500	778	1178	–	–

Per phase armature resistance is 0.2 Ω. Calculate the voltage regulation at full load current of 500 A at 0.8 p.f. lagging by the following methods. (i) EMF method (ii) MMF method (iii) ASA method.

[B.Sc Engg III, AMU, 1998
B.E. III, AMU, 1999]

Q.3 Two similar alternators operating in parallel have the following data:

Alternator I : Capacity 2 MW, frequency drops from 50.5 Hz at no load to 48 Hz at no load

Alternator II : Capacity 2 MW, frequency drops from 50 Hz at no load to 48.5 Hz at full load

(i) Calculate how a total of 3.6 MW is shared by each alternator. Also calculate the operating bus frequency at this load.

(ii) Compute the maximum load the two units in parallel can deliver without overloading either of them.

[BSc. Engg III, 1998, AMU]

Q.4 Two star connected generators are connected in parallel and supply a balanced load of 1500 kVA at 11 kV line voltage and 0.8 lagging power factor. The synchronous reactances of the two machines respectively are 35 Ω and 4 Ω. The prime mover governors of the two machines are adjusted so as to equally share the load. The phase current in one machine is 43 A, at a lagging power factor. Calculate:

(a) the phase current in the second machine

(b) the induced emf in each machine

(c) the power factor at which each machine operates.

[B.E. III, 1999, AMU]

Q.5 Two three-phase, 60 Hz alternators are operating in parallel. One of them has a capacity of 1000 kVA and a linear governor characteristic drooping from 61 Hz at no load to 59.6 Hz at full load. The other has a capacity of 1500 kW and governor characteristic drooping linearly from 61.4 Hz at no load to 59.2 Hz at full load.

[B.E. III, AMU, 1999]

Q.6 A 3 ϕ, 6.6 kV alternator with R_a = 1 Ω/phase and X_s = 5 Ω/phase is supplying 300 A at 0.8 p.f. lagging to infinite bus bar.

(i) If the excitation is raised by 30% (with steam input constant), find the current and power factor.

(ii) If the steam input is increased by 20% (keeping excitation constant) what will be the current and power factor.

[B.E. II, AMU, 1992]

Q.7 Two three-phase star connected alternators supply a load of 3000 kW at 0.8 p.f. lagging the two machines share load equally. The excitation of second machine is so adjusted that it supplies 150 A at a lagging power factor. The synchronous impedances of the two machines are 0.4 + j 12 Ω and 0.5 + j 10 Ω respectively. Determine the current, power factor, induced emf and load angle of each machine. The terminal voltage is 6.6 kV.

[B.E. II, AMU, 1991]

Q.8 A 1000 kVA, 6.6 kV, 3 phase star connected synchronous generator has a synchronous reactance of 25 Ω/phase. It supplies full load current of 0.8 lagging power factor to the bus bar.

(a) If the excitation is raised by 25%, find the current and power factor.

(b) If steam input is increased by 25% keeping excitation constant, what is the current and power factor?

[B.E., AMU, 1991]

Q.9 A 3 ϕ, 16 pole, synchronous machine has Y-connected winding with 144 slots and 10 conductors per slot. The flux per pole is 0.04 wb and speed is 375 rpm. Find the phase and line induced emf's. The total turns/phase may be assumed to be series connected.

[B.E., AMU, 1995]

Q.10 A 3 ϕ, Y-connected alternator is rated at 1500 kVA, 12 kV. The armature effective resistance and synchronous reactance are 2 Ω and 35 Ω respectively per phase. Calculate the percentage regulation for a load of 1200 kW at power factor of (i) 0.8 lagging (ii) 0.8 leading.

[B.Tech II, UPTU, 2002]

Q.11 A 1200 kVA, 66 kV 3 ϕ, Y-connected alternator with a resistance of 0.4 Ω and reactance of 6 Ω per phase delivers full load current at p.f. of 0.8 lagging and normal rated voltage. Estimate the terminal voltage for the same excitation and load current at 0.8 p.f.

[B.Tech II, UPTU, 2003]

Q.12 A full load current of 100 A requires an excitation current of 3 A when a single phase 1200 V alternator is short circuited by an ammeter of resistance 0.01 Ω. On open circuit the same excitation produces 350 V. The resistance of the armature is 0.5 Ω. Calculate the regulation of the alternator at 0.8 power factor leading.

[B.Tech II, UPTU, 2001]

Q.13 A 1100 V, three-phase alternator is running at 300 rpm and has 24 poles. Find the number of conductors in the stator winding, if the magnetic flux is 5×10^{-2} wb/pole by assuming the breadth factor equal to 0.96 and coil to be full pitched.

[B.Tech II, UPTU, 2001]

Q.14 The stator of a 3 ϕ synchronous generator has 9 slots per pole and carries a balanced double layer winding. Determine the winding factor if each coil spans 7 slots.

[B.Tech III, JMI, 2005]

Q.15 A 3 ϕ, *Y*-connected synchronous generator rated at 10 kVA, 230 V, has a synchronous reactance of 1.2 Ω per phase and an armature resistance of 0.5/*ph*. Calculate the voltage regulation at full load 0.8 p.f. lagging.

[B.Tech III, JMI, 2005]

Q.16 A 3 ϕ synchronous generator has synchronous reactance of 0.2 per unit and is delivering full load at 0.8 power factor lagging. Calculate the maximum load in per unit, which it can supply under steady state without losing synchronism.

[B.Tech III, JMI, 2006]

Q.17 A 3 ϕ, *Y*-connected round rotor synchronous generator rated at 10 kVA, 230 V has a synchronous reactance of 1.2 Ω per phase and an armature resistance of 0.5 Ω per phase. Determine the power factor at which the full load voltage regulation is zero.

[B.Tech III, JMI, 2006]

Q.18 A 3 MVA, 6 pole synchronous generator runs at 1000 rpm in parallel with other machines on 3.3 kV bus bars. If the generator is on no load, calculate the synchronizing power and synchronizing torque for one mechanical degree of displacement when its synchronous reactance is 25%.

[B.Tech III, JMI, 2006]

Q.19 The speed regulations of two 50 MW three-phase alternators operating in parallel are 2% and 3% respectively from full load to no load. If each machine is fully loaded, when the total load is 100 MW. What would be the load on each machine, when the total load is 60 MW?

[B.Tech III, JMI 2006]

Q.20 A 50 Hz alternator has single turn full pitched coils. The emf induced in each coil is constant over the middle $2/3^{rd}$ of each half cycle and varies uniformly over the remainder its rms value being 20 V. Calculate flux per pole.

[B.Tech III, I.U.. 2006]

Q.21 Find the voltage regulation at full load 0.8 p.f. lagging for a 3 phase 1500 kVA, 6.6 kV, Y-connected alternator having an armature resistance of 0.09 Ω per phase and a synchronous reactance of 8.5 Ω per phase.

[B.Tech II, I.U., 2005]

REVIEW QUESTIONS

Q.1 Describe the constructional features of cylindrical and salient pole machines.

[B.Tech III, I.U., 2006]

Q.2 Derive an expression for distribution factor of a winding having 'q' slots per pole per phase and a slot angle 'β'.

[B.Tech III, I.U., 2006]

Q.3 What are harmonics? How does the harmonics affect the voltage generated in an alternator?

[B.Tech III, I.U., 2006]

Q.4 State two-reaction theory. Derive the expression of power for cylindrical pole machine.

[B.Tech III, I.U., 2006]

Q.5 What is winding factor? How does it affect the emf induced in a coil of synchronous generator?

[B.Tech III, I.U., 2007]

Q.6 What is open circuit characteristic of synchronous generator? How is it obtained?

[B.Tech III, I.U., 2007]

Q.7 What is short circuit ratio? Discuss its importance.

[B.Tech III, I.U., 2007]

Q.8 Explain two reaction theory.

[B.Tech III, I.U., 2007]

Q.9 Write the different conditions for parallel operation of alternators. Explain with a neat diagram one synchronizing method of an alternator.

[B.Tech II, I.U., 2005]

Q.10 Explain the term "Voltage Regulation" as applied to synchronous generator. Discuss the graphical method of determination of voltage regulation by MMF method. Neglect armature resistance. Why is MMF method called optimistic method?

[B.Tech III, JMI, 2005]

Q.11 Explain with suitable sketches the method of synchronization of three-phase alternators by "Rotating lamps method". How is it determined that the incoming machine is faster or slower than the running machine?

[B.Tech III, JMI, 2005]

Q.12 Explain Blondel's two-reaction theory in relation to synchronous machines. Draw the phasor diagram for an underexcited synchronous generator based on this theory and justify the presence and position of each phasor.

[B.Tech III, JMI, 2005]

Q.13 Determine the full load voltage regulation of a salient pole synchronous generator operating at 0.85 p.f lag if $X_d = 1.6$ p.a and $X_q = 0.96$ p.a. Ignore armature resistance.

[B.Tech III, JMI, 2005]

Q.14 Draw OCC and ZPF lagging (cat full load) characteristics of a cylindrical rotor synchronous generator and explain the graphical method of determination of voltage regulation by Potier method.

[B.Tech III, JMI, 2006]

Q.15 Discuss the need of parallel operation of synchronous generators and explain the conditions that must be satisfied for synchronizing a synchronous generator with the infinite bus.

[B.Tech III, JMI, 2006]

Q.16 Draw phasor form factor, chording factor and breath factor. Also derive emf equation of an alternator.

[B.Tech II, UPTU, 2004]

Q.17 Define diagram of a non-salient pole machine supplying full load at

(i) Lagging power load

(ii) Leading power load

[B.Tech II, UPTU, 2004]

Q.18 Discuss the main constructional features of a cylindrical rotor type alternator.

[B.Tech II, UPTU, 2000]

Q.19 Discuss the synchronous impedance and mmf methods of calculating regulation of an alternator. Why do these methods give different results? Explain.

[B.Tech II, UPTU, 2000]

Q.20 Explain power angle characteristic of synchronous machine.

[B.Tech II, UPTU, 2002]

Q.21 What is voltage regulation? Discuss the synchronous impedance method of calculating voltage regulation of a synchronous machine with the help of short circuit and open circuit tests.

[B.Tech II, UPTU, 2001]

Q.22 Write a short note on any one of the following:

(a) Phasor diagram of cylindrical rotor alternator.

[B.Tech II, UPTU, 2001]

(b) Power angle characteristic of synchronous machine

[B.Tech II, UPTU, 2003]

Q.23 By means of a neat diagram, describe the main parts of an alternator with their functions.

[B.Tech II, UPTU, 2003]

Q.24 Derive the emf equation of an alternator incorporating distribution and pitch factor in it. Discuss the effects of these factors on output and performance.

[B.Tech II, UPTU, 2004]

Q.25 Explain power angle diagram.

[B.Tech II, UPTU, 2004]

Chapter

10

Synchronous Motors

10.1 Introduction
10.2 Constructional Features
10.3 Principle of Operation
10.4 Starting Methods
10.5 *V*-Curve
10.6 Torque and Power
10.7 Hunting of Synchronous Motor
10.8 Braking
10.9 Applications
- Solved Questions
- Unsolved Questions
- Review Questions

10.1 INTRODUCTION

The three-phase motors are employed in every industry to carry out a large number of operations. But the polyphase synchronous motor has a typical application in the power industries, that is, power factor improvement. That too by using the synchronous motor for power factor improvement or correction. The synchronous motor behaves as a variable condenser and provides leading kVAr to the system, whose power factor is lagging.

The synchronous motor is available with single-phase and three-phase supply. This motor is constant speed motor and employed for high power ratings of the order of mega watts.

The synchronous motors are always connected across the lines in order that it behaves as a shunt capacitor bank connected in wye. So this motor when operated with high field excitation and under no load is employed for power factor improvement thus increasing the active power taken by the system hence the efficiency increases and also the cost of the active power used by the load is also decreased accordingly. So in most of the large electricity consumers use this method at the supply mains and thus saves the electricity cost being charged by the power industries.

10.2 CONSTRUCTIONAL FEATURES

The construction of a synchronous motor is similar to that of salient pole synchronous generators as shown below. The only difference is that the synchronous motor does not use prime mover, as the synchronous motor uses electrical energy and changes it into mechanical power respectively, which is obtained at the shaft of the machines.

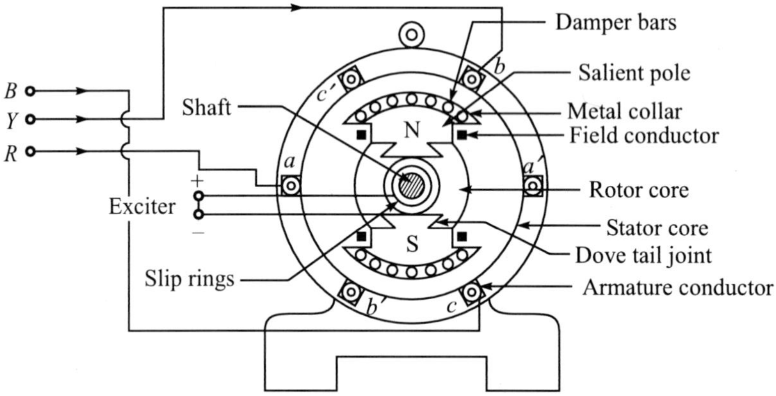

Fig. 10.1 Synchronous motor.

The constructional features are similar and is discussed in the chapter on Synchronous Generators. The damper bars are squirrel cage winding and are short-circuited by the metal collar and as a result the rotor damper bars behaves as a squirrel cage three phase induction motor. The motor gets started by induction motor action and after the magnetic locking of the stator field and rotor poles, the rotor also starts rotating at the same speed i.e. 'N_s'. After that the damper bars become inactive i.e., has no influence on the torque responsible for rotation of the rotor.

10.3 PRINCIPLE OF OPERATION

The synchronous motor has a stationary winding called armature winding which is mounted on the slots of armature so the armature winding is also stationary. The other winding called field winding, the purpose of which is to produce field flux respectively. The field winding is wound over the rotor pole core, which results in north and south pole respectively, when d.c. current passes through field winding.

When the armature windings carry three-phase balanced alternating currents then a rotating magnetic field is produced in the air gap of the machine as in the case of three-phase IM. The north field of rotor gets attracted with the south pole of the rotating field so a torque is produced in one direction. In a very short interval of time the south pole of stator comes in front of the south pole of the rotor as a result repulsive force is exerted on the rotor. So when the poles are alike the torque exerted is clockwise then with pole changing the torque exerted counterclockwise. So the rotor will be stationary and hence the synchronous motor is non self-starting. To start the synchronous motor some auxiliary devices should be used.

So we see from the Fig. (10.2) that the rotor is exerted by equal and opposite forces which results in no motion of the rotor as the rotor is not able to respond to such changes due to large inertia of the motor rotor and the rotor will be at rest.

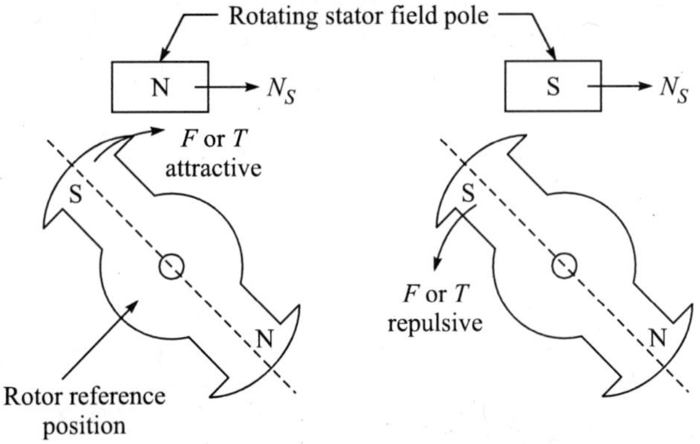

Fig. 10.2 Forces and torque on rotor of synchronous motor.

10.4 STARTING METHODS

As the 3ϕ synchronous motor is not self-starting so the motor is started by using same auxiliary devices which provides motion to the motor rotor externally.

In the synchronous motors which do not have damper windings are started by using a d.c. motor coupled to the shaft of synchronous motor. The d.c. motor gives prime motion to the synchronous motor rotor and afterwards it is decoupled. The motor rotor poles get locked with the stator field pole which is rotating. The rotating field continues to drive the rotor at the same speed.

In the second method, the stator is held by clutch gear and slowing the stator is released such that it starts rotating with synchronous speed. After that it is tightened with rubber footing such that the rotor starts rotating. The stator is tightly held such that the stator is brought to rest and the rotor acquires the full synchronous speed.

In the third method, the synchronous motor is driven by an induction motor having pole pairs less by one. Such that the synchronous motor rotor is moved at its rated speed and then the inductor motor is removed once the synchronous speed is achieved by the motor.

In the last and best method, the damper windings which is also called amortisseur windings are used which are placed on the rotor pole shoes slots and are shorted by metal collar made up of steel such that the rotating stator field flux 'ϕ_s' cuts these rotor bars or windings and emf is included. Since the windings are shorted so the rotor current starts flowing through it. The rotor currents produces rotor flux 'ϕ_r'. The interaction of these two fluxes leads to a torque which results in the movement of rotor by induction motor action. At no load the rotor acquires about 3–4% of synchronous speed of the motor for small motors and 1–2% large motors. The rotor north and south poles get attracted with the opposite poles of the stator field. Once the opposite poles get locked magnetically. The rotors start moving with the speed of rotating magnetic field 'B_s', which is the synchronous speed 'N_s'. The motor remains moving till the locking of the two fields are maintained.

10.5 *V*-CURVE

The curve between the armature current 'I_a' and the field current 'I_f' has a shape of '*V*' and so it is called *V*–curve of synchronous motor. The curve between power factor (cos ϕ) and field current 'I_f' has shape 'Λ' and is called inverted '*V*' curve respectively.

The '*V*' and 'Λ' curve are shown in Fig. (10.3).

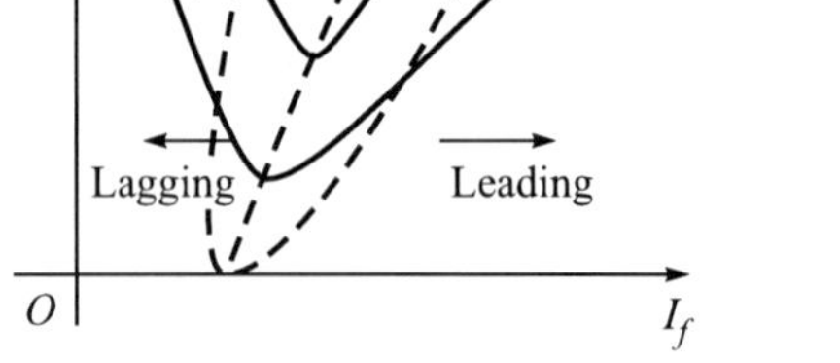

Fig. 10.3 *V*-Curve of 3ϕ Synchronous motor.

The power developed 'P_d' by the motor phase is equal to $V_{ph} \cdot I_{a_{ph}} \cdot \cos\theta$. So if the excitation to the field winding 'I_f' is changed at a given load torque 'T_L', the torque angle 'δ' must change so as to keep the power developed constant

$$T \propto \phi_s \cdot \phi_r \cdot \sin\delta$$

and $$P_d = \omega T$$

or, $$T = \frac{P_d}{\omega} = \frac{P_d}{2\pi n}, \text{ Nm}$$

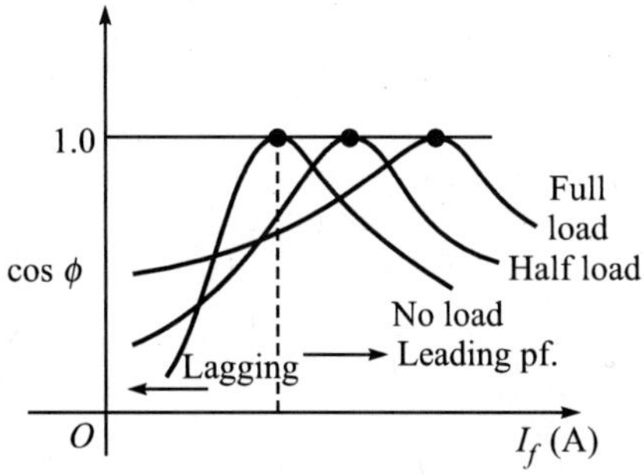

Fig. 10.4 Λ–Curve of 3ϕ synchronous motor.

We see from graph that as 'I_f' is increased the load increases i.e., 'I_a' increases and more power is developed.

If the excitation is decreased, then the load decreases accordingly. We are talking about the electrical load of the motor. Similarly, we can consider that if the torque angle 'δ' is kept constant then the increase in excitation current 'I_f' will result in increase of 'ϕ_r' so to keep torque constant the flux 'ϕ_s' should decrease and hence the motor power factor is increased and that too leading power factor. The motor behaves as a variable capacitor for high field excitation or over excited condition.

When the motor is under-excited then the lagging reactive kVAr is drawn by the motor to maintain the torque so the motor behaves as lagging load or inductive load. At one value of field excitation when both the fluxes are equal then the motor behaves as a resistor having unity power factor.

Thus, where large value variable capacitance, having larger power rating is required then over-excited synchronous motors are used. If large rating variable inductance is required, then under-excited synchronous motors are used.

The dotted lines on the graph shows the locus of power factor equal to 0.8 leading lagging, and the unity power factor respectively.

The 'V' curves of three-phase synchronous motor can be obtained as per the connections shown in Fig. (10.5).

The input of the synchronous motor is connected using two wattmeters which will give the power at the input of motor for different values of field excitation 'I_f'. The power factor ($\cos\phi$) and field current (I_f) at any value of loads can be determined. Thus, V curve can be plotted between 'I_a' and 'I_f'. By measuring '$\cos\phi$' by two-wattmeter method and plotted against 'I_f' the inverted V curve can be drawn for different loads applied at shaft of motor. It can be observed that for higher mechanical loads more armature current 'I_a' is required by the motor phase windings to provide the requisite torque to move the rotor at synchronous speed 'N_s'.

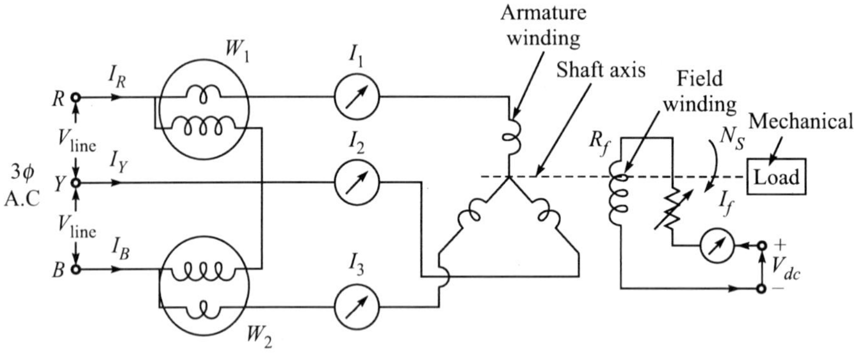

Fig. 10.5 Connection diagram for 'V' curves of 3ϕ synchronous motor.

Most of the synchronous capacitors are having ratings from 20 MVAr to 200 MVAr. The larger ratings are hydrogen cooled. Sometimes pony motor are used to start the synchronous capacitors (over-excited synchronous motor).

10.6 TORQUE AND POWER

The power per phase of stator windings which are connected in *wye* is given by

$$P_{ph} = V_{ph} \cdot I_{ph} \cos\theta$$

where,

P_{ph} = power per phase, watts

V_{ph} = phase voltage = $\dfrac{V_L}{\sqrt{3}}$, volts

I_{ph} = phase current = I_L, Amps.

$\cos\theta$ = power factor of motor

The power developed by the motor is less than the power input to the motor. The developed power per phase is given by

$$(P_{ph})_{dev} = \frac{V_{ph} \cdot E_b}{X_s} \cdot \sin\delta$$

Where,

E_b = Back emf of each winding due to self-inductance of windings, volts

X_s = Synchronous reactance, ohms.

δ = Angle between 'V_{ph}' and 'E_b' is called torque angle

Where, $(P_{ph})_{dev} = \omega T$, watts

Where, $\omega = \dfrac{2\pi N}{60}$ rad/s

T = developed torque, Nm

The torque developed per phase by synchronous motor is given by

$$\boxed{[T_{dev}]_{ph} = \frac{(P_{ph})_{dev}}{\omega}}, \text{ Nm}$$

The torque speed characteristic of the synchronous and induction motor is given by

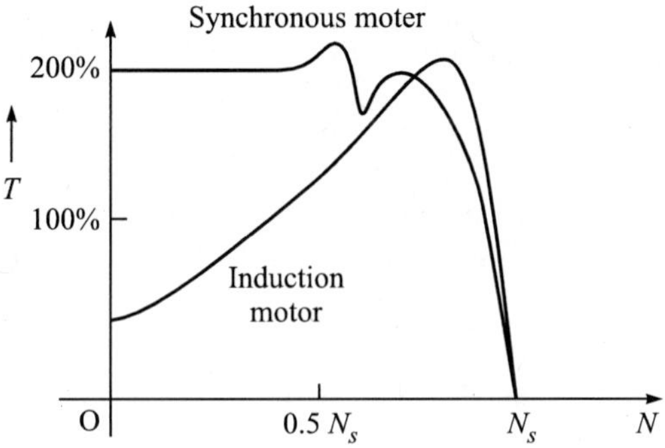

Fig. 10.6 T/N characteristics of synchronous and induction motor (>1000 hp).

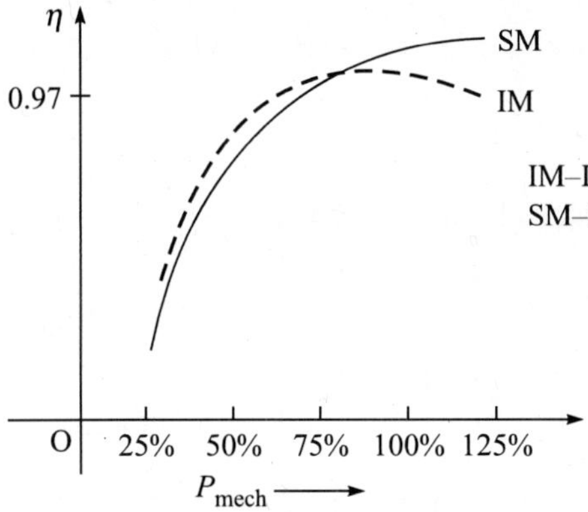

Fig. 10.7 η/P_{mech} characteristic of IM & SM.

The mechanical power developed by synchronous motor 'P_m' is expressed by

$$P_m = \frac{V_{ph} \cdot E_b}{X_s} \cdot \sin\delta$$

The curve between power 'P_m' and torque angle 'δ' is given by Fig. (10.8). The maximum power per phase is

$$(P_m)_{max} = \frac{V_{ph} \cdot E_b}{X_s}$$

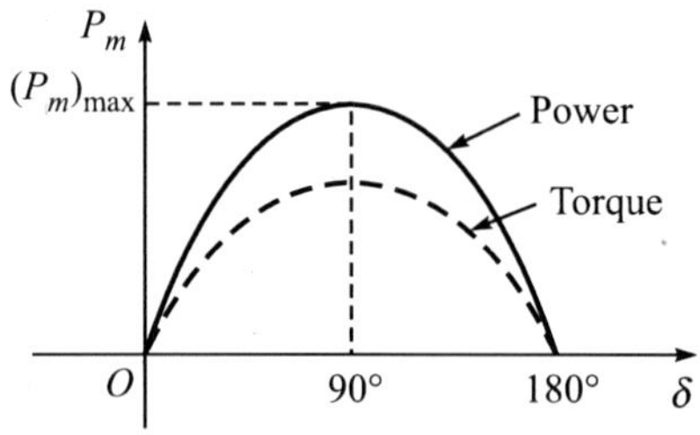

Fig. 10.8 Power and torque variation with torque angle 'δ'.

The rotor speed is constant

So, $$\boxed{T = \frac{9.55 \cdot P_m}{N_s}}, \text{ Nm}$$

where,

T = Torque per phase, Nm

P_m = Mechanical power per phase, watts

N_s = Synchronous speed = $\frac{120f}{p}$, rpm

$9.55 = \text{Constt} = \frac{60}{2\pi}$

The maximum torque the motor can develop is called pull out torque which occurs at $\delta = 90°$.

The mechanical and electrical angles are related by

$$\delta = \frac{p\theta_m}{2}$$

P = No. of poles of motor = $120f/N_s$

θ_m = Mechanical angle from no load position.

10.7 HUNTING OF SYNCHRONOUS MOTOR

The torque angle 'δ' increases whenever the load at the shaft of motor increases. So the motor rotor hunts or oscillates whenever sudden load changes in order to search the equilibrium position of the motor, that is, when both the stator field and rotor field magnetic axis are same. It takes some time to achieve the equilibrium position and during this transient period there are oscillations of motor rotor about the equilibrium position. This is called hunting of synchronous motor. When, the induced voltage 'E_b' and stator phase voltage 'V_{ph}' are lined up and $E_b = V_{ph}$ by adjusting 'I_f', then the motor is said to be floating on the busbar and the motor current 'I_L' or 'I_a' is practically zero or very very small and that to overcome the friction and windage losses incurred by the motor.

If we apply load to the motor by increasing 'T_L' at the shaft then the speed of the motor decreases, i.e, $N < N_s$. So the rotor poles lag behind the stator field poles by an angle 'δ'. Due to this the induced emf in the stator phase windings 'E_b' reaches its maximum value a little later than before. The induced emf 'E_b' lags behind 'V_{ph}' by δ (electrical). The mechanical displacement 'θ_r' produces an electrical phase shift of 'δ' between 'V_{ph}' and 'E_b'. This phase shift produces a potential difference across 'X_s'.

$$V_x = V_{ph} - E_b$$

So, current flow in the stator circuit given by

$$j\,IX_s = V_x$$

So, $$I = -j\,\frac{V_x}{X_s} = -j\frac{(V_{ph} - E_b)}{X_s}$$

The equivalent of motor is given by

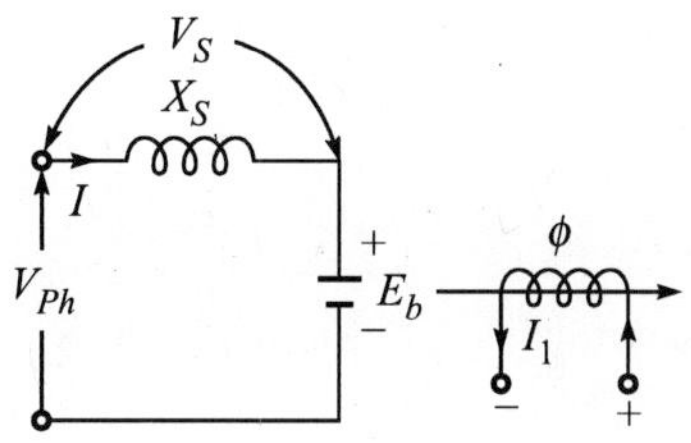

Fig. 10.9 Synchronous motor equivalent circuit.

At no load, V_{ph}, E_b are in phase

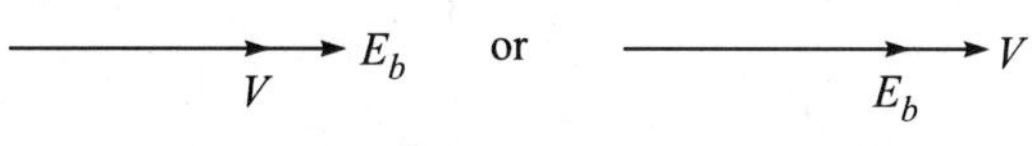

Fig. Phasor diagram of SM (at NL).

At load, there is a phase difference between V_{ph} and E_b.

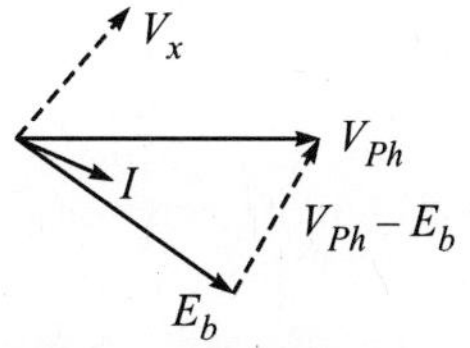

Fig. 'E_b' lags 'V_{ph}'.

If 'I' is in phase with 'V_{ph}' then motor absorbs active power and is transformed completely to mechanical power except for the copper and iron losses in stator.

The value of 'E_b' depends on the power output of motor and desired power factor.

If the mechanical load exceeds the pull out or maximum torque of motor then the rotor poles get separated from the stator poles and the magnetic locking breaks and the motor comes to rest. The stopping of the motor immediately causes the line drop due to large inductive load so it causes line disturbances and the CB's trip immediately. If the motor speed differs from the synchronous speed the motor armature windings and damper windings get heated rapidly. The pull out torque of the motor is 50% to 150% times greater than the nominal full load torque of the motor.

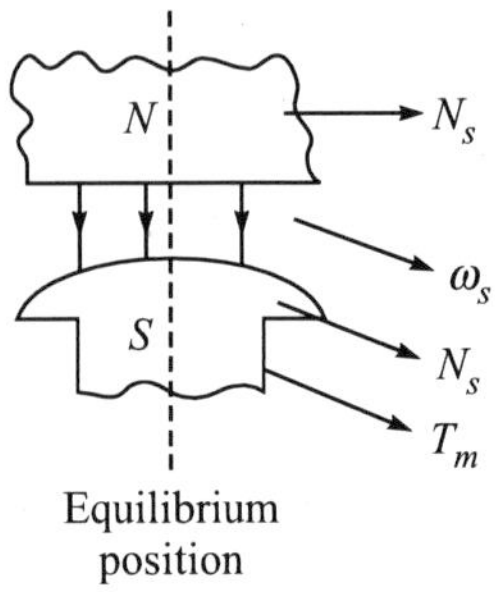

Fig. 10.10 Without load/constant load on synchronous motor.

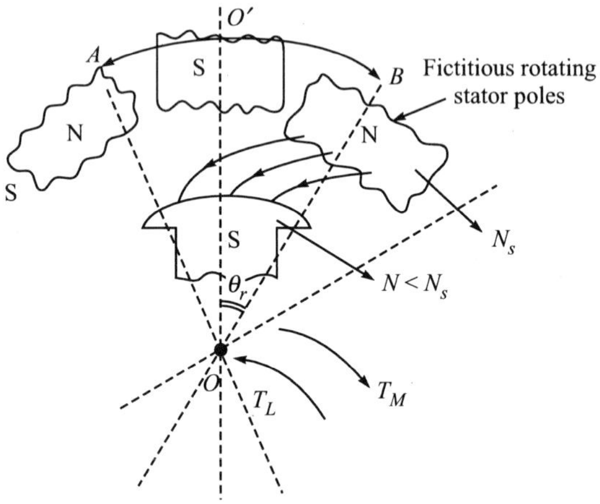

Fig. 10.11 With sudden increase of load on synchronous motor.

From the Fig. (10.11), it is clear that as the motor at the shaft of motor increases the motor rotor oscillates about 'OO''during the hunting till it gets locked with the opposite pole of stator coming behind the previous one with which it gets detached.

10.8 BRAKING

The synchronous motor is needed to be brought to rest in most of the industrial applications. But due to extremely high inertia of the rotor of synchronous motor it takes several hours before it comes to rest. So to brake or stop the motor the following methods are implemented.

1. By short circuit the armature windings thereby generating braking torque and dissipating the energy in resistors of armature circuit. The field excitation is not changed and it is kept at its maximum value.

2. By connecting external resistors to the armature circuits thereby reducing the voltage across the phase windings 'V_{ph}' below the induced emf. 'E_b'. The energy is dissipated in the external resistors. The field excitation 'I_f' = s kept at its maximum value.
3. By applying mechanical braking when the motor speed reaches half its rated value. Thus, it prevents the brake shoes.

10.9 APPLICATIONS

The salient pole types SM are generally available in industrial applications.

The PMSM are used in applications which need large duty cycles such as pumps, fans and compressors.

The hysteresis SM are used in gyrocompasses and small centrifuges.

The SM is also used for high power applications such as blowers, conveyors, steel rolling mills, main line traction, ship propulsion, aircraft testing, mine hoists, reversing mills, etc.

It is used in gas turbine starting and pumped hydro turbine starting. The synchronous motor is used as a frequency changer by coupling it with an alternator such that $\frac{f_{\text{alternator}}}{f_{\text{motor}}} = \frac{P_{\text{alternator}}}{P_{\text{motor}}}$ where, frequency ratio is equal to poles ratio.

SOLVED QUESTIONS

Q.1 The synchronous reactance per phase of a three-phase star connected 6.6 kV synchronous motor is 20 Ω. For a certain load input, the input is 915 kW at normal voltage and the induced line emf is 8.94 kV. Neglecting resistance, determine:

(i) line current (ii) power factor

[C.E. 1996]

A.1 $X_s = 20\ \Omega,\ P_{in} = 915\ \text{kW};\ V_L = 6.6\ \text{kV} = 6600\ \text{V}$

$$V_{Ph} = \frac{V_L}{\sqrt{3}}\ (Y)$$

$$V_{Ph} = 3810\ \text{V}$$

$$E_{b/\text{ph}} = \frac{8.94\ \text{kV}}{\sqrt{3}} = 5.16\ \text{kV}$$

$$R_a = 0\ \Omega$$

$E_b > V_p$, so the motor operates with leading power factor.

$$P_{in} = \sqrt{3}\ V_L I_L \cos\theta = \sqrt{3}\ V_L \cdot I_p \cos\theta$$

$$I_p \cdot \cos\theta = \frac{P_{in}}{\sqrt{3}\,V_L} = \frac{915000}{\sqrt{3}\times 6600} = 80 \text{ A}$$

$$\theta_i = \tan^{-1}\frac{X_s}{R_a} = \tan^{-1}\frac{20}{0} = \tan\infty = 90°$$

$$IZ_s = I\sqrt{R_a^2 + X_s^2} = I\sqrt{O^2 + 20^2} = 20I = E_r$$

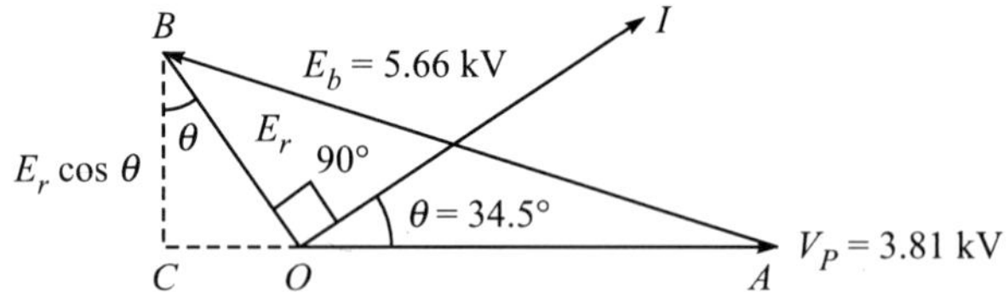

$$\therefore \quad AB^2 = BC^2 + CA^2$$

$$[5.16 \times 10^3]^2 = (E_r \cos\theta)^2 + CA^2 = 80^2 + CA^2$$

$$\therefore \quad CA = 4.9 \text{ kV}$$

$$\therefore \quad E_r = OB = \sqrt{OC^2 + CB^2} = \sqrt{(AC - OA)^2 + CB^2}$$

$$= \sqrt{(1090)^2 + (20\times 80)^2} = 1941 \text{ V} = 1.941 \text{ kV}$$

(i) Line current $I_L = I_p = \dfrac{E_r}{Z_s} = \dfrac{1941}{20} = 97.14$

(ii) Power factor $\cos\theta = \dfrac{I\cos\theta}{I} = \dfrac{80}{97.1} = 0.824$ (lead)

$$\therefore \quad \theta = 34.5°$$

Q.2 A three-phase, 415 V, 6 pole, 50 Hz, *Y* connected synchronous motor has e.m.f. of 520 V (L–L). The stator winding has a synchronous reactance of 2 Ω per phase and motor develops a torque of 220 Nm. The motor is operating at 415 V, 50 Hz bus.

(i) Calculate the current drawn from the supply and its power factor.

(ii) Draw the phasor diagram showing all the relevant quantities.

(GATE 1992)

A.2 $V_L = 415$ V, $P = 6$, $f = 50$ Hz, $E_{b_L} = 520$ V

$X_s = 2$ Ω/ph, $T_d = 220$ Nm.

$$\therefore \quad V_p = \frac{V_L}{\sqrt{3}} (Y) = \frac{415}{\sqrt{3}} = 239.6 \text{ V}$$

$$E_{b/\text{ph}} = \frac{520}{\sqrt{3}} = 300 \text{ V}$$

(i) 'I' and '$\cos\phi$'

$$N_s = \frac{120f}{p} = \frac{120 \times 50}{6} = 1000 \text{ rpm.}$$

$$\text{Total power developed} = 3\,P_m = T_d \times \frac{2\pi N_s}{60}$$

$$= 23038 \text{ W}$$

$$\therefore \quad P_{d/\text{ph}} = 23038 \text{ W}/3 = 7679 \text{ W}$$

Power developed/phase

$$P_m = \frac{E_b \cdot V}{Z_s} \cos(\theta - \alpha) - \frac{E_b^2}{Z_s} \cos\theta$$

$$= \frac{E_b V}{Z_s} \sin\alpha$$

$$\frac{E_b V}{Z_s} \cdot \sin\alpha = 7679.$$

$$\therefore \quad \sin\alpha = \frac{7679 \times 2}{300 \times 239.6} = 0.2137$$

$$\alpha = 12.34^\circ$$

$$E_r = \sqrt{E_b^2 + V^2 - 2E_b V \cos\alpha}$$

$$= \sqrt{300^2 + 239.6^2 - 2 \times 300 \times 239.6 \times \cos 12.34^\circ}$$

$$= 83.48 \text{ V}$$

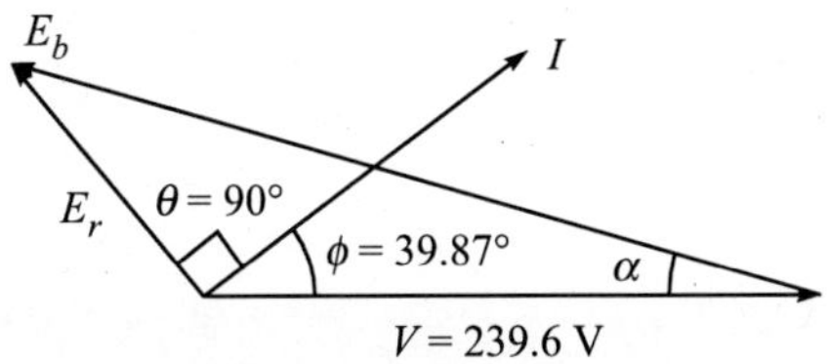

Fig. 2 Phasor diagram of SM.

∴ Current drawn per phase

$$I = \frac{E_r}{Z_s} = \frac{83 \cdot 48}{2} = 41.74 \text{ A}$$

From Fig. 2

$$\frac{E_b}{\sin(\theta + \phi)} = \frac{E_r}{\sin \alpha}$$

$$\therefore \quad \sin(\theta + \phi) = \frac{E_b \sin \alpha}{E_r} = \frac{300 \sin(12.34^\circ)}{83.48} = 0.768$$

$$\theta + \phi = \sin^{-1}(0.768) = 129.87^\circ$$

∵ $E_b > V$, power factor is leading.

$$\phi = 129.87^\circ - \theta = 129.87 - 90$$

$$\phi = 39.87^\circ \text{ (leading)}$$

∴ Power factor $\cos \phi = 0.768$ (lead)

Q.3 A 3ϕ synchronous motor absorbing 60 kW is connected in parallel with a factory load of 240 kW having a lagging power factor of 0.8. If the combined load has a p.f. of 0.9, what is the value of the leading kVAR supplied by the motor and at what power factor is it working.

A.3 $P_L = 240$ kW, $\cos \phi_L = 0.8$, $P_m = 60$ kW,

$\cos \phi = 0.9 \Rightarrow \phi = 25.84^\circ$

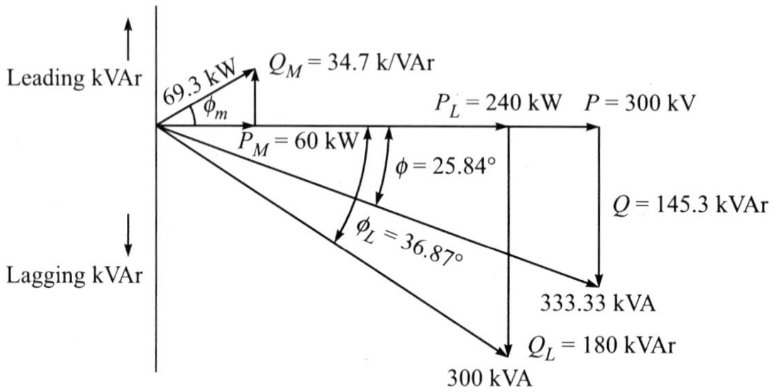

Fig. 3

Value of leading kVAr

Factory load kVAr $Q_L = 240 \tan \phi = 240 \tan(\cos^{-1} 0.8) = 180$ kVAr

Total load $P = P_L + P_M = 240 + 60 = 300$ kW

Combined kVAr $= Q = P \tan \phi = 300 \tan \phi (\cos^{-1} 0.9)$

$= 145.3$ kVAr (lag)

Leading kVAr supplied by motor $Q_M = Q_L - Q$

$$= 180 - 145.3$$

$$= 34.7 \text{ kVAr.}$$

kVA supplied by motor $= \sqrt{P_M^2 + Q_M^2} = \sqrt{60^2 + 34.7^2}$

$$= 69.3 \text{ kVAr}$$

Power factor of motor $\cos \phi_m = \dfrac{60}{69.3} = 0.866$ (lead)

UNSOLVED QUESTIONS

Q.1 A 100 h.p., 400 V, three-phase, star connected synchronous motor has a resistance and leakage reactance per phase of 0.4 ohm and 0.4 ohm respectively. Calculate for full load 0.8 p.f. leading the open circuit emf per phase and mechanical power developed. Assume an efficiency of 92%.

[B.Tech II, UPTU, 2002]

Q.2 A 2 kV, three-phase, four-pole, star connected synchronous motor runs at 1500 rpm. The constant excitation voltage is 2 kV, while its synchronous impedance is $(0 + j3)\ \Omega$ per phase. Determine the power input, power factor and torque developed for an armature current of 200 A.

[B.Tech III, JMI, 2005]

Q.3 A 2.5 kV, three-phase, star connected synchronous motor has synchronous impedance of $(0 + j5)\ \Omega$ per phase. The motor input is 1 MW at rated voltage and at an excitation emf of 3.6 kV. Calculate the line current and power factor.

[B.Tech III, JMI, 2005]

Q.4 A 6.6 kV, star connected, three-phase synchronous motor works at constant voltage and constant excitation. Its synchronous reactance is 20 Ω per phase and armature resistance is negligible. When the input power is 1 MW, the power factor is 0.8 leading. Find the power angle and the power factor when the input is increased to 1.5 MW.

[B.Tech III, JMI, 2006]

Q.5 A three-phase, 3.3 kV star connected synchronous motor has an effective resistance and synchronous reactance of 2 Ω and 18 Ω per phase respectively. If the excitation voltage is 3.8 kV between lines, calculate the maximum power that the motor can develop and also determine the corresponding current and power factor.

[B.Tech III, JMI, 2006]

Q.6 A 200 h.p., 400 V, three-phase star connected synchronous motor has impedance of $(0.06 + j\,0.5)\ \Omega$ respectively. Calculate open circuit emf and the mechanical power developed for full load 0.8 p.f. lagging. Assume 90% efficiency of motor.

Q.7 A 750 kW, 11 kV, three-phase, star connected synchronous motor has a synchronous reactance of 35 ohms/phase and 5 ohms/phase resistance. Determine the excitation emf per phase when the motor is operating on full load at 0.8 p.f. leading. Its efficiency under this condition is 93%.

[B.E. II, AMU, 1992]

REVIEW QUESTIONS

Q.1 Explain the operation of synchronous motors. Also describe their industrial applications.

[B.Tech II, UPTU, 2004]

Q.2 (a) Explain the sequence of events that take place when the load on a synchronous motor is changed.

(b) What is hunting and how can it be avoided? Explain.

(c) Draw the phasor diagram of a salient pole synchronous motor and explain.

[B.Tech II, UPTU, 2000]

Q.3 What do you understand by synchronous condenser? Explain.

[B.Tech II, UPTU, 2002]

Q.4 (a) Draw and explain the *V*–curve of a synchronous motor for different resistances.

(b) Discuss the principle of operation of a synchronous motor.

[B.Tech II, UPTU, 2001]

Q.5 (a) A synchronous motor is not self-starting. Explain.

(b) State the application of synchronous motors. Compare synchronous motors with induction motor drives.

[B.Tech II, UPTU, 2003]

Q.6 Write short notes on synchronous motor *V*–curve.

[B.Tech II, UPTU, 2003]

Q.7 Explain the operation of synchronous motors describe their industrial applications.

[B.Tech II, UPTU, 2004]

Q.8 Is it possible to vary the speed of synchronous motor by varying the field excitation or by any other method? Explain what happens when the field current is (i) increased (ii) decreased.

[B.Tech II, UPTU, 2001]

Q.9 Explain effects of varying excitations on the armature current and power factor in a three-phase synchronous motor. Draw the curves for armature current and power factor variation with excitation for two loads 'P_1' and 'P_2' with $P_1 < P_2$.

[B.Tech II, UPTU, 2002]

Q.10 (a) Draw and explain the:

(i) *V* curve of a synchronous motor for different resistances.

(ii) Power angle diagram.

(b) Discuss the principle of operation of a synchronous motor.

[B.Tech II, UPTU, 2001]

Q.11 Derive an expression for the power developed in a synchronous motor of cylindrical rotor construction, when armature resistance is not negligible. Find the condition for maximum power developed and draw a corresponding power angle characteristic.

[B.Tech III, JMI, 2005]

Q.12 Derive an expression for the torque angle of an over-excited salient pole synchronous motor considering armature resistance. Also draw a phasor diagram for these conditions.

[B.Tech III, JMI, 2005]

Q.13 Explain the operation of a synchronous motor connected to infinite bus and operating at constant load with variable excitation. Draw appropriate phasor diagram and justify the presence of position of each phasor. Also indicate the locus of armature current and excitation voltage phasors.

[B.Tech III, JMI, 2006]

Q.14 Develop an expression for the power developed by an under-excited synchronous motor having salient pole structure. Ignore armature resistance.

[B.Tech III, JMI, 2006]

Q.15 Using a neat diagram describe one starting method of synchronous motor.

[B. Tech III, I.U, 2006]

Q.16 Draw the phasor diagram for non-salient pole motor. Derive the expression for mechanical power.

[B.Tech III, I.U, 2007]

Q.17 Write short notes on

(a) Starting of synchronous motor

(b) Hysteresis motor.

[B.Tech III, I.U, 2007]

Q.18 Justify the following statements:

(a) An over-excited synchronous motor has more still synchronous coupling as compared to an under-excited motor.

(b) Damper windings are not needed in cylindrical solid rotor synchronous machines.

(c) For a constant electrical power output, excitation of a synchronous machine cannot be decreased below certain value.

(d) For same kVA load, efficiency of an alternator at leading p.f. if more than that at lagging p.f.

(e) A synchronous machine with small air gap has less synchronizing power coefficient as compared to a machine with large air gap.

[B.Tech III, AMU, 1998]

Q.19 A synchronous motor connected to an infinite bus is supplying a constant load, is supplying a constant torque load. Its field current is adjusted to operate it at unity power factor. Now if the field current is increased, explain what happens to the following:

(a) The magnitude of the resultant flux

(b) The magnitude of the armature current and its power factor

(c) The magnitude of armature current in phase with the terminal voltage

(d) The load angle 'δ'.

Justify your answer with the help of suitable phasor diagram whenever required.

[B.Tech III, AMU, 1999]

Q.20 The full load torque angle of a synchronous motor at rated voltage and frequency is 30° (elect.). The stator resistance is negligible. How would the torque angle be affected by the following changes:

(i) The load torque and terminal voltage remaining constant, the excitation and frequency are raised by 10%.

(ii) The load power and terminal voltage remaining constant, the excitation and frequency are reduced by 10%.

(iii) The load torque and excitation remaining constant, the terminal and frequency are raised by 10%.

[B.Tech III, AMU, 2000]

Q.21 (a) Explain the phenomenon of hunting in synchronous machines. Explain how damper winding reduces hunting.

(c) Explain why synchronous motor is not self-starting.

[B.Tech III, AMU, 2000]

Q.22 Explain the operation of a synchronous condenser. Give phasor diagram.

[B.E. II, AMU, 1992]

Q.23 (a) Explain why a synchronous motor is not self-starting. Discuss a method to start the motor.

(b) Discuss *V*–curves of synchronous motors.

[B.E II, AMU, 1991]

Chapter

11

Single Phase Induction Motor

11.1 Introduction
11.2 Principle of Operation
11.3 Types of Single-Phase IM
11.4 Starting Methods
11.5 Equivalent Circuit of SPIM
11.6 Testing of SPIM
11.7 Stepper Motor
11.8 Hysteresis Motor
11.9 Linear Induction Motors
- Solved Problems
- Unsolved Problems
- Review Questions

11.1 INTRODUCTION

The induction motors which operate on single phase supply are known as single-phase induction motors or in short, SPIM. These single phase induction motors find applications from fractional horsepower motors used in toys to few horse-power range used in the pump motors that are used in houses.

As the rating of the SPIM increases, the cost increases sharply and so it is not preferred in ratings greater than 5 hp operated on power lines. The size also becomes larger in comparison to the equivalent three-phase induction motor.

In this chapter, the most commonly used SPIMs are discussed. Their performance with loading is also covered. The single phase induction motors are used for a large number of domestic and commercial uses. In the domestic purpose it is used in the ceiling fans, pumps, refrigerators, dryers, machine tool operation, blowers, etc.

They are noisy as compared to the equivalent three-phase motors so they are not preferred in the industries. Mostly the squirrel cage induction motor operated on three-phase supply at a line voltage of 400 V is preferred.

11.2 PRINCIPLE OF OPERATION

The single phase motor to has a rotor similar in appearance as that of squirrel cage rotor. There is winding on the stator and no winding is present on the rotor only conducting bars are present which are shorted. The winding of stator is distributed in the stator slots as shown in the Fig. (11.1) below.

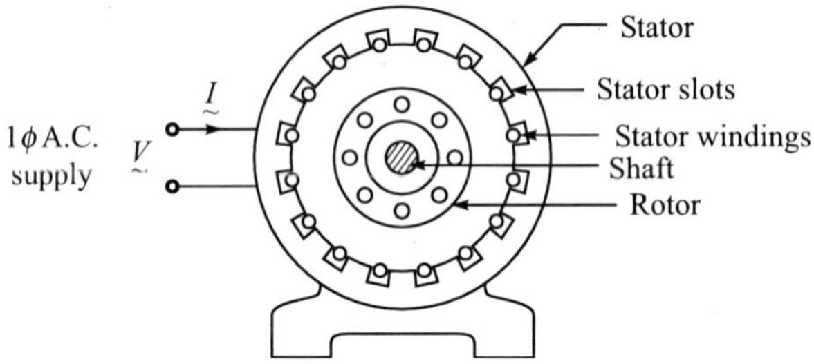

Fig. 11.1 Single phase induction motor.

For understanding the principle of operation of the motor, consider Fig. (11.2) in which the stator winding is shown concentrated for simplicity.

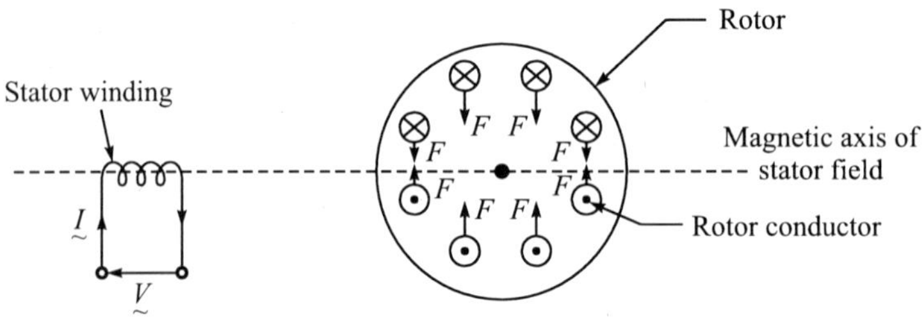

Fig. 11.2 Induced current in rotor conductor due to stator field.

We see from Fig. (11.2) that the rotor conductors on the upper half is exerted by a force 'F' in the lower direction (downward as shown) and the lower half rotor conductors are exerted by same force 'F' in the upward direction as shown in Fig. (11.2). So the torque 'T_u' exerted on the upper half rotor conductors is projected towards the anticlockwise direction as per Flemming's left hand rule whereas the lower half rotor conductors are exerted by a torque 'T_l' in the clockwise direction. So the upper half conductors will try to rotate the rotor in the anticlockwise direction and the lower half rotor conductors will try to rotate the rotor in the clockwise direction. As a result, the rotor will not rotate and will vibrate at a frequency of 50 Hz, so we cannot predict the vibration with naked eye since the resolution of the eye is about $\frac{1}{10}$ whereas, the resolution needed is $\frac{1}{50}$ which is much lesser than the resolution of the eye. So the single phase induction motor fails to start with only single winding in the stator.

The principle of operation of 1ϕ induction motor is explained by revolving field theory in which the stator field ampere turn or m.m.f. is expressed as sum of two m.m.f's of half the amplitude as given by the expression (11.1) below.

$$\mathcal{F} = \mathcal{F}_m \cos \theta \cos \omega t \tag{11.1}$$

where, '$\mathcal{F}_m \cos \theta$' is the amplitude of the stator mmf '$\mathcal{F}$', 'θ' is the angle between stator winding w.r.t. magnetic axis of it. The expression (11.1) can be resolved as follows:

$$\mathcal{F} = \frac{\mathcal{F}_m}{2} (2 \cos \theta \cos \omega t) \tag{11.2}$$

or,

$$\mathcal{F} = \frac{\mathcal{F}_m}{2} [\cos(\theta - \omega t) + (\cos \theta + \omega t)]$$

or,

$$\mathcal{F} = \frac{\mathcal{F}_m}{2} \cos(\theta - \omega t) + \frac{\mathcal{F}_m}{2} \cos(\theta + \omega t) \tag{11.3}$$

or,

$$\mathcal{F} = \mathcal{F}_f + \mathcal{F}_b \tag{11.4}$$

where,

$$\mathcal{F}_f = \phi_f \cdot \mathcal{R} \tag{11.5}$$

$$\mathcal{F}_b = \phi_b \cdot \mathcal{R} \tag{11.6}$$

where,

$\mathcal{F}_f$ = Forward mmf, AT

$\mathcal{F}_b$ = Backward mmf, AT

$\mathcal{R}$ = Reluctance of magnetic circuit path of flux, AT/Wb

There are two fluxes 'ϕ_f' and 'ϕ_b' in which 'ϕ_f' is rotating at a speed 'ω' in the anticlockwise direction whereas 'ϕ_b' is rotating at the same speed in the clockwise direction. Thus, two torques are produced due to the interaction of the fluxes and the rotor currents which are produced due to the induction process. Thus, these torques are directed oppositely so the net torque on the rotor of the motor is equal to zero. As a result the single phase induction motor is non self-starting.

$$T = T_f + T_b = 0 \tag{11.7}$$

$$T_b = -T_f$$

so,

$$T_f + T_b = 0$$

11.3 TYPES OF SINGLE PHASE IM

The single phase induction motor is classified as follows:

(a) Split phase resistance start

(b) Split phase capacitance start
(c) Split phase capacitance start-capacitance run
(d) Shaded pole SPIM
(e) Permanent capacitor SPIM
(f) Reluctance start IM
(g) Repulsion start IM.

The various types of SPIM is made as per the requirements for different purposes which depends on the ratings and size of machine. The SPIM are available from very small size of few centimeters to large size requiring foundation. The ratings are from $\frac{1}{200}$ hp to about 2 hp range. The classification is on the basis of construction and method of starting respectively.

11.4 STARTING METHODS

The single phase induction motors are started by using an auxiliary winding called starting winding. The purpose of this winding is to provide the starting torque to the motor rotor. The starting winding is split into two windings called the main winding and auxiliary winding respectively. Both the windings are space displaced by 90° and both windings have different '$\frac{R}{X}$' ratio. Thus, the current drawn by the motor is split, one part 'I_m' passes through the main winding whereas other part 'I_a' passes through the auxiliary winding respectively. Since the impedance of the main winding is different from that of auxiliary winding. So the current through main winding 'I_m' will have some phase angle which will be different from that of the phase angle of auxiliary winding current 'I_a'. As a result, there will be phase difference between these two currents i.e., 'I_m' and 'I_a' respectively. As a result an unbalanced rotating field will present in the air gap between the stator and rotor. Since the rotor conductors are cutting this rotating field flux lines so an emf is induced in the rotor conductor. As the rotor conductors are short circuit, so rotor current will flow through the rotor conductors which is resulted as a result of the induced emf in rotor conductors. The interaction of the fields produced by the stator current and rotor current which has some phase difference will result in a torque as a result of which the motor starts rotating. This method is called phase splitting. The motor which works on this principle is called split phase induction motor.

In the split phase induction motor, the $\left(\frac{R}{X}\right)$ ratio of the auxiliary winding is kept higher with respect to the main winding. That is, the main winding is more inductive whereas the auxiliary winding is more resistive. In the stator slots, the main windings occupy about $\frac{2}{3}$rd of the total number of slots whereas the rest

$\frac{1}{3}$ rd slot is occupied by the auxiliary winding. The auxiliary windings significant only during starting. After that, it gets out of the circuit in the SPIM employing centrifugal switch or clutch. The centrifugal switch opens when the rotor speed reaches about 75% of the synchronous speed 'n_s'.

Resistance Start Split phase IM

The phase sum of 'I_a' and 'I_m' is equal to 'I' and it is expressed mathematically as follows:

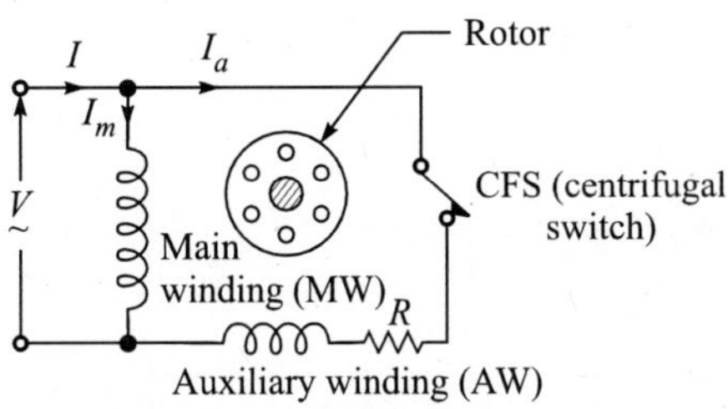

Fig. 11.3 Resistance start SPIM.

$$\bar{I} = \bar{I}_a + \bar{I}_m \tag{11.8}$$

The phasor diagram is given by considering $\bar{V}$ as reference phasor i.e, $\bar{V}\angle 0°, = V$

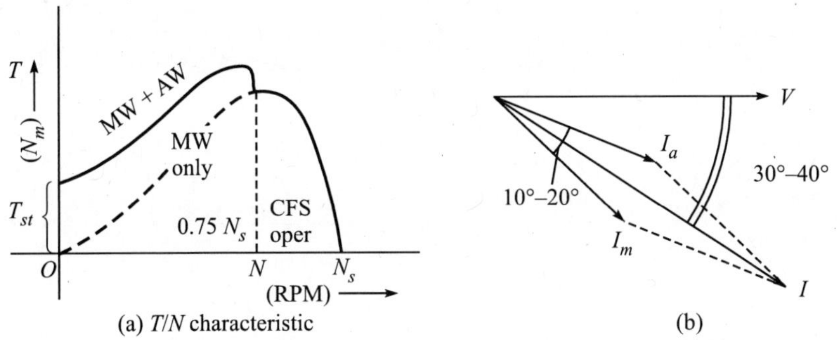

Fig. 11.4 Phasor diagram of voltage and current for resistance start SPIM.

This type of motor has low power factor during starting. This type of motor is preferred for the applications where low starting torque and constant load torque is applied at the shaft of motor. But due to the higher power consumption the motor efficiency is poor. So this type of motor is not used generally.

Uses: Ceiling fans, desk fans etc.

Demerits: Small starting torque

i.e., $$T_{st} = 0.6\ T_{fl}$$

Reactance Start Split Phase IM

In this type of SPIM the motor draws a lagging current from the supply. So the power factor of this type of motor is lower than that of the SPIM previously discussed.

This lagging nature of the motor affects the operation of other devices which are attached to the same supply lines. The voltage regulation of the supply line will become poorer. So, this type of starting is seldom used practically.

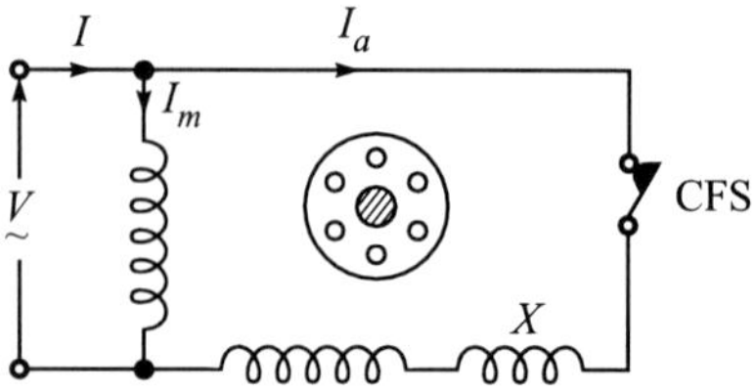

Fig. 11.5 Reactor start SPIM.

Split Phase Capacitor Start IM

In this type of starting of SPIM, a capacitor of high value of the order of 'μF' is placed in series with auxiliary winding. Thus, the phase angle of the auxiliary current 'I_a' is made leading with respect to that of supply voltage 'V'. The main winding current 'I_m' is lagging with reference to the supply voltage 'V'. To determine the phase angle of the current drawn by the motor from lines 'I', we use the phasor representation of the current and voltage as shown in the Fig. (11.7).

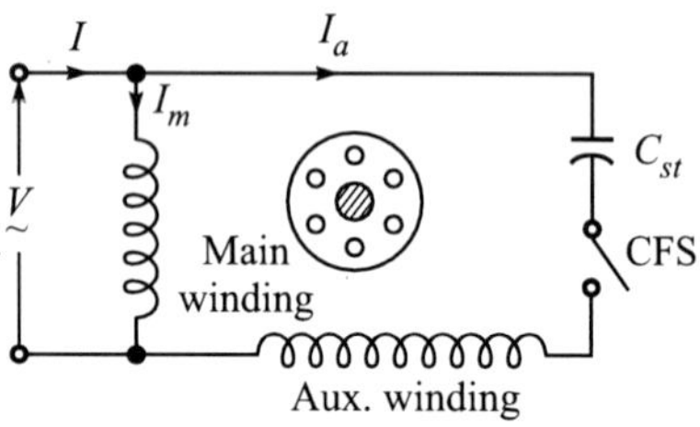

Fig. 11.6 Capacitance Start SPIM.

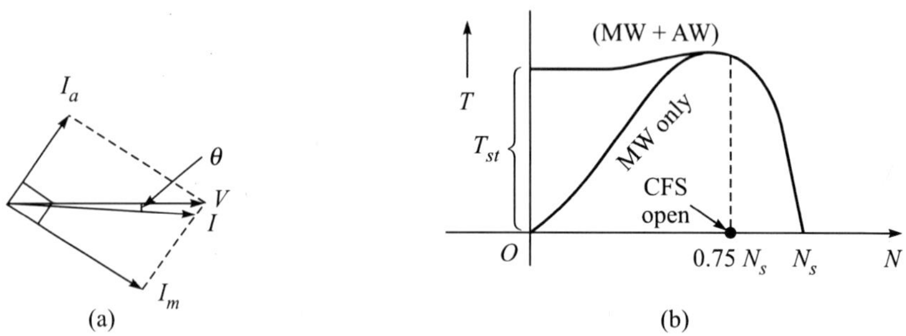

Fig. 11.7 (a) Phasor diagram of 'V' and 'I' for capacitor start SPIM. (b) T/N characteristic of SPIM (single capacitor start).

Thus, from Fig. (11.7) it is clear that the phase angle 'θ' between 'V' and 'I' is decreased. So the 'cos θ' value which is known as power factor is increased. So by employing a capacitor of sufficient value the starting torque of the motor can be improved, on the other hand, the power factor of the motor is made high at starting. So the motor will not affect other loads attached to the same supply lines during starting.

By this method, the motor can be brought to its full speed quickly. But the drawback of this method is that the power factor of motor decreases during running condition.

Advantages:

(1) High starting torque.
(2) Quiet operation
(3) High starting torque ($T_{st} = 4\ T_{fl}$)
(4) Better efficiency at starting

Appls: Pumps, CF fans, etc.

Capacitance Start–Capacitance Run SPIM

This method of starting provides the improvement of power factor of motor during starting and during running condition. It employs two capacitors one having higher value which is called starting capacitor, 'C_{st}' and the other is of lower value called the running capacitor, 'C_{run}'. The capacitor 'C_{st}' is present in the circuit during starting condition whereas it is cut off the circuit after the motor speed reaches about 75% n_s. (where n_s synchronous speed) due to the centrifugal switch which gets opened at a particular speed about 75% n_s due to centrifugal force. The connection diagram of this type of motor is shown in the Fig. (11.8).

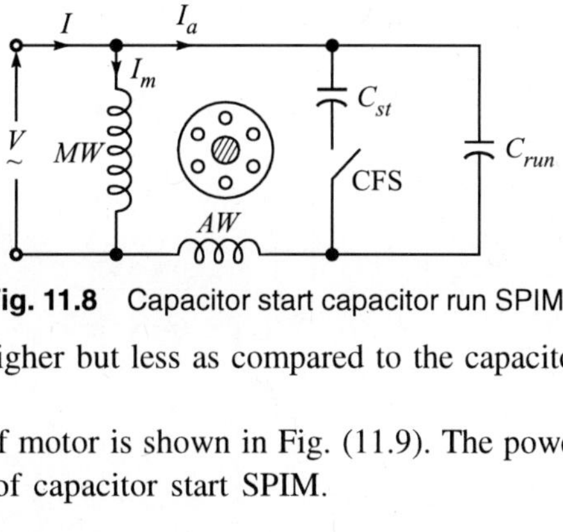

Fig. 11.8 Capacitor start capacitor run SPIM.

The torque of this SPIM is also higher but less as compared to the capacitor start SPIM.

The phasor diagram of this type of motor is shown in Fig. (11.9). The power factor at starting is better than that of capacitor start SPIM.

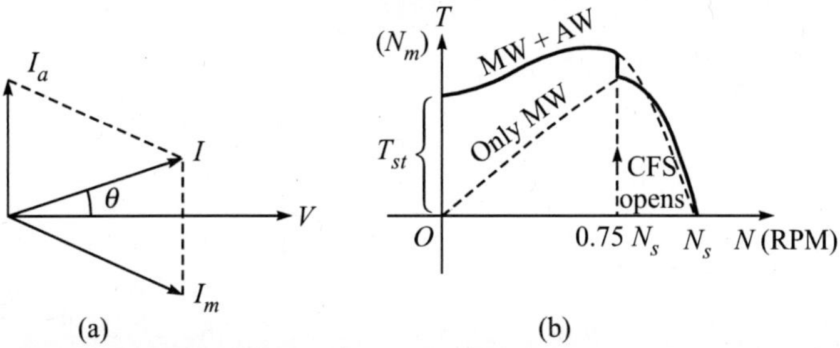

Fig. 11.9 (a) Phasor diagram of capacitor start capacitor run SPIM. (b) *T*/*N* characteristic of two capacitor SPIM.

Such type of SPIM is used in the pumps for lifting water in tanks of houses and industries.

Advantages:

(1) High power factor
(2) High efficiency at starting and running
(3) Low noise
(4) High starting torque $T_{st} = (2.5 - 3.5)T_{fe}$

Applications: Pumps, blowers, refrigerators, fans exhaust type, air-conditioning units etc.

Shaded Pole Induction Motor

The shaded pole induction motor uses a shading band made up of copper which is mounted on one end of pole piece of the motor. There is a division of the main flux. 'ϕ' into two, few flux lines pass through the unshaded portion of the pole say 'ϕ_{ns}' whereas few flux lines pass through the shaded pole say 'ϕ_s'. This could be visualized easily as shown below.

Due to the shading band an emf is induced in this copper band and so current circulates through it. The field under the pole which appears to be moving from unshaded portion of pole to shaded portion of the pole as a result the field inside the air gap will appear to be rotating in the clockwise direction as a result the rotor will also move in the same direction i.e., clockwise as shown in Fig. (11.10). Since the flux lines have an effect similar to stretching of band.

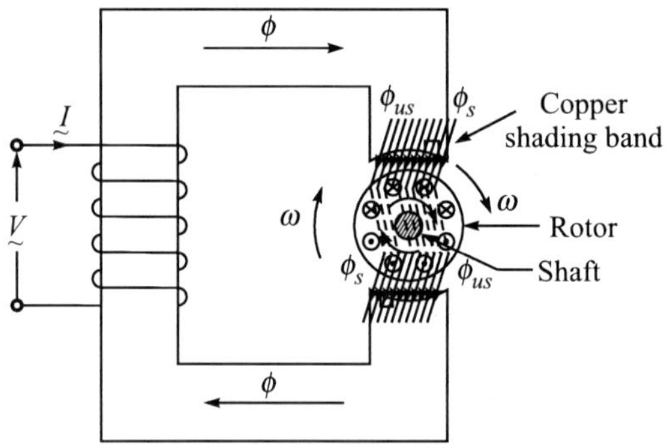

Fig. 11.10 Shaded pole IM.

But the torque produced by this type of motor is very small, that is, about 40% of $T_{\text{full load}}$. So it is used for applications where small starting torque is required.

Advantages:

(1) No need of centrifugal switch or clutch
(2) Self-starting

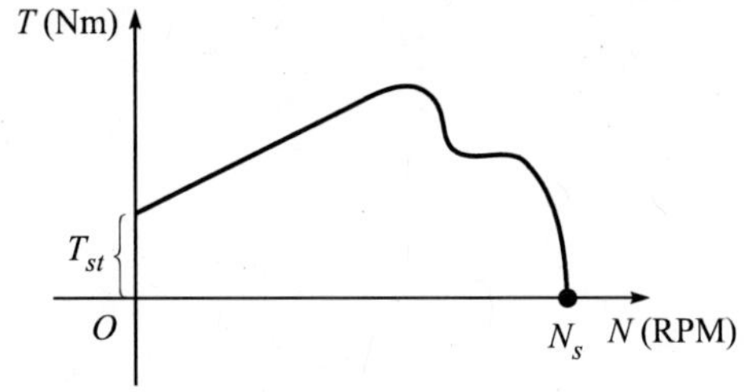

Fig. 11.11 *T/N* characteristic of shaded pole motor.

(3) Need no maintenance
(4) Long life

Disadvantages:

(1) Low starting torque
(2) Low efficiency
(3) Slow
(4) Small rating

Rating: $\frac{1}{3000}$ kW – $\frac{1}{8}$ kW, $I_{st} = 1.5\ I_{\text{Roted}}$

Applications: Toys, blowers, phono motors, projectors, bending machine, etc.

Permanent Capacitor SPIM

In this type of IM the capacitor of the order of micro farads are connected in the auxiliary circuit during starting and running condition. Hence, a small starting torque will lead to starting of the IM.

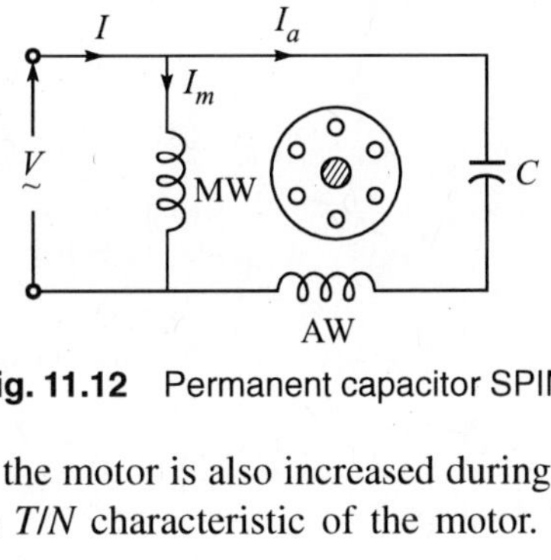

Fig. 11.12 Permanent capacitor SPIM.

The power factor of the motor is also increased during running operation. The Fig. (11.13) shows the *T/N* characteristic of the motor. The advantages of this motor are as follows:

(i) Self-starting
(ii) Low noise
(iii) Need less maintenance
(iv) Low cost

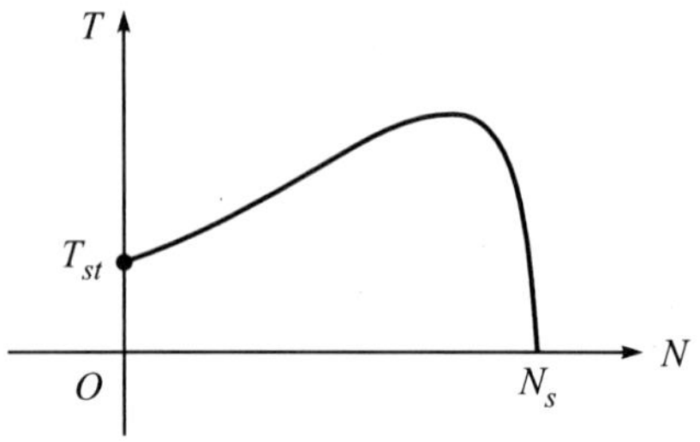

Fig. 11.13 *N/T* characteristic of permanent capacitor SPIM.

The demerits of this type of motor are

(i) Low starting torque ($T_{st} = 0.5\ T_{FL}$)
(ii) Use only for constant loads

Uses:

(i) Ceiling fans
(ii) Intake fans
(iii) Blowers.

Repulsion Start SPIM

In this type of motor the armature consists of d.c. winding and commutator, the brushes are placed such that they are short circuited along an axis displaced through a small angle from the axis of stator winding. The repulsion motor has characteristic similar to series motor. So the starting toque 'T_{st}' is high. When the motor acquires 70% of synchronous speed, the centrifugal switch opens. The short circuited path between the brushes thus makes it as a squirrel cage SPIM. In same types, the brushes are lifted.

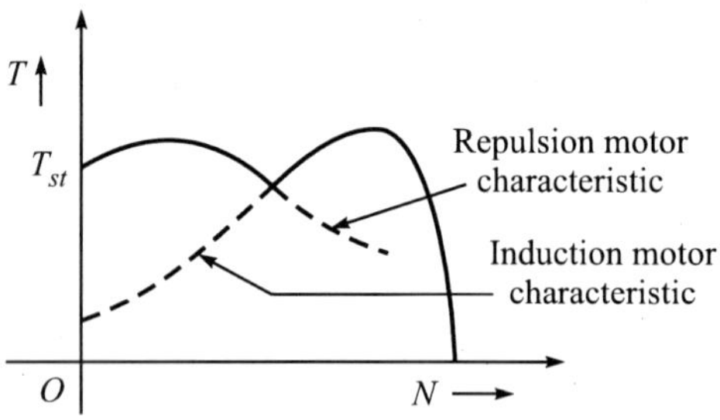

Fig. 11.14 *N/T* characteristic of Repulsion start SPIM.

Advantages:

(i) T_{st} is high
(ii) Low starting current

(iii) Heavy loads can be driven

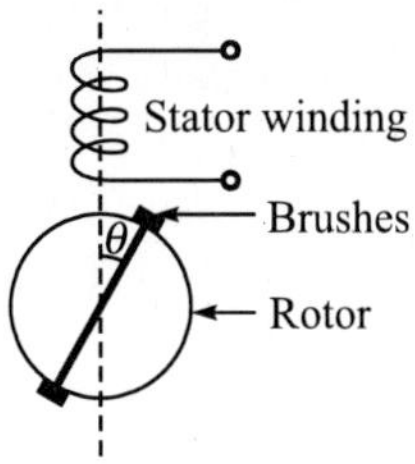

Disadvantages:

(i) Costly
(ii) Noisy
(iii) Reversal not easy

Applications: Machine tools, pumps, fans, etc.

Repulsion Induction SPIM

The rotor has two different types of windings, the outer slots have commutator winding while the inner slots has squirrel cage winding made up of aluminium. The stator winding is similar as SPIM.

During starting the outer commutator winding provides the torque for starting the operation of SPIM. During running the inner winding contributes to the rotor motor.

Advantages:

(1) High T_{st}
(2) Good speed regulation
(3) Ability to withstand heavy loads for longer periods

Disadvantages:

(1) Costly
(2) Needs maintenance

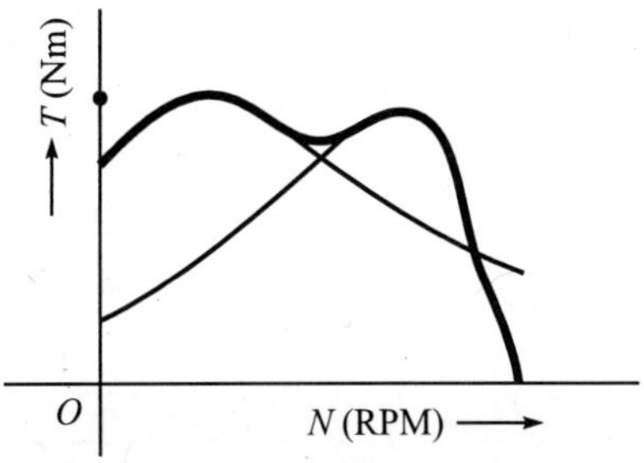

Fig. 11.15 *T/N* characteristic of repulsion induction SPIM.

Applications:

(1) Mixing machines
(2) CF pumps
(3) Large fans
(4) Hoist

Single Phase Reluctance Motor

It starts by induction motor action but when it approaches synchronous speed 'n_s' the salient pole of the rotor is locked magnetically with the rotating field and so the motor acquires synchronous speed.

The reluctance motor only synchronise for small slip speed '$s \cdot n_s$' and the mass of the rotating member is small.

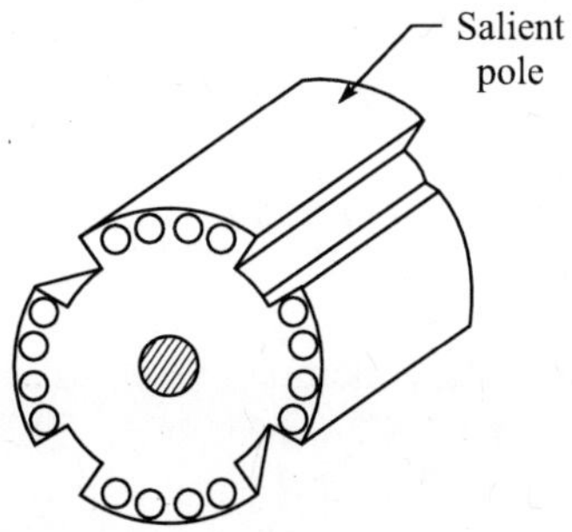

Fig. 11.16 Rotor of reluctance motor.

Advantages:

(1) Cheaper when compared to other constant speed a.c. motor.
(2) For variable frequency speed control in electronic circuitry.

Ratings: From FHP to 100's of horse powers.

11.5 EQUIVALENT CIRCUIT OF SPIM

In this Fig. (11.17) the 'T_f' and 'T_b' are the forward and backward torque respectively. The forward torque will rotate the rotor in the anticlockwise direction and the backward torque will move the rotor in the clockwise direction. The actual torque which will be exerted on the motor is the sum of these two torques. The forward rotating field 'B_f' and the backward rotating field 'B_b' are rotating at synchronous speed 'n_s' respectively. The slip of the rotor with respect to the forward rotating field is called forward slip 'S_f' and the slip of the rotor with respect to the backward rotating field is called backward slip 'S_b'. The relation between the forward and backward slip is given by

$$S_f + S_b = 2$$

$$S_f = \frac{n_s - n}{n_s} = S \text{ (say)}$$

Then,
$$S_b = \frac{n_s - (-n)}{n_s} = \frac{2n_s - n_s + n}{n_s}$$

$$S_b = 2 - \left(\frac{n_s - n}{n_s}\right) = 2 - S_f = 2 - S$$

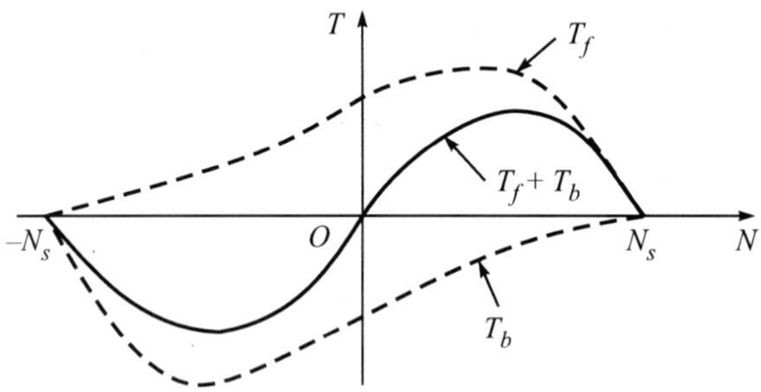

Fig. 11.17 *T/N* characteristic of single phase IM.

The rotor equivalent circuits for the forward field and the backward fields differs with the condition of rotor (i.e., standstill and when in motion). When the rotor is at standstill or rest both the forward and backward slips are equal to '1'. Since the rotor speed is equal to zero at standstill. But as the value of speed 'n'

of rotor increases then $S_f = \dfrac{n_s - n}{n_s}$ decreases and the backward slip $S_b = \dfrac{n_s + n}{n_s}$ increases correspondingly. So from the equivalent circuit of rotor circuit as shown in Fig. (11.18).

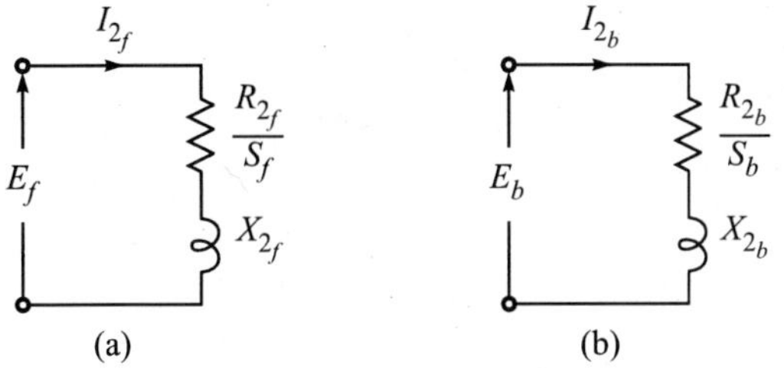

Fig. 11.18 Rotor equivalent circuit of SPIM.

We have,
$$I_{2f} = \frac{E_f}{\left(\dfrac{R_{2f}}{S_f} + X_{2f}\right)} = \frac{E_f}{\left(\dfrac{R_{2f}}{S_f} + X_{2f}\right)} = \frac{E_f}{Z_{2f}}$$

and
$$I_{2b} = \frac{E_b}{\left(\dfrac{R_{2b}}{S_b} + X_{2b}\right)} = \frac{E_b}{\left(\dfrac{R_{2b}}{(2 - S)} + X_{2b}\right)} = \frac{E_b}{Z_{2b}}$$

As the speed of motor increases, the value of slip $S = \dfrac{n_s - n}{n_s}$ decreases, so, the forward rotor current decreases and the backward rotor current increases respectively.

Due to the stator current 'I_1', a pulsating stator field is produced in the air gap which is visualized as two fields 'B_f' and 'B_b' respectively moving in opposite directions in the air gap of motor with a speed 'n_s' respectively. Each of these fields induces e.m.f's in the rotor circuit, namely, 'E_f' and 'E_b'. These two emf's will drive currents 'I_f' and 'I_b' through the rotor circuits respectively. If the rotor resistance is equal to 'R_2' and the rotor leakage reactance is 'X_2'. Then the equivalent circuit of rotor is given as in Fig. (11.19) which combines the effect of both fields as follows.

The parameters of the equivalent circuit as shown in Fig. (11.19) are as explained below:

V_1 = Stator applied voltage, volts
I_1 = Stator current, amperes
R_1 = Stator resistance, ohms.
X_1 = Stator leakage reactance, ohms.
X'_{2f} = Rotor forward leakage reactance, referred to stator side, ohms.

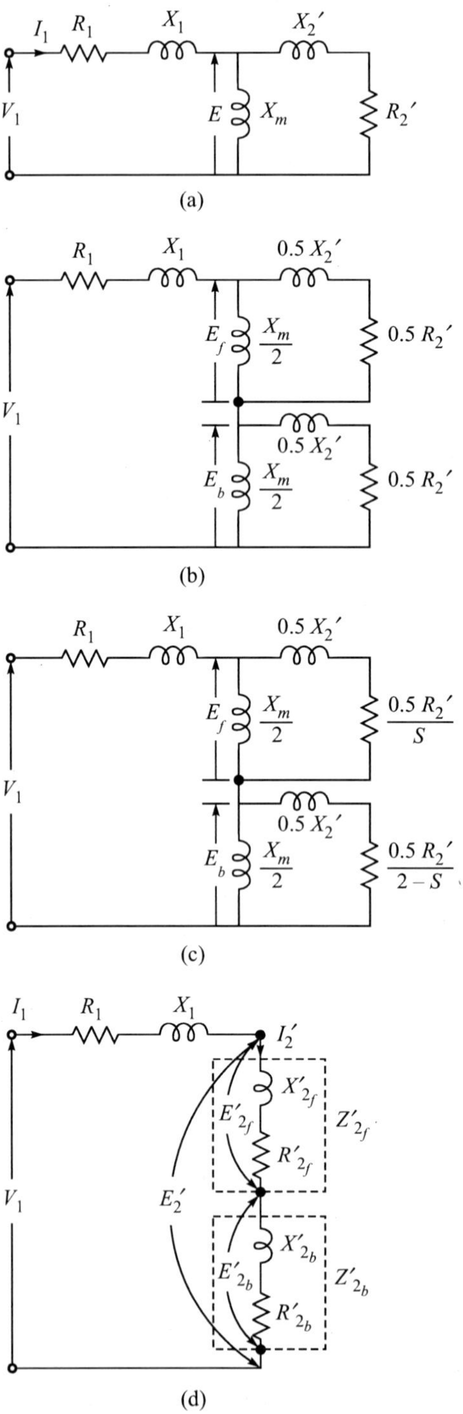

Fig. 11.19 Equivalent circuit of SPIM at standstill and at slip 'S'.

X'_{2b} = Rotor backward leakage reactance referred to the stator side, ohms.

R'_2 = Rotor resistance referred to stator side.

Z'_{2f} = Rotor impedance (forward) referred to the stator side, ohms.

Z'_{2b} = Rotor impedance (backward) referred to the stator side, ohms.

E'_2 = Total rotor induced emf referred to the stator side, volts.

During stand still, $S = 1 \therefore n = 0$

So, $E_f = E_b$

During running condition $n \neq 0$ so, $S < 1$

So, $E_f > E_b$ since $Z_{2f} > Z_{2b}$ (for same rotor current)

Since $E_f \propto \phi_f$ and $E_b \propto \phi_b$

So, the flux in the forward direction will exceed that of backward flux during running condition hence if a small torque is exerted in the forward direction then the induction motor rotor continues to move in that direction as the torque in that direction exceeds by much more amount when compared with the backward torque.

11.6 TESTING OF SPIM

The single phase induction motors are tested under no load condition (i.e., without actually loading the machine shaft) and under full load condition (by blocking the machine's shaft by tightening the belt of the pulley). Thus, two types of test are performed on the motor to determine the parameters of the equivalent circuit of the induction motor which corresponds to the electrical analog of the mechanical equivalent (i.e., motor). These tests are no load test and blocked rotor test respectively.

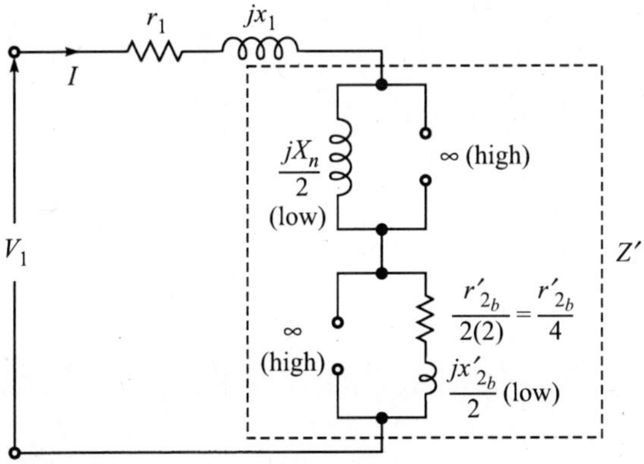

Fig. 11.20 Equivalent circuit of SPIM (under no load).

By the no load test, 'S' is small since 'n' is high as per the relation

$$S_f = \frac{n_s - n}{n_s} = S \text{ say, } S \to 0$$

So, $\left(\frac{r'_{2f}}{2s_f}\right)$ is high. To an approximation, it can be regarded as open circuit since current through it is very small.

If a wattmeter is attached at the stator side of SPIM, then it will measure a power 'P_0' given by the expression as follows:

$$P_0 = P_{\text{stray}} + I_0^2\left(r_1 + \frac{r'_{2b}}{4}\right)$$

Excluding stray power loss 'P_{stray}', then power measured by the wattmeter 'P_0' is given by

$$P_0 \approx I_0^2\left(r_1 + \frac{r'_{2b}}{4}\right) \approx P_{\text{iron}}$$

Thus, no load power corresponds to iron loss. The impedance of the motor equivalent circuit is equal to Z_0 (say) which is given by

$$Z_0 = (r_1 + jx_1) + Z'_2 = Z_1 + Z'_2$$

$$Z_0 = \frac{V_0}{I_0} \text{ (known)}$$

where, V_0, I_0 are the voltmeter and ammeter readings which are attached at the stator side.

Then, the reactance is given by

$$X_0 = \sqrt{Z_0^2 - \left(\frac{4r_1 + r'_{2b}}{4}\right)^2}$$

$$= \sqrt{\left[r_1 + j\left(x_1 + \frac{x_0}{2}\right) + \left(\frac{r'_{2b}}{4} + j\frac{x'_{2b}}{2}\right)\right]^2 - \left(r_2 + \frac{r'_{2b}}{4}\right)^2}$$

$$\simeq x_1 + \frac{x'_2}{2} + \frac{x_0}{2}$$

where,

$$\left(r_1 + \frac{r'_{2b}}{4}\right) = \frac{4r_1 + r'_{2b}}{4} = \frac{P_0}{I_0^2}\text{; from above expression}$$

where, P_0, I_0 are the wattmeter and ammeter readings under NL condition and which are known.

If $\quad x_1 = x'_{2b}$ and $x_1 = x'_{2b} = \dfrac{x_1 + x'_{2b}}{2}$

$$X_0 = \sqrt{\left[r_1 + \frac{r'_{2b}}{4} + j\left(x_1 + \frac{x_0}{2} + \frac{x'_{2b}}{2}\right) + r_1 + \frac{r'_{2b}}{4}\right]\left[\cancel{r_1} + \cancel{\frac{r'_{2b}}{4}} + j\left(x_1 + \frac{x_0}{2} + \frac{x'_{2b}}{2}\right) - \cancel{r_1} - \cancel{\frac{r'_{2b}}{4}}\right]}$$

$$X_0 = \sqrt{\left[2\left(r_1 + \frac{r'_{2b}}{4}\right) + j\left(x_1 + \frac{x_0}{2} + \frac{x'_{2b}}{2}\right)\right]\left[j\left(x_1 + \frac{x_0}{2} + \frac{x'_{2b}}{2}\right)\right]}$$

which gives magnetizing branch reactance as $X_0 \approx 2\left(X_0 - x_1 - \frac{x'_2}{2}\right)$

Blocked Rotor Test

Since, rotor is blocked, so, $n = 0,\ S = \frac{n_s - n}{n_s} = 1 = S_f$

since, $\frac{r'_{2f}}{2s_f}$ is small, x'_{2f} is also small and X_m is high so, Z'_{2f} is small $\Rightarrow I'_{2f}$ is high

As $\qquad I'_{2f} = I'_{2b}$ (at $n = 0$ or blocked rotor)

The equivalent circuit of the induction motor reduces to

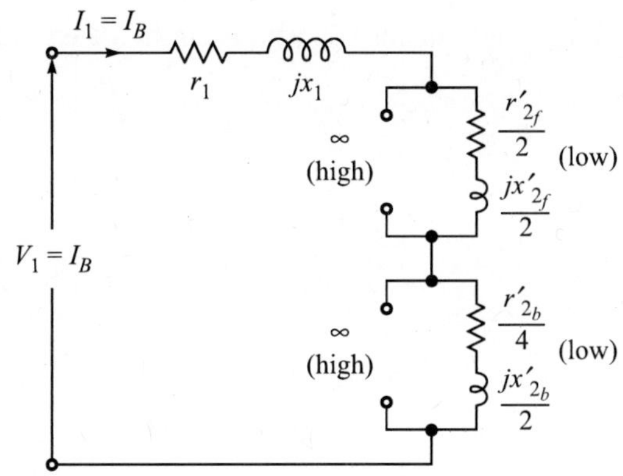

Fig. 11.21 Equivalent circuit of SPIM (Blocked rotor).

So, the total impedance of SPIM is equal to

$$Z_{\text{blocked}}\ (\text{say}) = Z_B = \left(r_1 + \frac{r'_{2f}}{2} + \frac{r'_{2b}}{4}\right) + j\left(x_1 + \frac{x'_{2f}}{2} + \frac{x'_{2b}}{2}\right)$$

$$= R_B + jX_B$$

If 'V_B' and 'I_B' are the voltmeter and ammeter readings with rotor blocked, then the impedance 'Z_B' is given by $Z_B = \frac{V_B}{I_B}$.

If a wattmeter is attached at the stator side then the wattmeter readings 'P_B' is given by

$$P_B = I_B^2\, R_B = I_B^2 \left(r_1 + \frac{r'_{2f}}{2} + \frac{r'_{2b}}{4} \right), \text{ watts.}$$

So, $$r_1 + \frac{r'_{2f}}{2} + \frac{r'_{2b}}{4} = \frac{P_B}{I_B^2}, \text{ ohms.}$$

'r_1' can be determined by Voltmeter-Ammeter method with d.c. current.

$$r_{1_{a.c}} = r_{1_{dc}} \times 1.6.$$

The rotor resistance referred to stator side is given by

$$r'_2 = \frac{P_B}{I_B^2} - r_1$$

$$x_1 + x'_2 = \sqrt{Z_B^2 - R_B^2}$$

If $$x_1 = x'_2, \text{ then } x_1 = x'_2 = \frac{x_1 + x'_2}{2}$$

Moreover, the $$P_B = P_{cu}$$

So, the sum of the powers 'P_0' and 'P_B' gives account of the total losses incurred in the machine. So the motor efficiency can be determined as per the relation

$$\eta = \frac{P_0}{P_{in}} = \frac{P_{in} - \text{losses}}{P_{in}}$$

$$\eta - 1 - \frac{\text{losses}}{P_{in}}$$

The 'P_{in}' is given by the exp.

$$P_{in} = V \cdot I, \text{ watts.}$$

where, 'V' and 'I' are the r.m.s. value of the applied voltage and the current drawn by the motor respectively.

11.7 STEPPER MOTOR

Stepper motor as the name implies is a motor in which the motor rotor moves in steps. The step angle depends on the construction and design of the machine. The most commonly used steps are 2.5°, 5°, 7.5°, 15° or higher depending on the pulse applied to the windings.

This type of stepper motors are used for precision operation with reference to position and speed. If the pulse rate or frequency is increased then the step angle can be reduced and vice versa. The pulses are digital type and are applied to the stator windings of the motor. The rotor of this motor also available in number of types. The applications are in printers, tape drives, *XY* plotters, valves, robots, machine tools, clocks, and control operations. The ratings are from FHP range to several H.P.

The most popular stepper motors are:

(i) Permanent magnet stepper motor (PMSM)
(ii) Variable reluctance stepper motor (VRSM)

These two types of stepper motor are distinguished on the basis of the rotor constructional differences as shown in the Fig. below.

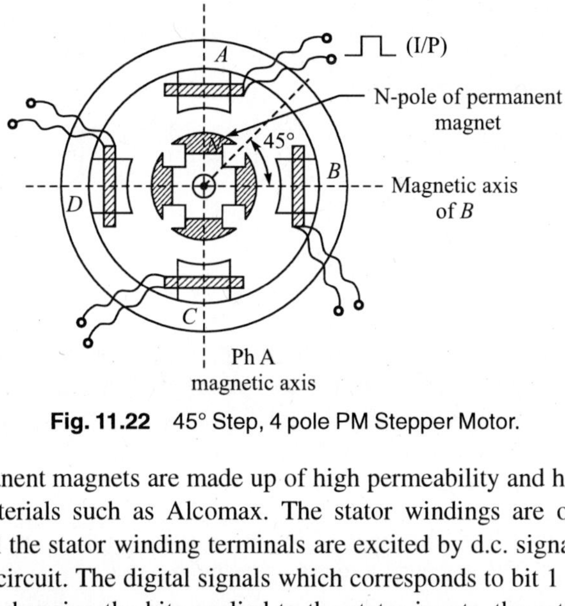

Fig. 11.22 45° Step, 4 pole PM Stepper Motor.

The permanent magnets are made up of high permeability and high retentivity magnetic materials such as Alcomax. The stator windings are of multi-phase windings. All the stator winding terminals are excited by d.c. signals given from a low power circuit. The digital signals which corresponds to bit 1 (high) or bit 0 (low). So by changing the bits applied to the stator inputs, the rotor motion can be moved accordingly.

That is, consider the digital pulses applied to the four-phase windings *A*, *B*, *C* and *D* in the sequence as follows.

Table: Showing the position of rotor in degrees with reference to different excitations w.r.t. magnetic axis of '*A*' initially.

phD	*phC*	*phB*	*phA*	θ	*Position of Rotor*
0	0	0	1	0°	Along *A* (magnetic axis)
0	0	1	1	22.5°	Between *A* & *B*
0	1	1	1	45.0°	Along *B* magnetic axis
1	1	1	1	57.5°	Between *B* and *C*

Thus, several other positions could be obtained by different values of signals applied to the phase windings of the stator. The position of the rotor will be so aligned in the direction in which the magnetic poles of the rotor and stator are in the same direction as shown in the Fig. (11.22).

Variable Reluctance Stepper Motor

It uses rotor having low retentivity. The reluctance of the rotor periphery is made different by making slots and teeth respectively as shown in Fig. (11.23). Thus, the motor rotor takes that position in which the reluctance of the magnetic path is minimum.

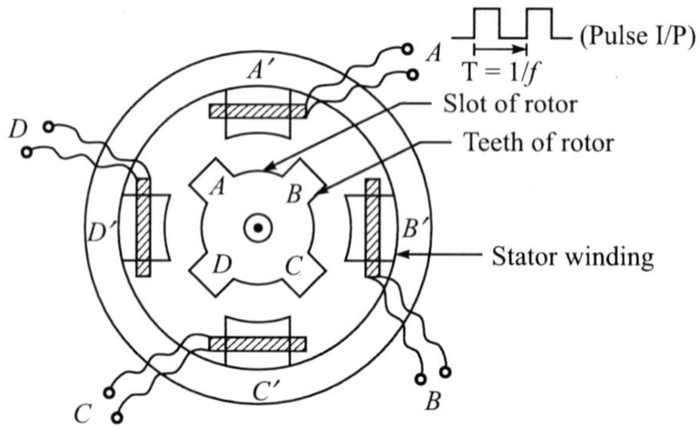

Fig. 11.23 VR stepper motor.

By increasing the *I/P* pulse frequency the speed of the rotor could be increased accordingly. The torque produced in the reluctance stepper motor is less with reference to the PM stepper motor.

Step size are 1.8°, 5°, 7.5°, 15°, etc. Used in the digital clocks. Comparison of the two stepper motors are as follows:

(1) The mass of rotor of VR stepper motor is small.
(2) The speed of VR stepper motor rotor is high.
(3) It requires less time to produce step change them PM stepper motor.
(4) No oscillations at end of each step with respect to PM stepper motor.
(5) Step angle can be made smaller them PM stepper motor.
(6) It is available in ratings higher than PM stepper motor.

11.8 HYSTERESIS MOTOR

The hysteresis motor can have salient pole type or non-salient pole type of rotor. The rotor is composed of a ceramic material which has high resistance so the eddy currents will be almost absent due to high resistance of rotor. So the only losses that could be induced in the rotor are the hysteresis loss. The rotor is made up of high retentivity steel produces a high loss due to hysteresis. The hysteresis torque will lead to the motion of rotor. The hysteresis torque will lead to high speed of the rotor which is independent of voltage and frequency. As the rotor approaches synchronous speed, the magnetic field which is revolving in the air gap will get locked with the field of rotor as a result the rotor also start rotating at synchronous speed. The torque of rotor is almost constant. The torque produced by the hysteresis motor is steady so it is used for a number of applications such as CD players, tape recorders, etc.

Advantages:

(1) Low inertia
(2) High pick-up speed
(3) Gyroscope
(4) Quiet operation

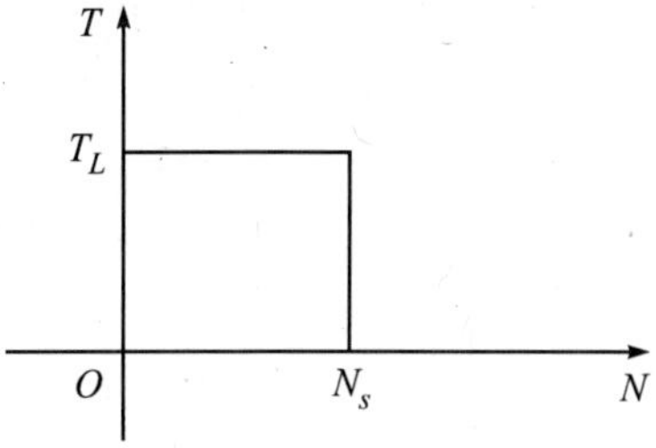

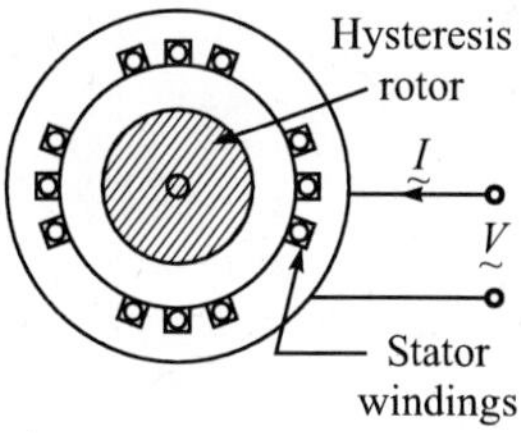

Fig. 11.24 *T/N* characteristic of hysteresis motor.

11.9 LINEAR INDUCTION MOTOR

As the name of the motor implies that this IM is used for linear motion. The construction of the LIM is shown as follows

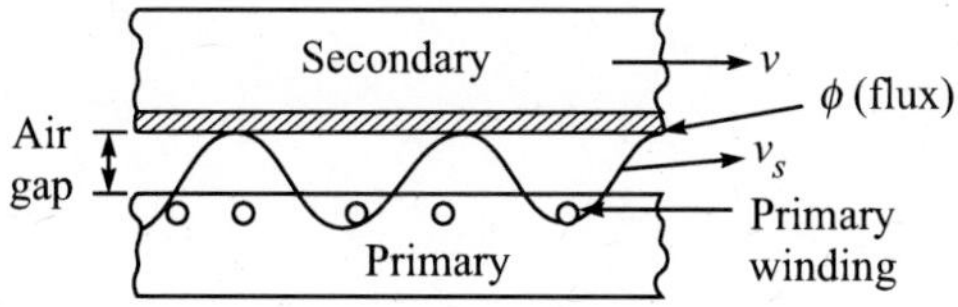

Fig. 11.25 LIM (single sided).

The air gap of LIM is large ~10 mm. So the current required by primary is higher to provide the required flux 'ϕ' in the air gap. The characteristic of LIM between force 'F' and velocity 'v_s' is given by Fig. (11.25)

$$V_s = 2\text{ wf, m/s}$$
$$V_0 = \text{linear synchronous speed, m/s}$$
$$\omega = \text{Width of pole pitch, m.}$$
$$f = \text{Frequency of flux in air gap, Hz}$$

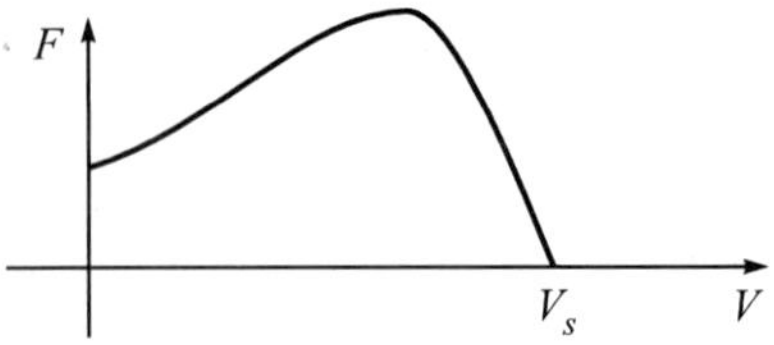

Fig. 11.26 Characteristic of LIM.

The velocity 'V_s' is independent of pole of machine. The slip of motor is given by

$$S = \frac{V_s - v}{v_s}$$

where, 'v' is the speed of secondary member.

Applications: Transport, material handling, pumping, sliding door closer, curtain pullers, etc.

SOLVED QUESTIONS

Q.1 The following test data of a $\frac{1}{2}$ hp, 1ϕ, 230 V, 50 Hz, 2500 rpm induction motor

Blocked Rotor Test: $V_B = 80$ V, $I_B = 5A$, $P_B = 150$ W

No Load Test: $V_0 = 230$ V, $I_0 = 2A$, $P_0 = 100$ W

If stator resistance is 3 Ω, obtain the equivalent circuit parameters. Also obtain rotational loses.

A.1 (i)

$$R_1 = 3\ \Omega$$

$$P_B = I_B^2 (R_1 + R'_{2b}) \Rightarrow R_1 + R'_{2b} = \frac{P_B}{I_B^2} = \frac{150}{25} = 6\ \Omega$$

$$\therefore \quad R'_{2b} = 6 - 3 = 3\ \Omega$$

$$Z_B = \frac{V_B}{I_B} = \frac{80}{5} = 16\ \Omega$$

$$\therefore \quad X_1 + X'_2 = \sqrt{Z_B^2 - (R_B + R'_2)^2} = \sqrt{16^2 - 6^2}$$

$$= \sqrt{220} = 14.83\ \Omega$$

If $\quad X_1 = X'_2$, then $X_1 = X'_2 = \dfrac{X_1 + X'_2}{2} = 7.42\ \Omega$

$$P_0 = I_0^2 \left(R_1 + \frac{R'_{2b}}{4} \right) = I_0^2\, R_0$$

$$\therefore \quad R_0 = \frac{P_0}{I_0^2} = \frac{100}{4} = 25\ \Omega$$

$$\therefore \quad Z_0 = \sqrt{R_0^2 + \Bigg(\underbrace{\frac{X_m}{2} + X_1 + \frac{X'_{2b}}{2}}_{X_0}\Bigg)^2} = \frac{V_0}{I_0}$$

$$\frac{230}{2} = 115 = \sqrt{25^2 + X_0^2}$$

$$X_0^2 = 115^2 - 25^2 = 12600$$

$$X_0 = 35.6\ \Omega$$

$$\because \quad X_1 = 7.42\ \Omega = X'_{2b}$$

$$\therefore \quad \frac{X_m}{2} = 7.42 + 3.71 = 35.6$$

$$\Rightarrow \quad \boxed{X_m = 48.94\ \Omega}$$

Thus, the equivalent circuit of the SPIM is given by

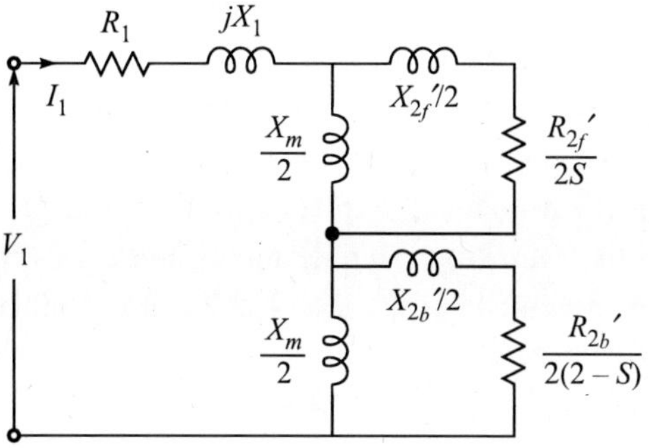

(ii) The rotational losses are given by

$$P_0 = I_0^2 \left(R_1 + \frac{R'_{2b}}{2}\right) + P_{\text{rotational}}$$

$$P_{rot} = P_0 = I_0^2\left(R_1 + \frac{R'_{2b}}{2}\right)$$

$$= 100 - 4\left(3 + \frac{3}{2}\right)$$

$$= 100 - 4\ (3 + 1.5) = 100 - 18 = 82\ \text{W}$$

UNSOLVED QUESTIONS

Q.1 The following tests are performed on a SPIM of rating 200 W, 220 V, 50 Hz, 6 pole having 12 Ω stator resistance. If the test results are:

BRT: 110 V, 3*A*, 220 W
NLT: 220 V, 1.35*A*, 70 W

Determine the parameters of equivalent circuit. Also determine the current, power factor and efficiency of motor.

[12.44 Ω, 13.67 Ω, 0.74 lag, 60%]

Q.2 A test on 1/4 hp, 120 V, 60 Hz, 1725 rpm, 1ϕ, IM are as follows:

$r_1 = 2\ \Omega,\ r_2' = 4\ \Omega,\ x_1 = 3\ \Omega,\ x_2' = 3\ \Omega,\ x_m = 60\ \Omega,\ r_{\omega+fi} = 600\ \Omega$

Draw the equivalent circuit of SPIM. Determine the power factor of motor.

[0.57 lag]

REVIEW QUESTIONS

Q.1 Explain double revolving field theory of single phase IM.
Q.2 Why the SPIM is non-self-starting?
Q.3 Discuss the starting methods of SPIM.
Q.4 Compare the various starting methods of SPIM.
Q.5 Derive the equivalent circuit of SPIM.
Q.6 Why testing of SPIM is done? Explain with a neat diagram.
Q.7 Write notes on the following: Stepper motor, hysteresis motor, LIM.
Q.8 A test on the main winding of a 1 kW, 4 pole, 215 V, 50 Hz, 1ϕ, IM gave the following results:

No load test: 215 V, 3.9*A*, 185 W
Blocked Rotor test: 85 V, 9.8*A*, 390 W

Resistance of the main winding is given to be 1.6 Ω.

(a) Calculate the parameters of the circuit model assuming that the magnetizing reactance is connected at the input terminals.

(b) Determine the line current, power factor and shaft torque, when the motor is running at a speed of 1440 rpm.

[$x_m = 55.1\ \Omega,\ r_1 = 1.6\ \Omega,\ r_2' = 3.12\ \Omega,\ I_L = 6.27\ \angle 43°\ A,\ O$]

[B.E II, AMU, 93]

Q.9 (a) Explain the following motors with neat diagrams and also give their torque speed characteristics

(i) Capacitor Start Motors

(ii) Capacitor Run Motors
(iii) Shaded Pole Motors

(b) The following test data are obtained from a $\frac{1}{4}$ hp, 120 V, 60 Hz, 1730 rpm, induction motor. Stator winding (main) resistance = 2.9 Ω Blocked rotor (standstill) test: $V = 43$ volts, $I = 5$ A, $P = 140$ W; No load test: $V = 120$ volts, $I = 3.5$ A, $P = 125$ W.

(i) Obtain the double revolving field equivalent circuit for the motor.
(ii) Determine the rotational loss, input current, input power, power factor and torque developed.

[B.Tech IV, JMI, 2005]

Q.10 (i) Explain double revolving field theory with neat diagrams.
(ii) The currents in the main and auxiliary windings are as follows:

$$i_m = \sqrt{2}\, I_m \cos \omega t;\; i\,a = \sqrt{2}\, I_a \cos(\omega t + \theta_a)$$

The effective numbers of turns for the main and auxiliary windings are 'N_m' and 'N_a'. The windings are placed in quadrature. Obtain expressions for the stator rotating mmf wave.

[B.Tech IV, JMI, 2005]

Q.11 Explain the single phase repulsion motor, give its characteristic and applications.

[B.Tech IV, JMI, 2005]

Q.12 Explain why a single phase induction motor with only one single phase winding on the stator and connected across a single phase supply does not develop starting torque but continues to rotate in a direction in which the rotor is given an initial rotation.

[B.Tech IV, JMI, 2006]

Q.13 (a) Discuss why single phase induction motors do not have a starting torque.
(b) Draw the circuit diagram of a capacitor start capacitor run single phase induction motor and explain its working.

[B.Tech II, UPTU, 2003]

Q.14 Write short notes on stepper motors.

[B.Tech II, UPTU, 2003, 2001]

Q.15 With the help of double revolving field theory explain that a single phase induction motor is not a self-starting machine.

[B.Tech II, UPTU, 2001]

Q.16 (a) A single phase induction motor does not develop a starting torque whereas a three-phase induction motor develops, explain. Why?

(b) Briefly discuss the different methods used for the starting of a one phase induction motor.

[B.Tech II, UPTU, 01]

Q.17 Write notes on any one of the following single phase induction motors
(i) Resistance split phase motor
(ii) Capacitance split phase motor

[AMU, B Tech II, 91]

Q.18 Explain the need of auxiliary winding in single phase induction motor.

[B.Tech II, AMU, 92]

Q.19 Explain briefly the construction and working of (a) Reluctance motors (b) Hysteresis motor.

[B.E. III, AMU, 99]

Q.20 (a) Write with the help of neat sketch, explain the working of a resistance split phase IM.

(b) A test on the main winding of a 1 kW, 4 pole, 215 V, 50 Hz. Single phase IM gave the following test results:

No load test: 215 V, 3.9*A*, 185 W

Blocked rotor test: 85 V, 9.8, 390 W

Calculate the parameters of the circuit model of the motor assuming that the magnetizing reactance hangs at the input terminals of the motor. Main winding resistance is given to be 1.6 Ω.

[B.E II, AMU, 2006]

Q.21 What is a stepper motor? Explain the working of PM stepper motor.

[B.Tech III, AMU, 2000]

Q.22 Using first principles develop the equivalent circuit of single phase induction motor.

Q.23 Why testing of single phase IM done? Discuss the various tests performed on 1ϕ IM.

Chapter

12

Solid State Motor Control

12.1 Introduction
12.2 Solid State Control Devices
12.3 Types of Drive
12.4 Chopper Control
12.5 Cycloconverter Control
12.6 Inverter Control
- Solved Questions
- Unsolved Questions
- Review Questions

12.1 INTRODUCTION

The d.c. motor plays a very important role in the industrial and domestic use. The importance of d.c. motor lies in the speed control applications such as CD drives, CD-ROM drivers, audio cassette recorders employing small power to traction motor drives employing very high power. Previously, the speed control is done by employing high power rheostat or by using contactors and practices which consume large power. It also occupies large spaces due to which there is a need of new innovative methods employing small power for its operation and at the same time it is cost effective.

With the advent of the thyristors, MOSFETS, IGBT, GTO and other power semiconductor devices these problems have been solved. They not only proved reliable but also effective methods for speed control of the d.c. motors. In this chapter both the d.c. and a.c. motor speed control methods are discussed which are used practically.

Though now the microcontroller and the computers have further enhanced the speed control methods. But this is out of scope of this book.

12.2 SOLID STATE CONTROL DEVICES

The most commonly used semiconductor device used for power control is called thyristors which is a common name of silicon controlled rectifiers (SCR). Other devices are diac and triac respectively which are used for the speed control of fans type load and also as light dimmers. Thus, it is very much necessary to know about the operation of this particular device i.e., SCR's.

The thyristor is a four layer device as shown in Fig. (12.1).

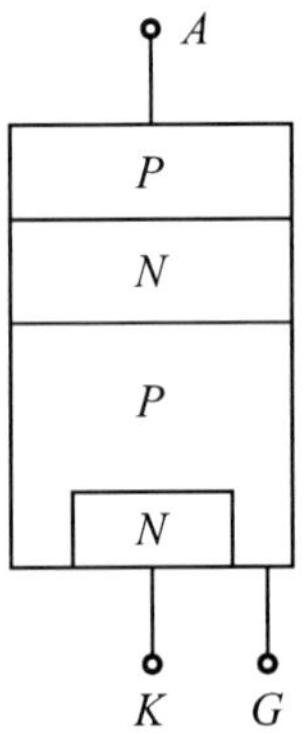

Fig. 12.1 Structure of SCR or Thyristor.

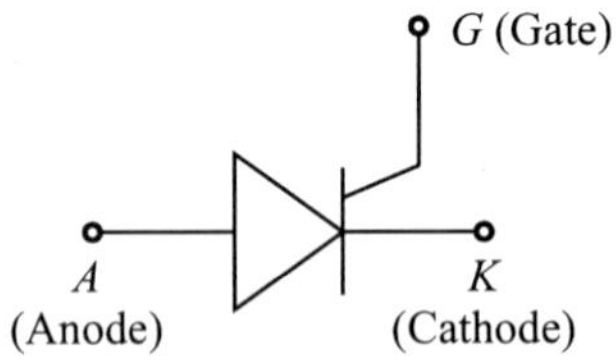

Fig. 12.2(a) Symbolic representation of SCR or Thyristor.

Symbolically, it is represented as given in Fig. 12.2(a). The ratings of the thyristor differs depending upon the applications. The thyristor of higher power ratings are mounted on a heat sink to permit heat dissipated in the air and helps in the heat balancing. The thyristor characteristic is shown below.

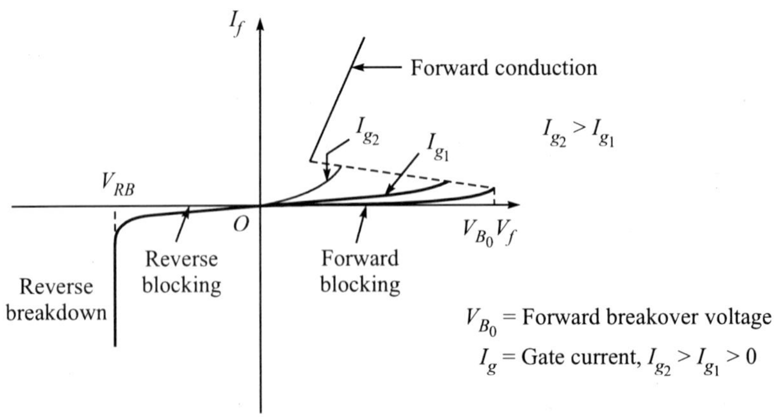

Fig. 12.2(b) *V-I* characteristic of thyristor.

Thus, from the characteristic we see that the thyristor works on various modes of operation, namely, forward conduction similar to forward biased condition of ordinary diode. In the reverse blocking condition the thyristor is reverse biased and very small current of the order of few milliamperes flows through it. So we can regard it as off state of SCR. To avoid damage of thyristor, care should be taken to ensure that the maximum reverse voltage across a thyristor does not exceed the breakdown voltage 'V_{RB}'. Positive gate voltage to reverse biased SCR should be limited to prevent the junction due to thermal runaway.

In the forward blocking state only small leakage current flows. In the forward conduction state the thyristor operates if $V_f > V_{BO}$, by gate triggering, $\frac{dV}{dt}$ turn on, and irradiation of *G-K* junction.

If $V_f > V_{BO}$, then large current passes through thyristor from anode to cathode depending on the load impedance. The thyristor is said to be on. The drop across device is of the order of 1.0 volt and increases with current.

When this current decreases below 'I_{BO}' then the thyristor returns to blocking state called holding current.

The second method is an important method employed for turning on the thyristor by applying low voltage pulse to the gate terminal (i.e., between '*G*' and '*K*'). Gate layer disappears and the 'SCR' behaves as a forward biased diode. A positive '*G*' to '*K*' pulse decreases 'V_{BO}' and helps in conduction. The current is of the order of few milliamperes in the gate. Once the thyristor starts conducting then there is no need of gate signal, so a gate pulse is sufficient for turning on the SCR or thyristor. The gate current is so chosen that the anode current exceeds latching current.

The thyristor goes to its off state only when the gate signal is removed and the forward current decreases below the holding current value. The latching current is about three times that of holding current.

The time required for the SCR to turn on is of the order of 1 to 3 μsec., provided the load inductance does not limit the anode current buildup. This time can be changed by changing gate current and voltage wave shapes. Higher current gate pulse will turn on the thyristor quickly by injecting the charge carriers in *G-K* region.

The turning off of thyristor is called commutation. The large voltage transients are prevalent during the commutation process which can damage the thyristor. So snubber circuit consisting of '*R*' and '*C*' in series is placed across the thyristor. The turn off time is in the range of 5 to 200 μsec.

The gate pulse width is of the order of 0.1 m sec. of d.c. supply and voltage and current of it is increased. That is, power of pulse is increased. The SCRs are protected by keeping the '*I*' below its rated value, the temperature should not be greater than 120°C, heat sinks should be used. With 180° conduction, the ratio of the rms current to the average current is $\frac{\pi}{2}$. The average current is therefore restricted to $\frac{2I_{dc}}{\pi}$. By using fuses transients are produced. The snubber circuit is used to protect the device from voltage transients.

Other thyristor devices are namely

(i) DIAC

(ii) TRIAC

(iii) GTO (Gate turn off thyristor)

(iv) RCT (Reverse conducting thyristor)

(v) AGT (Amplifying gate thyristor)

Many new devices are added to the thyristor family. Every devices has its own advantages and disadvantages. The symbolic representations of the devices are shown in Fig. (12.3) below.

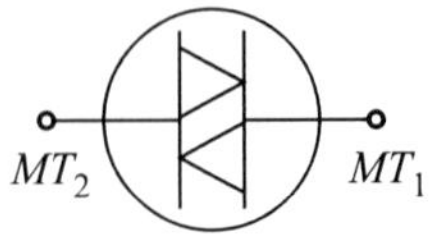

(a) Diac

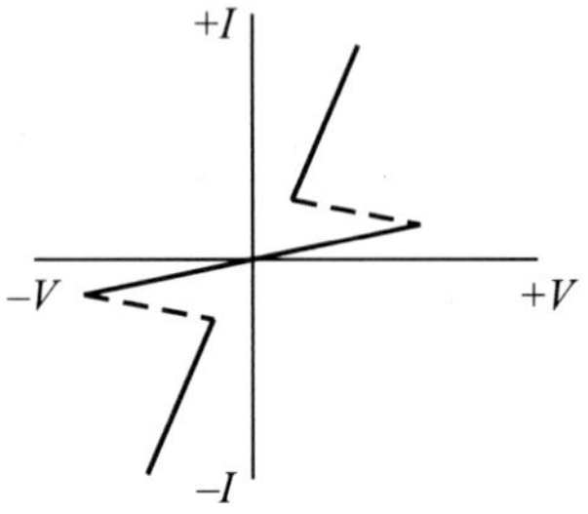

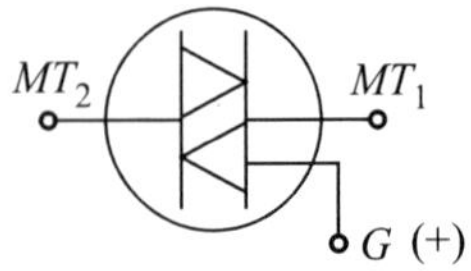

(b) Triac

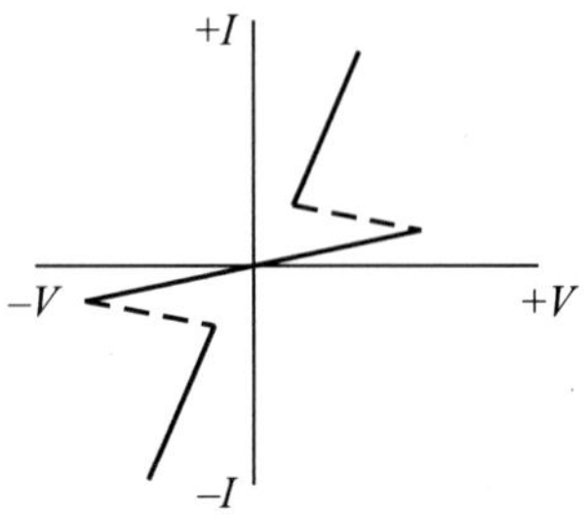

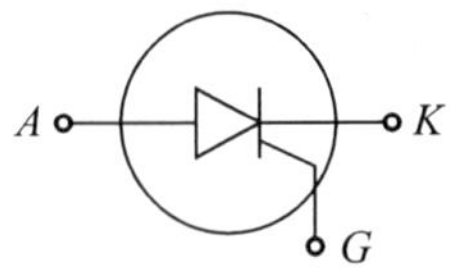

(c) GTO

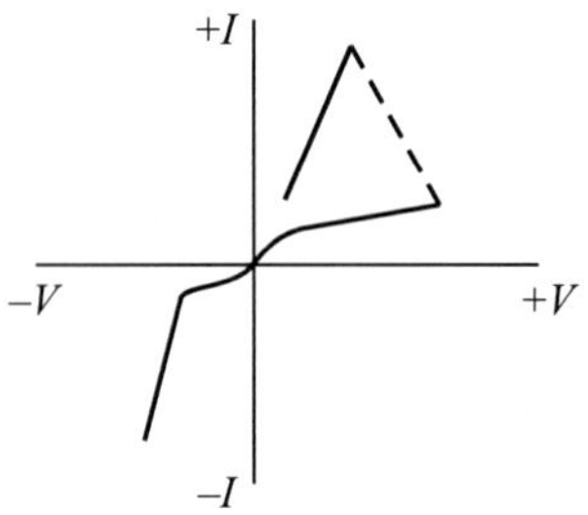

Fig. contd.

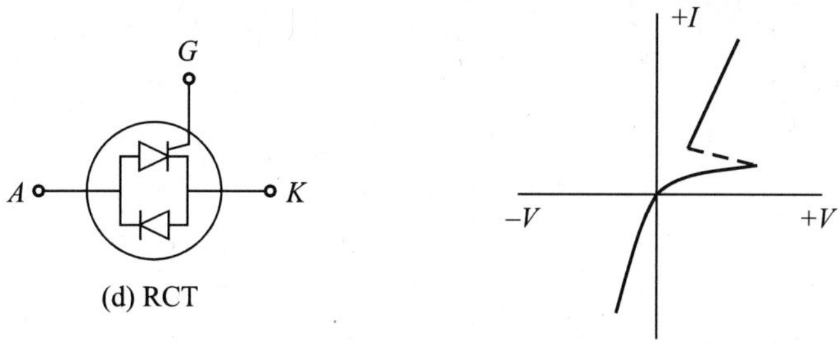

Fig. 12.3 Devices and their *V-I* characteristics.

The power circuits employing choppers and inverters employ GTO's. There is no need of forced commutation in these devices.

The device type is increasing day by day. So for the selection of the device a strategy should be made. They are as follows.

(i) Voltage and current ratings
(ii) Switching frequency (slew rate)
(iii) Cost
(iv) Protection of device used
(v) Ease of triggering
(vi) Ruggedness
(vii) Availability
(viii) Incidental dissipation.

But then also the device selection as a power switch is becoming very difficult with the advent of new power semiconductor devices.

12.3 TYPES OF DRIVE

A system used for the control of load is known as drive. The drives may consist of a motor and load. The motor output characteristic is changed with the load requirements and this needs change in the speed of driving motor which is coupled to the load directly or indirectly.

On the basis of type of motor used, the drives are basically classified as

(a) d.c. drives
(b) a.c. drives.

The d.c. drives employs d.c. motor and the a.c. drives employs synchronous and induction motors. The stepper motor also uses d.c. pulses so it is preferred in the electronic devices driver circuitry. Pulse of current leads to precise increment of angular shaft position. The duration of pulse is set as per the requirement of motor and its load.

The d.c. drives employ the chopper control and converter control. The chopper control is used where d.c. is available and where a.c. is available and there is a need of converting the a.c. into d.c. then converters are used.

The d.c. drives are discussed in the coming sections. The a.c. drives employs generally induction motors and in few applications synchronous motors are used. The drives employing induction motors control the frequency or slip. The Fig. (12.4) depicts the control methods.

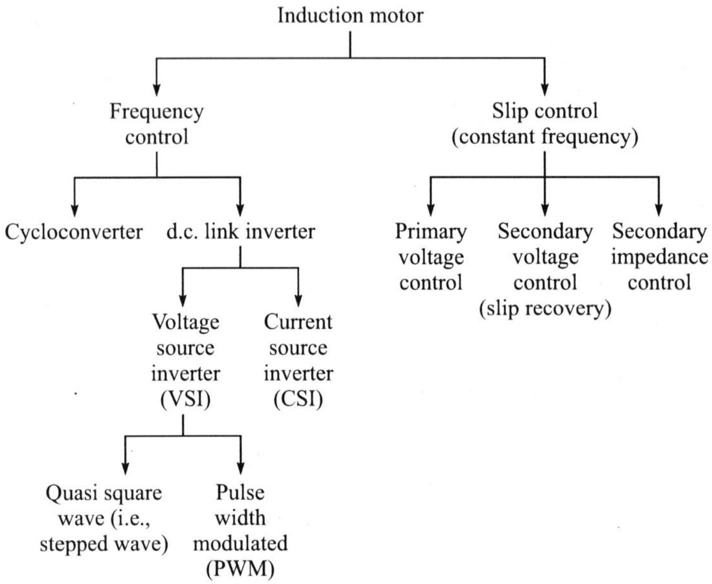

Fig. 12.4 Speed control methods of IM.

The inductor motor drives are generally preferred for constant speed operation. The switched reluctance drive is becoming popular where wide speed variation with precise control is required. In most of the industries induction motor is used due to low cost and ruggedness. The synchronous motor drives are used in the servo applications.

Bridge Rectifier (Controlled) with Motor Load

Semi Converter or Half bridge Converter (2 Pulse Converter)

The motor control in the single quadrant, is possible using this converter by varying firing angle 'α'. The average armature voltage is given by the expression given below.

$$V_{a(av)} = \frac{1}{2\pi} \int_0^{2\pi} v_a(\theta)\, d\theta; \ \theta = \omega t \qquad (12.1)$$

Since, '$v_a(\theta)$' is a.c. in the armature which is given by.

$$= \frac{1}{\pi} \int_\alpha^\pi V_m \sin \omega t \, d(\omega t)$$

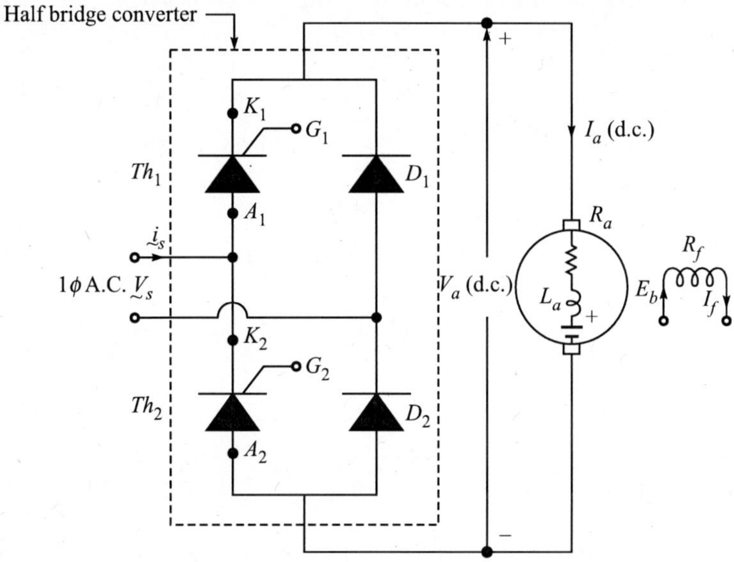

Fig. 12.5 Single phase half bridge converter (with d.c. motor load).

$$= \frac{V_m}{\pi} \int_{\alpha}^{\pi} \sin\theta \, d\theta$$

$$= \frac{V_m}{\pi} \left| -\cos\theta \right|_{\alpha}^{\pi}$$

$$= \frac{V_m}{\pi} \left[-\cos\pi - (-\cos\alpha) \right]$$

$$= \frac{V_m}{\pi} [-(-1) + \cos\alpha]$$

$$\boxed{V_{a(av)} = \frac{V_m}{\pi}(1 + \cos\alpha)} \tag{12.2}$$

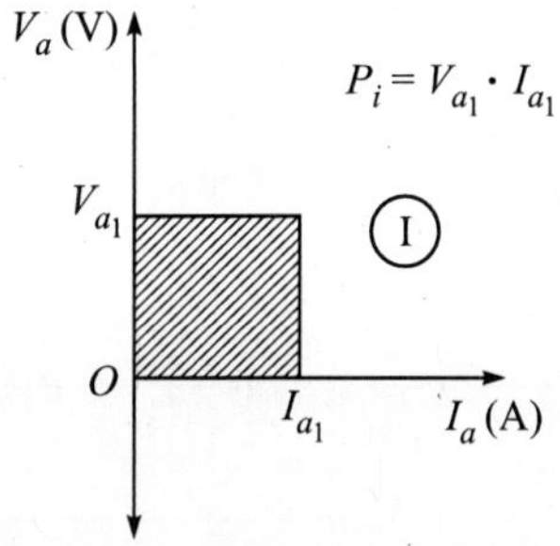

Fig. 12.6 One quadrant operation of d.c. motor.

where, 'α' is the firing angle of the converter. The average load current or motor current is given by

$$I_{a(av)} = \frac{V_{a_{(av)}} - E_b}{R_a}\ ;\ E_b \propto N\phi,\ \text{For}\ \phi = \text{constt},\ E_b \propto N \tag{12.3}$$

where, 'E_b' is the back emf, in volts. 'R_a' is the armature resistance, ohms. On substitution, we have

$$\boxed{I_{a_{(av)}} = \frac{V_m}{\pi R_a}(1 + \cos\alpha) - \frac{E_b}{R_a}} \tag{12.4}$$

Comparison of Converter Under Different Modes of Operations

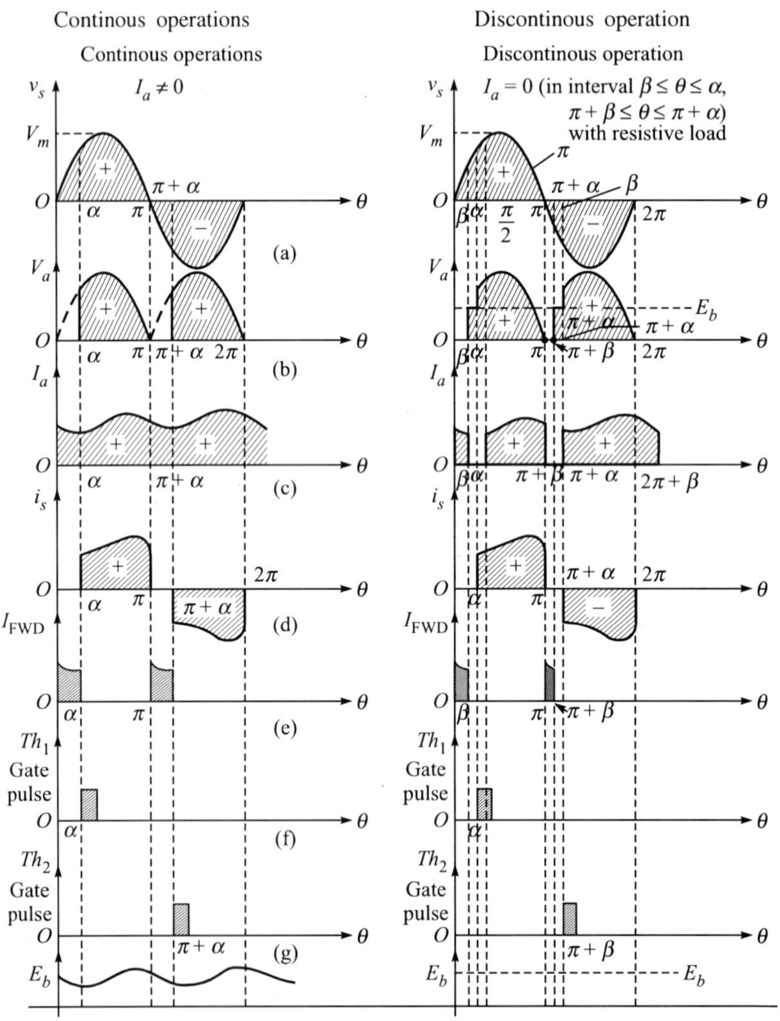

Fig. 12.7 Waveforms of half bridge converter (single phase).

Discontinuous Operation of Single Phase Semi Converter (with Inductive Load)

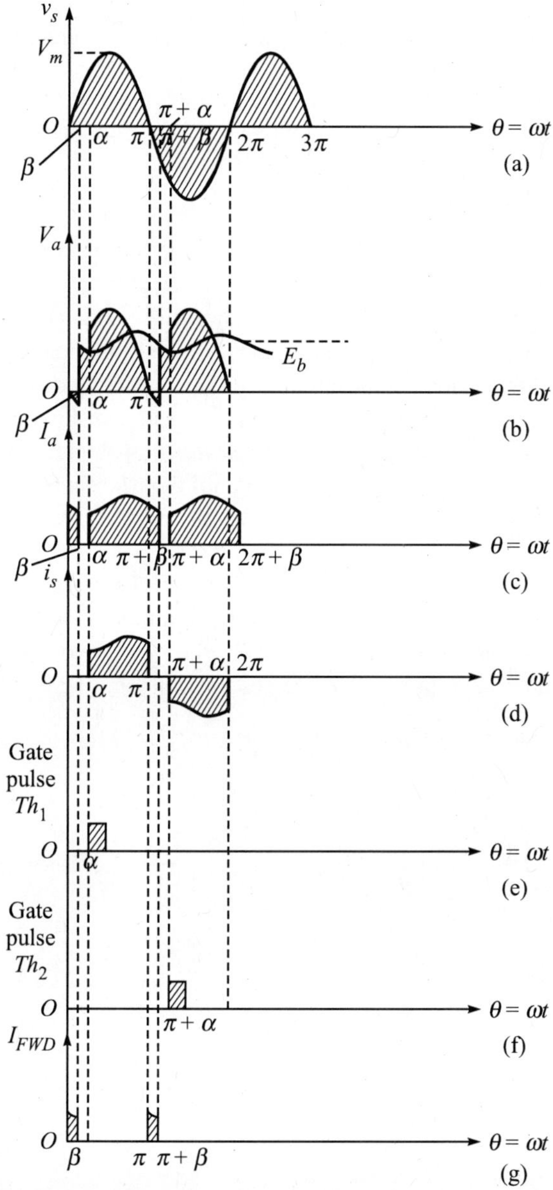

Fig. 12.8 Waveforms of 1ϕ half bridge converter with inductive load (discontinuous mode of operation).

when, $\frac{V_m}{\pi R_a}(1 + \cos\alpha) > \frac{E_b}{R_a}$

$$I_{a(av)} = + \text{ ve (continuous operation)}$$

when, 'α' is high, $\frac{V_m}{\pi R_a}(1 + \cos\alpha) < \frac{E_b}{R_a}$ (discontinuous operation)

For α (lesser than 90°) and small load torque, the armature current is continuous, for larger load torque and higher 'α' the 'I_a' is discontinuous.

Single Phase Full Bridge Converter

The single phase half bridge converter is unsuitable for two quadrant operation as the single quadrant (I quad.) operation corresponds to motoring operation taking power from the supply but in (IV quad.) operation the motor performs braking operation drawing large braking current which will lead to damage of thyristors unless some protection is taken in the circuit. The free wheeling diode is generally connected across the motor armature terminals in such a fashion that the back emf 'E_b' of the motor forward biases the free wheeling diode and the current 'I_{FWD}' flows during the negative cycle of input until the next thyristor is fixed. So where frequent braking operation is needed the semiconverter is unsuitable and the full bridge converter is suitable as shown in Fig. (12.9).

The full bridge converter employs four thyristors in the bridge circuit in such a fashion that two thyristor operate during positive half cycle whereas other two remain inoperative. During negative half cycle the other two operate and the previous two become inoperative. The dashed line in Fig. (12.9) illustrates the same.

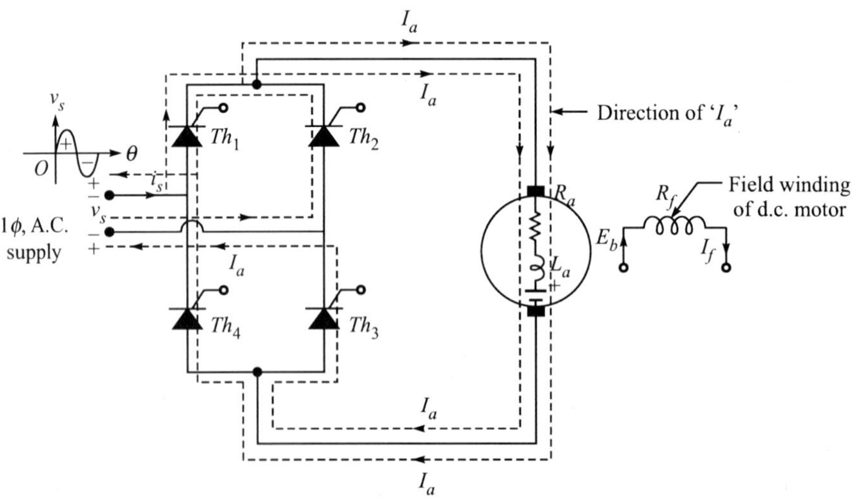

Fig. 12.9 Single phase full bridge converter supplying separately excited d.c. motor.

Comparison of Full Bridge Converter (Under Different Modes of Operation

Continuous Operation ($I_a \neq 0$) | Dis Continuous Operation ($I_a = 0$) in some intervals

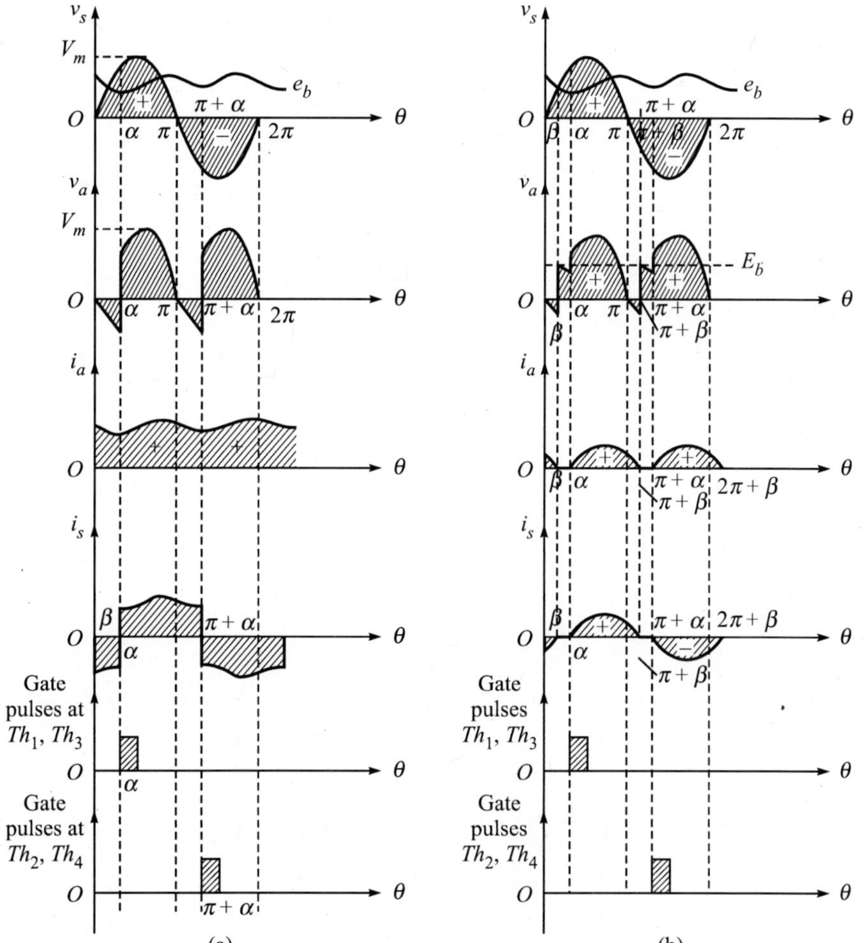

Fig. 12.10 Waveforms of full bridge converter (single phase) with motor load.

Using KVL across the motor equivalent (RLE load) circuit, we have

The armature voltage 'v_a' available across the motor armature terminals is given by

$$V_a = e_b + i_a R_a + L_a \frac{di_a}{dt} \tag{12.5}$$

At firing angle 'α', the expression (12.5) get modified as

$$v_a(\alpha) = e_b(\alpha) + i_a R_a + L_a \frac{di_a}{dt}$$

or,
$$v_a(\alpha) - e_b(\alpha) = i_a R_a + L_a \frac{di_a}{dt}$$

Net voltage across the motor load is

$$v_{R-L}(t) = v_a(\alpha) - e_b(\alpha) = i_a R_a + L_a \frac{di_a}{dt}$$

$$= V_m \sin\alpha - e_b(\alpha) = i_a R_a + L_a \frac{di_a}{dt} \qquad (12.6)$$

If $V_m \sin\alpha > e_b(\alpha)$, the armature current '$i_a$' is continuous. Otherwise, discontinuous.

Average value of the armature voltage is given by

$$V_a(av) = \frac{1}{2\pi}\int_0^{2\pi} v_a\, d\theta \qquad (12.7)$$

$$= \frac{2}{2\pi}\int_\alpha^{\pi+\alpha} V_m \sin\theta\, d\theta$$

$$= \frac{2V_m}{2\pi}\left[-\cos\theta\right]_\alpha^{\pi+\alpha}$$

$$= \frac{2V_m}{2\pi}\left[-\cos(\pi+\alpha) - (-\cos\alpha)\right]$$

$$= \frac{2V_m}{2\pi}\left[\cos\alpha - \cos(\pi+\alpha)\right]$$

$$= \frac{2V_m}{2\pi}\cdot 2\cos\alpha = \frac{2V_m}{\pi}\cos\alpha$$

So,
$$\boxed{V_a(av) = \frac{2V_m}{\pi}\cos\alpha}, \text{ volts} \qquad (12.8)$$

The armature current $i_a\,(av) = \dfrac{V_{L(av)} - E_b}{R_a}$

$$= \frac{2V_m}{\pi R_a}\cos\alpha - \frac{E_b}{R_a} \qquad (12.9)$$

The load current 'i_a' and load voltage 'V_a' have ripple frequency twice that of supply frequency.

Since harmonics are present in the output thus it affects the voltage and current waveforms as per the relations as given below by the exp. (12.10). For *n*th harmonic in the output,

Voltage across the load for *n*th harmonic is given by

$$V_{a_n} = \frac{V_m}{2\pi}\sqrt{\frac{1}{(2n-1)^2}+\frac{1}{(2n+1)^2}-\frac{2\cos 2\alpha}{(2n-1)(2n+1)}} \tag{12.10}$$

where, $n = 1, 2, 3, \ldots$

and armature current corresponding to the *n*th harmonic is given by

$$I_{an} = V_{an}/Z_{an} \tag{12.11}$$

$$I_{an} = \frac{V_{an}}{\sqrt{R_a^2 + (n\omega L_a)^2}}; \; n = 1, 2, 3,$$

The operating power factor $(\cos\phi)_0$ is given by

$$(\cos\phi)_0 = \frac{(I_a)_{av}^2 R_a + E_b (I_a)_{av}}{(V_a)_{av}(I_a)_{av}} = \frac{\text{Active power taken by load}}{\text{Input power to motor}} \tag{12.12}$$

The operating efficiency of the converter is given by

$$[\eta]_0 = \frac{[E_b (I_a)_{av} = \omega T_a]}{(I_a)_{av}^2 R_a + E_b (I_a)_{av}} = \frac{\text{Output power}}{\text{Input power}} \tag{12.13}$$

Thus, $(\cos\phi)_0$ and $(\eta)_0$ decreases with increase of 'α' which is an important parameter of the converter.

If armature current falls to zero, both the thyristors which are conducting in the positive half cycles (i.e., Th_1, Th_3) as shown in Fig. (12.9) are turn off by line commutation with the negative cycle of supply voltage but due to 'L_a' of armature circuit the current 'i_a' passes through the motor even with the starting of negative half cycle. At an angle '$\pi + \beta$'; $V_a = E$ and the motor goes in discontinuous mode. At an angle of '$\pi + \alpha$' the (Th_2, Th_4) are fixed as a result it starts conducting and the next cycle starts. Thus, the negative cycle of input is changed into positive output signal. Only that part of negative cycle which is available after angle '$\pi + \alpha$'. The output current which passes through the motor armature 'i_a', is fluctuating '*d.c*'. The frequency of the fluctuating d.c. is twice the supply frequency which is 50 Hz. As the torque of the motor is directly proportional to the armature current as given by the relation

$$T \propto i_a \tag{12.14}$$

so, the moment of inertia of motor '*J*' is high so these fluctuations of armature current 'i_a' does not affect the output of motor. The output of the motor is governed by the shaft power

$$P_{sh} = \omega T = 2\pi n T \tag{12.15}$$

Where, 'n' is the speed of rotor in 'rpm'. If the shaft power is constant, then it implies that the speed of the motor is constant, if the torque is constant. Thus, we can imagine that due to large mass of armature or rotor of motor small current change does not affect the speed of the motor and so we do not have to bother much about it. In the discontinuous mode of operation, the motor intervals are as follows:

(a) The interval in which the motor armature draws current from the source as a result the motor rotor rotates and motoring operation results. This mode of operation is in the interval ($\alpha \le \theta \le \pi + \beta$) which is called the duty interval of the motor as shown in Fig. 12.10(b).

(b) The other interval when the armature current 'I_a' is equal to zero. The armature voltage 'V_a' is equal to the induced e.m.f. 'E'. This occurs in the interval $[(\pi + \beta) \le \theta \le (\pi + \alpha)]$; this is called zero current interval as shown in Fig. 12.10(b).

The motor armature voltage 'v_a' is given by the equation.

$$V_a = E_b + L_a \frac{di_a}{dt} + i_a R_a:$$

The average armature current 'I_a' is given by

$$I_a = \frac{V_a - E_b}{R_a}$$

$$V_a = \frac{1}{\pi}\left[\int_{\alpha}^{\pi+\beta} v_s \, d\theta + \int_{\pi+\beta}^{\pi+\alpha} E_b \, d\theta\right]$$

$$= \frac{1}{\pi}\left[\int_{\alpha}^{\pi+\beta} V_m \sin\theta \, d\theta + \int_{\pi+\beta}^{\pi+\alpha} E_b \, d\theta\right]$$

$$= \frac{V_m \cdot \left|-\cos\theta\right|_{\alpha}^{\pi+\beta} + E_b \left|\theta\right|_{\pi+\beta}^{\pi+\alpha}}{\pi}$$

$$= \frac{V_m [-\cos(\pi+\beta) + \cos(\alpha)] + E_b [\pi + \alpha - (\pi - \beta)]}{\pi}$$

The average armature voltage,

$$V_a = \frac{V_m (\cos\alpha - \cos\beta) + E_b (\alpha - \beta)}{\pi} \tag{12.16}$$

The 'E_b' is the average value of 'e_b'.
Substituting the value of 'V_a' in exp. of 'I_a'

$$I_a = \left[\frac{V_m (\cos\alpha - \cos\beta) + E_b (\alpha - \beta)}{\pi} - E_b\right] \Big/ R_a \tag{12.17}$$

where, 'R_a' is the armature resistance, in ohms.

The back emf $E_b \propto N\phi$; 'N' is the speed of the motor, RPM

If ϕ = constant, then $E_b = K_e \cdot N$

Also, the torque of motor $T \propto I_a$

or, $T = K_t \cdot I_a$ So, $I_a = T/K_t$

Substituting the values of 'E_b' and 'I_a' in expression, we get

$$\frac{T}{K_t} = \frac{V_m}{\pi R_a}(\cos\alpha - \cos\beta) + \frac{K_e \cdot N(\alpha-\beta)}{\pi R_a} - \frac{K_e N}{R_a}$$

or, $$N\left[\frac{K_e(\alpha-\beta)}{\pi R_a} - \frac{K_e \cdot \pi}{\pi R_a}\right] = \frac{V_m}{\pi R_a}(\cos\alpha - \cos\beta) - \frac{T}{K_t}$$

or, $$N\left[\frac{K_e(\alpha-\beta) - K_e \cdot \pi}{\pi R_a}\right] = \frac{K_t V_m(\cos\alpha - \cos\beta) - \pi R_a T}{K_t}$$

or, $$K_e N\left[\frac{(\alpha-\beta-\pi)}{\pi R_a}\right] = \frac{K_t V_m(\cos\alpha - \cos\beta) - \pi R_a T}{K_t}$$

or, $$K_e N = \frac{\pi R_a [K_t V_m(\cos\alpha - \cos\beta) - \pi R_a T]}{(\alpha-\beta-\pi) K_t}$$

or, $$N = \left(\frac{1}{K_t K_e}\right)\left[\frac{\pi R_a V_m(\cos\alpha - \cos\beta) - \pi^2 R_a^2 T}{(\alpha-\beta-\pi)}\right]$$

or, $$N \propto \frac{\pi R_a V_m(\cos\alpha - \cos\beta)}{[\alpha-(\beta+\pi)]} - \frac{\pi^2 R_a^2}{[\alpha-(\beta+\pi)]} \cdot T \qquad (12.18)$$

Boundary condition between the continuous and discontinuous operation is given by substituting,

$$\pi + \alpha = \pi + \beta \text{ or } \alpha = \beta.$$

i.e, $$N_c \propto \frac{\pi R_a V_m(\cos\alpha - \cos\alpha)}{[\alpha - \alpha - \pi)]} - \frac{\pi^2 R_a^2}{(\alpha - \alpha - \pi)} \cdot T$$

$$Nc \propto -\pi R_a^2 T \qquad (12.19)$$

Where, 'N_c' is the critical speed given by

$$\boxed{N_c = K_e \cdot \pi R_a^2 \cdot T} \qquad (12.20)$$

The *N/T* characteristic of a full bridge converter is obtained by exp. (12.18) which is shown in Fig. (12.11).

In the continuous mode of operation, the speed drop with loading is small as compared to that of motor operating in the discontinuous mode of operation. The motor speed corresponding to critical is 'N_c' which is dependent over the armature resistance 'R_a'. The higher the armature resistance the higher will be the critical

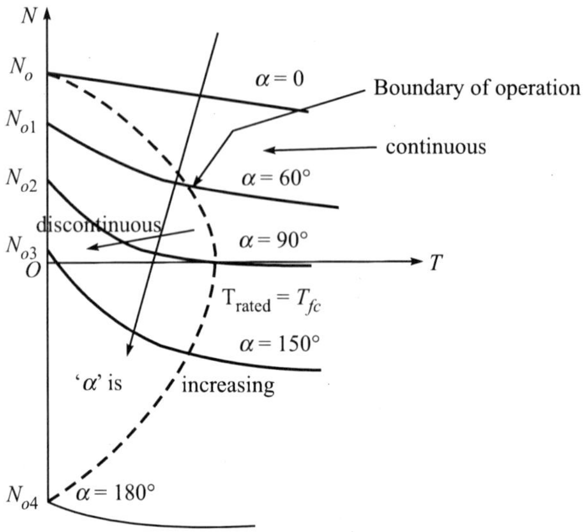

Fig. 12.11 *N/T* characteristic of 1ϕ full bridge converter.

value of speed for a given firing angle 'α'. For load torque 'T_L' > T_{fL} (of motor), the motor goes in the discontinuous mode of operation for low power drive operation. If the torque 'T' is increased then from exp. (12.18), the speed of motor 'N' decreases and the armature current 'I_a' increases it causes the angle ($\pi + \beta$) to increase and so the average armature voltage 'V_a' decreases respectively as per eqn. (12.16). This is the case, when the armature current is not continuous.

In the continous mode of operation,

$$V_a = \frac{2V_m}{\pi} \cos \alpha$$

and,

$$I_a = \frac{V_a - E_b}{R_a} = \frac{2V_m}{\pi R_a} \cos\alpha - \frac{E_b}{R_a}$$

$$T \propto I_a$$

$$K_t \cdot T = \frac{2V_m}{\pi R_a} \cos \alpha - \frac{K_e \cdot \phi \cdot N}{R_a}$$

For ϕ = constant, the above expression reduces to

$$\frac{N}{R_a} \cdot K_e = \frac{2V_m}{\pi R_a} \cos \alpha - K_t \cdot T$$

$$N \propto \frac{2V_m}{\pi} \cos \alpha - K'T$$

$$N = \frac{2V_m}{\pi} \cdot K \cos \alpha - K \cdot K'T$$

$$N = \frac{2V_m}{\pi} K \cos \alpha - K''T$$

$$\boxed{N = N_0 - K''T} \tag{12.21}$$

Thus, $$N_0 = \frac{2V_m}{\pi} \cdot K \cos \alpha \tag{12.22}$$

For $$\alpha = 0°,\ N_0 = \frac{2V_m}{\pi} \cdot K$$

$$\alpha = 60°,\ N_{0_1} = \frac{2V_m}{\pi} \cdot K \cdot \frac{1}{2} = \frac{V_m}{\pi} \cdot K$$

$$\alpha = 90°,\ N_{0_2} = 0$$

$$\alpha = 150°,\ N_{0_3} = \frac{2V_m}{\pi} \cdot K \cdot \left(-\frac{\sqrt{3}}{2}\right) = -\frac{\sqrt{3}V_m}{\pi} K$$

$$\alpha = 180°,\ N_{0_4} = -\frac{2V_m}{\pi} \cdot K$$

We see that, as $\alpha > 90°$, the '*N*' reverses the sign i.e., negative due to large mass of rotor the motor can't rotate instantly so the regenerative action of the motor takes place and the motor operates in the IV quadrant. So the motor operation in this quadrant is only for loads in which the speed of the motor is increases rapidly as in the case of lifts, hoist, etc. So, for low value of 'α' the motor operation is continuous for $T_L < T_{fl}$ and also for $T_L > T_{fl}$.

But if the firing angle is increased, if $T_L > T_{fl}$, so even for $\alpha < 90°$, the motor operation is in discontinuous mode. So the factors on which the continuous or stable operation of motor depends is the power rating of the motor, load torque and the firing angle respectively.

Three-Phase Half Bridge Converter

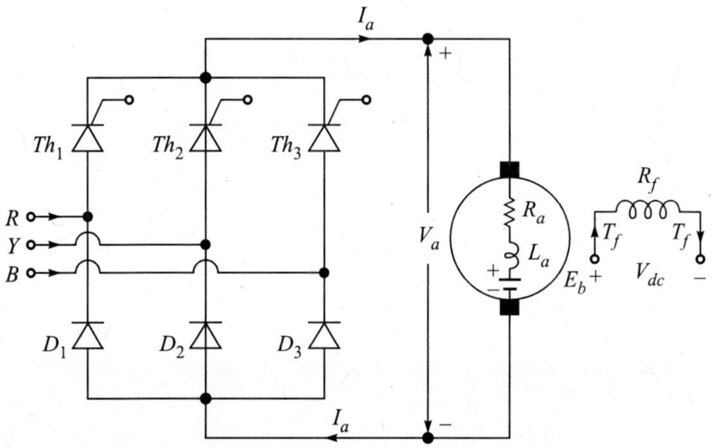

Fig. 12.12 Three-phase half bridge converter (connected to d.c. motor load).

A three-phase half bridge converter employs three thyristors Th_1, Th_2, Th_3 and three diodes D_1, D_2, and D_3 respectively in the upper and lower arms of the bridge converter. During a cycle of period (0 – 2π radians) there are six pulses at the output of the converter. So this type of converter is called six pulse converter. For the continuous mode operation, that is, when armature current flows continuously through motor, the output waveforms are as shown below.

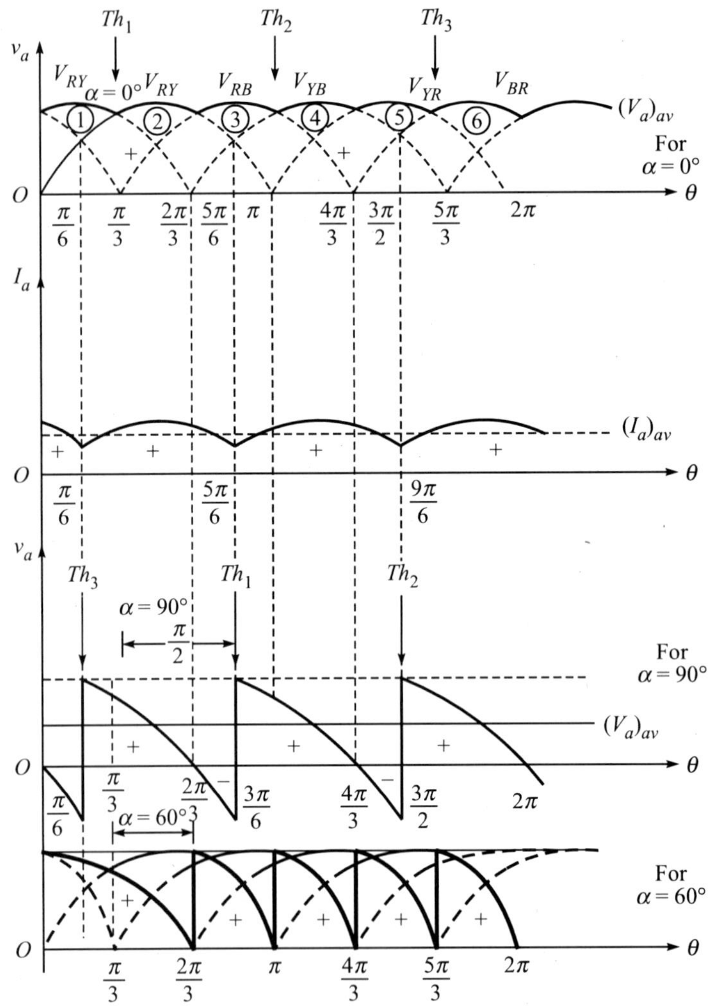

Fig. 12.13 Output waveforms of 3 ϕ semiconverter (continuous conduction mode).

The thyristors are fired at $\alpha = 0°$ (line commutation), $\alpha = 60°$ and $\alpha = 90°$ from the instant of start of negative cycle of input supply which reverses the biases of the thyristors. For firing angle of $\alpha < 60°$, the output voltage 'V_a' has only positive values but as $\alpha > 60°$, the output voltage 'V_a' has both positive and negative values. So we can say that the motor terminal voltages have both the values i.e., it reverses with the increase of firing angle. The armature current 'I_a' remains positive i.e., in the same direction through the motor.

The 3ϕ semiconverter can perform two quadrants operation of d.c. motor as shown in the Fig. (12.14) below.

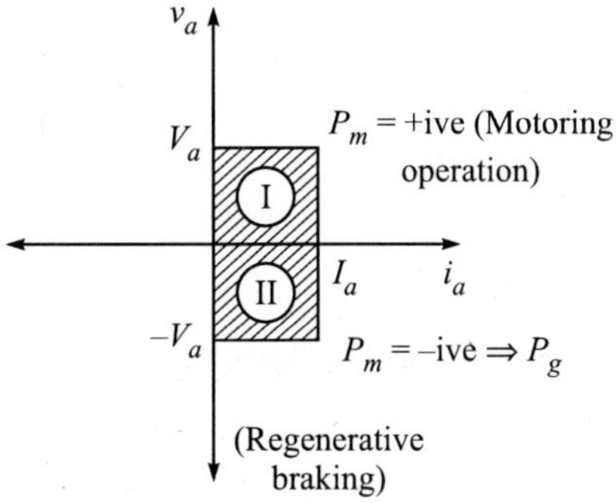

Fig. 12.14 Two quad. operation of 3 ϕ semiconverter with d.c. motor load.

For $\alpha < 60°$, the rectifier action of semiconverter and for $\alpha > 60°$ the inverting action of semiconverter, the average voltage output of the converter is given by

$$\boxed{(V_a)_{av} = \frac{3\sqrt{3}\,V_m}{2\pi}(1+\cos\alpha)} \tag{12.23}$$

Thus, $$(I_a)_{av} = \frac{(V_a)_{av} - E_b}{R_a} = I_a(\text{say}) \tag{12.24}$$

$$E_b \propto N\phi \text{ or, } E_b = KN \text{ if } \phi = \text{constant}$$
$$T \propto (I_a)_{av} \text{ or, } T = K'(I_a)_{av}$$

Substituting the values of $(I_a)_{av}$ and (E_b) in exp. we get

$$\frac{T}{K'} = \frac{\frac{3\sqrt{3}\,V_m}{2\pi}(1+\cos\alpha) - KN}{R_a}$$

or, $$KN = \frac{3\sqrt{3}\,V_m}{2\pi}(1+\cos\alpha) - \frac{R_a}{K'}\cdot T$$

or $$\boxed{N = \frac{3\sqrt{3}\,V_m(1+\cos\alpha)}{2\pi\cdot K} - \frac{R_a}{K\cdot K'}\cdot T} \tag{12.25}$$

The *N/T* characteristic of 3ϕ semiconverter is given in Fig. (12.15) below.

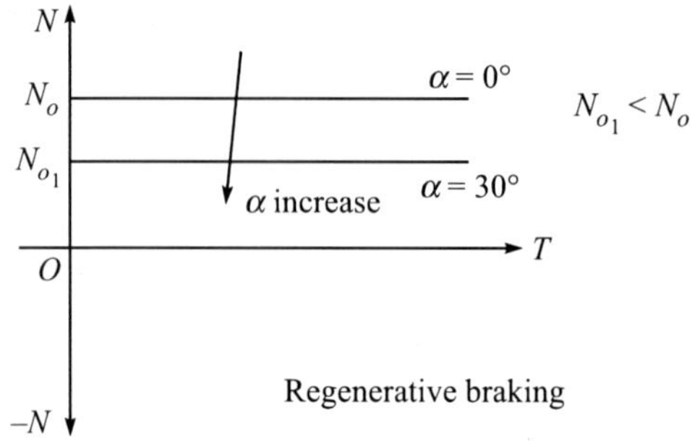

Fig. 12.15 *N/T* characteristic of 3 ϕ semiconverter.

The d.c. obtained on d.c. side is better and with smoothing inductor the discontinuous conduction can be reduced. After $\alpha = 60°$, the need of FWD arises.

Three-Phase Full Bridge Converter

The thyristor are fired sequentially with a delay angle 'α' and the thyristors are conducting for a period $\frac{\text{'}2\pi\text{'}}{3}$. Two thyristors conduct at a time. The output voltage 'V_a' is given by

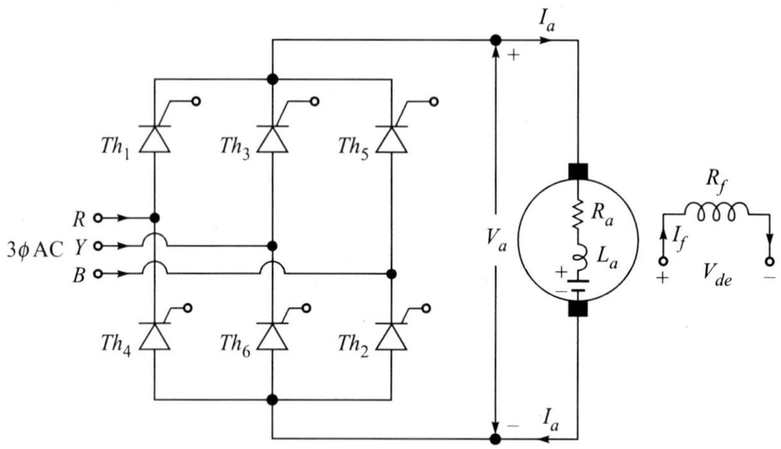

Fig. 12.16 3 ϕ full bridge converter.

$$V_a = \frac{3\sqrt{3}\,V_m}{\pi} \cos \alpha \tag{12.26}$$

and,

$$I_a = \frac{V_a - E_b}{R_a}$$

The speed 'N' is given by an expression similar as

$$\boxed{N = \frac{3\sqrt{3}\,V_m}{\pi K}\cos\alpha - \frac{R}{K \cdot K'} \cdot T} \tag{12.27}$$

When the continuous mode of operation is considered, we have N/T characteristic as follows.

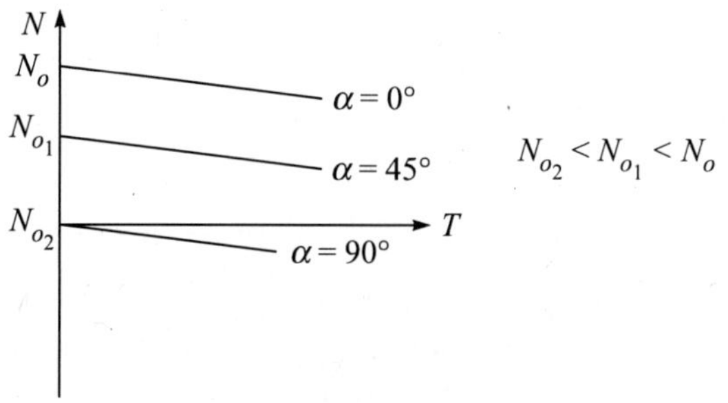

Fig. 12.17 N/T characteristic.

The load side voltage contains harmonics in multiples of 'ix'.
For n^{th} harmonic,

$$V_{Ln} = \frac{3\sqrt{3}\,V_m}{\pi}\sqrt{\frac{1}{(6n+1)} + \frac{1}{(6n-1)} - \frac{2\cos 2\alpha}{(6n+1)(6n-1)}} \tag{12.28}$$

$$i_{Ln} = \frac{V_{Ln}}{Z_{Ln}} \tag{12.29}$$

'Z_{Ln}' is in impedance offered by load to 'i_{Ln}'. The armature current and armature voltage are positive for all values of 'α' between 0° and 90°.

The torque speed relation can be expressed as

$$T = N_0(\alpha) - K_N N \tag{12.30}$$

where, no load speed 'N_0' is given by

$$N_0 = \frac{3\sqrt{3}\,K''V_m \cdot V_f}{\pi\, R_a R_f} \cdot \cos\alpha \tag{12.31}$$

$$K_N = \frac{K'''\,K_f^2\,V_f^2}{R_a\,R_f^2}$$

so the N/T characteristic of 3ϕ, full bridge converter is given by Fig. (12.18).

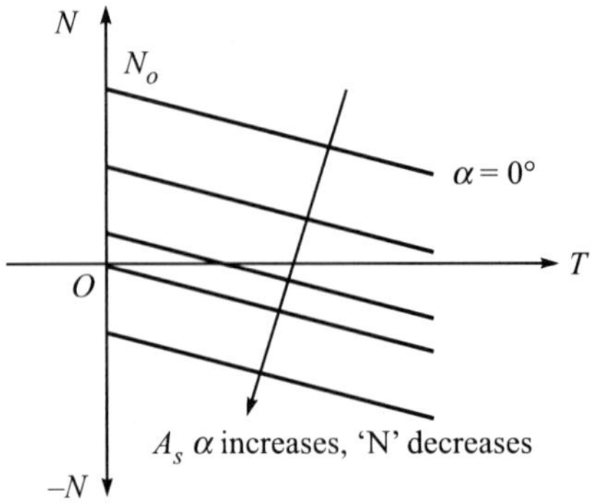

Fig. 12.18 *N*/*T* characteristic of 3ϕ full bridge converter.

Separately Excited DC Motor with Rectified Three-Phase Supply

Single phase supply is employed for low power motors below 5 hp. For high power applications the three-phase power supply is used up to 1000 hp motors can be driven using full bridge converter. The additional inductance in the armature circuit enhances the continuous armature current mode. The free wheeling diode maintains continuous load current over larger range. The average output voltage of the converter is given by

$$\boxed{(V_a)_{av} = \frac{3\sqrt{3}\,V_m}{2\pi}(1+\cos\alpha)}$$

For α = 0°, corresponds to uncontrolled three-phase bridge converter.
The average armature current 'I_{Lav}' given by

$$\boxed{(V_a)_{av} = \frac{V_{Lav} - E_b}{R_a} = \frac{3\sqrt{3}\,V_m}{2\pi\,R_a}(1+\cos\alpha) - \frac{E_b}{R_a}}$$

since,
$$(V_a)_{av} = E_b + (I_a)_{av}\,R_a \quad \text{(for separately excited motor)}$$
$$E_b = (V_a)_{av} - (I_a)_{av}\,R_a \quad ; E_b \propto N\phi \text{ and } T \propto I_a$$
$$K_e\,N\phi = (V_a)_{av} - \frac{T}{K_t}\cdot R_a$$

If
$$\phi = \text{constt}$$
$$K'_e\,N = (V_a)_{av} - \frac{T}{K_t}\cdot R_a$$
$$N \propto (V_a)_{av} - \frac{T}{K_t}\cdot R_a$$
$$N \propto \frac{3\sqrt{3}\,V_m}{2\pi}(1 + \cos\alpha) - \frac{T}{K_t}\cdot R_a$$

$$N = \frac{3\sqrt{3}\,V_m}{2\pi\,Ke'}(1 + \cos\alpha) - \frac{T}{K_t \cdot Ke'} R_a$$

$$\boxed{N = N_0(\alpha) - K \cdot T}$$

The no load speed is given by

$$\boxed{N_0(\alpha) = \frac{3\sqrt{3}\,V_m}{2\pi\,Ke'}(1 + \cos\alpha)}$$

For $\alpha = 0°$

$$N_0(0°) = \frac{3\sqrt{3}\,V_m}{\cancel{2}\pi\,Ke'} \cdot \cancel{2} = \frac{3\sqrt{3}\,V_m}{\pi\,Ke'}$$

This corresponds to the no load speed of uncontrolled converter. The *N*/*T* characteristic for different 'α' is shown below.

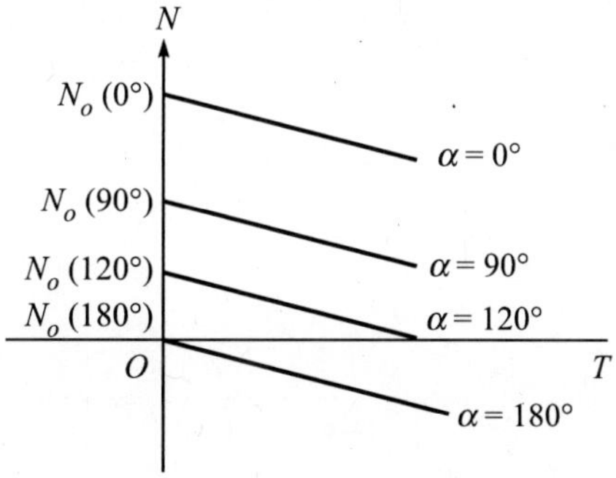

Fig. 12.19 *N*/*T* characteristic of 3 ϕ semiconvertor.

For $\alpha = 90°$

$$N_0(90°) = \frac{3\sqrt{3}\,V_m}{2\pi\,Ke'} \cdot 1 = \frac{3\sqrt{3}\,V_m}{2\pi\,Ke'}$$

For $\alpha = 120°$

$$N_0(120°) = \frac{3\sqrt{3}\,V_m}{2\pi\,Ke'}(1 - 0.5)$$

$$= \frac{3\sqrt{3}\,V_m}{2\pi\,Ke'} \cdot \frac{1}{2} = \frac{3\sqrt{3}\,V_m}{4\pi\,Ke'}$$

For $\alpha = 180°$

$$N_0(180°) = \frac{3\sqrt{3}\,V_m}{2\pi\, Ke'}\,(1 - 1)$$
$$= 0°$$

At $\alpha > 90°$, current flows through free wheeling diode in the anticlockwise direction as shown.

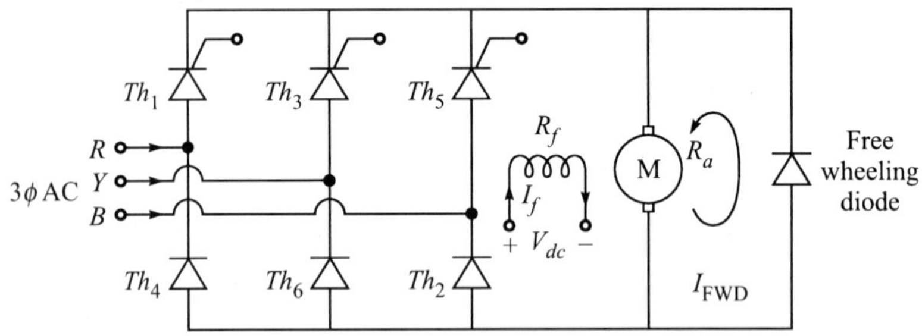

Fig. 12.20 3 ϕ semiconverter (with free wheeling diode).

The output voltage of the three-phase semiconverter contains 3 pulse per cycle, so it is known as three pulse converter. The output contains larger harmonics with respect to full bridge converter. The one quantant operation can be obtained with good efficiency.

Three-Phase Full Bridge Converter

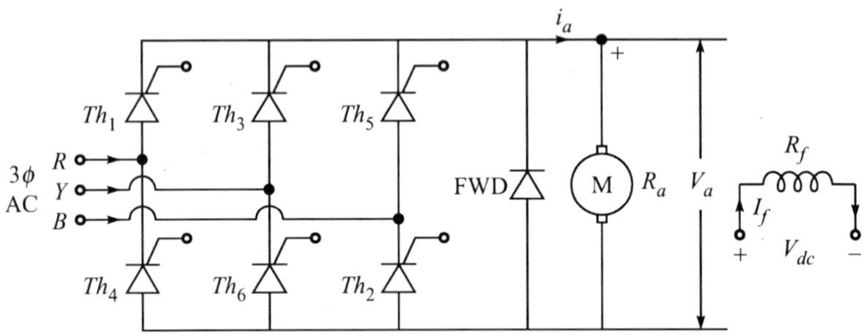

Fig. 12.21 3 ϕ full bridge converter (with free wheeling diode).

If an inductance 'L' is connected in series with the motor armature circuit such that the total inductance of the armature circuit is high. Then the armature current 'i_a' flows continuously even if 'E_b' is high.

For the continuous current mode, the average armature voltage is given by

$$(V_a)_{av} = \frac{3\sqrt{3}\,V_m}{\pi}\cos\alpha$$

$$(I_a)_{av} = \frac{(V_a)_{av} - E_b}{R_a}$$

$$E_b = (V_a)_{av} - (I_a)_{av} \cdot R_a$$

$$K_e N\phi = (V_a)_{av} - \frac{T}{K_t} \cdot R_a$$

For $I_f = \text{constant}, \ \phi = \text{constant}$

So, the expression reduces to

$$K_e' N = (V_a)_{av} - \frac{T}{K_t} \cdot R_a$$

$$N = \frac{3\sqrt{3}\, V_m}{\pi \cdot K_e'}(\cos \alpha) - \frac{T}{K_t \cdot K_e'} R_a$$

$$\boxed{N = N_0(\alpha) - kT}$$

The no load speed of the motor is dependent on the firing angle 'α', such that

$$N_0(\alpha) = \frac{3\sqrt{3}\, V_m}{\pi K_e'} \cos \alpha$$

For $\alpha = 0°$

$$N_0(0°) = \frac{3\sqrt{3}\, V_m}{\pi K_e'} \cdot 1 = \frac{3\sqrt{3}\, V_m}{\pi K_e'}$$

For $\alpha = 60°$

$$N_0(60°) = \frac{3\sqrt{3}\, V_m}{\pi K_e'} \cdot \frac{1}{2} = \frac{1}{2}\left(\frac{3\sqrt{3}\, V_m}{\pi K_e'}\right)$$

For $\alpha = 90°$

$$N_0(90°) = \frac{3\sqrt{3}\, V_m}{\pi K_e'} \cdot 0 = 0$$

For $\alpha = 120°$

$$N_0(120°) = \frac{3\sqrt{3}\, V_m}{\pi K_e'} \cdot \left(-\frac{1}{2}\right) = -\frac{1}{2}\left(\frac{3\sqrt{3}\, V_m}{\pi K_e'}\right)$$

For $\alpha = -\ 180°$

$$N_0(180°) = \frac{3\sqrt{3}\, V_m}{\pi K_e'} \cdot (-\ 1) = -\left(\frac{3\sqrt{3}\, V_m}{\pi K_e'}\right)$$

The speed torque characteristic at different 'α' is given by

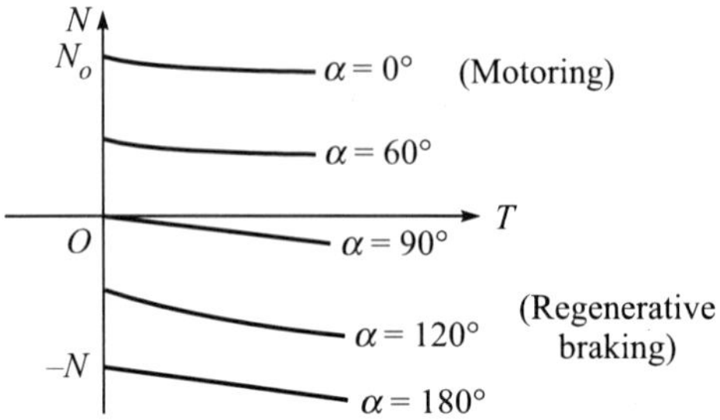

Fig. 12.22 *N/T* characteristic of 3 ϕ full bridge converter.

Since in the output six pulses are present per cycle of a.c. voltage, this type of converter is sometimes termed as six pulse converter. For $\alpha < 90°$ the $(V_a)_{av}$ and $(I_a)_{av}$ are positive so the motor operates in the *I* quadrant and the motor speed can be decreased over a wide range smoothly. But for $\alpha > 90°$, the $(V_a)_{av}$ is negative and $(I_a)_{av}$ is positive so the power taken by motor is negative and the motor operates in the IV quadrant. The armature current becomes discontinuous, so the power factor of the converter decreases and also the efficiency decreases as the harmonics increase in the output. The harmonic voltage and current is given by

$$V_{an} = \frac{3\sqrt{3}\,V_m}{\pi}\sqrt{\frac{1}{(6n+1)} + \frac{1}{(6n-1)} - \frac{2\cos 2\alpha}{(6n+1)\,(6n-1)}}$$

and

$$I_{an} = \frac{V_{an}}{Z_{an}}\ ;\ Z_{an} = r_a + j\,\omega_n\,L_a$$

$$\omega_n = 2\pi\,f_n$$

where, 'Z_{an}' is the impedance of armature circuit of motor offered to the harmonic armature current 'I_{an}'.

If the motor speed ripple is small, then E_b = constant.

For Discontinous Current 'i_a' Mode

'E_b' is less than 'V_m' and for low value of armature inductance 'L_a', the discontinuity occurs for less firing angle say $\alpha = 60°$.

Addition of free wheeling diode across the motor increases the power factor of the bridge converter and it is given by

$$\boxed{(\cos\phi)_{\text{bridge converter}} = P_a / 3V_m\,I_a} \qquad (12.32)$$

where, 'P_a' is the active power consumed by armature circuit resistance 'R_a', in watts

'V_m' is the maximum value of supply voltage, volts
'I_a' is the armature current, in amperes.

The average value of the armature current conduction is increased using free wheeling diode by the relation given below.

$$(V_a)_{av} = \frac{3\sqrt{3}\,V_m}{\pi}[1 + \cos(\alpha + 60°)] \tag{12.33}$$

Thus, the control of motor using converter firing angle 'α' control is increased to 120° by using free wheeling diode.

Dual Converter

The half bridge and full bridge converters fail to operate in all the four quadrants. So under such cases, dual converters are used. The dual converter as the name implies consists of two converter bridges connected in antiparallel fashion as shown in the Fig. (12.23).

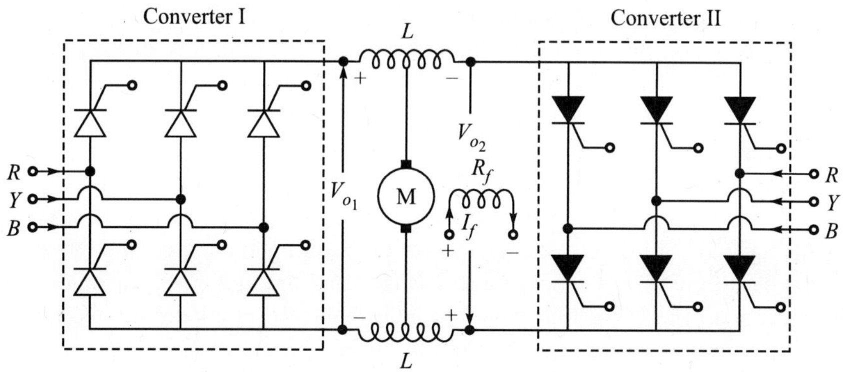

Fig. 12.23 Dual converter (with dc motor load).

The reactors 'L' are used between the two bridges which limits the circulating current from one bridge to another to pass through the motor 'M'.

Let 'α_1' and 'α_2' be the firing angles of the bridge converter (I) and (II) respectively. The sum of the firing angles should be 'π' radians for dual operation i.e., conversion and inversion respectively.

The averages output voltages of the converter is given by

$$V_{0_1} = -\,V_{0_2} \tag{12.34}$$

or, $$\frac{2V_m}{\pi}\cos\alpha_1 = -\frac{2V_m}{\pi}\cos\alpha_2$$

or, $$\cos\alpha_1 = -\cos\alpha_2 = \cos(\pi - \alpha_2)$$

or, $$\alpha_1 = \pi - \alpha_2$$

or, $$\boxed{\alpha_1 + \alpha_2 = \pi} \tag{12.35}$$

The dual converter operates in two modes respectively, namely, non-circulating current mode and circulating current mode.

Non-Circulating Current Mode: In this mode, the two bridges operate simultaneously so that conduction occurs in only one converter bridge at a time with the other non-operating. This type of operation is realized for continuous 'i_a'. A definite time lag is provided between the two converters. The motor control by this method is slow with respect to circulating current mode operation. Moreover, the difficulty arises when 'i_a' becomes discontinuous.

Circulating Current Mode: A circulating current flows from one converter bridge to another converter bridge whenever there is a difference in the output average voltage of the bridge (i.e., $V_{0_1} \neq -V_{0_2}$). One bridge acts as a rectifier converting a.c. to d.c. whereas the other bridge acts as a inverter converting d.c. into a.c. This circulating current makes the operation of the bridge in the continuous armature current mode. The motor reversal obtained by this method is faster with respect to non-circulating current mode.

This is used for motors up to 10 kW with 1ϕ a.c. supply and up to 1000 kW or more with three-phase a.c. supply.

It should be noted that if the circulating current is equal to 30% of the rated current value of the motor the discontinuous 'i_a' to the motor could be restricted and the motor will operate with continuous armature current. Thus, the speed regulation improves the drive.

The speed reversal of the motor can be done by increasing the firing angle 'α_1' of the converter (I) which is operating in the rectifying mode and thereby decreasing the firing angle of the converter (II) which is operating in the inverting mode. Thus, the back emf 'E_b' of the motor is changed and the motor will operate in the 2nd quadrant. The current control loop adjusts the 'α' continuously and the motor speed drops to zero and accelerates in the reverse direction. In the circulating current mode the inductor 'L' is a must and it should be connected between the two converters.

Demerits of Circulating Current Mode:

(i) The inductor 'L' increases cost.
(ii) Weighs high
(iii) Losses are increased.

In the non-circulating current mode, one converter operates at one time. So the inductor 'L' is not used. So there is an advantage of low cost, low weight and low losses with respect to circulating current mode.

When operating in the *I* quadrant, only the converter (1) is operating as a rectifier changing a.c. to d.c. and supplying it to the d.c. motor.

The firing angle 'α_1' is kept high. So that the armature current 'i_a' is equal to zero for the case when '$(V_a)_{av}$' is equal to 'E_b' of the motor.

i.e, $$\boxed{(i_a)_{av} = \frac{(V_a)_{av} - E_b}{R_a}} \quad ; \quad \text{For continuous '}i_a\text{' mode.}$$

where, $$(V_a)_{av} = E_b \text{ for some '}\alpha_1\text{'}$$

$$(V_a)_{av} = \frac{2V_m}{\pi} \cos \alpha_1$$

After the zero armature current is sensed, firing pulses are applied to converter (2) at 'α_2'. The firing pulses to gates of thyristors of converter (1) are removed.

The current control loop adjusts 'α_2' of converter (2) and thus decreases the motor speed and after stopping again accelerates in the reverse direction.

By field current 'I_f' reversal to the different quadrant operation is possible using dual converter to the field winding circuit of separately excited d.c. motor. When the field current equals to zero after a delay time the other converter operates at a firing angle 'α_2' which is kept at the lowest value. When the field current value is such that 'E_b' reverses, at firing angle 'α_1' (high) are applied to the converter (1). By such method, fast reversal, stopping and braking operation could be obtained.

The output waveforms of dual converter is depicted in Fig. (12.26)

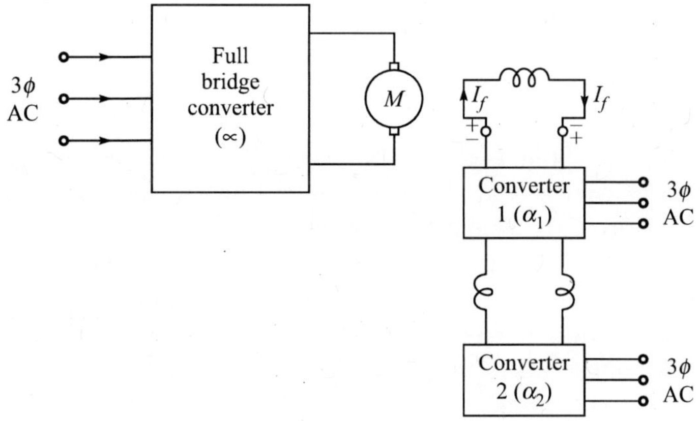

Fig. 12.24 4 quadrant operation (using field reversal).

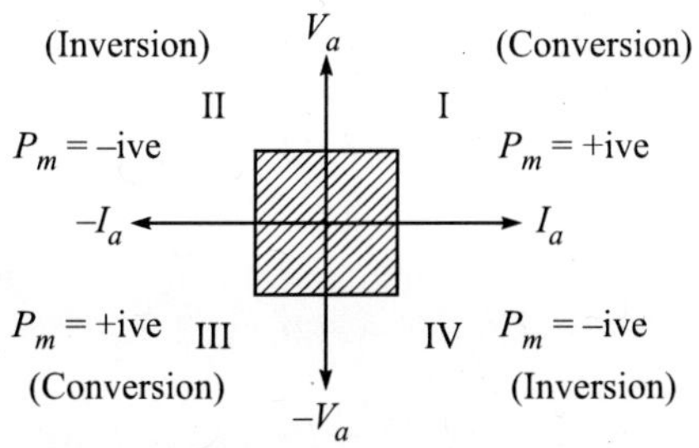

Fig. 12.25 4 quadrant operation of dual converter.

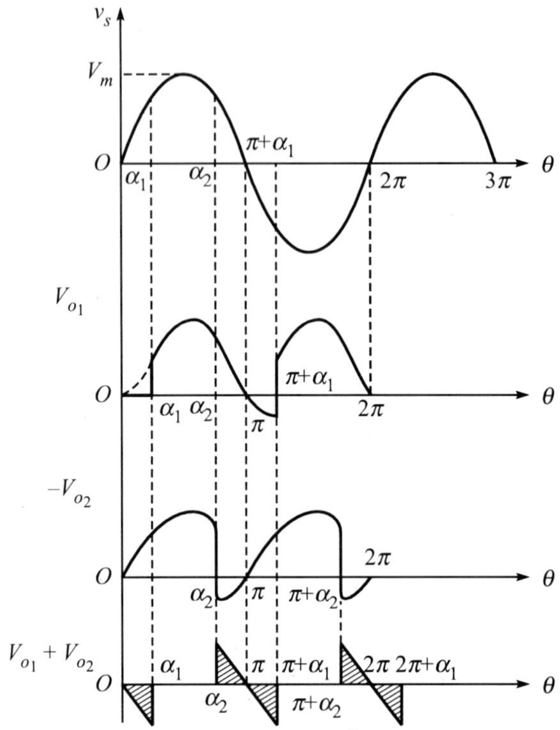

Fig. 12.26 Output waveforms of dual converter.

By controlling the firing angles of the converters, the average values of the voltages of the converters across the motor i.e., V_{0_1} and V_{0_2} are made equal but of opposite polarity. Even though this condition is satisfied, the instantaneous values of the d.c. output voltages are not equal. This difference in voltages (V_{0_1} – V_{0_2}) is the main cause of circulating current 'i_c'. This current leads to fast operation of drives which results in changing the motor operation from one mode to other i.e., motoring to braking and vice versa.

The circulating current 'i_c' is restricted from flowing through the motor by using an inductor 'L_s' called smoothing inductor. The circulating current passes from one converter to another and thus one converter operates as converter whereas the other works as inverter. Above certain firing angle 'α' the current becomes discontinuous in the case of previous mode that is, non-circulating mode. But the circulating current mode eliminates this problem. This current 'i_c' can be controlled by controlling 'α'. The sum of 'α_1' and 'α_2' is equal to 180°.

The dual converter is available in a number of configurations. There is a time lag of a few milliseconds between the converting and inverting operation. This is provided by a control circuitry.

12.4 CHOPPER CONTROL

The d.c. chopper is employed where variable d.c. supply is required and where the supply of d.c. is readily available. The chopper control has the merit of having almost negligible losses. The choppers can be used for two or four quadrant operation.

The basic methods of chopper control are:

(a) Time ratio control
(b) Current limit control.

By regulating the duty cycle 'D' of the conducting pulses the average load voltage or output voltage 'V_0' can be controlled.

That is,

$$V_{\text{Lav}} = V_0 = \frac{T_{\text{ON}}}{(T_{\text{ON}} + T_{\text{OFF}})} V_i = (D)\ V_i \tag{12.36}$$

where,
V_i = Input voltage, volts.
T_{ON} = ON period, seconds
T_{OFF} = OFF period, seconds
D = Duty cycle,

With (T_{ON} + T_{OFF}) kept constant, and by varying 'T_{ON}', the pulse width modulated (PWM) wave is obtained at the output.

If T_{ON} = constant and 'T_{OFF}' is varied, then the resultant wave is frequency modulated wave (FMW).

Basic Chopper Circuit

This type of chopper circuit is sometimes called buck converter since the output of the converter 'V_0' is always less than 'V_i'. That is 'V_0' can't exceed 'V_i'.

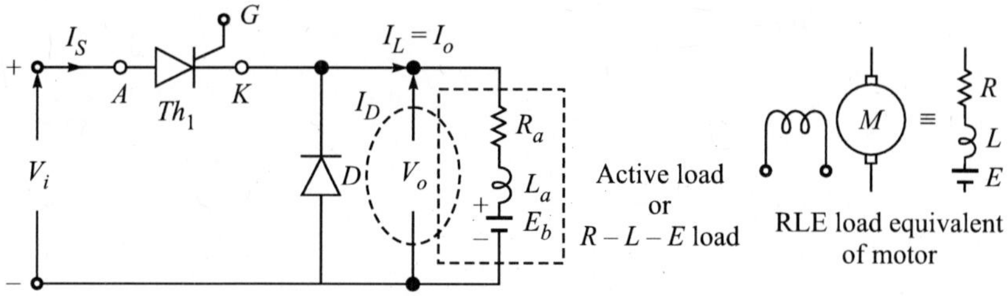

Fig. 12.27 Chopper circuit (Class A).

When thyristor switch 'Th_1' conducts then the load gets supply and vice versa. When the thyristor is in 'OFF' state the output voltage 'V_0' is zero by the action of diode 'D' if 'I_D' flows that too will occur if 'L_a' presents.

For all loads, the current 'I_L' or 'I_0' remains in the same direction. Operation takes place with positive output voltage '$+V_0$'. So this type of chopper performs one quadrant operation

i.e, $$P_m = V_0 I_0 = \text{positive (Motoring operation)} \tag{12.37}$$

With constant 'α' (firing angle) of thyristor. 'Th_1' the load voltage '$v_L(t)$' waveform is fixed or same. The load current $i_L(t)$ waveform depends on the nature of load (resistive or inductive) and its impedance

$$(Z_L = \sqrt{R_a^2 + (\omega L_a)^2}) \tag{12.38}$$

For resistive loads, such as incandescent lamps, the load current $i_L(t)$ waveforms are identical. But for inductive loads, such as motor loads, the load current magnitude varies but is continuous as shown in Fig. (12.26).

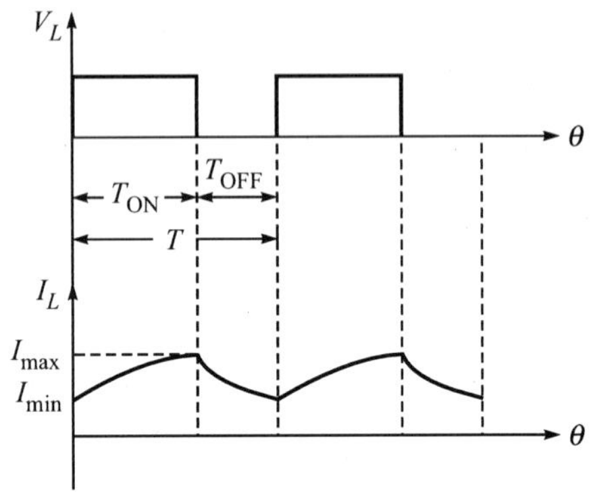

Fig. 12.28 Output waveforms of chopper for high 'D' (continuous load current).

When the 'T_{ON}' (on period) of the thyristor is decreased by decreasing 'α' (delay angle) of the thyristor by giving the gate pulse by the control circuitry. Thus, decreasing the duty cycle 'D'. We have discontinuous current 'i_L' available to the motor load as shown in Fig. (12.29).

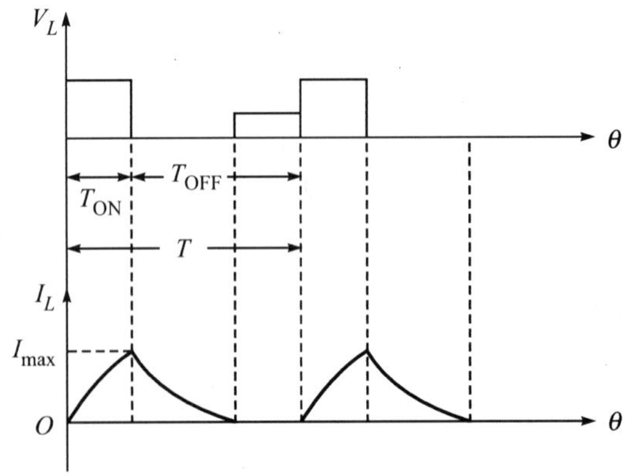

Fig. 12.29 Output waveforms of chopper for low D (discontinuous load current).

We see that for low 'D' and low 'L_a' the current 'i_L' may fall to zero during 'off' period of thyristor switch. For shunt and separately excited motors 'E_b' is not varying with 'I_L'. But in the series motors, since $\phi \propto I_L$ and $E_b \ \alpha \ N\phi$. So, the 'E_b' varies with 'I_L'. For simplicity consider 'E_b' to be constant in series motor as well.

We have from Figs. (12.28) and (12.29)

$$T_{ON} + T_{OFF} = 2\pi = T \tag{12.37a}$$

$$\text{Periodic time} = T = \frac{2\pi}{\omega}$$

So, the frequency of the chopper $= f = \dfrac{1}{T} = \dfrac{\omega}{2\pi}$

For thyristor or SCR choppers, the chopping frequency lies between 100 Hz and 1 kHz.

In low power applications, the MOSFET based chopper, this frequency is greater than 200 kHz.

On time of switch is given by

$$T_{ON} = D \cdot T = D \cdot 2\pi = 2\pi D, \text{ seconds} \tag{12.38a}$$

$$\text{Switch on time} = \frac{2\pi D}{\omega} \tag{12.39}$$

The r.m.s. value of the load voltage 'v_L' is

$$V_{L_{rms}} = \sqrt{D} \cdot V_i \tag{12.40}$$

$$\text{The ripple factor} = \frac{\text{a.c. components}}{\text{Average value}} = \frac{\text{a.c. component}}{\text{Av. d.c. output}}$$

$$\text{Ripple factor, } R \cdot F = \sqrt{\frac{V_{L_{rms}}^2 - V_{L_{av}}^2}{V_{L_{av}}}} = \sqrt{\frac{1-D}{D}} \tag{12.41}$$

For full conduction, $T_{ON} = T$, $D = 1$ so, $R \cdot F = 0$ i.e., output is pure d.c.

Two Quadrants Chopper

With Th_1 and Th_2 conducting motor current 'i_L' is positive and load voltage 'V_L' is positive. The motor is in motoring mode taking power from supply to drive the load at its shaft.

When Th_1 and Th_2 are in OFF state i.e., *RB*. The positive motor current '$+i_L$' passes through 'D_1' and 'D_2'. The load voltage is negative that is, $v_L = -\text{ve}$ and so the current direction reverses when duty cycle > 0.5; $T_{ON} > T_{OFF}$

$$V_{L_{av}} = +\text{ve}$$

When $D = 0.5;\ T_{ON} = T_{OFF}$

$$V_{L_{av}} = 0$$

When $D < 0.5;\ T_{ON} < T_{OFF}$

$$V_{L_{av}} = -\text{ve}$$

Thus, we see that the motor works in the I and IV quadrant. That is, both motoring and regenerating mode of operation is possible with such type of converter.

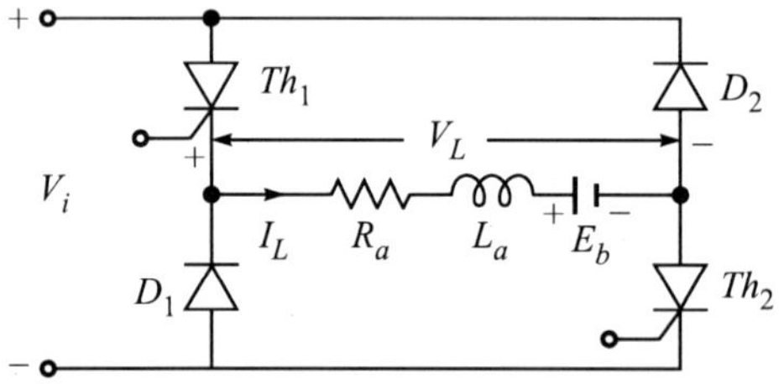

Fig. 12.29(a) Two quadrant chopper.

For 4 quadrant operation, the diodes D_1 and D_2 are replaced by SCRs.

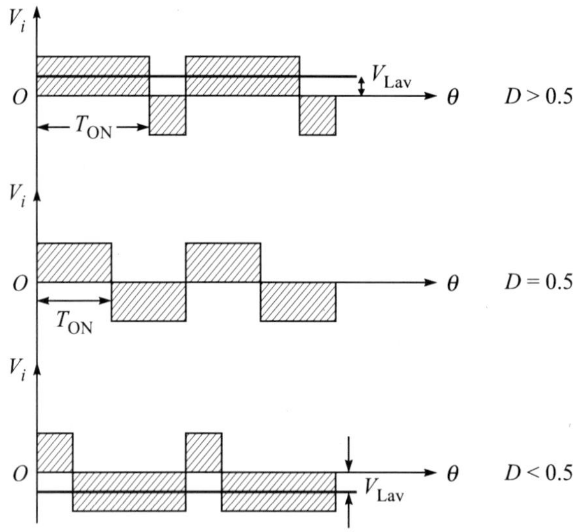

Fig. 12.30 *O/P* waveforms of 2 Quad. chopper (at different duty cycles).

In the boost converter, the circuit given below suffice in understanding the operation.

In such converter V_L or V_0 exceeds 'V_i' so they are used for boosting voltages hence, are called boost converter.

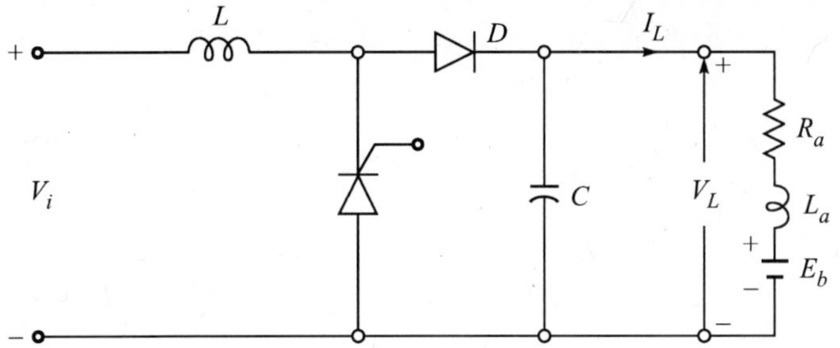

Fig. 12.31 Boost converter.

12.5 CYCLOCONVERTER CONTROL

It is seen that the d.c. motors are controlled by changing the armature voltage or the armature current. Whereas, the a.c. motors are controlled by varying the voltage and frequency.

Though the d.c. machines suffice the needs of high efficiency at all torques and speeds there are some demerits of d.c. machines i.e., it uses commutator and brushes which need replacement regularly after a period of time, the cost is high, operating voltage is low i.e., about 1000 V, size is comparatively small 2000 kW speed limitation of 2000 r/min.

The static frequency changers convert the line frequency (50 Hz) directly into the desired load frequency. The cycloconverters fall in this category. They are used for both induction and synchronous motors.

Fig. (12.32) shows the cycloconverter based drive

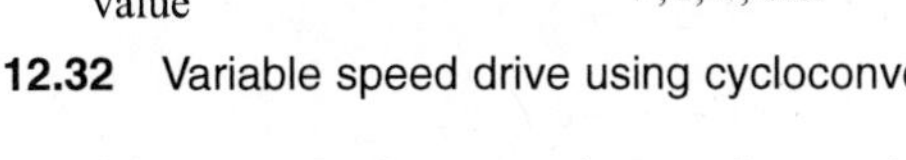

Fig. 12.32 Variable speed drive using cycloconverter.

The cycloconverter drives are for low speed operation such as ship propulsion motors which are rated in 'MW'.

The firing control should be carried out by microprocessors to obtain precision control so that the output voltage of the converter follows the input waveform. The thyristors are switched off naturally by line commutation.

The thyristors are connected in antiparallel. At low frequency, the power factor is nearly unity. The cycloconverter absorbs active power from line. Delayed triggering of gate is required to generate sinusoidal low frequency voltage at output. The power factor at the input is about 85% when the motor runs at the rated power and the rated speed.

The motor is stopped by changing 'α' of the gate pulse. To reduce the speed of motor both frequency and voltage have to be reduced in the same proportion. By using cycloconverter, 4 quad operation is possible. The stator flux of three-phase SCIM is automatically adjusted in relation to the frequency to maintain constant flux in the air gap between stator and rotor. The *N/T* characteristic remains have the same shape as $\frac{V}{f}$ = constant.

For a 4 pole, 1440 rpm, IM, the cycloconverter frequency must be slightly less than $\frac{1440 \times 4}{12030}$ = 48 Hz in order to fed power to line. That is, the regenerating mode.

During operation, one thyristor conducts while the other is blocked. Due to the zigzag sine waves due to constant switching between output and input the cycloconverter fed motors are slightly hotter than normal. So blowers are needed at low speed operation of drive. Reactive power is drawn from line so the power factor is poor about 0.85 even for purely resistive loads.

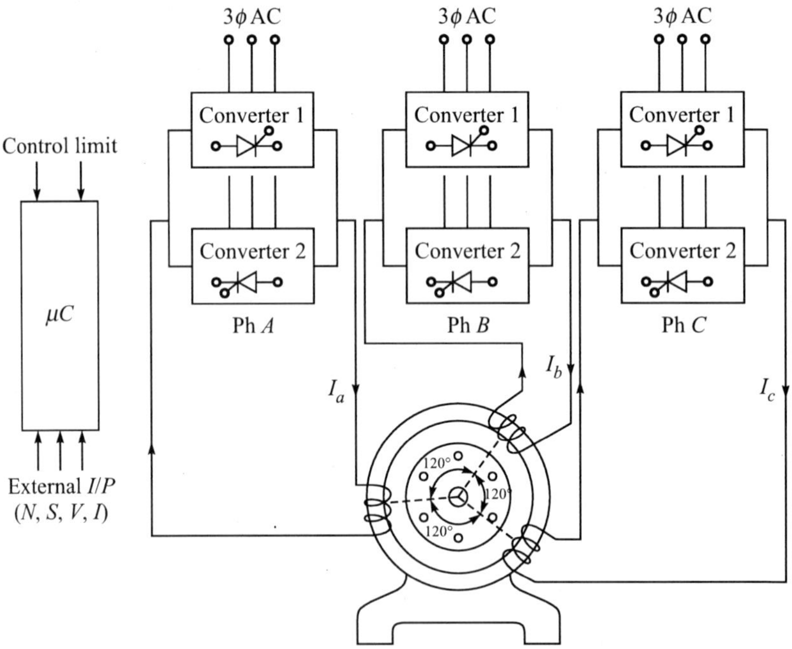

Fig. 12.33 Cycloconverter based SCIM drive.

12.6 INVERTER CONTROL

The inverter performs the inverse operation of a converter. It changes d.c. power into a.c. There are two types of inverters, namely:

1. Self commutated inverters, in which the commutation means are provided in the inverter itself. Sometimes it is referred as forced commutated inverters.
2. Line commutated inverters in which the commutation is by virtue of line voltages on the a.c. side of inverter.

Line Commutated Inverter

The current flows only when the thyristor 'Th_1' is forward biased. In the Fig. (12.34)

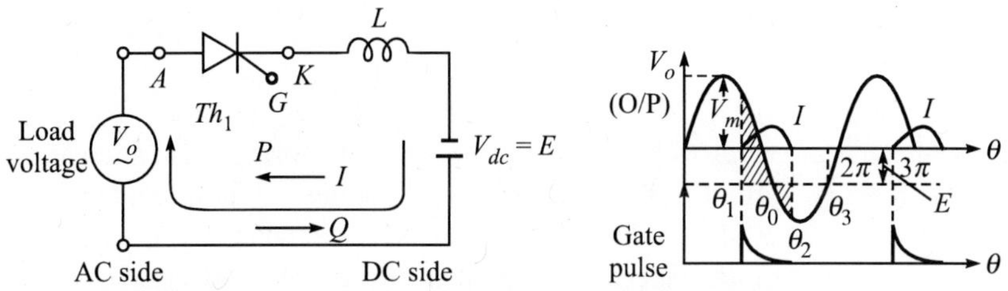

Fig. 12.34 (a) Inverter (line commutated) (b) Waveforms at *O*/*P* and *I*/*P*.

The a.c. side must be existing a.c. system that generates its own a.c. voltage whether or not the inverter is in operation. The a.c. system could be a grid supply that can easily absorb the active power 'P' supply by the inverter.

To achieve power converter the peak a.c. voltage "V_m' has to be greater than the d.c. voltage 'E' and the thyristors must be fired within a precise defined range i.e., before 'θ_0' or after 'θ_3'.

From Fig. (12.34) (b)

At 'θ_0' ; $\qquad I = I_{max}$

At 'θ_2' ; $\qquad I = 0$

The shaded area of lower is maximum if 'θ_2' is shifted to 'θ_3'. This is done by changing the firing angle 'α' of the thyristor, in this case $\alpha = \theta_1$. That is, by increasing the value of 'α'.

Under normal inverter operation, the current 'I' peak lags behind 'V_0'. So the a.c. source still has to supply the reactive power 'Q' to the inverter. $P = E \cdot I_{av}$, watts where 'E' is the voltage of d.c. source and 'I_{av}' is the average value of d.c. current 'I'.

The current pulses flowing into the a.c. terminals are non-sinusoidal and a low impedance a.c. system is necessary so as not to distort the sinusoidal voltage of 'I'. The pulses contain fundamental component that is in phase with sinusoidal voltage 'V_0'. The effective value of this component is given by

$$I_{rms}(f) = \frac{E_{dc}\, I_{dc}}{V_{rms}} \tag{12.42}$$

where, $\qquad E_{dc} = E$

In practice, filters are used to ensure that the current flowing into the a.c. line are more or less sinusoidal. It should be noted that line commutated inverter are meant for 3 phase systems and not single phase systems. It is seen from Fig. 12.34(b) that the gate pulse goes away but the thyristor remains conducting until the line negative voltage commutates it or reverse biases it and drives it in blocking state. The commutation of the inverters requires reactive power which d.c. supply can't provide. Therefore, auxiliary or special circuit suffice this need of 'Q'. This type of inverter is called forced commutated inverter. This type of inverter is employed for IM drives. Whereas, there are some loads which provide 'Q' such as over-excited synchronous motor which can be used as line commutated inverter.

A forced commutated inverter employs thyristors, capacitors, etc. The turn off time of thyristor limits the upper frequency of the inverter. So for this type of inverter the turn off time of thyristor should be as low as possible. To get an inverter which has an output voltage not depending on the load current and which can provide reactive current to the load, diodes are connected antiparallel to the thyristors. Thus, these diodes enable one to use inductive load at the output of the inverter. Commutating inductances are provided to prevent short circuits when both the thyristors work simultaneously. They also prevent the discharge of the capacitor when commutation takes place through the feedback diodes. Sometimes diodes are used in series with the thyristors to prevent discharging of 'C' due to 'E_b' of motor or variation of 'T_L'.

When the filter inductor is large then the line current assumes rectangular wave. Using phasor diagram the inverter operation could be visualized as follows.

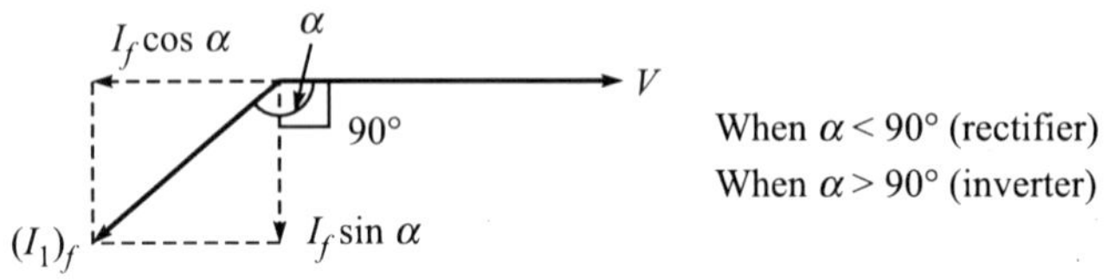

Fig. 12.35 V, I_{1_f} are O/P phasor of inverter

$$\text{The displacement power factor } \cos\alpha = \frac{P}{\sqrt{P^2+Q^2}}$$

$$= \frac{V_{av}}{(V_{av})_f} \tag{12.42a}$$

For single phase inverter

$$P = V \cdot (I_1)_f \cdot \cos\alpha' \text{ (a.c. side)} \tag{12.43}$$

$$= V_{dc}\, I_{dc} \text{ (d.c. side)} \tag{12.44}$$

Power dissipation factor in the filter (external) is equal to

$$I_{dc}^2\, R_f \tag{12.45}$$

$$\therefore \qquad \cos \alpha' \text{ (DPF)} = \frac{V_{dc} I_{dc}}{V \cdot (I_1)_f} \tag{12.46}$$

For d.c. link inverter; at $\alpha = 0$
(single phase)

$$V_{dc} = \frac{2V_m}{\pi} |\cos \alpha|_{\alpha=0} = \frac{2V_m}{\pi} \cdot \frac{\sqrt{2}}{\sqrt{2}} = \frac{2\sqrt{2}V}{\pi}$$

$$\therefore \qquad \cos \alpha' = \frac{2\sqrt{2}V}{\pi \cdot V} \cdot \frac{I_{dc}}{(I_0)_f}$$

$$\cos \alpha' = 0.9 \frac{I_{dc}}{(I_0)_f}$$

$$\therefore \text{ Displacement power factor} = 0.9\left(\frac{I_{dc}}{(I_0)_f}\right) \tag{12.47}$$

The full bridge voltage source inverter with load is shown in Fig. 12.36(a).

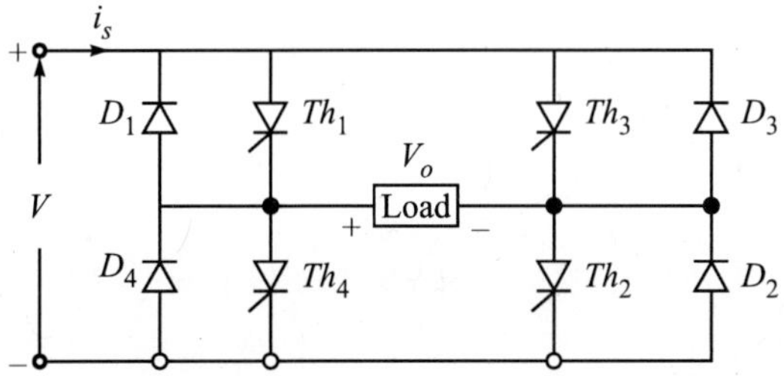

Fig. 12.36(a) Full bridge VSI.

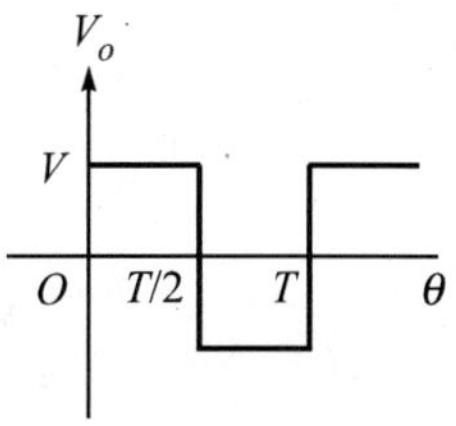

Fig. 12.36(b) Output waveform.

Th_1 and Th_2 operate during positive half cycle of input whereas Th_3 and Th_4 operates during negative half cycle of input. The output is a square wave of amplitude 'V'. The frequency of firing pulses decides the frequency of output of inverter.

The commutation circuit is employed for turning off of the thyristor. One commutation circuit is shown below.

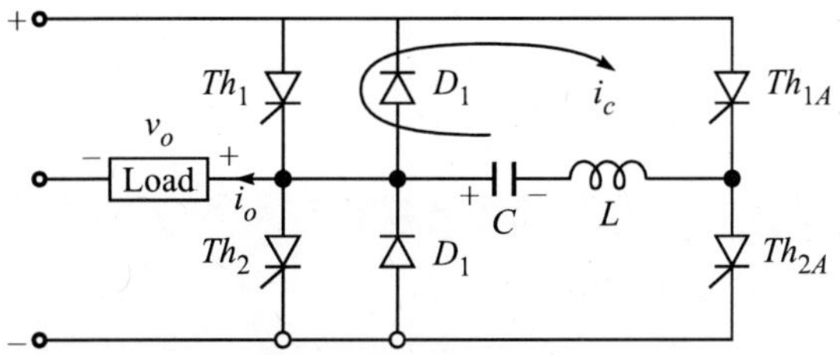

Fig. 12.37 Mc Murray commutation circuit.

For switching off the main thyristor 'Th_1' at instant $T/2$ prior to turning on the other main thyristor 'Th_2' the auxiliary thyristor 'Th_{1A}' is fired. As a result current 'i_c' flows and thus 'Th_1' is reversed biased and it is turned off. When $i_0 = i_c$ then, the current through 'Th_1' is equal to zero and hence it is turned off.

SOLVED QUESTIONS

Q.1 A single phase converter is used to control the speed of a 3 hp, 220 V, 1500 rpm, separately excited d.c. motor. The converter is connected to single phase, 230 V, 50 Hz supply. The armature resistance is $R_a = 1\ \Omega$ and armature inductance is $L_a = 10$ mH. The motor constant is equal to 0.05 V/rpm. In the rectifying mode (or motoring operation) the d.c. machine operates as the motor runs at 1400 rpm and carries an armature current of 12 amperes. Assume that the motor current contains no ripples.

Determine (a) Firing angle 'α'.
(b) Determine power to the motor.
(c) Determine the power factor of supply.

In the inverting mode (regenerating operation). The polarity of motor back emf 'E_b' is reversed by reversing field excitation.

(a) Determine the firing angle to keep the motor current at 12 amperes. When speed is 1400 rpm.
(b) Determine the power fed back to the supply at 1400 rpm.

A.1 1. (a) $E_b = 0.08 \times 1400 = 112$ V.

$$V_a = E_b + I_a R_a = 112 + 12 \times 1 = 124 \text{ V}$$

$$124 = \frac{2\sqrt{2} \times 230}{\pi} \cos \alpha$$

$$\cos \alpha = \frac{389.36}{650.44} = 0.5986 \text{ rad.}$$

$$\alpha = \cos^{-1} 0.5986 = 53.230°$$

(b) $P = V_a I_a$

$$= 124 \times 12 = 1488 \text{ W}$$

(c) Consider the supply current waveform as square wave

With amplitude 12 A $= I_a$

The r.m.s. value = 12 A

Thus, supply VA is

$$S = VI = 230 \times 12 = 2760 \text{ VA}$$

If losses neglected then,

$$P = (P_{in}) \text{ motor} = 1488 \text{ W}$$

$$\therefore \text{ Supply power factor } = \frac{P}{S} = \frac{1488}{2760}$$
$$= 0.5391$$

2. (a) $E_b = -112$ V

$$V_a = E_b + I_a R_a = -112 + 12 \times 1$$
$$= -100 \text{ V}$$

$$\therefore \quad V_a = \frac{2\sqrt{2} \times 230}{\pi} \cos \alpha = -100$$

$$\cos \alpha = -0.483$$

$$\Rightarrow \quad \alpha = 118.90°$$

(b) Power from d.c. machine

$$P_{dc} = 112 \times 12 = 1344 \text{ W}$$
$$P_{Ra} = (12)^2 \times 1 = 144 \text{ W}$$

Power lost in 'R_a'

$$P_{s(\text{regenerated})} = 1344 - 144 = 1200 \text{ W}$$

Q.2 For a 3 pulse converter due to large R_a and L_a the output current is continuous and ripple free. For $\alpha = 30°$ and $\alpha = 120°$, determine the output voltage. If $V_m = 260$ V

A.2 For $\alpha = 30°$

$$V_0 = \frac{3\sqrt{6}}{2\pi} V_m \cos \alpha$$

$$V_0 = \frac{3\sqrt{6}}{2\pi} \times 260 \times \cos 30$$

$$V_0 = 263.5 \text{ Volt.}$$

For $\alpha = 120°$

$$V_0 = \frac{3\sqrt{6}}{2\pi} \times 260 \times \cos 120°$$

$$V_0 = -152.1 \text{ volt.}$$

For $\alpha = 30°$, the operation of motor load will be in I quadrant and for $\alpha = 120°$ it will be in IV quadrant.

Q.3 A three-phase full bridge converter is used to control the speed of a 3 hp, 220 V, 1500 rpm, separately excited d.c. motor. The converter is fed by three-phase, 400 V, 50 Hz supply. The motor parameters are $R_a = 0.1\ \Omega$, $L_a = 5$ mH, $K\phi = 0.09$ V/rpm. The rated current is 12 A.

1. Rectifier (or motoring) operation. The machine operates as the a motor draws rated current and runs at 1500 rpm. Assume motor current ripple free.

   (a) Determine the firing angle.

   (b) Determine the supply power factor.

2. Inverter operation. The d.c. machine is operated in regenerative braking mode at 1200 rpm and rated current.

   (a) Determine firing angle.

   (b) Determine the power fed back to the supply and the supply power factor.

A.3 1 (a) $V_m = \dfrac{400}{\sqrt{3}} = 231 \text{ V}$

$$E_b = 0.09 \times 1500 = 135 \text{ volt.}$$

$$V_a = E_b + I_a R_a$$

$$= 135 + 12 \times 0.1 = 136.2 \text{ volt.}$$

$$V_a = \frac{3\sqrt{6} \times 231}{\pi} \cos \alpha = 136.2 \text{ V}$$

$$\alpha = \cos^{-1} 0.252 = 75.41°$$

(b) $I = \sqrt{\dfrac{1}{\pi} \times (12)^2 \times \dfrac{2\pi}{3}} = 9.798 \text{ A.}$

Supply VA = 3 VI = 6789.99 VA.

Assuming no losses, $P_s = V_a I_a = 136.2 \times 12 = 1634.4 \text{ W}$

$\therefore$ Supply p.f. $\cos \phi = \dfrac{P_s}{S} = 0.24$

2 (a) Inverting mode, $E_b = -135 \text{ V}$

$\therefore$ $V_a = E_b + I_a R_a$

$$= -135 + 12 \times 0.1 = -133.8 \text{ V}$$

$$V = \frac{3\sqrt{6} \times 277}{\pi} \cos \alpha = -133.8 \text{ V}$$

$$\cos \alpha = -0.21 \Rightarrow \alpha = 102.1°.$$

(b) Power from the machine

$$P_{dc} = 135 \times 12 = 1620 \text{ W}$$

$$P_{Ra} = (12)^2 \times 0.1 = 14.4 \text{ W}$$

$$P_s = P_{dc} - P_{Ra} = 1605.6$$

$$(\text{VA})_s = 6790 \text{ VA}$$

$$\therefore \quad (\text{P.F.})_s = 0.24$$

Q.4 A 2 quadrant chopper is used to control the speed of a d.c. motor and for regenerative braking. The motor constant is 0.1 V/rpm. The chopping frequency is 100 Hz and the motor armature resistance is 0.5 Ω. The inductance 'L_a' is very large and the armature current is ripple free. The supply voltage is 230 V.

(a) Chopper Th_1 and diode D_1 are operated to control the speed of motor. At N = 1500 rpm and I_a = 10A (ripple free)

(i) Determine the turn on time of the chopper.

(ii) Determine the power developed by the motor, power absorbed by 'R_a' and power from the source.

(b) When Th_2 and D_2 operate (regeneration).

At N = 1400 rpm and I_0 = –10A (ripple free)

A.4 (a) (i) $V_a = E_b + I_a R_a$

$$= 0.1 \times 1500 + 10 \times 0.5$$

$$= 150 + 5 = 155 \text{ V}$$

$$V_a = \frac{t_{ON}}{T} \cdot V = \frac{t_{ON}}{T} \times 230 = 155$$

$$t_{ON} = 0.674\, T$$

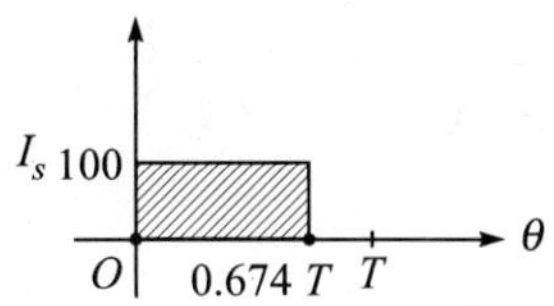

(ii) $P_{\text{motor}} = E_b \cdot I_a = 155 \times 10 = 1550 \text{ W} = 1.55 \text{ kW}$

$$P_{R_a} = (I_a)^2 \times R_a = (10)^2 \times 0.5 = 50 \text{ W}$$

$$P_s = V(I_s)_{av} = 230 \times \frac{100 \times 0.676}{1}$$

$$= 23000 \text{ W} = 23 \text{ kW}$$

(iii) $P_{\text{motor}} = E_b\, I_a$

$$= -140 \times 10$$

$$= -1400 \text{ W} = -1.4 \text{ kW}$$

$$P_g = 1.4 \text{ kW}$$

$$P_R = (10)^2 \times 0.5 = 50 \text{ W}$$

$$\therefore \quad P_s = 1400 - 50 = 1350 \text{ W} = 1.35 \text{ kW}$$

(b) $V_a = E_b + (-I_a R_a)$

$= 140 - 10 \times 0.5 = 135 \text{ V}$

$$V_a = \frac{T - t_{ON}}{T} \cdot V$$

$$135 = \left(1 - \frac{t_{ON}}{T}\right) \cdot 230$$

$$\frac{t_{ON}}{T} = 0.59,$$

If $T = 5$ ms then $t_{ON} = 2.9$ ms (approx.)

Q.5 In a single bridge inverter the load current is $I_L = 10\sqrt{2} \sin\left(\omega t - \frac{\pi}{4}\right)$. The d.c. voltage is equal to 110 V. Determine the average value of supply current and the power from the d.c. supply. Also determine the power delivered by the load.

A.5 (a) $I_{s(av)} = \frac{1}{\pi}\int_0^{\pi} 10\sqrt{2} \sin\left(\omega t - \frac{\pi}{4}\right) \cdot d\theta$

$$= \frac{10\sqrt{2}}{\pi}\left[-\cos\left(\omega t - \frac{\pi}{4}\right)\right]_0^{\pi}$$

$$= \frac{10\sqrt{2}}{\pi}\left[-\cos\left(\pi - \frac{\pi}{4}\right) + \cos\left(-\frac{\pi}{4}\right)\right]$$

$$= \frac{10\sqrt{2}}{\pi}[+0.707 + 0.707]$$

$= 6.367$ A

(b) $P_s = 110 \times 6.367 = 700.43$ W

(c) From fourier analysis of square wave 'V_a', the r.m.s. value of fundamental output voltage is

$$V_{a_1} = \frac{4V_a}{\pi\sqrt{2}} = \frac{4 \times 110}{\pi\sqrt{2}} = 99.1 \text{ V}$$

$$P_{out} = V_{a_1} \cdot I_a \cos\theta$$

$$= 99.1 \times \frac{10\sqrt{2}}{\sqrt{2}} \times \cos 45°$$

$$= 700.6 \text{ W}$$

UNSOLVED QUESTIONS

Q.1 For a semiconverter if the SCR's are fixed at 60°. The motor current is 15A and is assumed ripple free. The supply voltage is 120 V. Draw the output waveforms, power taken by motor, supply VA, power factor, $I_{rms(SCR)}$, I_D. Assume converter is loss less.

Q.2 The speed of 110 V, 5 hp d.c. motor is controlled by single phase full bridge converter. The a.c. supply is 120 V, 60 Hz. Consider the d.c. motor and converter to be ideal and lossless. Assume high 'L_a' in series with the motor. The motor voltage constant is 0.055 V/rpm.

(a) For a speed of 1000 rpm and rated motor current. Determine firing angle of converter, r.m.s. value of supply current, thyristor current, supply power factor.

Q.3 The three-phase full converter is used to control speed of motor (d.c). If E_b = 0.1 V/rpm. The V_{L-L} = 110 V, R_a = 0.2 Ω. For α = 50°, N = 900 rpm.

(a) Determine the average value of the motor current, assuming ripple free
(b) Determine the r.m.s. value of thyristor current, line current.
(c) Supply power factor
(d) Draw the *O/P* waveforms.

Q.4 A one quadrant chopper is used to control the speed of a d.c. motor D.C. voltage = 120 V, R_a = 0.15 Ω, K = 0.05 V/rpm, chopper frequency = 250 Hz.

At a speed of 1200 rpm, the motor current is 125A. The motor current can be assumed ripple free.

Determine (a) Duty cycle (b) T_a (c) P_{motor} = ? (d) P_s = ?

Q.5 Power is transferred from a 300 V battery to a 3 ph, 230 V, 50 Hz a.c. supply via a inverter. A large 'L' and 'R' of 10 Ω is included on the d.c. side. Calculate the power transferred and the power factor if (a) α = 90° (b) α = 120° (c) α = 150°.

[OW, 2247.2 W, 833.9 W; O, 0.477, 0.827]

Q.6 A 230 V, 960 rpm and 200 A separately excited d.c. motor has an armature resistance of 0.02 Ω. The motor is fed from a chopper which provides both monitoring and braking operation. The source has voltage of 230 V. Assuming continuous current conduction.

(i) Calculate duty ratio of the chopper for motoring operation at rated torque and 350 rpm.
(ii) Calculate duty ratio of chopper for braking operation at rated torque and 350 rpm.

(B. Tech, JMI, 2004)

Q.7 Uncontrolled full wave bridge rectifier with a series *RL* load has an a.c. source with $V_m = 100$ V at 60 Hz and load with $R = 10\ \Omega$ and $L = 10$ mH. Determine the average current in the load, power absorbed by the load and the power factor of the circuit and also estimate the peak to peak variation in load current base on the first a.c. term is the fourier series.

(B. Tech, JMI, 2006)

Q.8 What are the different chopper configurations available for speed control of d.c. motors?

(JMI, 2007)

Q.9 Discuss what are various types of inverters used in control of three-phase induction motors.

(JMI, 2007)

Q.10 Explain principles of speed control of a separately excited d.c. motor and obtain normalized equations and characteristics of a variable speed d.c. motor.

REVIEW QUESTIONS

Q.1 (i) What is an electric drive? Discuss its advantages over other types of drive.

(ii) Discuss the different types of load.

(JMI, 2004)

Q.2 Discuss the dynamic braking and natural and rheostatic speed torque characteristic of a d.c. shunt motor.

(JMI, 2004)

Q.3 Discuss the cycloconverter fed drive with schematic diagram and show the voltage waveforms for half cycle output.

(JMI, 2004)

Q.4 Discuss the chopper control of a separately excited d.c. motor with its speed torque curves.

(JMI, 2004)

Q.5 Discuss the principle of vector control of induction motor with phasor diagram.

(JMI, 2004)

Q.6 Write down the algorithms for vector control scheme.

(JMI, 2004)

Q.7 Derive the expression of field angle and rotor flux linkage using the rotor flux based calculator of direct vector control method.

(JMI, 2004)

Q.8 What is an electric drive? Explain and classify electric drives.

(JMI, 2007)

A

Appendix

SI UNITS

Metor – Length equal to 1650763.73 wavelengths in vacuum of the radiation corresponding to the transition between levels 2 p_{10} and 5 d_5 of the Krypton-86 atom.

Kilogram – Unit of mass equal to the mass of the international prototype of the kilogram in the custody of the International Bureau of Weights and Measures, Serres, Paris, France.

Second – Duration of 9192631770 periods of the radiation corresponding to the transition between the two hyperfine levels of the ground state of the cesium-133 atom

Ampere – That constant current which if maintained in two straight parallel conductors of infinite length yet negligible cross-section. placed 1 metre apart in a vacuum would produce between these conductors a force of 2×10^{-7} newtons per metre of length.

Kelvin – That thermodynamic temperature which is the fraction 1/273.16 of the triple point of water.

Radiam – Unit of measure of plane angle with its vertex at the centre of a circle and subtended by an arc equal to the radius.

Conversion Factors

Length (m)

Larger Units (commonly used)	*Smaller Units* (commonly used)
Giga $\times 10^9$	milli $\times 10^{-3}$
Mega $\times 10^6$	micro $\times 10^{-6}$
Kilo $\times 10^3$	nano $\times 10^{-9}$
	pieo $\times 10^{-12}$

Area (m^2)

$$1\ \text{mm}^2 = 10^{-6}\ \text{m}^2$$
$$1\ \text{cm}^2 = 10^{-4}\ \text{m}^2$$

Speed

$$1\ \text{cm/s} = 10^{-2}\ \text{m/s.}$$
$$1\ \text{km/hr} = \frac{5}{18}\ \text{m/s} = 0.2778\ \text{m/s}$$
$$1\ \text{rps} = 1\ \text{Hertz. (Hz).}$$
$$1\ \text{rpm} = \frac{1}{60}\ \text{Hertz} = 0.0167\ \text{rps.}$$

Force

$$9.807\ \text{N/Kg f} = \frac{1\ \text{kg}\cdot\text{m/s}^2}{\text{N}}$$

Energy

$$1\ \text{J/Nm} = \frac{1\ \text{W}\cdot\text{s}}{\text{Nm}} = 1.602 \times 10^{-12}\ \frac{\text{J}}{\text{eV}} = 1\ \frac{\text{J}}{\text{kg m/s}^2}$$
$$= 3.6 \times 10^6\ \text{J/kW hr}$$

Power

$$745.7\ \frac{\text{W}}{\text{hp}} = \frac{1\ \text{W}}{\text{J/s}} = 1.3405\ \frac{\text{hp}}{\text{kW}} = 4.186\ \frac{\text{W}}{\text{g}\cdot\text{cal/s}}$$
$$1\ \text{BHP} = 735.5\ \text{W},\ 1\ \text{HP} = 746\ \text{W}$$

Resistivity 'ρ'

$$10^{-8}\ \frac{\Omega\ \text{m}}{\mu\,\Omega\cdot\text{cm}} = 0.01\ \frac{\Omega\cdot\text{m}}{\Omega\cdot\text{cm}}$$

Potential Difference 'V'

$$1\ V = 1\ \frac{\text{J}}{\text{C}} = 1\ \frac{\text{N}\cdot\text{m}}{\text{C}} = 1\ \frac{\text{Wb/s}}{\text{V}}$$

Magnetic flux 'ϕ' and Flux density 'β'

$$10^8\ \frac{\text{Maxwell}}{\text{Wb}} = 10^5\ \frac{\text{K}\cdot\text{lines}}{\text{Wb}} = \frac{1\ \text{Wb}}{\text{A}} = 10^{-4}\ T = \frac{1\ M_x}{a_n^2}$$

Magnetomotive Force 'F'

$$\text{AT} = 0.7958\ \frac{\text{At}}{\text{Gb}}$$

Magnetic field strength '*H*'

$$10^2 \frac{\text{AT/m}}{\text{AT/cm}} = 79.58 \frac{\text{AT/m}}{\text{Oe}}; \ 1 \text{ Oe} = 79.6 \frac{\text{AT}}{\text{m}}$$

Permeability 'μ'

$$\mu = \mu_0 \, \mu_r \, ; \qquad \mu_0 = 4\pi \times 10^{-7}$$

$$\mu_{\text{cgs}} = \mu_{\text{mks}} \times \frac{10^7}{4\pi}$$

B

Appendix

Full Load Current in DC Motors (at N_{base}).

	'V_a' (volts)		
HP	120 V	240 V	500 V
$\frac{1}{4}$	3.1 A	1.6 A	
$\frac{1}{3}$	4.1 A	2.0 A	
$\frac{1}{2}$	5.4 A	2.7 A	
$\frac{3}{4}$	7.6	3.8 A	
1.0	9.5	4.7 A	
1.5	13.2	6.6 A	
2.0	17	8.5 A	
3.0	25	12.2 A	
5	40	20 A	
7.5	58	29 A	
10	76	38 A	
15		55 A	27 A
20		72 A	34 A
25		89 A	43 A
30		106 A	51 A
100		341 A	164 A
125		425 A	205 A
150		506 A	246 A
200		675 A	330 A

Full Load Current of Single Phase A.C. Motors

HP	115 V	230 V
$\frac{1}{6}$	4.4 A	2.2 A
$\frac{1}{4}$	5.8 A	2.9 A
$\frac{1}{3}$	7.2 A	3.6 A
$\frac{1}{2}$	9.8 A	4.9 A
$\frac{3}{4}$	13.8 A	6.9 A
1	16.0 A	8 A
1.5	20.0 A	10 A
2.0	24.0 A	12 A
3.0	34 A	17 A
5.0	56 A	28 A
7.5	80 A	40 A
10.0	100 A	50 A

Full Load Current 3 ϕ A.C. Motors

HP	*SCIM/SRIM V*					*Synchronous Motor at UPF (Amp)*			
	115 V	*230 V*	*460 V*	*575 V*	*2300 V*	*230 V*	*460 V*	*575 V*	*2300 V*
1/2	4	2	1	0.8					
3/4	5.6	2.8	1.4	1.1					
1	7.2	3.6	1.8	1.4					
1.5	10.4	5.2	2.6	2.1					
2.0	13.6	6.8	3.4	2.7					
3.0	–	9.6	4.8	3.9					
5.0	–	15.2	7.6	6.1					
7.5	–	22	11	9					
10.0	–	28	14	11					
15	–	42	21	17					
20	–	54	27	22					
25	–	68	34	27	–	53	26	21	
100	–	248	124	99	26	202	101	81	20

C

Appendix

Maximum Rating of S.C. and Ground Fault Protective Devices

Motor Type	% F L current			
	No delay fuse	Time delay duse	Plain C.B.	Inverse C.B.
1 ϕ (all types)	300	175	700	250
All 1 ϕ and 3 ϕ IM and SM full voltage starting				
letter code F to V	300	175	700	250
letter code B to E	250	175	700	200
letter code A	150	150	700	150
All 1 ϕ, 3 ϕ IM and SM (with reduced voltage starting) I_r < 30 Amp				
letter code	250	175	700	200
For I_r > 30 A				
code F to V	250	175	700	200
code B to E	200	175	700	200
code A	150	150	700	150
High reactance SCIM				
I < 30 A	250	175	700	250
I > 30 A	200	175	700	200

Wound roter IM				
No code	150	150	700	150
D.C. (constant voltage)				
P < 50 H.P	150	150	250	150
P > 50 H P	150	150	175	150

D

Appendix

Rated, Starting Currents and Torques of Various Classes of 220 V Induction Motors

Power (HP)	Irated (Amp)	Starting 'I_{st}' Current (A) Class (B, C, D)	Class (F)	Starting torque, T_{st}% rated torque at rated voltage: Class A & B 4 pole	6 pole	8 pole	Class C 4 pole	6 pole	8 pole
0.5	2.0	12	_			150			
1.0	3.5	24	_	275	175	150			
1.5	5.0	35	_	265	175	150			
2.0	6.5	45	_	250	175	150			
3.0	9.0	60	_	250	175	150	_	250	225
5.0	15	90	_	185	160	130	250	250	225
7.5	22	120	_	175	150	125	250	225	200
10.0	27	150	_	175	150	125	250	225	200
25.0	64	365	_	150	135	125	200	200	200
50.0	125	725	450	150	135	125	200	200	200
100.0	246	1450	900	125	125	125	200	200	200
200.0	480	2900	1800	125	125	125	200	200	200

(a) Starting current $I'_{st} = \dfrac{V_{\text{applied}}}{220} \times I_{st}$

(b) Starting torque $T'_{st} = \left(\dfrac{V_{\text{applied}}}{220}\right)^2 \times T_{st}$.

where, I_{st}, T_{st} → are values given in table.

– I_{starting} of class 'A' motor are usually higher than corresponding class B, C, D motors

- T_{starting} of class '*D*' motor are usually higher than corresponding class *A*, *B*, *C* motors.
- T_{st} of class '*F*' motor are usually lower than corresponding *A*, *B*, *C*, motors.

E

Appendix

IMPORTANT TRIGNOMETRIC RELATIONS

$$\cos 2\theta = 2\cos^2\theta - 1 = 1 - 2\sin^2\theta$$

$$\sin 2\theta = 2\sin\theta\cos\theta$$

$$\sin 2\theta + \cos^2\theta = 1$$

$\sin 0^\circ = 0$	$\cos^\circ = 1$	$\tan 0^\circ = 0$
$\sin 30^\circ = \frac{1}{2}$	$\cos 30^\circ = \frac{\sqrt{3}}{2}$	$\tan 30^\circ = \frac{1}{\sqrt{3}}$
$\sin 45^\circ = \frac{1}{\sqrt{2}}$	$\cos 45^\circ = \frac{1}{\sqrt{2}}$	$\tan 45^\circ = 1$
$\sin 60^\circ = \frac{\sqrt{3}}{2}$	$\cos 60^\circ = \frac{1}{2}$	$\tan 60^\circ = \sqrt{3}$
$\sin 90^\circ = 1$	$\cos 90^\circ = 0$	$\tan 90^\circ = \infty$
$\sin 180^\circ = 0$	$\cos 180^\circ = -1$	$\tan 180^\circ = 0$
$\sin 270^\circ = -1$	$\cos 270^\circ = 0$	$\tan 270^\circ = -\infty$
$\sin 360^\circ = 0$	$\cos 360^\circ = 1$	$\tan 360^\circ = 0$

Rule 1:

(a) **No change:** sin, cos θ, tan of $180^\circ - \theta$ or $180^\circ + \theta$ or $360^\circ - \theta$ or $360^\circ + \theta$ remains the same trignometric ratio.

(b) **Change:** sin, cos, tan of $90^\circ - \theta$, $90^\circ + \theta$ or $270^\circ - \theta$ or $270^\circ + \theta$ **changes**, sin becomes cos, tan becomes cot.

Rule 2 of sign: The sign of the resulting trignometric ratio depends upon the quadrant of the angle.

$\sin(-\theta) = -\sin\theta$; $\cos(-\theta) = +\cos\theta$; $\tan(-\theta) = -\tan\theta$

$\sin(90 - \theta) = \cos\theta$; $\cos(90 - \theta) = \sin\theta$; $\tan(90 - \theta) = \cot\theta$

$\sin(90 + \theta) = \cos\theta$; $\cos(90 + \theta) = -\sin\theta$; $\tan(90 - \theta) = -\cot\theta$

$\sin(180 - \theta) = \sin\theta$; $\cos(180 - \theta) = -\cos\theta$; $\tan(180 - \theta) = -\tan\theta$

$\sin(180 + \theta) = -\sin\theta$; $\cos(180 + \theta) = -\cos\theta$; $\tan(180 + \theta) = \tan\theta$

$\sin(270 + \theta) = -\cos\theta$; $\cos(270 - \theta) = -\sin\theta$; $\tan(270 - \theta) = \cot\theta$

$\sin(270 + \theta) = -\cos\theta$; $\cos(270 + \theta) = +\sin\theta$; $\tan(270 - \theta) = -\cot\theta$

$\sin(360 - \theta) = -\sin\theta$; $\cos(360 - \theta) = \cos\theta$; $\tan(360 - \theta) = -\tan\theta$

Formulae

$$\sin(A + B) = \sin A \cos B + \cos A \sin B$$

$$\sin(A - B) = \sin A \cos B - \cos A \sin B$$

$$\cos(A + B) = \cos A \cos B - \sin A \sin B$$

$$\cos(A - B) = \cos A \cos B + \sin A \sin B$$

$$\sin C + \sin D = 2\sin\left(\frac{C+D}{2}\right)\cos\left(\frac{C-D}{2}\right)$$

$$\cos C + \cos D = 2\cos\left(\frac{C+D}{2}\right)\cos\left(\frac{C-D}{2}\right)$$

$$\sin C - \sin D = 2\cos\left(\frac{C+D}{2}\right)\sin\left(\frac{C-D}{2}\right)$$

$$\cos C - \cos D = -2\sin\left(\frac{C+D}{2}\right)\sin\left(\frac{C-D}{2}\right)$$

$$2\sin A \cos B = \sin(A + B) + \sin(A - B)$$

$$2\cos A \sin B = \sin(A + B) - \sin(A - B)$$

$$2\cos A \cos B = \cos(A + B) + \cos(A - B)$$

$$2\sin A \sin B = \cos(A - B) - \cos(A + B)$$

$$\sin^{-1}(\sin\theta) = \theta$$

$$\cos^{-1}(\cos\theta) = \theta$$

$$\tan^{-1}(\tan\theta) = \theta$$

Other Important Formulae Used

- $(1 + a)^n = 1 + na + \frac{n(n-1)}{2!}a^2 + \frac{n(n-1)(n-2)}{3!}a^3 + \ldots$
- $a^\circ = 1$

– $a^m \times a^n = a^{m+n}$

– $a^m / a^n = a^{m-n}$

– If $ax^2 + bx + c = 0$ then $x = -\dfrac{b \pm \sqrt{b^2 - 4ac}}{2a}$

Sum of roots $= -\dfrac{b}{a}$, Product of the roots $= \dfrac{c}{a}$

Roots are equal if $b^2 - 4\,ac = 0$

Roots are real if $b^2 - 4\,ac > 0$

Roots are imagine any if $b^2 - 4\,ac < 0$

– $\log 1 = 0$; $\log a^a = 1$, $\log m + \log n = \log mn$,

$\log m - \log n = \log \dfrac{m}{n}$; $\log m^n = n \log m$; $e^{\log ex} = x$

$\log_a m = \log_b m \cdot \log_{ab}$

– $\dfrac{d}{dx}(\sin x) = \cos x$; $\dfrac{d}{dx}(\cos x) = -\sin x$; $\dfrac{d}{dx}(\ln x) = \dfrac{1}{x}$

$\dfrac{d}{dx}(x^n) = nx^{n-1}$

– $\int_a^b f(x)dx = -\int_b^a f(x)dx = \int_b^c f(x)dx + \int_c^b f(x)dx,\ a < c < b$

$\int_0^{2a} f(x)dx = 2\int_0^a f(x)dx$ if $f(2a - x) = f(x)$

$= 0$ if $f(2a - x) = -f(x)$

Euler's theorem

$$\frac{\partial z}{\partial t} = \frac{\partial z}{\partial x}\cdot\frac{dx}{dt} + \frac{\partial z}{\partial y}\cdot\frac{dy}{dt}; \qquad \text{if } Z = f(x, y)$$

$$\frac{\partial y}{\partial x} = -\frac{\partial\phi/\partial x}{\partial\phi/\partial y}; \qquad \text{if } \phi(x, y) = c$$

Maximum or minimum

$$\frac{\partial f}{\partial x} = 0, \qquad \frac{\partial f}{\partial y} = 0$$

For maxima: $\dfrac{\partial^2 f}{\partial x^2} < 0$ For minima: $\dfrac{\partial^2 f}{\partial x^2} < 0$

Vectors

$$\bar{r} = x\hat{i} + y\hat{j} + z\hat{k} \; ; \quad (\bar{r}) = \sqrt{x^2 + y^2 + z^2}$$

Salar product

$$\bar{a} \cdot \bar{b} = |\bar{a}||\bar{b}| \cos\theta ; \qquad \text{W.D.} = \text{W} = \bar{F} \cdot \bar{r}$$

$$\theta = \cos^{-1} \frac{\bar{a} \cdot \bar{b}}{|\bar{a}||\bar{b}|}$$

Vector product

$$\bar{a} \times \bar{b} = |\bar{a}||\bar{b}| \sin\theta \; \hat{\eta}$$

Moment of force or Torque = $T = \bar{F} \times \bar{r}$

Laplace

$$L(1) = \frac{1}{s} \; ; \; L^{-1}\left(\frac{1}{s}\right) = 1$$

Experiments

EXPERIMENT NO. 1

Object: To obtain magnetization characteristic of a d.c. shunt generator. Determine critical field resistance of d.c. shunt generator using O.C.C.

Apparatus used: Write detailed specifications of machine and instruments used.

Theory: Refer this book for this part related to the given object.

Circuit Diagram:

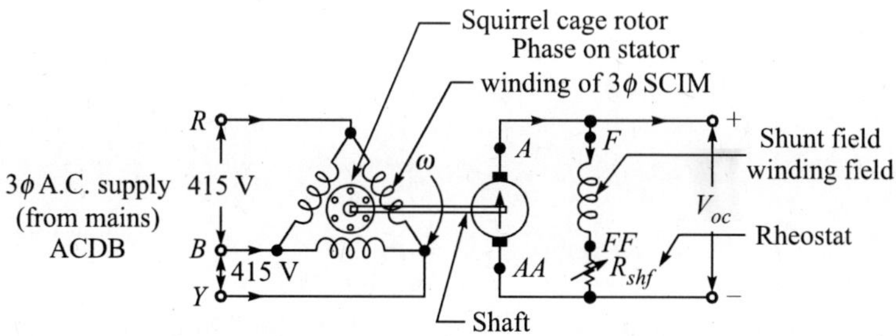

Fig. 1 Connection diagram of d.c. shunt generator for obtaining O.C.C.

Observation Table: For open circuit characteristic of the d.c. generator

Table I

S. No.	*Field Current I_f(A)*	*Terminal Voltage V_0(V) (open circuit voltage)*
1.		
2.		
3.		
⋮		
⋮		
10.		

Graph: As shown in Fig. 2.

Result: Plot graph between field current and terminal voltage. Write the value of 'R_{cf}' obtained from graph.

Discussions: Discuss the magnetization characteristic and the limitations of a d.c. shunt generator.

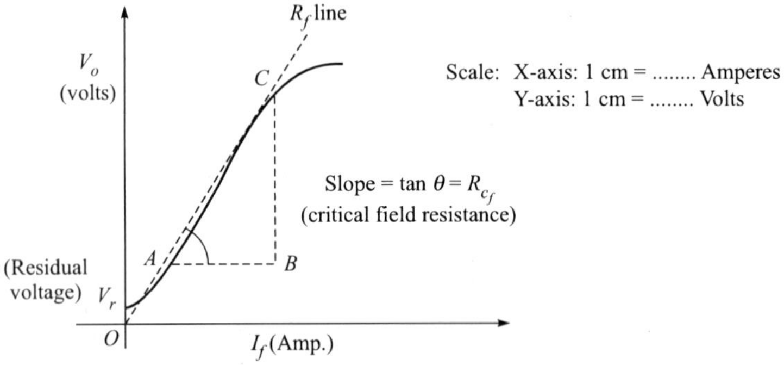

Fig. 2 O.C.C. of d.c. shunt generator

Viva Questions:

Q.1 What is the significance of O.C.C?

Q.2 What is critical speed?

Q.3 What is critical field resistance?

Q.4 What is prime mover? Where it is used?

Q.5 What is the difference between self excited and separately excited d.c. generators?

EXPERIMENT NO. 2

Object: To obtain load characteristic of a d.c. compound generator.

Apparatus used: Write detailed specifications of machine and instruments used.

Theory: Refer this book for this part related to the given object.

Circuit Diagram:

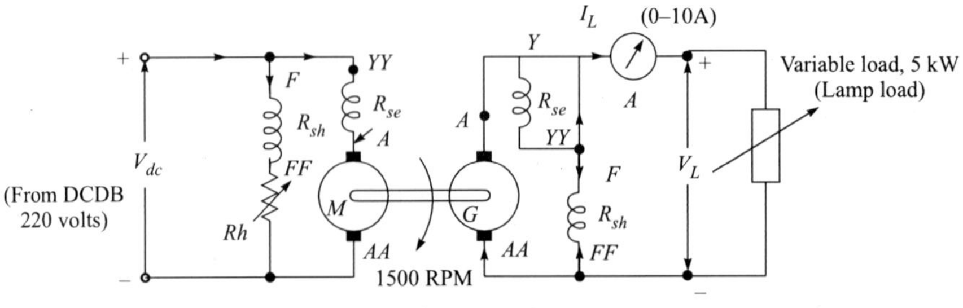

Fig. 1 Connection diagram of a d.c. compound generator.

Observation Table: For load characteristic of d.c. compound generator

Table I

S. No.	*Load Current I_L(A)*	*Load or Terminal Voltage V_L(V)*
1.		
2.		
3.		
⋮		
⋮		
10.		

Graph: As shown in Fig. 2.

Result: Plot graph between load current and terminal voltage.

Discussions: Discuss the load characteristic and the limitations of this machine.

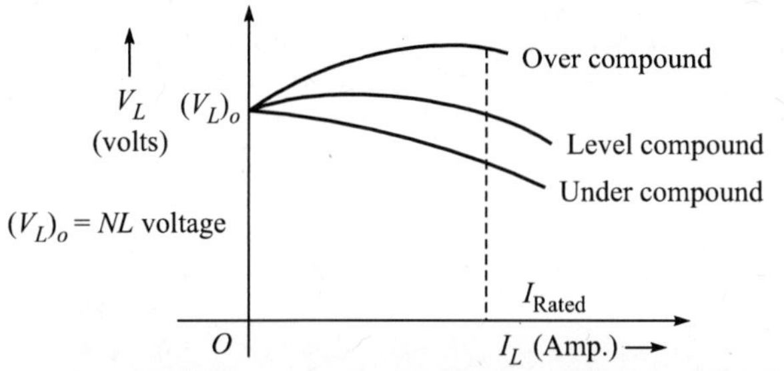

Scale: X-axis; 1 cm = ... A
Y-axis; 1 cm = ... V

Fig. 2 Load characteristic of a d.c. compound generator.

Viva Questions:

Q.1 What is load? What are the different types of load practically used?
Q.2 What is diverter resistor?
Q.3 What are the different types of commutative compound generator?
Q.4 Why the load characteristic droops at heavy loads?
Q.5 Where does level compound generators used?

EXPERIMENT NO. 3

Object: To study the speed control of a d.c. shunt motor and draw the speed variation curves with respect to

(a) Change of field current (field control)
(b) Change of resistance in armature circuit (armature control)

Apparatus used: Write detailed specifications of machine, equipment and instruments used.

Theory: Refer this book for this part related to the given object.

$$\boxed{N \propto \frac{V - I_a R_a}{\phi}}; \quad \text{where} \quad \phi \propto I_f$$

$$N \propto \frac{V_a}{\phi}; \text{ where } V - I_a R_a = V_a \text{ (armature voltage)}$$

If ϕ = constant, then $\boxed{N \propto V_a}$

If V = constant and $\phi \propto I_f$, then $N \propto \dfrac{(V - I_L R_a)}{I_f} + R_a$

or,
$$\boxed{N = K_a + K\frac{(V - I_L R_a)}{I_f}}$$

At no load, $N = N_o$ (on shaft)

Circuit Diagram:

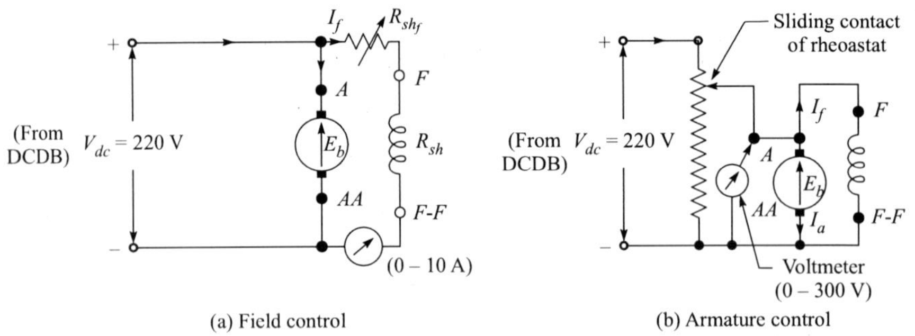

Fig. 1 Connection diagram for speed control of a d.c. shunt motor.

Observation Table: For speed control of a d.c. motor

Table I

S. No.	Field Control		Armature Control	
	Ammeter I_f(A)	Speed of the motor (rpm)	'Va'(V)	Speed of the motor (rpm)
1.				
2.				

Contd.

3. ⋮ ⋮ 10.		

Graph: Field control graph is shown Fig. 2 and armature control graph shown in Fig. 3.

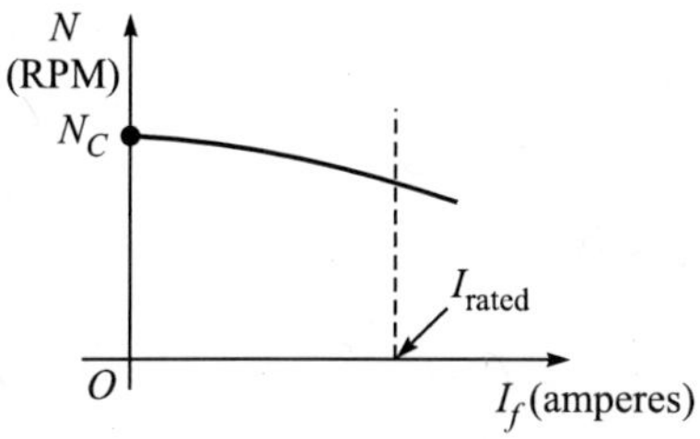

Fig. 2 N/I_f characteristic of a d.c. shunt motor.

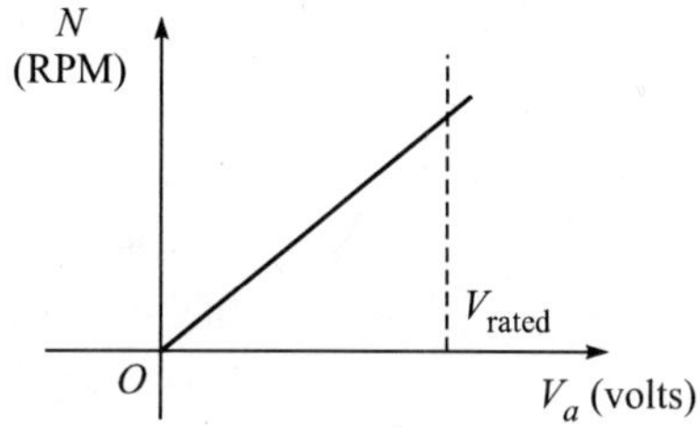

Fig. 3 N/V_a characteristics of a d.c. shunt motor.

Result: Comment on speed variations for the two controls used.

Discussion:

(a) In the field control method discuss the effect of field current on the speed of motor.

(b) In the armature control method discuss the effect of armature current on the speed of motor.

Viva Questions:

Q.1 Where field control of d.c. motor is required?

Q.2 Why do the speeds of motor decreases with increase of load at the shaft?

Q.3 What is rated current?

Q.4 What is base speed?

Q.5 What is armature reaction?

EXPERIMENT NO. 4

Object: Study of d.c. series motor and two-point starter and to draw the speed versus armature current characteristic.

Apparatus used: Write detailed specifications of machine, equipment and instruments used.

Theory: Refer this book for this part related to the given object.

Circuit Diagram:

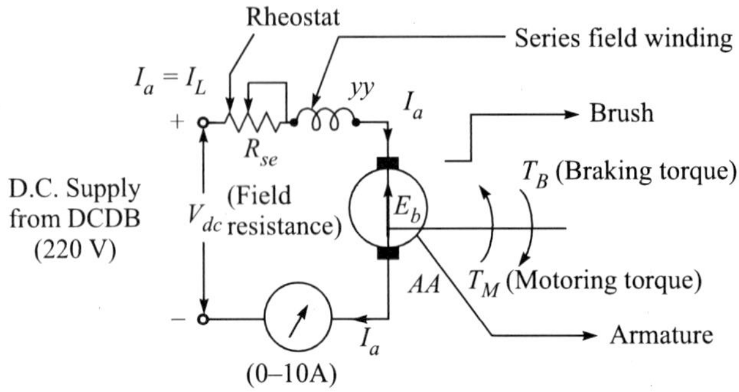

Fig. 1 Connection diagram for obtaining 'N/I_a' characteristic of d.c. series motor.

Observation Table: For operating characteristic of d.c. series motor

Table I

S. No.	*Armature Current 'I_a'(A)*	*Speed of Motor 'N' (rpm)*
1.		
2.		
3.		
⋮		
⋮		
10.		

Graph: As shown in Fig. 2 below.

Result: Plot graph between armature current and speed.

Discussion: Discuss the characteristic and write the motor applications.

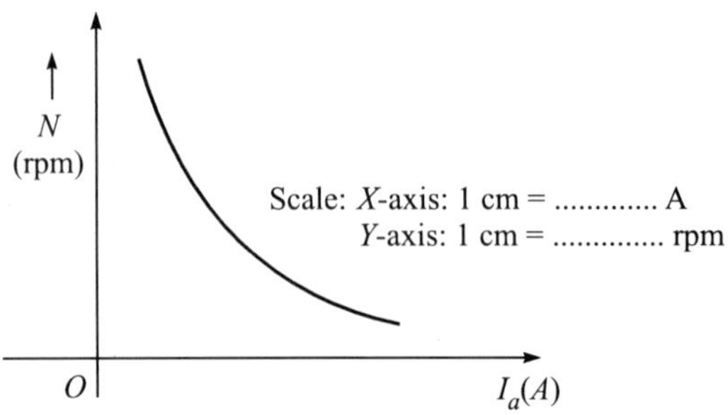

Fig. 2 N/I_a characteristic of d.c. series motor.

Viva Questions:

Q.1 What is the need of a starter in a d.c. motor?

Q.2 What type of loads are used in d.c. series motor?
Q.3 Why the shape of N/I_a curve is hyperbolic?
Q.4 Why the load torque opposes motoring torque?
Q.5 Why the series motor is not used on no load?

EXPERIMENT NO. 5

Object: To draw the V-curves of a synchronous motor. Plot armature current (I_a) versus field current (I_f) at different values of load.

Apparatus Used: Write detailed specifications of machine, equipment and instruments used.

Theory: Refer this book for this part related to the given object.

Circuit Diagram:

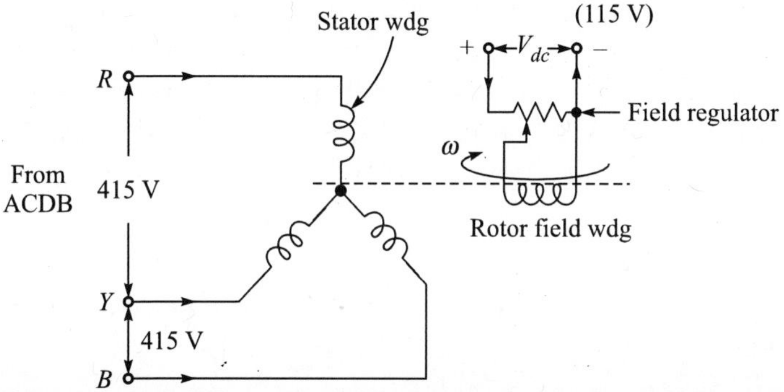

Fig. 1 Connection diagram of obtaining V-curve of a synchronous motor.

Observation Table: For V-curves of synchronous motor

Table I

S. No.	Field Current $I_f(A)$	Armature Current $I_a(A)$
No load		
Half load		
Full load		

Graph: As shown in Fig. 2 below.

Result: The V-curves shift for different loads on shaft ....

Discussion:

(a) The significance of V-curve.
(b) Applications of over- and under-excited synchronous motor.

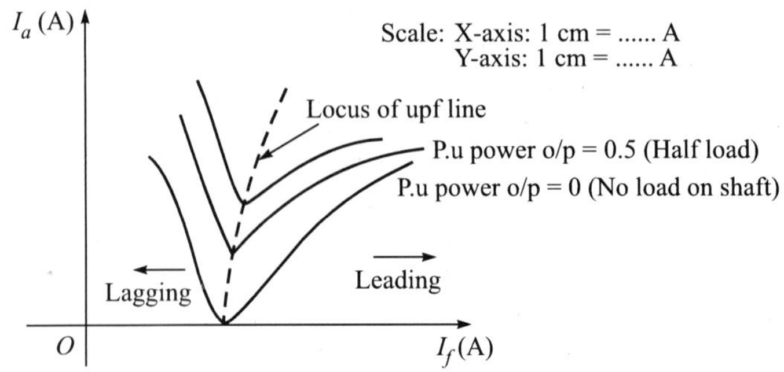

Fig. 2 V-curves of a synchronous motor at different loads.

Viva Questions:

Q.1 What is the practical importance of V-curve?

Q.2 Define the various portions of V-curve?

Q.3 Why does the field excitation kept high in a synchronous condenser?

Q.4 Why the field excitation is given after the a.c. supply to stator windings in synchronous motor?

Q.5 Why the synchronous motor does not operate as asynchronous motor?

EXPERIMENT NO. 6

Object: To obtain efficiency and regulation of a single phase transformer by Sumpners test.

Apparatus Required: Write detailed specifications of machine, equipment and instruments used.

Theory: Refer this book for this part related to the given object.

Observation Table: For determination of efficiency

Table I

S. No.	Voltage(V)	Current(A)	P_i (W_1)	P_{Cu} (W_2)	$(P_i + P_{Cu})/2$ (W)	Efficiency (%) 'η'
1.						
2.						
3.						
⋮						
⋮						
10.						

Circuit Diagram:

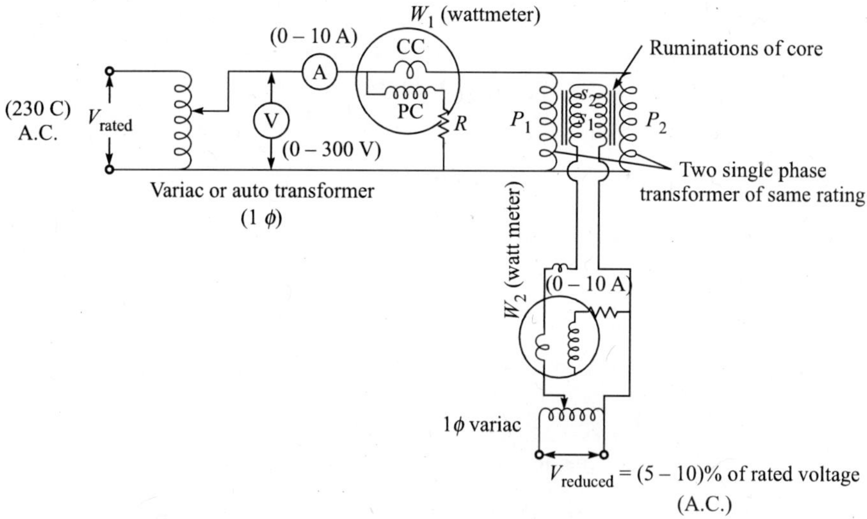

Fig. 1 Sumpners test connection diagram.

Formula used:

$$\eta = \frac{\text{Output power}}{\text{Output power} + \left(\dfrac{W_1 + W_2}{2}\right)} \times 100\%$$

W_1, W_2 = Wattmeter readings, Watts.

Table II (For Voltage Regulation)

S. No.	$V_{NL}(V)$	$V_{FL}(V)$	*V.R. (in %)*
1.			
2.			
3.			

Result:

(a) Efficiency of Transformer = ...

(b) Voltage Regulation = ...

Discussion:

(a) Why Sumpners test is called heat run test?
(b) What is the practical utility of Sumpners test?

Viva Questions:

Q.1 What is voltage regulation?
Q.2 Why is it called regenerative test?

Q.3 Why is it called back to back test?

Q.4 Why identical transformers are used?

Q.5 On what factors the efficiency of a transformer depends?

EXPERIMENT NO. 7

Object: To perform no-load test and short-circuit test on a three-phase alternator and to find voltage regulation by synchronous impedance method.

Apparatus Used: Write detailed specifications of machine, equipment and instruments used.

Theory: Refer this books for this part related to the given object.

Circuit Diagram:

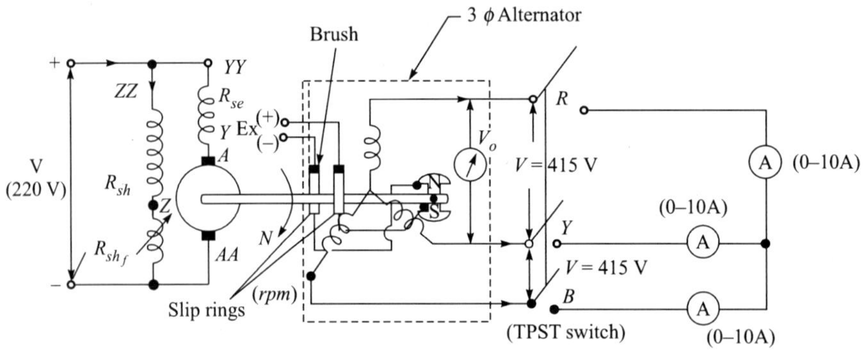

Fig. 1 For performing NL test and S.C. test on 3φ alternator

Formulae used:

$$Z_s = \frac{(V_0)_{AD}}{(I_{SC})_{AB}}; \quad X_s = \sqrt{Z_s^2 - R_a^2}; \quad R_a = 1.6\, R_{dc}$$

$$\%\text{V.R.} = [V_{NL} - V_{FL})/V_{FL}] \times 100\%$$

Observation Table: For Voltage Regulation

Table I For different loads (Resistive, inductive and capacitive)

S. No.	$V_{NL}(V)$	$V_{FL}(V)$	V.R. (in %)
1.			
2.			
3.			

For S.C.C

Table II

S. No.	$I_f(A)$	$I_{sc}(A)$
1. 2. 3. ⋮ ⋮ 10.		

For O.C.C

Table III

So. No.	$I_f(A)$	$V_0(Volts)$
1. 2. 3. ⋮ ⋮ 10.		

Graph: Using Table II and Table III, the OCC and SCC are obtained as shown in graph in Fig. 2.

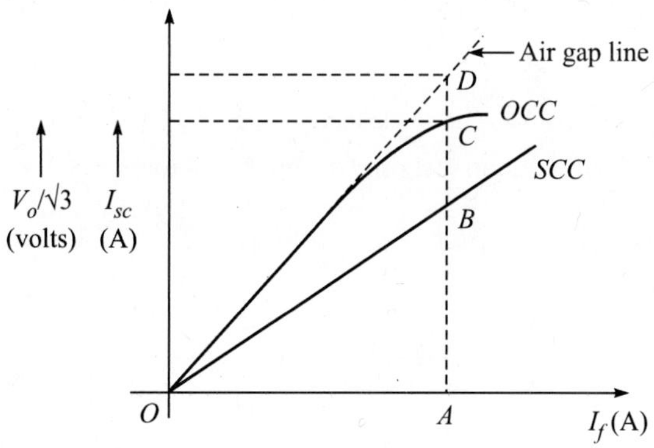

Fig. 2 O.C.C. and S.C.C. of an 3ϕ alternator.

Result:

(a) Synchronous Impedance 'Z_s' = ...

(b) Voltage Regulation 'V.R.' = ...

Discussion:

(a) Write the significance of air gap line.
(b) Why synchronous impedance method is called a pessimistic method? Discuss.

Viva Questions:

Q.1 What are the two main types of 3ϕ alternator?
Q.2 What are the other methods of determining voltage regulation?
Q.3 Why the a.c. and d.c. resistance differ in magnitude?
Q.4 What is the significance of air gap line of an alternator?
Q.5 How many slip rings are used in an alternator?

EXPERIMENT NO. 8

Object: To obtain load characteristic of a d.c. shunt generator.

Apparatus Used: Write detailed specifications of machine and instruments used.

Theory: Refer this book for this part related to the given object.

Circuit Diagram:

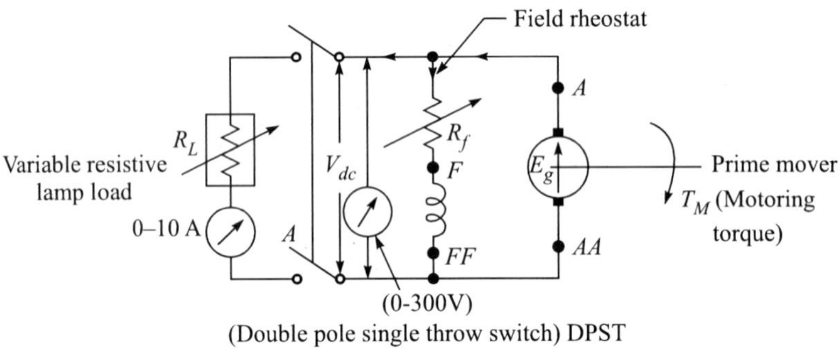

Fig. 1 Load test on d.c. shunt generator.

Observation Table:

Table I (For external characteristic)

S. No	Load Current $I_L(A)$	Terminal Voltage $V_T(V)$
1.		
2.		
3.		
⋮		
⋮		
10.		

Result: Plot graph between load current and terminal voltage.

Discussion: Discuss the load characteristic and the limitations of this machine.

Viva Questions:

Q.1 What is external characteristic?
Q.2 Why does the terminal voltage differ from the generated voltage in armature?
Q.3 Is the generator load characteristic depends on the power factor of load?
Q.4 How can one know the direction of generated emf practically?
Q.5 What is used as a prime mover in d.c. shunt generator?

EXPERIMENT NO. 9

Object: Determination of losses in a single phase transformer by open circuit and short circuit test. Obtain the equivalent circuit of transformer.

Apparatus Required:

S. No.	*App/Comp. used*	*Specification*	*Make*
1.			
2.			
3.			
⋮			
⋮			
10.			

Circuit Diagram:

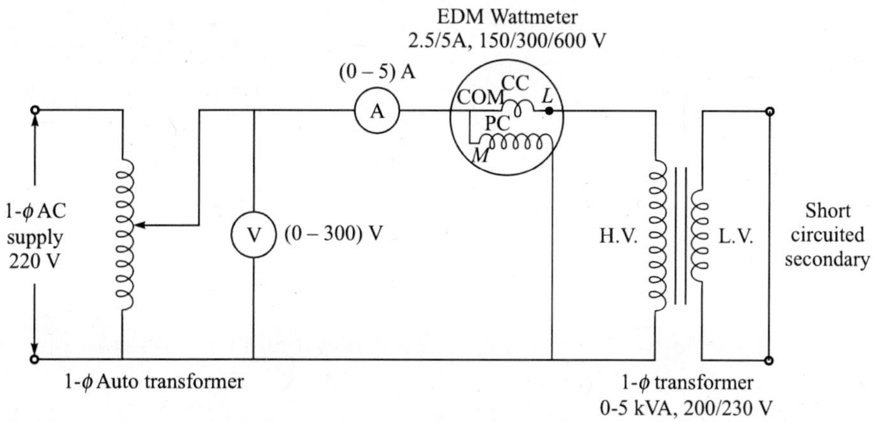

Fig. 1 SCT on 1 ϕ transformer.

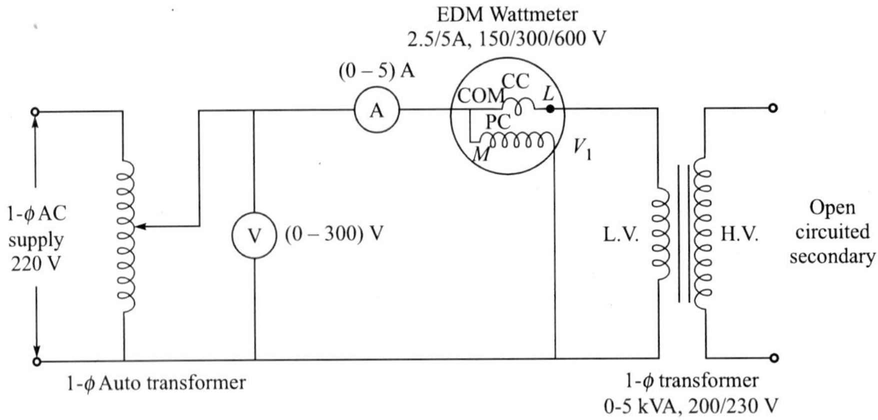

Fig. 2 O.C.T. on 1 ϕ transformer.

Theory: Refer this book for this part.

Observation Table:

Table-I For (O.C.T.)

S. No.	V_0*(Volts)*	I_0*(Amps)*	W_0*(Watts)*

Table II For (S.C.T.)

S. No.	V_{SC}*(Volts)*	I_{SC}*(Amps)*	W_{SC}*(Watts)*

Calculations: Iron loss (W_0) = ... W, Copper loss (W_{SC}) = ... W,

Result: The losses of transformer (0.5 kVA) provided are ... W and ... W respectively.

Discussion:

(a) Mention other losses of a transformer.
(b) Mention how the losses of a transformer can be minimized.

Viva Questions:

Q.1 What is power transformer?
Q.2 Why no load test on a transformer is performed?
Q.3 What are the types of loss occurs in a transformer?
Q.4 Why the transformer is rated in kVA?
Q.5 Why equivalent circuit is determined?

EXPERIMENT NO. 10

Object: Study of Ward Leonard speed control for a d.c. shunt motor.

Apparatus Required:

S. No.	Equipment	Rating
1.		
2.		
3.		
⋮		
⋮		
10.		

Circuit Diagram:

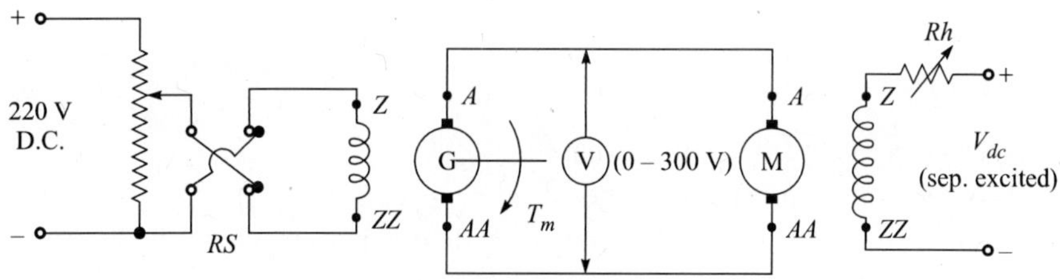

Fig. 1 Ward Leonard speed control.

Theory: As given in the book.

Formula used:

$$N \propto \frac{V - I_a R_a}{\phi} \qquad \text{(I)}$$

where,

N = Speed, rpm

V = Applied voltage to motor 'M', volts.
I_a = Armature current, amperes.
R_a = Armature resistance, ohms.
ϕ = Flux per pole, Wb.

For shunt motor, ϕ = Constant

$$\boxed{N \propto V - I_a R_a} \tag{II}$$

Observation Table:

Table I

S. No.	No Load (Forward Direction)		No Load (Reverse Direction)	
	V(volts)	*N(rpm)*	*V(volts)*	*N(rpm)*
1.				
2.				
3.				
⋮				
⋮				
10.				

Graph: N/V_{applied} in both cases.

Result: Write about the speed variation in both cases.

Discussion:

(a) Merits of this method.
(b) Applications

Viva Questions:

Q.1 What is the merit of Ward Leonard method?
Q.2 What is static Ward Leonard method?
Q.3 Why the conventional Ward Leonard method is not used nowadays?
Q.4 How can the motor direction be reversed?
Q.5 How can the brush drop be minimized?

EXPERIMENT NO. 11

Object: Study of speed control of d.c. shunt motor using thyristor controlled drive.

Apparatus used:

S. No.	Equipment	Ratings
1.		
2.		
3.		
⋮		
⋮		
10.		

Theory:

Formula used: $V_0 = K \cos \alpha$

where, V_0 = Average output voltage, volts.

α = Firing angle of thyristor

Circuit Diagram:

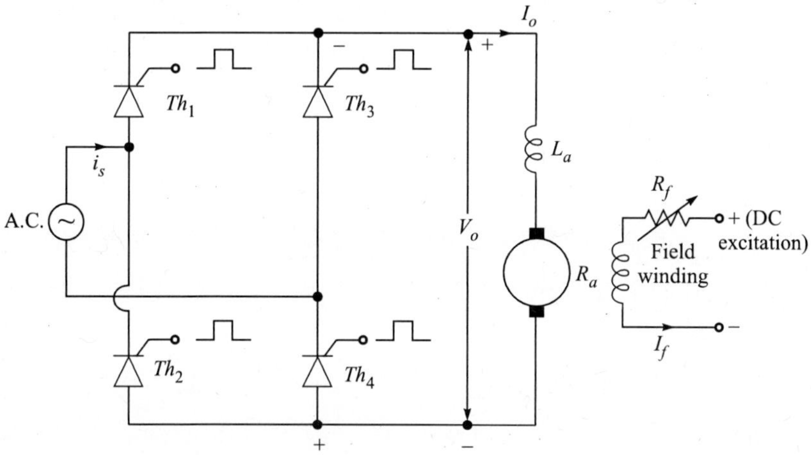

Fig. 1 Circuit diagram showing thyristor controlled d.c. drive.

Observation Table: For speed control of d.c. motor

S. No.	α (degree)	V_0 (volts)	N (rpm)
1.			
2.			
3.			
⋮			
⋮			
10.			

Graph: Draw the graph between 'N' and 'α'.

Result: Write about the speed variation with the changes in 'α' (firing angle).

Discussion:

(a) Advantages of solid state control over conventional methods.

(b) Practical applications.

Viva Questions:

Q.1 What is a firing angle?

Q.2 Why solid state speed control is preferred?

Q.3 What are the other solid state devices that can be used?

Q.4 How can the device protection be achieved?

Q.5 What is the effect of 'L_a' (armature inductance) on the output of converter?

Bibliography

1. Fitzgerald A.E. et al., *Electric Machinery* McGraw-Hill International, New Delhi, 1983.
2. Gupta B.R., and V. Singhal, *Fundamentals of Electric Machines*, New Age International (P) Ltd., New Delhi, 2005.
3. Kosov I.L., *Electric Machinery and Transformers*, Prentice-Hall, N.J, 1972.
4. Langsdorf, A.S., *Theory of A.C. Machinery* McGraw Hill, New York, 1955.
5. Say M.G., *Performance and Design of A.C. Machines* Pitmal, London, 1983.
6. Praker and Smith, *Problems in Electrical Engineering*, CBS Publishers, New Delhi.
7. P.K. Mukherjee and S. Chakrravorti, *Electrical Machines*, Dhanpat Rai Publications (Pvt.) Ltd., New Delhi, 2003.
8. G.K. Dubly, *Fundamentals of Electric Drives*, Narosa Publishing House, New Delhi.

Index

A

A.C. quantities,
  phasor approach to, 17
A.C. voltage waveform, 15
All dark lamp method, 364
All day efficiency, 151
Alternating current circuits, 15
Alternators,
  characteristics of, 348
  parallel operation of, 362
Armature reaction, 214, 343
  cross magnetizing effect of, 217
  demagnetizing effect of, 216
Auto transformers, 156
  starting of, 302

B

Balanced three-phase source, 19
Bar magnet, 71
Basic chopper circuit, 457
Blocked rotor test, 417
Blondel's two reaction theory, 356
Bridge rectifier, 432
Broken magnet, 71
Brush assembly, 187
Brushless motor, 265
  advantages of, 265

C

Capability curve, 355
Capacitance start, 407
Chopper circuit, 457
Chopper control, 457
Circulating current mode, 454
  demerits of, 454
Commutation, 210
  types of, 212
Commutator assembly, 191
Compound generator, 197
  load characteristic of, 207
Constant horse-power connections, 310
Copper losses, 219
Core losses, 218
Core type transformer, 128
Counter current braking, 262
Cummulative compound generator, 197
Cumulative compound D.C. motors, 267
Current transformer, 165
Cycloconverter control, 461
Cylindrical rotor synchronous machines, 334

D

D.C. cummulative compound generator, 223
D.C. differential compound generators, 223
D.C. generators,
  applications of, 222
  characteristics of, 201
  efficiency of, 221
  EMF equation of, 208
  voltage regulation of, 220
D.C. machines,
  construction features of, 185
  losses in, 218
  types of, 192

D.C. motor,
  applications of, 266
  characteristics of, 237
  efficiency of, 245
  speed control of, 253
  starting of, 246
  types of, 237
D.C. separately excited generator, 222
D.C. series generator, 222
  load characteristic of, 206
D.C. series motor, 267
D.C. shunt generator, 222
  load characteristic of, 204
  voltage build up in, 195
D.C. shunt motor, 254, 266
Delta connected system, 24
Δ/Y connections, 161
Diamagnetic materials, 73
Dielectric loss, 220
Differentially compound motors, 267
Diode,
  non-linear characteristic of, 4
Direct on line starters, 298
Direct-current circuits, 2
Double layer winding, 190
Doubly excited system, 120
Drive,
  types of, 431
Drum type armature winding, 114
Dual converter, 453
Dynamic braking (Rheostatic braking), 263
Dynamic rheostatic braking, 314

E

Electric circuit, 78
Electric field as a medium, 107
Electromagnetic relay, 110
Electromechanical systems,
  dynamical equations of, 122
EMF commutation, 213
EMF equation, 342
EMF rule, 6
End shields, 188
Equivalent circuit, 287

F

Faraday's first law, 54
Faraday's law of electromagnetic induction, 54
Faraday's second law, 54
Ferromagnetic materials, 74
Field assembly, 186
Flemming's left hand rule, 61
Flemming's right hand rule, 60, 63
Force, 121
Four-point starter, 251
Four-pole synchronous generator, 332
Fringing effect, 86
Fringing, 85

G

Gauss's law for magnetic fields, 51
Gramme ring winding, 113
Grounding transformer, 166

H

Harmonics, 164
High impedance transformers, 166
Hunting, 368
Hysteresis motor, 420
Hysteresis, 86

I

Ideal transformer, 131
  phasor diagram of, 133
Induction furnace, 167
Induction generator, 315
Induction heating transformers, 167
Induction motor,
  braking of, 314
  cogging of, 313
  crawling of, 313
Instrument transformers, 165
Inverter control, 462

K

Kirchoff's current law (KCL), 4, 5

Kirchoff's junction rule, 5
Kirchoff's voltage law (KVL), 6

L

Lap winding, 190
Lenz's law, 55
Level compound generator, 198
Line commutated inverter, 463
Linear induction motor, 421

M

Magnetic circuit, 78
Magnetic equivalent circuit, 79, 80, 83
Magnetic field as a medium, 110
Magnetic flux intensity, 76
Magnetic flux, 76
Magnetic materials, 72
  types of, 73
Magnetization force, 76
Magnetization,
  intensity of, 77
Magnetomotive force, 75
Maximum power transfer theorem, 13
Maximum voltage regulation, 149
Maxwell's equations, 50
Mechanical losses, 220
MMF method, 353
Motional EMF, 53
Motor losses, 244
Motor,
  operating principle of, 234
  performance characteristics of, 243
  torque equation of, 235
Motoring action, 116
Mutual inductance, 51
Mutual induction, 52

N

No load test, 152
Non-circulating current mode, 454

O

Ohm's law, 3
Open circuit characteristic, 348
Open circuit test, 152
Over compound generator, 198

P

Parallel magnetic circuit, 83
Paramagnetic materials, 74
Per unit system, 167
Permanent capacitor SPIM, 409
Permanent magnet D.C. motor, 264
Permanent magnets, 71
Permeability, 76
Permeance, 77
Plugging operation, 262, 314
Polarity test, 152
Pole changing, 307
Polyphase IM,
  characteristic of, 296
  construction of, 279
  types of, 278
Potential transformer, 166
Potier's method, 354
Potier's regulation diagram, 355
Power angle characteristic, 365
Power expressions, 290
Power flow diagram, 290
Power rating, 16
Practical transformer, 137, 141

R

Reactance start split phase IM, 406
Real transformer,
  phasor diagram of, 137
Reduced voltage starting, 300
Regenerative braking, 264
Regulation characteristic, 351
Relative permeability, 76
Reluctance, 77
Repulsion induction SPIM, 411
Repulsion start SPIM, 410
Resistance commutation, 213
Resistance in parallel, 8
Resistance rule, 6
Resistance start split phase IM, 405
Resistor,
  linear characteristic of, 4

Ring specimen (with air gap), 80
Ring specimen (without air gap), 79
Rotating magnetic field, 283
Rotor resistance control, 306

S

Self excited generator,
load characteristic of, 204
magnetization characteristic of, 202
Self inductances, 51
Self-excited generator, 194
Self-induction, 52
Semi converter of half bridge converter, 432
Semi-closed slots, 189
Separately excited D.C. generator,
load characteristics of, 203
magnetization characteristic of, 202
Separately excited D.C. motor speed control, 253
Separately excited D.C. motor (with rectified three-phase supply), 448
Series generator, 196
Series magnetic circuit, 78, 79, 81, 82
Series resistance, 9
Shaded pole induction motor, 408
Shell type transformer, 83, 129
Short circuit characteristic, 349
Short circuit equivalent, 154
Short circuit test, 153
Shunt generator, 194
Single layer winding, 190
Single phase full bridge converter, 436
Single phase IM,
starting methods of, 404
types of, 403
Single phase motor,
operation of, 402
Single phase reluctance motor, 411
Single phase wattmeter,
power measurement in, 29
Slip ring induction motor,
starting of, 304
Solid state control devices, 428
Special transformers, 165
Speed control (by changing frequency), 310
Speed control of IM by concatenation, 311
Speed regulation, 236
SPIM,
equivalent circuit of, 412
testing of, 415
Split phase capacitor start IM, 406
Square specimen (without air gap), 79
Star connected systems, 22
Star delta starting, 302
Starting resistor elements, 252
Stepper motor, 418
Stray load losses, 220
Sumpner's test (back to back test), 155
Synchronous generators, 333
constructional features of, 332
parallel operation of, 361
Synchronous impedance method, 351
Synchronous machine dynamics, 370
Synchronous machines,
transients in, 369
Synchronous motor,
braking of, 392
construction of, 384
hunting of, 390
operation of, 384
starting methods of, 385

T

Thevenin's theorem, 10
Three phase induction motor, 298
speed control of, 306
Three-phase a.c. circuit,
single line diagram of, 32
Three-phase balanced system,
requirements for, 21
Three-phase circuits, 19
Three-phase connections, 22
Three-phase full bridge converter, 446, 450
Three-phase half bridge converter, 443
Three-phase IM,
applications of, 316

operation of, 282
starting of, 298
Three-phase load, 25
properties of, 25
Three-phase synchronous motor,
V-curve of, 386
Three-phase transformer, 130, 158
Three-point starter, 249
Three-winding transformers, 165
Thumb rule, 73
Torque expressions, 292
Torques, 121
Transformer EMF, 51
Transformer testing, 151
Transformer,
constructional features of, 128
efficiency of, 149
equivalent circuit of, 141
types of, 130
voltage regulation of, 145
Two quadrants chopper, 459
Two-point starter, 248
Two-turn wave winding, 191
Two-wattmeter method, 30
phasor diagram of, 31

U

Unbalanced three-phase system, 22
Uniform magnetic field,
moving coil in, 112
Universal motors, 266

V

V/*V* connection, 164
Variable reluctance stepper motor, 420
Variable torque connections, 310

W

Ward Leonard method, 261
Wide band transformer, 166
Winding factor, 336

Y

Y/Δ connections, 162
Yoke, 185

Z

Zero power factor characteristic, 350
Zero voltage regulation, 148
Zigzag connections, 163